Food Additives Data Book

Food Additives Data Book

Edited by

Jim Smith

and

Lily Hong-Shum

Blackwell
Science

Editorial Offices:
9600 Garsington Road, Oxford OX4 2DQ, UK
Tel: +44 (0)1865 776868
Blackwell Science, Inc., 350 Main Street, Malden, MA 02148-5018, USA
Tel: +1 781 388 8250
Iowa State Press, a Blackwell Publishing Company, 2121 State Avenue, Ames, Iowa 50014-8300, USA
Tel: +1 515 292 0140
Blackwell Publishing Asia Pty Ltd, 550 Swanston Street, Carlton South, Victoria 3053, Australia
Tel: +61 (0)3 9347 0300
Blackwell Wissenschafts Verlag, Kurfürstendamm 57, 10707 Berlin, Germany
Tel: +49 (0)30 32 79 060

First published 2003 by Blackwell Science Ltd

Library of Congress
Cataloging-in-Publication Data
is available

0-632-06395-5

A catalogue record for this title is available from the British Library

Set in 9 on 11pt Times New RomanPS
SNP Best-set Typesetter Ltd., Hong Kong
http://www.sparks.co.uk
Printed and bound in Great Britain by
TJ International Limited, Padstow

For further information on
Blackwell Science, visit our website:
www.blackwellpublishing.com

Contents

List of Contributors

Editors

Jim Smith, PhD, MBA, FIFST
Director, Strategic Research & Technical Services
Prince Edward Island Food Technology Centre
PO Box 2000
101 Belvedere Avenue
Charlottetown
Prince Edward Island
Canada C1A 7N8

Lily Hong-Shum, MSc
Food Scientist, R&D
Newlywed Foods
9110-23 Ave
Edmonton, Alberta
Canada T6N 1H9
formerly Food Scientist with
Prince Edward Island Food Technology Centre
PO Box 2000
101 Belvedere Avenue
Charlottetown
Prince Edward Island
Canada C1A 7N8

Contributors

Part 1 Acidulants

Stephanie Doores
Associate Professor of Food Science
The Pennsylvania State University
Department of Food Science
103 Borland Laboratory
University Park
PA 16802
USA

Part 2 Antioxidants

Fereidoon Shahidi, PhD, FCIFST, FACS
Professor of Food Biochemistry
Department of Biochemistry, Memorial University of Newfoundland
St John's
Newfoundland
Canada A1B 3X9

P.K.J.P.D. (Janitha) Wanasundara
Professional Research Associate
University of Saskatchewan
Department of Applied Microbiology and Food Science
College of Agriculture, Room 3E68
51 Campus Drive
Saskatoon
Saskatchewan
Canada S7N 5A8

Part 3 Colourings

Paul Collins
GNT UK Ltd
14, Stadium Business Court
Millennium Way
Pride Park
Derby
Derbyshire DE24 8HP
England
formerly with Overseal Foods Ltd
Park Road
Overseal
Derbyshire DE12 6JK
England

The late **Peter Rayner**
formerly with Overseal Foods Ltd
Park Road
Overseal
Derbyshire
England DE12 6JK

Part 4 Emulsifiers

Eric A. Flack
Consultant
Greenewood
The Park
Great Barton
Bury St Edmonds
Suffolk IP31 2SX
England

Part 5 Enzymes

Jim Smith

Part 6 Flavour Enhancers

Lily Hong-Shum

Part 7 Flour Additives

Catriona Crawford
formerly with Campden & Chorleywood Food Research Association
Chorleywood
Hertfordshire WD3 5SH
England

Part 8 Gases

Jim Smith
Lily Hong-Shum

Part 9 Nutritive Additives

Lori W. Léger
Food Scientist
Prince Edward Island Food Technology Centre
PO Box 2000
101 Belvedere Avenue
Charlottetown
Prince Edward Island
Canada C1A 7N8

G.G. Zawadzka
Chemistry Technologist
Prince Edward Island Food Technology Centre
PO Box 2000
101 Belvedere Avenue
Charlottetown
Prince Edward Island
Canada C1A 7N8

Part 10 Polysaccharides

Rachel Shepherd
Lecturer, Department of Cell & Molecular Biology
University of Technology, Sydney
PO Box 123, Broadway
Sydney, NSW 2007
Australia
formerly with Industrial Research Ltd
Gracefield Laboratories
Gracefield Road
PO Box 31–310
Wellington
New Zealand

Part 11 Preservatives

Jim Smith

Part 12 Sequestrants

Jean Chaw-Kant, MSc
University of Alberta
6826 40th Avenue
Edmonton
Alberta
Canada T6K 1B4

Part 13 Solvents

Lynn M. McMullen
Associate Professor, Food Microbiology
Department of Agricultural, Food & Nutritional Science
University of Alberta
4–10 Agriculture Forestry Centre
Edmonton, Alberta
Canada T6G 2P5

Part 14 Sweeteners

David K. MacKinnon, BSc (Math), BSc (EE), EIT (APEGNB)
Visiting Worker
NRC Institute for Information Technology
Room 359 Build M-50
1200 Montreal Road
Ottawa, Ontario
Canada K1A 0R6

How to Use This Book

The prudent use of approved food additives continues to be important in the food industry in order to provide safe, convenient, quality food products with useful shelf-lives. Developments in processing technologies will minimise the use of additives but their advantages continue in certain product formats.

The *Food Additives Data Book* contains practical information about a wide range of food additives. They are organised by functional category such as acidulants, antioxidants, colourings, etc. The priority is to provide useful information for the practising food technologist and student. The 'function in foods' and 'technology of use in foods' sections are likely to be the most useful ones for each additive in the data book. These illustrate why the additive finds application in certain food products and how it is used from a practical point of view. Chemical and physical data are provided to help the food technologist in using the additives. Some brief information on legislation is provided also.

To find information about a certain additive it can be searched for in the table of contents (if the functional category is known) or in the index. The index is extensive and can be used to locate information about certain foods, synonyms, alternative additives and other topics of interest.

Part 1

Acidulants

NAME:	Acetic acid
CATEGORY:	Emulsifiers/stabilisers/ Chelating agents/ pH control agents/ Preservatives/ Flavour enhancers and modifiers/ Solvents/ Firming agents
FOOD USE:	Baked goods/ Cereals and cereal products/ Dairy products/ Edible oils and fats/ Fish, seafoods and products/ Meat, poultry and eggs and products/ Fruits, vegetables and nuts and products/ Sugars, sugar preserves, and confectionery/ Vinegar, pickles and sauces
SYNONYMS:	Ethanoic acid/ Ethylic acid/ Glacial acetic acid/ Methane carboxylic acid/ Monocarboxylic acid/ Vinegar/ CAS 64-19-7/ DOT 2789/2790/ FEMA No. 2006
FORMULA:	CH3-COOH
MOLECULAR MASS:	60.05
ALTERNATIVE FORMS:	Calcium acetate; Hydroxyacetic acid; Manganese acetate; Methyl acetate; Potassium acetate; Sodium acetate; Zinc acetate
PROPERTIES AND APPEARANCE:	Clear, colourless liquid
BOILING POINT IN °C AT VARIOUS PRESSURES (INCLUDING 760 mm Hg):	117.9
MELTING RANGE IN °C:	16.6
FLASH POINT IN °C:	43 (closed cup), 57 (open cup)
IONISATION CONSTANT AT 25°C:	1.76×10^{-5}
DENSITY AT 20°C (AND OTHER TEMPERATURES) IN g/l:	1.0492 @ 20°/4°C
HEAT OF COMBUSTION AT 25°C:	209.02 kg carlories/gram molecular weight

VAPOUR PRESSURE AT VARIOUS TEMPERATURES IN mm Hg:	−17.2	(1 mm)
	17.5	(10 mm)
	43	(40 mm)
	63	(100 mm)
	99	(400 mm)
	118.1	(760 mm)
PURITY %:	≥99.5 by weight	
HEAVY METAL CONTENT MAXIMUM IN ppm:	10	
ARSENIC CONTENT MAXIMUM IN ppm:	3	

SOLUBILITY % AT VARIOUS TEMPERATURE/pH COMBINATIONS:

in water:	**@ 25°C**	Acid and its salts are freely soluble
in ethanol solution:	**@ 25°C**	Freely soluble
	@ 25°C	Potassium and sodium salts are freely soluble
	@ 25°C	Calcium salt is slightly soluble

FUNCTION IN FOODS:

Acetic acid is GRAS for miscellaneous and general-purpose usage in the United States with no limitation other than good manufacturing practices. Acetic acid can be used as an acidifier and flavouring agent in condiments such as mustard, catsup, salad dressings, sauces, canned fruits, and mayonnaise. It can be used to alter the acidity of acidified milk, acidified low-fat milk, acidified skim milk, meat and poultry products. It can also be used to alter the acidity of cold pack cheese food such that the pH does not exceed 4.5, and pasteurised process cheese (pH 5.3), cheese food (pH 5.0), and cheese spread (pH 4.0). It is a pickling agent in products such as sausages and pigs' feet, sweet and sour pickles, marinades and vinaigrettes. It is used to separate fatty acids and glycerol in rendered fats at a level sufficient for purpose. It assists in caramelisation. It is used as an emulsifier in the manufacture of hydroxylated lecithin. It can also be used to sanitise equipment and as a boiler water additive. Maximum levels recommended in foods are 0.25% for baked goods and baking mixes, 0.5%

for chewing gum, fats and oils, 0.6% for meat products, 0.8% for cheese and dairy product analogs, 3.0% for gravies and sauces, 9.0% for condiments and relishes, and 0.15% for all other food categories when used in accordance with good manufacturing practices.

Calcium acetate is GRAS for miscellaneous and general-purpose usage in the United States with no limitation other than good manufacturing practices. Calcium acetate can be used as a firming agent, pH control agent, processing aid, sequestrant, texturiser, thickener, and as an antirope agent in bakery products. It acts as a stabiliser in sausage casings and as a stabiliser when salt migrates from food-packaging materials. Maximum levels recommended are 0.02% in cheese, 0.15% in sweet sauces, toppings and syrups, 0.2% in baked goods, baking mixes, gelatins, puddings and fillings, and 0.0001% in all other foods in accordance with good manufacturing practices.

Potassium acetate is used as a synthetic flavouring.

Sodium acetate is GRAS for miscellaneous and general-purpose usage in the United States with no limitation other than good manufacturing practices. Sodium acetate can be used as a pH control agent, flavouring agent and adjuvant, stabiliser and buffer in certain milk and meat products, and as a boiler additive for food-grade steam. Maximum levels recommended in foods are 0.007% in breakfast cereal, 0.05% in snack foods, soup mixes and sweet sauces, 0.12% in meat products and jams and jellies, 0.15% in hard candy, 0.2% in soft candy, 0.5% in fats and oils, 0.6% in grain products and snack foods. Sodium acetate is limited to 2 oz/100 lb to acidify artificially flavoured fruit jelly, preserves and jams.

Hydroxyacetic acid, manganese acetate, methyl acetate, and zinc acetate can be used as a component of adhesives.

TECHNOLOGY OF USE IN FOODS:

Acetic acid $pK_1 = 4.75$

Calcium and sodium acetate are hygroscopic.

Acetic acid is more effective in limiting bacterial and yeast growth than mould growth.

FOOD SAFETY ISSUES:

LD_{50} (mg/kg body weight) acetic acid. Mouse 3,530–4,960 oral route; 525 intravenous route. Rat 3,310–3,530 oral route. Rabbit 1,200 oral route; 1,200 subcutaneous route; 1,200 rectal route.

LD_{50} (mg/kg body weight) calcium acetate. Mouse 52 intravenous route. Rat 147 intravenous route.

LD_{50} (mg/kg body weight) sodium acetate. Mouse 3,310 oral route; 380 intravenous route. Rat 3,530–4,960 oral route.

Acceptable daily intake for humans of acetic acid and its calcium, potassium and sodium salts is not limited.

LEGISLATION:

USA:

Acetic acid: 9 CFR 318.7, 381.147; 21 CFR 73.85, 131.111, 131.136, 131.144, 133.124, 133.169, 133.173, 133.179, Part 145, 172.814, 178.1010, 182.1, 184.1005
Calcium acetate: 21 CFR 181.29, 184.1185
Hydroxyacetic acid: 21 CFR 175.105
Manganese acetate: 21 CFR 175.105
Methyl acetate: 21 CFR 175.105
Potassium acetate: 21 CFR 172.515
Sodium acetate: 21 CFR 150.141, 150.161, 173.310, 182.70, 184.1721
Zinc acetate: 21 CFR 175.105

REFERENCES:

Budavari, S., O'Neill, M. J., Smith, A., and Heckelman, P. E. (1989) *The Merck Index.* 11th edn., Merck, Rahway, NJ.

Code of Federal Regulations (CFR) (1995) Volumes 9 (U. S. Department of Agriculture [USDA]), 21 (Food and Drug Administration [FDA]), and 27 (Bureau of Alcohol, Tobacco, and Firearms [BATF]). Office of the Federal Register, National Archives and Records Administration, U. S. Government Printing Office, Washington, DC.

Deshpande, S. S., Salunkhe, D. K., and Deshpande, U. S. (1995) Food Acidulants. In: Maga, J. A., and A. T. Tu (Eds.), *Food Additive Toxicology*. Marcel Dekker, Inc., New York, Ch. 2.

Doores, S. (1990) pH Control Agents and Acidulants, pp. 477–510. In: Branen, A. L., Davidson, P. M., and S. Salminen (Eds.), *Food Additives.* Marcel Dekker, Inc., New York, Ch. 13.

Doores, S. (1993) Organic Acids. In: Davidson, P. M., and A. L. Branen (Eds.), *Antimicrobials in Foods*, 2nd edn., Marcel Dekker, Inc., New York, pp. 95–136.

Food Chemicals Codex (1981) 3rd edn., Committee on Food Chemicals Codex, Food and Nutrition Board, Institute of Medicine, National Academy of Sciences, National Academy Press, Washington, DC.

Food Chemicals Codex (1996) 4th edn., Committee on Food Chemicals Codex, Food and Nutrition Board, Institute of Medicine, National Academy of Sciences, National Academy Press, Washington, DC.

Lewis, R. J., Sr. (1989) *Food Additives Handbook.* Van Nostrand Reinhold, New York.

Lide, D. R. (Ed.) (1993) *Handbook of Chemistry and Physics.* 74th edn. CRC Press, Inc., Boca Raton, FL.

NAME:	Adipic acid
CATEGORY:	Chelating agents/ pH control agents/ Preservatives/ Flavour enhancers and modifiers/ Flour and baking additives
FOOD USE:	Baked goods/ Dairy products/ Edible oils and fats/ Fish and seafoods and products/ Meat, poultry and eggs and products/ Fruit, vegetables and nuts and products/ Beverages/ Vinegar, pickles and sauces
SYNONYMS:	Acifloctin/ Acinetten/ Adilactetten/ Adipinic acid/ Hexanedioic acid/ Molten adipic acid/ 1,4-Butanedicarboxylic acid/ 1,6-Hexanedioic acid/ CAS 124-04-9/ DOT 9077/ FEMA No. 2011
FORMULA:	HOOC-(CH2)4-COOH
MOLECULAR MASS:	146.14
ALTERNATIVE FORMS:	Calcium adipate; Magnesium adipate
PROPERTIES AND APPEARANCE:	White crystals or crystalline, non-hygroscopic powder
BOILING POINT IN °C AT VARIOUS PRESSURES (INCLUDING 760 mm Hg):	337.5 (closed cup)
MELTING RANGE IN °C:	151.5–154
FLASH POINT IN °C:	196
IONISATION CONSTANT AT 25°C:	3.71×10^{-5} 3.87×10^{-6}
DENSITY AT 20°C (AND OTHER TEMPERATURES) IN g/l:	1.360 @ 25°/4°
HEAT OF COMBUSTION AT 25°C:	668.29 kg calories/gram molecular weight
VAPOUR PRESSURE AT VARIOUS TEMPERATURES IN mm Hg:	159.5 (1 mm) 191 (5 mm) 205.5 (10 mm)

	222 (20 mm) 240.5 (40 mm) 265 (100 mm) 337.5 (760 mm)
PURITY %:	≥99.6
WATER CONTENT MAXIMUM IN %:	0.2
HEAVY METAL CONTENT MAXIMUM IN ppm:	10
ARSENIC CONTENT MAXIMUM IN ppm:	3
ASH MAXIMUM IN %:	≤0.002–0.02
SOLUBILITY % AT VARIOUS TEMPERATURE/pH COMBINATIONS:	
in water:	@ **100°C** 160 @ **20°C** 1.4
in ethanol solution:	Freely soluble
FUNCTION IN FOODS:	Adipic acid is GRAS for miscellaneous and general-purpose usage in the United States with no limitation other than good manufacturing practices. Adipic acid can be used as a pH control agent, neutralising agent, leavening agent and flavouring agent. It can be used to alter the acidity of acidified milk, acidified low-fat milk and acidified skim milk, margarine or oleomargarine. It can be used as a sequestrant in oils and improves melting characteristics and texture of process cheese and cheese spreads. It is used in puddings and gelatin to improve set and maintain acidities within the range of pH 2.5–3.0. It is an excellent, slow-acting, leavening agent that supports the even release of carbon dioxide in baked goods. It increases the whipping quality of products containing egg white. It can be used in the production of resinous and polymeric coatings. Maximum levels recommended are 0.0004% for frozen dairy desserts, 0.005% for non-alcoholic beverages, 0.05% for baked goods and baking mixes, 0.1% for gravies, 0.3% for fats, oils and meat products, 0.45% for dairy product analogues, 0.55% for gelatin, puddings, and fillings, 1.3% for snack foods, 5% for condiments and relishes and 0.02% for other food categories when used in accordance with good manufacturing practices.

Diisobutyl adipate can be used as a component of adhesives and a plasticiser for food packaging materials at levels not to exceed good manufacturing practices.

ALTERNATIVES:

Adipic acid is used as an acidulant in commercial baking powders to replace tartaric acid and cream of tartar and phosphates because adipic acid is non-hygroscopic.

TECHNOLOGY OF USE IN FOODS:

Adipic acid $pK_1 = 4.43$; $pK_2 = 5.41$

Adipic acid is four to five times more soluble than fumaric acid at room temperature and has the lowest acidity of any of the food acids. It imparts a slowly developing, smooth, mildly acid flavour and is essential in supplementing foods with delicate flavours. It is practically non-hygroscopic which is an advantage in prolonging the shelf life of powdered products.

SYNERGISTS:

Adipic acid combined with sodium metabisulfite can be used in the preservation of sausages and other meat products.

FOOD SAFETY ISSUES:

LD_{50} (mg/kg body weight) adipic acid. Mouse 1,900 oral route; 680 intravenous route; 275 intraperitoneal route. Rat 3,600 oral route. Rabbit 2,430–4,860 oral route; 2,430 intravenous route.

Acceptable daily intake for humans of adipic acid is limited conditionally up to 5 mg/kg body weight.

LEGISLATION:

USA:

Adipic acid: 9 CFR 318.7; 21 CFR 75.85, 131.111, 131.136, 131.144, 166.110, 172.515, 175.300, 175.320, 177.2420, 184.1009

Diisobutyl adipate: 21 CFR 175.105, 181.27

REFERENCES:

Budavari, S., O'Neill, M. J., Smith, A., and Heckelman, P. E. (1989) *The Merck Index.* 11th edn., Merck, Rahway, NJ.

Code of Federal Regulations (CFR) (1995) Volumes 9 (U. S. Department of Agriculture [USDA]), 21 (Food and Drug Administration [FDA]), and 27 (Bureau of Alcohol, Tobacco, and Firearms [BATF]). Office of the Federal Register, National Archives and Records Administration, U. S. Government Printing Office, Washington, DC.

Deshpande, S. S., Salunkhe, D. K., and Deshpande, U. S. (1995) Food Acidulants. In: Maga, J. A., and A. T. Tu (Eds.), *Food Additive Toxicology*. Marcel Dekker, Inc., New York, Ch. 2.

Doores, S. (1990) pH Control Agents and Acidulants. In: Branen, A. L., Davidson, P. M., and S. Salminen (Eds.), *Food Additives.* Marcel Dekker, Inc., New York, pp. 477–510.

Doores, S. (1993) Organic Acids. In: Davidson, P. M., and A. L. Branen (Eds.), *Antimicrobials in Foods*, 2nd edn., Marcel Dekker, Inc., New York, pp. 95–136.

Food Chemicals Codex (1981) 3rd edn., Committee on Food Chemicals Codex, Food and Nutrition Board, Institute of Medicine, National Academy of Sciences, National Academy Press, Washington, DC.

Food Chemicals Codex (1996) 4th edn., Committee on Food Chemicals Codex, Food and Nutrition Board, Institute of Medicine, National Academy of Sciences, National Academy Press, Washington, DC.

Lewis, R. J., Sr. (1989) *Food Additives Handbook.* Van Nostrand Reinhold, New York.

Lide, D. R. (Ed.) (1993) *Handbook of Chemistry and Physics.* 74th edn. CRC Press, Inc., Boca Raton, FL.

NAME:	Caprylic acid
CATEGORY:	Emulsifiers/stabilisers/ pH control agents/ Preservatives/ Flavour enhancers and modifiers/ Anti-caking agents/ Antifoaming agent/ Foaming agents/ Glazing and coating agents
FOOD USE:	Baked goods/ Dairy products/ Edible oils and fats/ Meat, poultry and eggs and products/ Fruit, vegetables and nuts and products/ Sugars, sugar preserves and confectionery
SYNONYMS:	C-8 acid/ Hexacid 898/ *n*-Caprylic acid/ *n*-Octoic acid/ *n*-Octylic acid/ neo-fat 8/ Octanoic acid/ Octic acid/ 1-Heptanecarboxylic acid/ CAS 124-07-2
FORMULA:	CH3-(CH2)6-COOH
MOLECULAR MASS:	144.22
ALTERNATIVE FORMS:	Aluminium caprylate; calcium caprylate; cobalt caprylate; iron caprylate; magnesium caprylate; manganese caprylate; potassium caprylate; sodium caprylate
PROPERTIES AND APPEARANCE:	Colourless, oily liquid
BOILING POINT IN °C AT VARIOUS PRESSURES (INCLUDING 760 mm Hg):	239.3
MELTING RANGE IN °C:	16.5–16.7
IONISATION CONSTANT AT 25°C:	1.28×10^{-5}
DENSITY AT 20°C (AND OTHER TEMPERATURES) IN g/l:	0.9088 @ 20°/4°
VAPOUR PRESSURE AT VARIOUS TEMPERATURES IN mm Hg:	92.3 (1 mm) 124 (10 mm) 150.6 (40 mm) 172.2 (100 mm) 213.9 (400 mm) 237.5 (760 mm)

PURITY %: 99.6

WATER CONTENT MAXIMUM IN %: 10

HEAVY METAL CONTENT MAXIMUM IN ppm: 3

ARSENIC CONTENT MAXIMUM IN ppm: 0.1

SOLUBILITY % AT VARIOUS TEMPERATURE/pH COMBINATIONS:

in water: **@ 20°C** 0.068

in ethanol solution: Freely soluble

FUNCTION IN FOODS:

Caprylic acid is GRAS for miscellaneous and general-purpose usage in the United States with no limitation other than good manufacturing practices. Caprylic acid can be used as a flavouring agent and adjuvant, lubricant, binder and defoaming agent. It can be used indirectly as an antimicrobial agent in cheese wrappers and as a coating on fresh citrus fruits. It can also be used to assist in lye peeling. It is an indirect antimicrobial food additive when migrating to food from paper and paperboard products. It can be used as a sanitising solution on food processing equipment. Maximum levels recommended for specific products are 0.005% for fats and oils, frozen dairy desserts, gelatins, puddings, meat products and soft candy, 0.013% for baked goods and baking mixes, 0.016% for snack foods, 0.04% for cheeses, and 0.001% or less for other food categories.

Aluminium, calcium, magnesium, potassium or sodium caprylate can be used as a binder, emulsifier and anti-caking agent when used in accordance with good manufacturing practices. The salts of caprylic acid can also be used in the production of resinous and polymeric coatings.

Cobalt, iron or manganese caprylate can be used as a drier when migrating from food packaging material, with no limitation other than good manufacturing practices.

TECHNOLOGY OF USE IN FOODS:

Caprylic acid $pK_1 = 4.89$

Caprylic acid imparts a 'sweat-like' or cheesy odour and buttery taste to foods. It is more inhibitory to a broad group of microorganisms at a lower concentration at around neutral pH than other acidulants.

FOOD SAFETY ISSUES:

LD_{50} (mg/kg body weight). Mouse 600 intravenous route.

Acceptable intake for humans of caprylic acid is not limited.

LEGISLATION:

USA:

Caprylic acid: 21 CFR 172.210, 172.860, 173.315, 173.340, 178.1010, 182.90, 184.1025, 186.1025
Aluminium caprylate: 21 CFR 172.863
Calcium caprylate: 21 CFR 172.863
Cobalt caprylate: 21 CFR 181.25
Iron caprylate: 21 CFR 181.25
Magnesium caprylate: 21 CFR 172.863
Manganese caprylate: 21 CFR 181.25
Potassium caprylate: 21 CFR 172.863
Sodium caprylate: 21 CFR 172.863

REFERENCES:

Budavari, S., O'Neill, M. J., Smith, A., and Heckelman, P. E. (1989) *The Merck Index*. 11th edn., Merck, Rahway, NJ.

Code of Federal Regulations (CFR) (1995) Volumes 9 (U. S. Department of Agriculture [USDA]), 21 (Food and Drug Administration [FDA]), and 27 (Bureau of Alcohol, Tobacco, and Firearms [BATF]). Office of the Federal Register, National Archives and Records Administration, U. S. Government Printing Office, Washington, DC.

Deshpande, S. S., Salunkhe, D. K., and Deshpande, U. S. (1995) Food Acidulants. In: Maga, J. A., and A. T. Tu (Eds.), *Food Additive Toxicology*. Marcel Dekker, Inc., New York, Ch. 2.

Doores, S. (1990) pH Control Agents and Acidulants. In: Branen, A. L., Davidson, P. M., and S. Salminen (Eds.), *Food Additives*. Marcel Dekker, Inc., New York, pp. 477–510.

Doores, S. (1993) Organic Acids. In: Davidson, P. M., and A. L. Branen (Eds.), *Antimicrobials in Foods*, 2nd edn., Marcel Dekker, Inc., New York, pp. 95–136.

Food Chemicals Codex (1981) 3rd edn., Committee on Food Chemicals Codex, Food and Nutrition Board, Institute of Medicine, National Academy of Sciences, National Academy Press, Washington, DC.

Food Chemicals Codex (1996) 4th edn., Committee on Food Chemicals Codex, Food and Nutrition Board, Institute of Medicine, National Academy of Sciences, National Academy Press, Washington, DC.

Lewis, R. J., Sr. (1989) *Food Additives Handbook.* Van Nostrand Reinhold, New York.

Lide, D. R. (Ed.) (1993) *Handbook of Chemistry and Physics.* 74th edn. CRC Press, Inc., Boca Raton, FL.

NAME:	Citric acid
CATEGORY:	Emulsifiers/stabilisers/ Chelating agents/ Nutritive additives/ Antioxidants/ pH control agents/ Preservatives/ Flavour enhancers and modifiers/ Solvents/ Flour and baking additives/ Anti-caking agents/ Firming agents/ Glazing and coating agents
FOOD USE:	Baked goods/ Dairy products/ Edible oils and fats/ Fish and seafoods and products/ Meat, poultry and eggs and products/ Fruit, vegetables and nuts and products/ Beverages/ Soft drinks/ Sugar, sugar preserves and confectionery/ Alcoholic drinks/ Vinegar, pickles and sauces
SYNONYMS:	Aciletten/ Boxylic acid/ Citretten/ Citro/ 2-Hydroxy-1,2,3-propanetricarboxylic acid/ β-Hydroxytricarballyic acid/ β-Hydroxytricarboxylic acid/ CAS 77-92-9/ FEMA No. 2306
FORMULA:	CH2-(COOH)-C-(OH)(COOH)-CH2-COOH
MOLECULAR MASS:	192.14
ALTERNATIVE FORMS:	Ammonium citrate; calcium citrate; dipotassium citrate; disodium citrate; ferric ammonium citrate; ferric citrate; ferrous citrate; iron ammonium citrate; isopropyl citrate; magnesium citrate; manganese citrate; monoglyceride citrate; monopotassium citrate; monosodium citrate; stearyl citrate; stearyl monoglyceridyl citrate; triethyl citrate; tripotassium citrate; trisodium citrate
PROPERTIES AND APPEARANCE:	Colourless, odourless, translucent crystals or as a white granular to fine crystalline powder
BOILING POINT IN °C AT VARIOUS PRESSURES (INCLUDING 760 mm Hg):	Decomposition
MELTING RANGE IN °C:	153 (anhydrous form) 100 (monohydrate)
FLASH POINT IN °C:	100
IONISATION CONSTANT AT 25°C:	7.10×10^{-4} 1.68×10^{-5} 4.10×10^{-7}

DENSITY AT 20°C (AND OTHER TEMPERATURES) IN g/l:	1.665 (anhydrous form) 20°/4° 1.542 (monohydrate) 20°/4°
HEAT OF COMBUSTION AT 25°C IN J/kg:	468.6 (anhydrous form)
PURITY %:	≥99.5
WATER CONTENT MAXIMUM IN %:	0.5 (anhydrous form) 8.8 (hydrous form)
HEAVY METAL CONTENT MAXIMUM IN ppm:	10
ARSENIC CONTENT MAXIMUM IN ppm:	3
ASH MAXIMUM IN %:	0.05

SOLUBILITY % AT VARIOUS TEMPERATURE/pH COMBINATIONS:

in water:	@ **10°C**	54.0	@ **20°C**	59.2	@ **30°C**	64.3	@ **40°C**	68.6	@ **50°C**	70.9
	@ **60°C**	73.5	@ **70°C**	76.2	@ **80°C**	78.8	@ **90°C**	81.4	@ **100°C**	84.0
in ethanol solution:	@ **25°C**	1 g/2 ml								

FUNCTION IN FOODS:	Citric acid is GRAS for miscellaneous and general-purpose usage in the United States with no limitation other than good manufacturing practices. Citric acid is used as a curing accelerator, dispersing agent, sequestrant, and synergist for antioxidants. Citric acid can be used to alter the acidity of acidified milk, acidified low-fat milk, acidified skim milk, cultured milk, cultured low-fat milk, acidified skim milk, cultured skim milk, margarine and oleomargarine. It can also be used to alter the acidity of cold pack cheese food such that the pH does not exceed 4.5, pasteurised process cheese (pH 5.3), cheese food (pH 5.0), cheese spread (pH 4.0), and dry curd cottage cheese (pH 4.5–4.7). It can be used to alter the acidity of canned fruits, corn, artificially sweetened canned figs, canned peaches, and canned prune juice, artificially sweetened fruit jelly, fruit preserves, jams, meat and meat food products, and poultry. It controls acidity in pectin and alginate gels and reduces heat processing requirements by lowering pH. It enhances the flavour of citrus-based foods including canned sweetened apricots, cherries, pears, and chili con carne. Citric acid is a precursor to diacetyl and indirectly improves the flavour and aroma of a variety of cultured dairy products. It is limited to 0.15% by weight of the milk used or equivalent amount of sodium citrate as a flavour precursor or at 0.1% in sour cream and sour half-and-half. It assists in caramelisation. It can be used a plasticiser and an emulsifying agent to provide texture and improve melting characteristics in pasteurised processed cheese at a level not to exceed 3%. Citric acid is limited at 0.01% alone or in combination with antioxidants for any product containing antioxidants. It is approved for use in ice-cream, sherbet and ices, beverages, and salad dressings. Citric acid singly or in combination with sodium acetate can delay discoloration on fresh cuts of beef, lamb, and pork at levels not to exceed 250 ppm or 0.9 mg/sq in of product surface or exceed 500 ppm of 1.8 mg/sq in of surface when in combination with ascorbic acid, erythorbic acid or sodium ascorbate. Citric acid may be used in cured products or in 10% solution used to spray surfaces of cured meats and meat food products prior to packaging to replace up to 50% of the ascorbic acid, erythorbic acid, sodium ascorbate, or sodium erythorbate that is used. It is used to preserve cured colour of pork cuts during storage at a level not to exceed 30% in water solution used to spray surfaces of cured cuts prior to packaging. It can be used at levels not to exceed 0.001% in dry sausage, 0.01% in fresh pork sausage, and 0.01% in dried meats. Its use is limited at 0.003% for dry sausage in combination with antioxidants. Citric acid may be used to replace up to 50% of the ascorbic acid or sodium ascorbate in poultry to accelerate colour fixing and is limited at 0.01% alone or in combination with antioxidants in poultry fats to increase effectiveness of antioxidants. Its use is limited for French dressing, mayonnaise and salad dressing at 25% or less of the weight of the acids of the vinegar or diluted vinegar calculated as acetic acid. Citric acid can be added to grapes after fermentation or in

combination with other acids after fermentation to correct deficiencies at a level not to exceed 9 g/l of finished wine. It may be added to adjust the acidity of citrus fruit juice or wine at a level not to exceed 9 g/l finished wine. It prevents cloudy precipitates and inhibits oxidation. Citric acid can be used to stabilise wine at a level not to exceed 5.8 lb/1000 gallons (*ca.* 120 mg/l).

Ammonium citrate is GRAS for miscellaneous and general-purpose usage in the United States with no limitation other than good manufacturing practices. Ammonium citrate can be used as a stabiliser when migrating from food packaging material. It is used in the manufacture of adhesives and in the production of resinous and polymeric coatings. It can be used as a flavour enhancer, pH control agent and is used in non-alcoholic beverages and cheeses.

Calcium citrate is GRAS for miscellaneous and general-purpose usage in the United States with no limitation other than good manufacturing practices. Calcium citrate can be used as a nutrient and dietary supplement, pH control agent, buffer, firming agent, and sequestrant. It improves the baking properties of flour. It can be used in infant formula. It can be used as an emulsifying agent in pasteurised process cheese, cheese food, and cheese spread at a limit of 3.0% and a gelling agent in artificially sweetened jelly, fruit preserves and jam. It is limited at 0.1% as a flavour precursor in sour cream and sour half-and-half.

Ferric ammonium citrate, ferric citrate and ferrous citrate are used as a nutrient supplement at levels not to exceed good manufacturing practices. It may also be used in infant formula.

Iron ammonium citrate is used as an anti-caking agent in salt for human consumption at a level not to exceed 25 ppm (0.0025%) in finished salt.

Isopropyl citrate is GRAS for miscellaneous and general-purpose usage in the United States with no limitation other than good manufacturing practices. Isopropyl citrate can be used as a sequestrant, solvent and vehicle. It is used to protect flavour in margarine at a level of 0.02%, in non-alcoholic beverages, fats and oils at levels not to exceed good manufacturing practices. It can be used to increase effectiveness of antioxidants in lard, shortening, oleomargarine, fresh pork sausage, and dried meats at a level of 0.02%.

Magnesium citrate is used in the production of resinous and polymeric coatings.

Manganese citrate is GRAS for miscellaneous and general-purpose usage in the United States with no limitation other than good manufacturing practices. Manganese citrate can be used as a dietary supplement in baked goods, non-alcoholic beverages, dairy product analogues, fish products, meat products, milk products, poultry products, and infant formula.

Monoglyceride citrate is used to increase effectiveness in lard, shortening, fresh pork sausage, and dried meats at a level not to exceed 0.02%. It is used to increase effectiveness of antioxidants in poultry fats at a level of 0.02%.

Mono-, di-, and tripotassium citrate, and mono-, di-, and trisodium citrate are classified as stabilisers when migrating from food-packaging materials.

Potassium citrate is GRAS for miscellaneous and general-purpose usage in the United States with no limitation other than good manufacturing practices. Potassium citrate can be used as sufficient for purpose in jelly, margarine, meat products, milk and wine. It can be used to acidify margarine or oleomargarine at a level sufficient for purpose. It is used in the production of resinous and polymeric coatings. It can be used as an emulsifying agent in pasteurised process cheese, cheese food and cheese spread at a limit of 3.0%. Its use is limited to 2 oz/100 lb (1.4 mg/kg) in artificially sweetened fruit jelly, fruit preserves and jam. It is used as a pH control agent and sequestrant in the treatment of citrus wines at a level not to exceed 25 lb/1000 gal wine (ca. 50 mg/l).

Sodium citrate is GRAS for miscellaneous and general-purpose usage in the United States with no limitation other than good manufacturing practices. Sodium citrate is used as a chelating agent in conjunction with phosphate buffers to prepare non-caking meat-salt mixtures and to provide heat and storage stability in condensed, evaporated and sterile concentrated milks. It can be used to accelerate colour fixing in cured and comminuted poultry or poultry products to replace up to 50% ascorbic acid or sodium ascorbate that is used. It can be used as an emulsifying agent in pasteurised process cheese, cheese food, and cheese spread at a limit of 3.0%. It can be used to acidify margarine or oleomargarine. It is a denuding agent in tripe. It is used in the production of resinous and polymeric coatings. Sodium citrate is limited at a concentration of 10% in solution to spray on the surface of cured cuts of meat prior to packaging. It is limited at 0.1% as a flavour precursor in sour cream and sour half-and-half. Its use is limited to 2 oz/100 lb (1.4 mg/kg) of artificially sweetened fruit jelly, fruit preserves and jams.

Stearyl citrate is GRAS for miscellaneous and general-purpose usage in the United States with no limitation other than good manufacturing practices. Stearyl citrate can be used to protect flavour in margarine at 0.15%.

Stearyl monoglyceridyl citrate can be used as an emulsion or stabiliser in or with shortenings containing emulsifiers at a level sufficient for purpose.

Triethyl citrate is GRAS for miscellaneous and general-purpose usage in the United States with no limitation other than good manufacturing practice. It is used as a flavouring agent, solvent and vehicle and surface-active agent.

TECHNOLOGY OF USE IN FOODS:

Citric acid $pK_1 = 3.14$; $pK_2 = 4.77$; $pK_3 = 6.39$

Citric acid is more hygroscopic than adipic or fumaric acid and can create storage problems in powdered products.

Citric acid is compatible with fruit and berry flavours; it imparts a clean, tart taste in hard candies.

The salts of citric acid are commonly used as sequestrants.

Sodium citrate is added to carbonated beverages to reduce the sharpness of acid taste and it imparts a cool, saline taste and aids in the retention of carbonation.

SYNERGISTS:

Citric acid acts with antioxidants to prevent rancidity by chelating metal ions.

Sodium citrate in conjunction with phosphate buffers is used to prepare non-caking meat/salt mixtures.

FOOD SAFETY ISSUES:

LD_{50} (mg/kg body weight) citric acid. Mouse 5,040–5,790 oral route; 42–960 intravenous route; 961 intraperitoneal route; 2,700 subcutaneous route. Rat 11,700 oral route; 725–884 intraperitoneal route; 5,500 subcutaneous route. Rabbit 330 intravenous route.

LD_{50} (mg/kg body weight) potassium citrate. Dog 167 intravenous route.

LD_{50} (mg/kg body weight) sodium citrate. Mouse 44, intravenous route; 1,460 intraperitoneal route. Rat 1,210 intraperitoneal route. Rabbit 338 intravenous route.

Acceptable daily intake for humans of citric acid and its calcium, potassium and sodium salts is not limited.

LEGISLATION:

USA:

Citric acid: 9 CFR 318.7, 381.147; 21 CFR 73.85, 131.111, 131.112, 131.136, 131.138, 131.144, 131.146, 133.124, 133.129, 133.169, 133.173, 133.179, Part 145, 146.113, 146.187, 150.141, 150.161, 155.130, 161.190, 166.110, 169.115, 169.140, 169.150, 173.165, 173.280, 182.1033, 182.6033, 184.1033; 27 CFR 24.182, 24.246

Ammonium citrate: 21 CFR 175.105, 175.300, 177.1350, 181.29, 184.1140

Calcium citrate: 21 CFR 133.169, 133.173, 133.179, 150.141, 150.161, 155.200, 182.1195, 182.5195, 184.1195

Dipotassium citrate: 21 CFR 181.29
Disodium citrate: 21 CFR 181.29
Ferric ammonium citrate: 21 CFR 184.1296
Ferric citrate: 21 CFR 184.1298
Ferrous citrate: 21 CFR 184.1307c
Iron ammonium citrate: 21 CFR 172.430
Isopropyl citrate: 9 CFR 318.7; 21 CFR 166.110, 182.1386, 184.1386
Magnesium citrate: 21 CFR 175.300
Manganese citrate: 21 CFR 182.5449, 184.1449
Monoglyceride citrate: 9 CFR 318.7, 381.147; 21 CFR 172.832
Monopotassium citrate: 9 CFR 318.7; 21 CFR 133.169, 133.173, 133.179, 150.141, 150.161, 175.300, 181.29, 182.1625, 184.1625; 27 CFR 246
Monosodium citrate: 9 CFR 318.7, 381.147; 21 CFR 131.112, 131.136, 131.138, 131.144, 131.146, 131.160, 131.185, 133.169, 133.173, 133.179, 150.141, 150.161, 175.300, 181.29, 182.1751, 184.1751
Stearyl citrate: 9 CFR 318.7; 21 CFR 166.110, 182.1851, 184.1851
Stearyl monoglyceridyl citrate: 9 CFR 318.7; 21 CFR 172.755
Triethyl citrate: 21 CFR 184.1911
Tripotassium phosphate: 21 CFR 181.29
Trisodium phosphate: 21 CFR 181.29

REFERENCES:

Budavari, S., O'Neill, M. J., Smith, A., and Heckelman, P. E. (1989) *The Merck Index*. 11th edn., Merck, Rahway, NJ.

Code of Federal Regulations (CFR) (1995) Volumes 9 (U. S. Department of Agriculture [USDA]), 21 (Food and Drug Administration [FDA]), and 27 (Bureau of Alcohol, Tobacco, and Firearms [BATF]). Office of the Federal Register, National Archives and Records Administration, U. S. Government Printing Office, Washington, DC.

Deshpande, S. S., Salunkhe, D. K., and Deshpande, U. S. (1995) Food Acidulants. In: Maga, J. A., and A. T. Tu (Eds.), *Food Additive Toxicology*. Marcel Dekker, Inc., New York, Ch. 2.

Doores, S. (1990) pH Control Agents and Acidulants. In: Branen, A. L., Davidson, P. M., and S. Salminen (Eds.), *Food Additives*. Marcel Dekker, Inc., New York, pp. 477–510.

Doores, S. (1993) Organic Acids. In: Davidson, P. M., and A. L. Branen (Eds.), *Antimicrobials in Foods*, 2nd edn., Marcel Dekker, Inc., New York, pp. 95–136.

Food Chemicals Codex (1981) 3rd edn., Committee on Food Chemicals Codex, Food and Nutrition Board, Institute of Medicine, National Academy of Sciences, National Academy Press, Washington, DC.

Food Chemicals Codex (1996) 4th edn., Committee on Food Chemicals Codex, Food and Nutrition Board, Institute of Medicine, National Academy of Sciences, National Academy Press, Washington, DC.

Lewis, R. J., Sr. (1989) *Food Additives Handbook.* Van Nostrand Reinhold, New York.

Lide, D. R. (Ed.) (1993) *Handbook of Chemistry and Physics.* 74th edn. CRC Press, Inc., Boca Raton, FL.

NAME:	**Dehydroacetic acid**
CATEGORY:	pH control agents/ Preservatives
FOOD USE:	Dairy products/ Fruit, vegetables and nuts and products
SYNONYMS:	DHA/ DHS/ Dehydracetic acid/ Methylacetopyronone/ Δ lactone/ 2-Acetyl-5-hydroxy-3-oxo-4 hexenoic acid/ 3-Acetyl-6-methyl-l, 2-pyran-2,4-dione/ 3-Acetyl-6-methyl-2,4-pyrandione/ 3-Acetyl-6-methyl-2H-pyran-2,4 (3*H*)-dione/ 3-Acetyl-6-methyl-pyrandione-2,4/ CAS 520-45-6
FORMULA:	C8 H8 O4 (cyclic structure)
MOLECULAR MASS:	168.16
ALTERNATIVES FORMS:	Sodium dehydroacetate
PROPERTIES AND APPEARANCE:	White or nearly white crystalline powder
BOILING POINT IN °C AT VARIOUS PRESSURES (INCLUDING 760 mm Hg):	269.9
MELTING RANGE IN °C:	109–111
DENSITY AT 20°C (AND OTHER TEMPERATURES) IN g/l:	5.8
VAPOUR PRESSURE AT VARIOUS TEMPERATURES IN mm Hg:	91.7°C (1 mm)
PURITY %:	≥98.0
HEAVY METAL CONTENT MAXIMUM IN ppm:	10
ARSENIC CONTENT MAXIMUM IN ppm:	3
ASH MAXIMUM IN %:	0.1

SOLUBILITY % AT VARIOUS TEMPERATURE/pH COMBINATIONS:

in water:	**@ 25°C**	<0.1
in vegetable oil:	**@ 25°C (olive oil)**	1.6
in propylene glycol:	**@ 25°C**	1 g/35 ml

FUNCTION IN FOODS:

Dehydroacetic acid and its sodium salt are GRAS when used in accordance with good manufacturing practices. Dehydroacetic acid is approved as a preservative for cut or peeled squash at levels not to exceed 65 ppm expressed as dehydroacetic acid remaining in or on squash. It is bacteriostatic at concentrations of 0.1–0.4% and fungistatic at concentrations of 0.005–0.1%. It can be used as a component of adhesives with no limitation other than good manufacturing practices. It can also be used as a fungistatic agent in cheese wrappers.

TECHNOLOGY OF USE IN FOODS:

Dehydroacetic acid $pK_1 = 5.27$

Dehydroacetic acid has one of the highest dissociation constants of the organic acids and remains effective at higher pH ranges than other acids.

It is twice as effective as sodium benzoate at pH 5.0 against *Saccharomyces cerevisiae*, and 25 times more effective against *Penicillium glaucum* and *Aspergillus niger*.

FOOD SAFETY ISSUES:

LD_{50} (mg/kg body weight). Rat 1,000 oral route.

LEGISLATION:

USA:
Dehydroacetic acid: 21 CFR 172.130, 175.105, 175.300

REFERENCES:

Budavari, S., O'Neill, M. J., Smith, A., and Heckelman, P. E. (1989) *The Merck Index.* 11th edn., Merck, Rahway, NJ.

Code of Federal Regulations (CFR) (1995) Volumes 9 (U. S. Department of Agriculture [USDA]), 21 (Food and Drug Administration [FDA]), and 27 (Bureau of Alcohol, Tobacco, and Firearms [BATF]). Office of the Federal Register, National Archives and Records Administration, U. S. Government Printing Office, Washington, DC.

Deshpande, S. S., Salunkhe, D. K., and Deshpande, U. S. (1995) Food Acidulants. In: Maga, J. A., and A. T. Tu (Eds.), *Food Additive Toxicology*. Marcel Dekker, Inc., New York, Ch. 2.

Doores, S. (1990) pH Control Agents and Acidulants. In: Branen, A. L., Davidson, P. M., and S. Salminen (Eds.), *Food Additives.* Marcel Dekker, Inc., New York, pp. 477–510.

Doores, S. (1993) Organic Acids. In: Davidson, P. M., and A. L. Branen (Eds.), *Antimicrobials in Foods*, 2nd edn., Marcel Dekker, Inc., New York, pp. 95–136.

Food Chemicals Codex (1981) 3rd edn., Committee on Food Chemicals Codex, Food and Nutrition Board, Institute of Medicine, National Academy of Sciences, National Academy Press, Washington, DC.

Food Chemicals Codex (1996) 4th edn., Committee on Food Chemicals Codex, Food and Nutrition Board, Institute of Medicine, National Academy of Sciences, National Academy Press, Washington, DC.

Lewis, R. J., Sr. (1989) *Food Additives Handbook.* Van Nostrand Reinhold, New York.

Lide, D. R. (Ed.) (1993) *Handbook of Chemistry and Physics.* 74th edn. CRC Press, Inc., Boca Raton, FL.

NAME:	Fumaric acid
CATEGORY:	Emulsifiers/stabilisers/ Nutritive additives/ Antioxidants/ pH control agents/ Preservatives/ Flavour enhancers and modifier/ Flour and baking additives/ Firming agent
FOOD USE:	Baked goods/ Dairy products/ Edible oils and fats/ Meat, poultry and eggs and products/ Fruit, vegetables and nuts and products/ Beverages/ Sugars, sugar preserves and confectionery/ Alcoholic drinks
SYNONYMS:	Allomaleic acid/ Boletic acid/ Butenedioic acid/ Lichenic acid/ *trans*-Butenedioic acid/ *trans*-1,2-ethylenedicarboxylic acid/ (E)-butenedioic acid/ (E) 1,2-ethylenedicarboxylic acid/ CAS 110-17-8/ DOT 9126/ NSC-2752/ U-1149/ USAF EK-P-583
FORMULA:	HOOC-(CH)2-COOH
MOLECULAR MASS:	116.07
ALTERNATIVES:	Calcium fumarate; ferrous fumarate; magnesium fumarate; potassium fumarate; sodium fumarate; sodium stearyl fumarate
PROPERTIES AND APPEARANCE:	White granules or crystalline powder
BOILING POINT IN °C AT VARIOUS PRESSURES (INCLUDING 760 mm Hg):	290 (sublimes)
MELTING RANGE IN °C:	287
IONISATION CONSTANT AT 25°C:	9.30×10^{-4} 3.62×10^{-5}
DENSITY AT 20°C (AND OTHER TEMPERATURES) IN g/l:	1.635 @ 20°/4°
HEAT OF COMBUSTION AT 25°C:	318.99 kg calories/gram molecular weight
PURITY %:	≥99.5

WATER CONTENT MAXIMUM IN %:	0.5							
HEAVY METAL CONTENT MAXIMUM IN ppm:	10							
ARSENIC CONTENT MAXIMUM IN ppm:	3							
ASH MAXIMUM IN %:	0.1							
SOLUBILITY % AT VARIOUS TEMPERATURE/pH COMBINATIONS:								
in water:	**@ 25°C**	0.63	**@ 40°C**	1.07	**@ 60°C**	2.4	**@ 100°C**	9.8
in vegetable oil:	**@ 25°C (olive oil)**	Almost insoluble						
in ethanol solution:	**@ 95%**		**@ 30°C**	5.76				

FUNCTION IN FOODS:

Fumaric acid is used at a level not in excess of the amount reasonably required to accomplish the intended effect. It can be found in pie fillings, refrigerated biscuit doughs, and maraschino cherries. Fumaric acid can be used as a pH control agent to alter the acidity of acidified milk, acidified low-fat milk, acidified skim milk, canned fruit, artificially sweetened fruit jelly, fruit preserves and jam. It increases the strength of gelatin gels and displays some antioxidant properties in fat-containing foods. It blends with certain flavouring compounds to intensify after-taste of a flavour in fruit juices. It is used as a component of adhesives and in the production of resinous and polymeric coatings at levels not to exceed good manufacturing practices. It can be used to accelerate colour fixing in cured, comminuted poultry, poultry products, meat or meat food products at a level of 0.065% (or 1 oz/100 lb) of the weight of the poultry, poultry by-products, meat or meat by-products before processing. Fumaric acid can be added to grapes after fermentation or in combination with other acids after fermentation to correct deficiencies at levels not to exceed 9 g/l finished wine. The fumaric acid can be added to stabilise wine at levels not to exceed 25 lb/1000 gal (*ca.* 50 mg/l) and acid content of finished wine shall not exceed 3 g/l.

Calcium, ferrous, magnesium, potassium and sodium fumarates are a source of iron in foods and nutrient supplement in infant formula at levels consistent with good nutrition practice.

Sodium stearyl fumarate can be used as a dough conditioner in yeast-leavened bakery products at a level not to exceed 0.5% by weight of flour. It is used as a conditioning agent in dehydrated potatoes and processed cereals for cooking at levels not to exceed 1% by weight, and as a conditioning agent in starch-thickened or flour-thickened foods at levels not to exceed 0.2% by weight of the food. It is used as a stabilising agent in non-yeast-leavened bakery products at a level not to exceed 1% by weight of flour used.

ALTERNATIVES: Fumaric acid can substitute for tartaric acid in beverages and baking powders and partially replace citric acid in fruit drinks.

TECHNOLOGY OF USE IN FOODS: Fumaric acid $pK_1 = 3.03$; $pK_2 = 4.44$

Fumaric acid is one of the most acidic of the solid acids. Its low solubility in water limits its widespread use. Its slow rate of moisture absorption aids in extending the shelf life of powdered products. It can eliminate excessive hardening and rubbery texture in alginate-based desserts.

It is compatible with fruit and berry flavour; imparts a tart, metallic taste, and blends well with other acidulants without imparting a 'burst' in flavour.

SYNERGISTS: Complements sodium benzoate as a preservative in green foods and fish products.

FOOD SAFETY ISSUES: LD_{50} (mg/kg body weight) fumaric acid. Rat 10,700 oral route.

LD_{50} (mg/kg body weight) disodium fumarate. Rat 3,600–4,800 oral route.

LD_{50} (mg/kg body weight) sodium fumarate. Rat 8,000 oral route.

Acceptable daily intake for humans is limited unconditionally at 0–6 mg/kg body weight and conditionally at 6–10 mg/kg body weight.

LEGISLATION:

USA:

Fumaric acid: 9 CFR 318.7, 381.147; 21 CFR 131.111, 131.136, 131.144, Part 145, 146.113, 150.141, 150.161, 172.350, 175.105, 175.300, 175.320; 27 CFR 24.182, 24.244

Calcium fumarate: 21 CFR 172.350

Ferrous fumarate: 21 CFR 172.350, 184.1307d

Magnesium fumarate: 21 CFR 172.350

Potassium fumarate: 21 CFR 172.350

Sodium fumarate: 21 CFR 172.350

Sodium stearyl fumarate: 21 CFR 172.826

REFERENCES:

Budavari, S., O'Neill, M. J., Smith, A., and Heckelman, P. E. (1989) *The Merck Index.* 11th edn., Merck, Rahway, NJ.

Code of Federal Regulations (CFR) (1995) Volumes 9 (U. S. Department of Agriculture [USDA]), 21 (Food and Drug Administration [FDA]), and 27 (Bureau of Alcohol, Tobacco, and Firearms [BATF]). Office of the Federal Register, National Archives and Records Administration, U. S. Government Printing Office, Washington, DC.

Deshpande, S. S., Salunkhe, D. K., and Deshpande, U. S. (1995) Food Acidulants. In: Maga, J. A., and A. T. Tu (Eds.), *Food Additive Toxicology*. Marcel Dekker, Inc., New York, Ch. 2.

Doores, S. (1990) pH Control Agents and Acidulants. In: Branen, A. L., Davidson, P. M., and S. Salminen (Eds.), *Food Additives.* Marcel Dekker, Inc., New York, pp. 477–510.

Doores, S. (1993) Organic Acids. In: Davidson, P. M., and A. L. Branen (Eds.), *Antimicrobials in Foods*, 2nd edn., Marcel Dekker, Inc., New York, pp. 95–136.

Food Chemicals Codex (1981) 3rd edn., Committee on Food Chemicals Codex, Food and Nutrition Board, Institute of Medicine, National Academy of Sciences, National Academy Press, Washington, DC.

Food Chemicals Codex (1996) 4th edn., Committee on Food Chemicals Codex, Food and Nutrition Board, Institute of Medicine, National Academy of Sciences, National Academy Press, Washington, DC.

Lewis, R. J., Sr. (1989) *Food Additives Handbook.* Van Nostrand Reinhold, New York.

Lide, D. R. (Ed.) (1993) *Handbook of Chemistry and Physics.* 74th edn. CRC Press, Inc., Boca Raton, FL.

NAME:	Glucono-delta-lactone
CATEGORY:	Chelating agents/ pH control agents/ Preservatives/ Flour and baking additives
FOOD USE:	Baked goods/ Dairy products/ Meat, poultry and eggs and products/ Fruit, vegetables and nuts and products
SYNONYMS:	D-gluconic acid-δ-lactone/ D-gluconolactone/ D-glucono-1,5-lactone
FORMULA:	$C_6H_{10}O_6$ (cyclic structure)
MOLECULAR MASS:	178.14
PROPERTIES AND APPEARANCE:	Fine, white, crystalline powder
MELTING RANGE IN °C:	153 (decomposition)
HEAVY METAL CONTENT MAXIMUM IN ppm:	<0.002
SOLUBILITY % AT VARIOUS TEMPERATURE/pH COMBINATIONS:	
in water:	59
in ethanol solution:	1
FUNCTION IN FOODS:	Glucono-δ-lactone is GRAS in the United States with no limitation other than good manufacturing practices. Glucono-δ-lactone can be used as a curing and pickling agent, leavening agent, pH control agent, and sequestrant and a binder. It can be used to alter the acidity of acidified milk, acidified low-fat milk, and acidified skim milk, meat and poultry products. It is used as a curing accelerator in cured, comminuted meat or meat food products at a level of 8 oz/100 lb (*ca.* 5 mg/kg) of meat or meat by-product where slowly lowered pH will increase the curing rate without shorting the emulsion. It serves as a replacement for bacterial starter cultures in fermented sausages. It is also used as a leavening agent and as a cleaning agent. It can be used as an ingredient in canned green beans and canned wax beans.

TECHNOLOGY OF USE IN FOODS:	Glucono-δ-lactone must undergo hydrolysis to gluconic acid before it can lower pH. Therefore, when used in the manufacture of cheese products the rate of pH is slowed.
FOOD SAFETY ISSUES:	LD_{50} (mg/kg body weight) glucono-δ-lactone. Rabbit 7,630 intravenous route.
LEGISLATION:	**USA:** Glucono-δ-lactone: 9 CFR 318.7, 381.147; 21 CFR 131.111, 131.136, 131.144, 155.120, 184.1318
REFERENCES:	Budavari, S., O'Neill, M. J., Smith, A., and Heckelman, P. E. (1989) *The Merck Index.* 11th edn., Merck, Rahway, NJ. *Code of Federal Regulations* (CFR) (1995) Volumes 9 (U. S. Department of Agriculture [USDA]), 21 (Food and Drug Administration [FDA]), and 27 (Bureau of Alcohol, Tobacco, and Firearms [BATF]). Office of the Federal Register, National Archives and Records Administration, U. S. Government Printing Office, Washington, DC. Deshpande, S. S., Salunkhe, D. K., and Deshpande, U. S. (1995) Food Acidulants. In: Maga, J. A., and A. T. Tu (Eds.), *Food Additive Toxicology*. Marcel Dekker, Inc., New York, Ch. 2. Doores, S. (1990) pH Control Agents and Acidulants. In: Branen, A. L., Davidson, P. M., and S. Salminen (Eds.), *Food Additives.* Marcel Dekker, Inc., New York, pp. 477–510. Doores, S. (1993) Organic Acids. In: Davidson, P. M., and A. L. Branen (Eds.), *Antimicrobials in Foods*, 2nd edn., Marcel Dekker, Inc., New York, pp. 95–136. *Food Chemicals Codex* (1981) 3rd edn., Committee on Food Chemicals Codex, Food and Nutrition Board, Institute of Medicine, National Academy of Sciences, National Academy Press, Washington, DC. *Food Chemicals Codex* (1996) 4th edn., Committee on Food Chemicals Codex, Food and Nutrition Board, Institute of Medicine, National Academy of Sciences, National Academy Press, Washington, DC. Lewis, R. J., Sr. (1989) *Food Additives Handbook.* Van Nostrand Reinhold, New York. Lide, D. R. (Ed.) (1993) *Handbook of Chemistry and Physics.* 74th edn. CRC Press, Inc., Boca Raton, FL.

NAME:	Lactic acid
CATEGORY:	Emulsifiers/stabilisers/ Nutritive additives/ pH control agents/ Preservatives/ Flavour enhancers and modifiers/ Solvents/ Flour and baking additives/ Humectants
FOOD USE:	Dairy products/ Edible oils and fats/ Fish and seafood and products/ Meat, poultry and eggs and products/ Fruit, vegetables and nuts and products/ Sugars, sugar preserves and confectionery/ Alcoholic drinks/ Vinegar, pickles and sauces
SYNONYMS:	Acetonic acid/ Ethylidenelactic acid/ Milk acid/ 1-Hydroxyethane 1-carbonylic acid/ 1-Hydroxyethane 1-carboxylic acid/ 2-Hydroxypropanoic acid/ 2-Hydroxypropionic acid/ alpha-Hydroxypropionic acid/ DL mixture CAS 598-82-3/ L isomer CAS 79-33-4/ D isomer CAS 10326-41-7
FORMULA:	COOH-CHOH-CH3
MOLECULAR MASS:	90.08
ALTERNATIVE FORMS:	Butyl lactate; calcium lactate; ferrous lactate; potassium lactate; sodium lactate
PROPERTIES AND APPEARANCE:	Colourless or yellowish, syrupy non-volatile, hygroscopic liquid
BOILING POINT IN °C AT VARIOUS PRESSURES (INCLUDING 760 mm Hg):	122 DL Lactic acid @ 15 mm Hg 103 D Lactic acid @ 15 mm Hg
MELTING RANGE IN °C:	52.8 D Lactic acid 16.8 DL Lactic acid 53 L Lactic acid
IONISATION CONSTANT AT 25°C:	8.4×10^{-4} @ 100°C
DENSITY AT 20°C (AND OTHER TEMPERATURES) IN g/l:	1.249 @ 15°C 1.2060 @ 21°/4° (DL lactic acid)
HEAT OF COMBUSTION AT 25°C:	326.8 kg carlories/gram molecular weight DL lactic acid
PURITY %:	≥95

HEAVY METAL CONTENT MAXIMUM IN ppm:	10		
ARSENIC CONTENT MAXIMUM IN ppm:	3		
ASH MAXIMUM IN %:	0.1%		
SOLUBILITY % AT VARIOUS TEMPERATURE/pH COMBINATIONS:			
in water:	**@ 25°C** Soluble	**@ 100°C** Quickly soluble	
in ethanol solution: 95%	**@ 25°C** Almost insoluble		
in propylene glycol:	**@ 25°C** Soluble		
FUNCTION IN FOODS:	Lactic acid is GRAS for miscellaneous and general-purpose usage in the United States with no limitation other than good manufacturing practices. Lactic acid is used as an antimicrobial agent, curing and pickling agent, flavouring agent and flavour enhancer, pH control agent, solvent and vehicle, dough conditioner, thickener, and yeast food in bread. It has been used in the manufacture of angel food cake, canned fruits, meat food sticks, meringues, dry milk powders, sausage, canned vegetables, imitation sausages, jams, jellies, sherbets, confectionery products and whipped toppings. Lactic acid can be used to alter the acidity of acidified milk, acidified low-fat milk, acidified skim milk, margarine, oleomargarine, artificially sweetened fruit jelly, fruit preserves and jam, meat products and in brines for pickles and olives at a level sufficient for use. It can also be used to alter the acidity of cold pack cheese food such that the pH does not exceed 4.5, pasteurised process cheese (pH 5.3), cheese food (pH 5.0), cheese spread (pH 4.0), and dry curd cottage cheese (pH 4.5–4.7). It can be used as a sanitising agent on food processing equipment and as an antimicrobial spray on meat carcasses. A mixture of sodium alginate, calcium carbonate, lactic acid, and calcium lactate can be used to bind ground and formed raw or cooked poultry pieces at a level not to exceed 1.55% of the product with the combination of lactic acid and calcium lactate at a level not to exceed 0.6% of product. Lactic acid and calcium lactate can be used in a mixture as a binder for restructured meat food products not to exceed a level of 0.3% of the product formulation. Lactic acid is used in the manufacture of hydroxylated lecithin used as an emulsifier. Lactic acid can be added to grapes after fermentation or in combination with other acids after fermentation to correct deficiencies at a level not to exceed 9 g/l in finished wine. It can be added to correct natural acid deficiencies in juice grape wine.		

Office of the Federal Register, National Archives and Records Administration, U. S. Government Printing Office, Washington, DC.

Deshpande, S. S., Salunkhe, D. K., and Deshpande, U. S. (1995) Food Acidulants. In: Maga, J. A., and A. T. Tu (Eds.), *Food Additive Toxicology*. Marcel Dekker, Inc., New York, Ch. 2.

Doores, S. (1990) pH Control Agents and Acidulants. In: Branen, A. L., Davidson, P. M., and S. Salminen (Eds.), *Food Additives*. Marcel Dekker, Inc., New York, pp. 477–510.

Doores, S. (1993) Organic Acids. In: Davidson, P. M., and A. L. Branen (Eds.), *Antimicrobials in Foods*, 2nd edn., Marcel Dekker, Inc., New York, pp. 95–136.

Food Chemicals Codex (1981) 3rd edn., Committee on Food Chemicals Codex, Food and Nutrition Board, Institute of Medicine, National Academy of Sciences, National Academy Press, Washington, DC.

Food Chemicals Codex (1996) 4th edn., Committee on Food Chemicals Codex, Food and Nutrition Board, Institute of Medicine, National Academy of Sciences, National Academy Press, Washington, DC.

Lewis, R. J., Sr. (1989) *Food Additives Handbook*. Van Nostrand Reinhold, New York.

Lide, D. R. (Ed.) (1993) *Handbook of Chemistry and Physics*. 74th edn. CRC Press, Inc., Boca Raton, FL.

HEAVY METAL CONTENT MAXIMUM IN ppm:	10		
ARSENIC CONTENT MAXIMUM IN ppm:	3		
ASH MAXIMUM IN %:	0.1%		
SOLUBILITY % AT VARIOUS TEMPERATURE/pH COMBINATIONS:			
in water:	**@ 25°C** Soluble	**@ 100°C** Quickly soluble	
in ethanol solution: 95%	**@ 25°C** Almost insoluble		
in propylene glycol:	**@ 25°C** Soluble		
FUNCTION IN FOODS:	Lactic acid is GRAS for miscellaneous and general-purpose usage in the United States with no limitation other than good manufacturing practices. Lactic acid is used as an antimicrobial agent, curing and pickling agent, flavouring agent and flavour enhancer, pH control agent, solvent and vehicle, dough conditioner, thickener, and yeast food in bread. It has been used in the manufacture of angel food cake, canned fruits, meat food sticks, meringues, dry milk powders, sausage, canned vegetables, imitation sausages, jams, jellies, sherbets, confectionery products and whipped toppings. Lactic acid can be used to alter the acidity of acidified milk, acidified low-fat milk, acidified skim milk, margarine, oleomargarine, artificially sweetened fruit jelly, fruit preserves and jam, meat products and in brines for pickles and olives at a level sufficient for use. It can also be used to alter the acidity of cold pack cheese food such that the pH does not exceed 4.5, pasteurised process cheese (pH 5.3), cheese food (pH 5.0), cheese spread (pH 4.0), and dry curd cottage cheese (pH 4.5–4.7). It can be used as a sanitising agent on food processing equipment and as an antimicrobial spray on meat carcasses. A mixture of sodium alginate, calcium carbonate, lactic acid, and calcium lactate can be used to bind ground and formed raw or cooked poultry pieces at a level not to exceed 1.55% of the product with the combination of lactic acid and calcium lactate at a level not to exceed 0.6% of product. Lactic acid and calcium lactate can be used in a mixture as a binder for restructured meat food products not to exceed a level of 0.3% of the product formulation. Lactic acid is used in the manufacture of hydroxylated lecithin used as an emulsifier. Lactic acid can be added to grapes after fermentation or in combination with other acids after fermentation to correct deficiencies at a level not to exceed 9 g/l in finished wine. It can be added to correct natural acid deficiencies in juice grape wine.		

Butyl lactate is used as a component of adhesives at levels not to exceed good manufacturing practices.

Calcium lactate is GRAS for miscellaneous and general-purpose usage in the United States with no limitation other than good manufacturing practices. Calcium lactate is a flavour enhancer, flavour agent or adjuvant, leavening agent, nutrient supplement, stabiliser and thickener. It can be used as a firming agent for apple slices and canned grapefruit at a level not to exceed 0.035% calcium by weight of the finished product, and to prevent discoloration of fruit prior to thermal processing. It is used as a gelling agent in artificially sweetened fruit jelly, preserves, and jam. It ensures clarity in the brine of Spanish olives. It is used in salad dressings and marinades and in baking products. A mixture of sodium alginate, calcium carbonate, lactic acid, and calcium lactate can be used to bind ground and formed raw or cooked poultry pieces at a level not to exceed 1.55% of the product with the combination of lactic acid and calcium lactate at a level not to exceed 0.6% of product. It can be used to bind and extend poultry products and is required at a rate of 10% of the binder if rennet-treated, calcium-reduced dried skim milk used or 25% of binder if rennet-treated sodium caseinate is used. Calcium lactate is used to protect flavour in cooked semi-dry and dry products including sausages, imitation sausage and non-specific meat food sticks at a level of 0.6% in product formulation.

Ferrous lactate is GRAS for miscellaneous and general-purpose usage in the United States with no limitation other than good manufacturing practices. Ferrous lactate can be used as a dietary and nutrient supplement.

Potassium lactate is GRAS for miscellaneous and general-purpose usage in the United States with no limitation other than good manufacturing practices. Potassium lactate is used as a flavour enhancer, flavouring agent or adjuvant, humectant and pH control agent. It can be used to acidify margarine and oleomargarine at a level sufficient for purpose. It can be used to flavour various poultry and poultry food products at a level not to exceed 2% of formulation.

Sodium lactate is GRAS for miscellaneous and general-purpose usage in the United States with no limitation other than good manufacturing practices. Sodium lactate is used as a flavour enhancer, flavouring agent or adjuvant, pH control agent, denuding agent, emulsifier, hog scald agent, humectant, lye peeling agent, washing agent and cooked-out juices retention agent. It is used in biscuits, fruits, hog carcasses, meat products, nuts, sponge cake, Swiss roll, vegetables, and bottled water. It can be used to acidify margarine and oleomargarine at a level sufficient for purpose. It is limited in meat products where allowed to 5% of phosphate in pickle at a 10% pump level; and 0.5% of phosphate in product.

It is used to flavour various poultry and poultry food products, meat and meat food products at a level not to exceed 2% of the formulation.

TECHNOLOGY OF USE IN FOODS:

Lactic acid $pK_1 = 3.08$

Manufactured in technical, food and USP grade in two concentrations – 50% and 88%

SYNERGISTS:

Lactic acid acts synergistically with acetic acid as an antimicrobial preservative

FOOD SAFETY ISSUES:

LD_{50} (mg/kg body weight) lactic acid. Mouse 4,875 oral route. Rat 3,730 oral route. Guinea pig 1,810 oral route.

LD_{50} (mg/kg body weight) ferrous lactate. Mouse 147 oral route.

LD_{50} (mg/kg body weight) sodium lactate. Rat 2,000 intraperitoneal route.

Acceptable daily intake for humans of lactic acid is not limited for lactic acid and its calcium, potassium, ammonium and sodium salts and is limited up to 100 mg/kg body weight for DL-lactic acid.

Lactic acid and the calcium, potassium, and sodium salt should not be used in infant foods and formulas in either the D(−) or DL form; acetic acid may be substituted.

LEGISLATION:

USA:

Lactic acid: 9 CFR 318.7, 381.129, 381.147; 21 CFR 131.111, 131.136, 131.144, 133.124, 133.129, 133.169, 133.173, 133.179, 150.141, 150.161, 166.110, 172.814, 178.1010, 184.1061; 27 CFR 24.182

Butyl lactate: 21 CFR 175.105

Calcium lactate: 9 CFR 318.7, 381.129, 381.147; 21 CFR 145.145, 150.141, 150.161, 155.200, 184.1207

Ferrous lactate: 21 CFR 182.5311, 184.1311

Potassium lactate: 9 CFR 318.7, 318.147; 21 CFR 184.1639

Sodium lactate: 9 CFR 318.7, 381.147; 21 CFR 184.1768

REFERENCES:

Budavari, S., O'Neill, M. J., Smith, A., and Heckelman, P. E. (1989) *The Merck Index.* 11th edn., Merck, Rahway, NJ.

Code of Federal Regulations (CFR) (1995) Volumes 9 (U. S. Department of Agriculture [USDA]), 21 (Food and Drug Administration [FDA]), and 27 (Bureau of Alcohol, Tobacco, and Firearms [BATF]).

Office of the Federal Register, National Archives and Records Administration, U. S. Government Printing Office, Washington, DC.

Deshpande, S. S., Salunkhe, D. K., and Deshpande, U. S. (1995) Food Acidulants. In: Maga, J. A., and A. T. Tu (Eds.), *Food Additive Toxicology*. Marcel Dekker, Inc., New York, Ch. 2.

Doores, S. (1990) pH Control Agents and Acidulants. In: Branen, A. L., Davidson, P. M., and S. Salminen (Eds.), *Food Additives.* Marcel Dekker, Inc., New York, pp. 477–510.

Doores, S. (1993) Organic Acids. In: Davidson, P. M., and A. L. Branen (Eds.), *Antimicrobials in Foods*, 2nd edn., Marcel Dekker, Inc., New York, pp. 95–136.

Food Chemicals Codex (1981) 3rd edn., Committee on Food Chemicals Codex, Food and Nutrition Board, Institute of Medicine, National Academy of Sciences, National Academy Press, Washington, DC.

Food Chemicals Codex (1996) 4th edn., Committee on Food Chemicals Codex, Food and Nutrition Board, Institute of Medicine, National Academy of Sciences, National Academy Press, Washington, DC.

Lewis, R. J., Sr. (1989) *Food Additives Handbook.* Van Nostrand Reinhold, New York.

Lide, D. R. (Ed.) (1993) *Handbook of Chemistry and Physics.* 74th edn. CRC Press, Inc., Boca Raton, FL.

NAME:	Malic acid
CATEGORY:	Antioxidants/ pH control agents/ Preservatives/ Flavour enhancers and modifiers
FOOD USE:	Baked good/ Dairy products/ Edible oils and fats/ Meat, poultry and eggs and products/ Fruit, vegetables and nuts and products/ Beverages/ Soft drinks/ Sugars, sugar preserves and confectionery/ Alcoholic drinks
SYNONYMS:	Hydroxybutanedioic acid/ Hydroxysuccinic acid/ Pomalous acid/ 1-Hydroxy-1,2-ethanedicarboxylic acid/ L form CAS 97-67-6/ DL form CAS 617-48-1
FORMULA:	HO-CH-(COOH)-CH2-COOH
MOLECULAR MASS:	134.09
PROPERTIES AND APPEARANCE:	White or nearly white, crystalline powder or granules
BOILING POINT IN °C AT VARIOUS PRESSURES (INCLUDING 760 mm Hg):	150 (DL malic acid) 140 (D or L malic acid)
MELTING RANGE IN °C:	128 (DL malic acid) 100 (D or L malic acid)
IONISATION CONSTANT AT 25°C:	3.9×10^{-4} 7.8×10^{-6}
DENSITY AT 20°C (AND OTHER TEMPERATURES) IN g/l:	1.601 (DL malic acid) 1.595 (D or L malic acid) @ 20°/4°
HEAT OF COMBUSTION AT 25°C IN J/kg:	317.37 (L malic acid)
PURITY %:	≥99
HEAVY METAL CONTENT MAXIMUM IN ppm:	20

ARSENIC CONTENT MAXIMUM IN ppm: 3

ASH MAXIMUM IN %: 0.1

SOLUBILITY % AT VARIOUS TEMPERATURE/pH COMBINATIONS:

in water:	**@ 20°C**		55.8 (DL), 36.3 (L)	**@ 25°C**	Very soluble
in ethanol solution:	**@ 95%**	**@ 20°C**	45.53 DL malic acid	**@ 20°C**	86.60 L malic acid

FUNCTION IN FOODS: Malic acid is GRAS for miscellaneous and general-purpose usage in the United States with no limitation other than good manufacturing practices. Malic acid may be used as a flavour enhancer, flavouring agent and adjuvant, and pH control agent. It can be used in sherbets and ices, candies, baked goods, canned fruits, fat, lard, shortening, fruit preserves, jams, jellies, gelatins, puddings, and beverages primarily for flavouring and acidification. Malic acid can be used to alter the acidity of acidified milk, acidified low-fat milk, acidified skim milk, and artificially sweetened fruit jelly, fruit preserves and jam. Maximum levels recommended are 0.8% in gelatin, pudding and fillings, 2.6% in jams and jellies, 3.0% in chewing gum and soft candy, 3.4% in non-alcoholic beverages, 3.5% in processed fruits and fruit juice, 6.9% in hard candy and 0.7% in all other foods to be used in accordance with good manufacturing practices. It is used to increase the effectiveness of antioxidants in poultry fats at a level of 0.01%. It is limited at 0.01% on the basis of total weight in combination with antioxidants in lard and shortening. It can be used in French dressing, mayonnaise, and salad dressing at a level not to exceed 25% of the weight of the acids of the vinegar or diluted vinegar calculated as acetic acid. It can be added to grapes prior to fermentation or in combination with other acids to correct deficiencies after fermentation at a level not to exceed 9 g/l. It can also be added to adjust the acidity in apples, apple juice or wine at a level not to exceed 9 g/l of finished wine. It can be used to correct natural acid deficiencies in juice or wine.

ALTERNATIVES: Malic acid imparts the same degree of acidity as citric acid, but in lower amounts.

TECHNOLOGY OF USE IN FOODS: Malic acid $pK_1 = 3.4$; $pK_2 = 5.11$

Malic acid does not have the same build-up of acid taste as other acids. It imparts a smooth, tart taste with no burst of flavour and is compatible with fruit and berry flavour in hard candies. It provides excellent anti-browning properties in fruits.

The US Internal Revenue Service has specified the addition of malic acid to volatile fruit flavour concentrates containing 6–15% alcohol to render these products non-potable during transportation from manufacturer to winery.

SYNERGISTS: Malic acid acts synergistically with antioxidants to prevent rancidity.

FOOD SAFETY ISSUES: LD_{50} (mg/kg body weight) L malic acid. Rat 1,600 oral route. Rabbit 5,000 oral route.

LD_{50} (mg/kg body weight) sodium malate. Dog 1,000 oral route.

Acceptable daily intake for humans of malic acid is not limited for malic acid but is limited at 100 mg/kg for DL-malic acid. D and DL malic acids are not to be used in baby foods.

LEGISLATION:

USA:

Malic acid: 9 CFR 318.7, 381.147; 21 CFR 131.111, 131.136, 131.144, Part 145, 146.113, 150.141, 150.161, 169.115, 169.140, 169.150, 184.1069; 27 CFR 24.182, 24.192, 24.246, 240.1051

REFERENCES:

Budavari, S., O'Neill, M. J., Smith, A., and Heckelman, P. E. (1989) *The Merck Index*. 11th edn., Merck, Rahway, NJ.

Code of Federal Regulations (CFR) (1995) Volumes 9 (U. S. Department of Agriculture [USDA]), 21 (Food and Drug Administration [FDA]), and 27 (Bureau of Alcohol, Tobacco, and Firearms [BATF]). Office of the Federal Register, National Archives and Records Administration, U. S. Government Printing Office, Washington, DC.

Deshpande, S. S., Salunkhe, D. K., and Deshpande, U. S. (1995) Food Acidulants. In: Maga, J. A., and A. T. Tu (Eds.), *Food Additive Toxicology*. Marcel Dekker, Inc., New York, Ch. 2.

Doores, S. (1990) pH Control Agents and Acidulants. In: Branen, A. L., Davidson, P. M., and S. Salminen (Eds.), *Food Additives.* Marcel Dekker, Inc., New York, pp. 477–510.

Doores, S. (1993) Organic Acids. In: Davidson, P. M., and A. L. Branen (Eds.), *Antimicrobials in Foods*, 2nd edn., Marcel Dekker, Inc., New York, pp. 95–136.

Food Chemicals Codex (1981) 3rd edn., Committee on Food Chemicals Codex, Food and Nutrition Board, Institute of Medicine, National Academy of Sciences, National Academy Press, Washington, DC.

Food Chemicals Codex (1996) 4th edn., Committee on Food Chemicals Codex, Food and Nutrition Board, Institute of Medicine, National Academy of Sciences, National Academy Press, Washington, DC.

Lewis, R. J., Sr. (1989) *Food Additives Handbook.* Van Nostrand Reinhold, New York.

Lide, D. R. (Ed.) (1993) *Handbook of Chemistry and Physics.* 74th edn. CRC Press, Inc., Boca Raton, FL.

NAME:	Phosphoric acid
CATEGORY:	Emulsifiers/stabilisers/ Chelating agents/ Nutritive additives/ Antioxidants/ pH control agents/ Preservatives/ Flavour enhancers and modifiers/ Flour and baking additives/ Anti-caking agents/ Firming agents
FOOD USE:	Baked goods/ Cereals and cereal products/ Dairy products/ Edible oils and fats/ Fish and seafood and products/ Meat, poultry and eggs and products/ Beverages/ Soft drinks/ Sugars, sugar preserves and confectionery/ Alcoholic drinks/ Vinegar, pickles and sauces
SYNONYMS:	Orthophosphoric acid/ CAS 7664-38-2/ DOT 1805
FORMULA:	P(OH)2-OOH
MOLECULAR MASS:	98.0
ALTERNATIVE FORMS:	Phosphoric acid; ammonium phosphate, dibasic; ammonium phosphate, monobasic; ammonium potassium hydrogen phosphate; ammonium potassium phosphate; calcium glycerophosphate; calcium hexametaphosphate; calcium phosphate, dibasic; calcium phosphate, monobasic; calcium phosphate, tribasic; calcium pyrophosphate; ferric phosphate; ferric pyrophosphate; ferric sodium pyrophosphate; magnesium glycerophosphate; magnesium phosphate, dibasic; magnesium phosphate, monobasic; magnesium phosphate, tribasic; manganese glycerophosphate; potassium glycerophosphate; potassium phosphate, dibasic; potassium phosphate, monobasic; potassium pyrophosphate; potassium tripolyphosphate; sodium acid pyrophosphate; sodium aluminium phosphate; sodium metaphosphate; sodium phosphate, dibasic; sodium phosphate, monobasic; sodium phosphate, tribasic; sodium polyphosphate; sodium pyrophosphate; sodium tripolyphosphate; tetrasodium pyrophosphate
PROPERTIES AND APPEARANCE:	Colourless solution
MELTING POINT IN °C:	42.35
FLASH POINT IN °C	42.4

IONISATION CONSTANT AT 25°C:	7.52×10^{-3} 6.23×10^{-8} 2.2×10^{-13} (@ 18°C)
DENSITY AT 20°C (AND OTHER TEMPERATURES) IN g/l:	1.8741 (100% solution) 1.6850 (85% solution) 1.334 (50% solution) 1.0523 (10% solution)
VAPOUR PRESSURE AT VARIOUS TEMPERATURES IN mm Hg:	0.0285 mm @ 20°
PURITY %:	75–85%
HEAVY METAL CONTENT MAXIMUM IN ppm:	10
ARSENIC CONTENT MAXIMUM IN ppm:	2
SOLUBILITY % AT VARIOUS TEMPERATURE/pH COMBINATIONS:	
in water:	Soluble
in ethanol solution:	Soluble
FUNCTION IN FOODS:	Phosphoric acid is GRAS for miscellaneous and general-purpose usage in the United States with no limitation other than good manufacturing practices. Phosphoric acid can be used to alter the acidity of acidified milk, acidified low-fat milk, and acidified skim milk, margarine and oleomargarine, poultry, meat, and meat food products. It can be used to acidify cold pack cheese food such that the pH of the finished products is not below pH 4.5, pasteurised process cheese food (pH 5.3), pasteurised process cheese food (pH 5.0), pasteurised process cheese spread (pH 4.0) and dry-curd cottage cheese (4.5–4.7). It can be used as a synergist in combination with antioxidants in lard, shortening and poultry fat at a limit of 0.01%. It assists in caramelisation. It can be used as a sanitiser on food processing equipment.

Ammonium phosphate, dibasic is GRAS for miscellaneous and general-purpose usage in the United States with no limitation other than good manufacturing practices. Ammonium phosphate, dibasic can be used as a pH control agent, processing aid, firming agent, and leavening agent. It is found in baked goods, alcoholic beverages, condiments, and puddings. It is used as a yeast nutrient in wine production and to start secondary fermentation of sparkling wines at a level not to exceed 8 lb/1000 gallon (*ca.* 16 mg/l).

Ammonium phosphate, monobasic is GRAS for miscellaneous and general-purpose usage in the United States with no limitation other than good manufacturing practices. Ammonium phosphate, monobasic can be used as a pH control agent, and dough strengthener. It is found in baked goods, baking powder, frozen desserts, margarine, whipped toppings and yeast food. It is used as a yeast nutrient in wine production and to start secondary fermentation of sparkling wines at a level not to exceed 8 lb/1000 gallon (16 mg/l).

Ammonium potassium hydrogen phosphate can be used as an indirect additive as a component of adhesives intended for use in packaging and in the production of resinous and polymeric coatings in packaging materials.

Ammonium potassium phosphate can be used as an indirect additive in the production of resinous and polymeric coatings in packaging materials.

Calcium glycerophosphate is GRAS for miscellaneous and general-purpose usage in the United States with no limitation other than good manufacturing practices. Calcium glycerophosphate can be used as a dietary and nutrient supplement. It is found in gelatins, puddings and fillings. It can be used as an indirect additive in the production of resinous and polymeric coatings in packaging materials.

Calcium hexametaphosphate is GRAS for miscellaneous and general-purpose usage in the United States with no limitation other than good manufacturing practices. Calcium hexametaphosphate is used as a sequestrant.

Calcium phosphate, dibasic is GRAS for miscellaneous and general-purpose usage in the United States with no limitation other than good manufacturing practices. Calcium phosphate, dibasic is used as a dietary and nutrient supplement, dough conditioner, stabiliser, yeast food in baked foods, cereal products, and dessert gels. It can be used as an indirect additive in the production of resinous and polymeric coatings in packaging materials.

Calcium phosphate, monobasic is GRAS for miscellaneous and general-purpose usage in the United States with no limitation other than good manufacturing practices. Calcium phosphate, monobasic is used as a dietary and nutrient supplement, dough conditioner, firming agent, leavening agent, sequestrant, stabiliser, yeast food in cereals, and dough. It can be used as an acidulant in baking powders and wheat flours or as a mineral supplement for foods. It can be used in artificially sweetened fruit jelly, fruit preserves and jams as a gelling agent. It is limited in bread, rolls, and buns to not more than 0.75 part for each 100 parts by weight of flour used. It can be used as an indirect additive in the production of resinous and polymeric coatings in packaging materials.

Calcium phosphate, tribasic is GRAS for miscellaneous and general-purpose usage in the United States with no limitation other than good manufacturing practices. Calcium phosphate, tribasic can be used as a dietary supplement, nutrient, stabiliser in cereals, desserts, flour, lard, and dry vinegar. It can be used in clarifying sugar syrups and as a anti-caking agent in table salt. It can be used to preserve colour during dehydration processing in mechanically deboned chicken to be dehydrated at a level not to exceed 2% of the mechanically deboned chicken prior to dehydration. It can be used as an aid in rendering animal fats. It can be used as an indirect additive in the production of resinous and polymeric coatings in packaging materials.

Calcium pyrophosphate is GRAS for miscellaneous and general-purpose usage in the United States with no limitation other than good manufacturing practices. Calcium pyrophosphate can be used as a dietary and nutrient supplement.

Ferric phosphate is GRAS for miscellaneous and general-purpose usage in the United States with no limitation other than good manufacturing practices. It is used as a dietary and nutrient supplement.

Ferric pyrophosphate is GRAS for miscellaneous and general-purpose usage in the United States with no limitation other than good manufacturing practices. Ferric pyrophosphate can be used as a dietary and nutrient supplement.

Ferric sodium pyrophosphate is GRAS for miscellaneous and general-purpose usage in the United States with no limitation other than good manufacturing practices. Ferric sodium pyrophosphate is used as a dietary supplement.

Magnesium glycerophosphate can be used as a stabiliser when migrating from food packaging material. It can be used as an indirect additive in the production of resinous and polymeric coatings in packaging materials.

Magnesium phosphate, dibasic is GRAS for miscellaneous and general-purpose usage in the United States with no limitation other than good manufacturing practices. Magnesium phosphate, dibasic is used as a pH control agent and a dietary and nutrient supplement. It can be used as an indirect additive in the production of resinous and polymeric coatings in packaging materials.

Magnesium phosphate, monobasic can be used as an indirect additive in the production of resinous and polymeric coatings in packaging materials.

Magnesium phosphate, tribasic is GRAS for miscellaneous and general-purpose usage in the United States with no limitation other than good manufacturing practices. Magnesium phosphate, tribasic is used as a pH control agent, stabiliser and nutrient supplement. It can be used as an indirect additive in the production of resinous and polymeric coatings in packaging materials.

Manganese glycerophosphate is GRAS for miscellaneous and general-purpose usage in the United States with no limitation other than good manufacturing practices. Manganese glycerophosphate can be used as a dietary supplement.

Potassium glycerophosphate is GRAS for miscellaneous and general-purpose usage in the United States with no limitation other than good manufacturing practices. Potassium glycerophosphate can be used as a dietary supplement and as a stabiliser when migrating from food packaging material.

Potassium phosphate, dibasic is GRAS for miscellaneous and general-purpose usage in the United States with no limitation other than good manufacturing practices. Potassium phosphate, dibasic can be used as a sequestrant in the preparation of non-dairy powdered coffee creamers. It is used as an emulsifier in pasteurised process cheese, cheese food, and cheese spread at a level not to exceed 3.0% of the weight of the finished products. It can be used to decrease the amount of cooked-out juices in meat food products at a level of 5% of phosphate in pickle, at 10% in pump level, and 0.5% of phosphate in product only (clear solution may be injected into product) or in poultry at a limit of 0.5% of total poultry product. It can be used to remove feathers from poultry carcasses.

Potassium phosphate, monobasic is GRAS for miscellaneous and general-purpose usage in the United States with no limitation other than good manufacturing practices. Potassium phosphate, monobasic can be used to decrease the amount of cooked-out juices in meat food products at a level of 5% of phosphate in pickle, at 10% in pump level, and 0.5% of phosphate in product only (clear solution may be injected into product) or in poultry at a limit of 0.5% of total poultry product.

Potassium pyrophosphate can be used to decrease the amount of cooked-out juices in meat food products at a level of 5% of phosphate in pickle, at 10% in pump level, and 0.5% of phosphate in product only (clear solution may be injected into product) or in poultry at a limit of 0.5% of total poultry product.

Potassium tripolyphosphate is used as a boiler water additive. It can be used to decrease the amount of cooked-out juices in meat food products at a level of 5% of phosphate in pickle, at 10% in pump level, and 0.5% of phosphate in product only (clear solution may be injected into product) or in poultry at a limit of 0.5% of total poultry product.

Sodium acid pyrophosphate is GRAS for miscellaneous and general-purpose usage in the United States with no limitation other than good manufacturing practices. Sodium acid pyrophosphate is used as an emulsifier in pasteurised process cheese, cheese food, and cheese spread at a level not to exceed 3.0% of the weight of the finished products. It can be used as a sequestrant. It can be used as a curing accelerator to accelerate colour fixing in frankfurters, wieners, Vienna, bologna, garlic bologna, knockwurst and similar products not to exceed alone or in combination with other curing accelerators 8 oz/100 lb (*ca.* 5 mg/kg) of meat or meat and meat by-products nor 0.5% in finished product. It can be used to decrease the amount of cooked-out juices in meat food products at a level of 5% of phosphate in pickle, at 10% in pump level, and 0.5% of phosphate in product only (clear solution may be injected into product) or in poultry at a limit of 0.5% of total poultry product. It can be used to remove hair from hog carcasses and feathers from poultry carcasses.

Sodium aluminium phosphate is GRAS for miscellaneous and general-purpose usage in the United States with no limitation other than good manufacturing practices. Sodium aluminium phosphate can be used as an emulsifier in pasteurised process cheese, cheese food, and cheese spread at a level not to exceed 3.0% of the weight of the finished products.

Sodium metaphosphate (hexametaphosphate) is GRAS for miscellaneous and general-purpose usage in the United States with no limitation other than good manufacturing practices. Sodium metaphosphate is used as a starch modifying agent, sequestrant, and texturiser. It is used in fish, lima beans, peanuts, and canned peas. It is used as an emulsifier in pasteurised process cheese, cheese food, and cheese spread at a level not to exceed 3.0% of the weight of the finished products. It is used as a stabiliser and thickener in French dressing. It can be used in artificially sweetened fruit jelly, fruit preserves and jam at a level not to exceed 8 oz/100 lb (*ca.* 5 mg/kg) of the finished food. In meat food products it can be used in combination with sodium tripolyphosphate to protect flavour in fresh beef, beef for further cooking, cooked beef, beef patties, meat loaves, meat toppings and similar products derived from pork, lamb, veal, mutton and goat meat which are cooked or frozen after processing at a level not to exceed 0.5% of total product. It can be used to decrease the amount of cooked-out juices in meat food products at a level of 5% of phosphate in pickle, at 10% in pump level, and 0.5% of phosphate in product only (clear solution may be injected into product) or in poultry at a limit of 0.5% of total poultry product. It can be used to remove hair from hog carcasses. It is an indirect additive when migrating to food from paper and paperboard products.

Sodium phosphate, dibasic is GRAS for miscellaneous and general-purpose usage in the United States with no limitation other than good manufacturing practices. Sodium phosphate, dibasic can be used as a boiler water additive, sequestrant, texturizer, a dietary and nutrient supplement and buffer in coffee whiteners, cream sauce, evaporated milk, instant pudding, and whipped products. It is used as an emulsifier in pasteurised process cheese, cheese food, and cheese spread at a level not to exceed 3.0% of the weight of the finished products. It can be used in artificially sweetened fruit jelly, fruit preserves and jam at a level not to exceed 2 oz/100 lb (*ca.* 1 mg/kg) of the finished food. It can be used to decrease the amount of cooked-out juices in meat food products at a level of 5% of phosphate in pickle, at 10% in pump level, and 0.5% of phosphate in product only (clear solution may be injected into product) or in poultry at a limit of 0.5% of total poultry product. It can be used to remove hair from hog carcasses and feathers from poultry carcasses. It can be used as an indirect additive in the production of resinous and polymeric coatings in packaging materials.

Sodium phosphate, monobasic is GRAS for miscellaneous and general-purpose usage in the United States with no limitation other than good manufacturing practices. Sodium phosphate, monobasic can be used as a boiler water additive, a sequestrant, dietary and nutrient supplement. It is used as an emulsifier in pasteurised process cheese at a level not to exceed 3.0% of the weight of finished products. It can be used in artificially sweetened fruit jelly, fruit preserves and jam at a level not to exceed 2 oz/100 lb (*ca.* 1 mg/kg) of the finished food. It can be used to decrease the amount of cooked-out juices in meat food products at a level of 5% of phosphate in pickle, at 10% in pump level, and 0.5% of phosphate in product only (clear solution may be injected into product) or in poultry at a limit of 0.5% of total poultry product. It can be used to remove feathers from poultry carcasses at a level sufficient for use. It can be used in the production of resinous and polymeric coatings and as a stabiliser when migrating from packaging materials.

Sodium phosphate, tribasic is GRAS for miscellaneous and general-purpose usage in the United States with no limitation other than good manufacturing practices. Sodium phosphate, tribasic is used as a boiler water additive, sequestrant, buffer, dietary and nutrient supplement. It is used in cereals and evaporated milk. It is used as an emulsifier in pasteurised process cheese, cheese food, and cheese spread at a level not to exceed 3.0% of the weight of the finished products. It can be used in artificially sweetened fruit jelly, fruit preserves and jam at a level not to exceed 2 oz/100 lb (*ca.* 1 mg/kg) of the finished food. It is a denuding agent in tripe at a level sufficient for purpose. It can be used to remove hair from hog carcasses and feathers from poultry carcasses. It can be used in the production of resinous and polymeric coatings and as a stabiliser when migrating from packaging materials. It can be used as an aid in rendering animal fats.

Sodium polyphosphate can be used in combination with sodium tripolyphosphate and sodium metaphosphate to protect flavour in fresh beef, beef for further cooking, cooked beef, beef patties, meat loaves, meat toppings and similar products derived from pork, lamb, veal, mutton and goat meat which are cooked or frozen after processing at a level not to exceed 0.5% of total product. It can be used to decrease the amount of cooked-out juices in meat food products at a level of 5% of phosphate in pickle, at 10% in pump level, and 0.5% of phosphate in product only (clear solution may be injected into product) or in poultry at a limit of 0.5% of total poultry product.

Sodium pyrophosphate is GRAS for miscellaneous and general-purpose usage in the United States with no limitation other than good manufacturing practices. Sodium pyrophosphate can be used as a sequestrant. It can be used to decrease the amount of cooked-out juices in meat food products at a level of 5% of phosphate in pickle, at 10% in pump level, and 0.5% of phosphate in product only (clear solution may be injected into product) or in poultry at a limit of 0.5% of total poultry product. It can be used to remove hair from hog carcasses and feathers from poultry carcasses. It can be used as an indirect additive in the production of resinous and polymeric coatings in packaging materials.

Sodium tripolyphosphate is GRAS for miscellaneous and general-purpose usage in the United States with no limitation other than good manufacturing practices. Sodium tripolyphosphate is used as a sequestrant, texturiser used in angel food cake mix, desserts, gelling juices, canned ham, lima beans, meringues, and canned peas. In meat food products it can be used alone or in combination with sodium metaphosphate and sodium polyphosphate to protect flavour in fresh beef, beef for further cooking, cooked beef, beef patties, meat loaves, meat toppings and similar products derived from pork, lamb, veal, mutton and goat meat which are cooked or frozen after processing at a level not to exceed 0.5% of total product. It can be used to decrease the amount of cooked-out juices in meat food products at a level of 5% of phosphate in pickle, at 10% in pump level, and 0.5% of phosphate in product only (clear solution may be injected into product) or in poultry at a limit of 0.5% of total poultry product. It can be used to remove hair from hog carcasses and feathers from poultry. It can be used as a substance migrating to food from paper and paperboard products.

Tetrasodium pyrophosphate is GRAS for miscellaneous and general-purpose usage in the United States with no limitation other than good manufacturing practices. Tetrasodium pyrophosphate can be used as a sequestrant and boiler water additive. It is used as an emulsifier in pasteurised process cheese, cheese food, and cheese spread at a level not to exceed 3.0% of the weight of the finished products. It can be used to remove feathers from poultry carcasses at a level sufficient for use. It is used as a substance when migrating to food from cotton and cotton fabrics used in dry packaging and can be used as an indirect additive in the production of resinous and polymeric coatings in packaging materials.

TECHNOLOGY OF USE IN FOODS: Phosphoric acid $pK_1 = 2.12$; $pK_2 = 7.21$; $pK_3 = 12.67$

SYNERGISTS: Phosphoric acid increases the effectiveness of antioxidants in lard and shortening.

FOOD SAFETY ISSUES: LD_{50} (mg/kg body weight) phosphoric acid. Rat 1,530 oral route.

Acceptable daily intake for humans of phosphoric acid is up to 70 mg/kg body weight.

LEGISLATION:

USA:

Phosphoric acid: 9 CFR 318.7, 381.147; 21 CFR 75.85, 131.111, 131.136, 131.144, 133.124, 133.129, 133.169, 133.173, 133.179, 178.1010, 182.1073
Ammonium phosphate, dibasic: 21 CFR 184.1141b; 27 CFR 24.246
Ammonium phosphate, monobasic: 21 CFR 184.1141a; 27 CFR 24.246
Ammonium potassium hydrogen phosphate: 21 CFR 175.105, 175.300, 181.29
Ammonium potassium phosphate: 21 CFR 175.300
Calcium glycerophosphate: 21 CFR 175.300, 181.29, 182.5201, 184.1201
Calcium hexametaphosphate: 21 CFR 182.6203
Calcium phosphate, dibasic: 21 CFR 175.300, 181.29, 182.1217, 182.5217, 182.8217
Calcium phosphate, monobasic: 21 CFR 136.150, 150.141, 150.161, 175.300, 181.29, 182.1217, 182.5217, 182.6215, 182.8217
Calcium phosphate, tribasic: 9 CFR 318.7, 381.147; 21 CFR 175.300, 182.1217, 182.5217, 182.8217
Calcium pyrophosphate: 21 CFR 182.5223, 182.8223
Ferric phosphate: 21 CFR 182.5301, 184.1301
Ferric pyrophosphate: 21 CFR 182.5304, 184.1304
Ferric sodium pyrophosphate: 21 CFR 182.5306
Magnesium glycerophosphate: 21 CFR 175.300, 181.29
Magnesium phosphate, dibasic: 21 CFR 175.300, 181.29, 182.5434, 184.1434
Magnesium phosphate, monobasic: 21 CFR 175.300, 181.29, 182.5434

Magnesium phosphate tribasic: 21 CFR 175.300, 182.5434, 184.1434
Manganese glycerophosphate: 21 CFR 182.5455
Potassium glycerophosphate: 21 CFR 182.5628
Potassium phosphate, dibasic: 9 CFR 318.7, 381.147; 21 CFR 133.169, 133.173, 133.179, 182.6285
Potassium phosphate, monobasic: 9 CFR 318.7, 381.147
Potassium pyrophosphate: 9 CFR 318.7, 381.47
Potassium tripolyphosphate: 9 CFR 318.7, 381.147; 21 CFR 173.310
Sodium acid pyrophosphate: 9 CFR 318.7; 21 CFR 133.169, 133.173, 133.179, 182.1087
Sodium aluminium phosphate: 21 CFR 133.169, 133.173, 133.179, 182.1781
Sodium metaphosphate: 9 CFR 318.7, 381.47; 21 CFR 133.169, 133.173, 133.179, 150.141, 150.161, 169.115, Part 176, 182.90, 182.6760, 182.6769
Sodium phosphate, dibasic: 9 CFR 318.7, 381.147; 21 CFR 133.169, 133.173, 133.179, 150.141, 150.161, 173.310, 175.300, 181.29, 182.1778, 182.5778, 182.6290, 182.6760, 182.6778, 182.8778
Sodium phosphate, monobasic: 9 CFR 318.7, 381.147; 21 CFR 133.169, 133.173, 133.179, 150.141, 150.161, 173.310, 182.1778, 182.5778, 182.6085, 182.8778
Sodium phosphate, tribasic: 9 CFR 318.7, 381.47; 21 CFR 133.169, 133.173, 133.179, 150.141, 150.161, 173.310, 182.1778, 182.5778, 182.6778, 182.8778
Sodium polyphosphate: 9 CFR 318.7, 381.147
Sodium tripolyphosphate: 9 CFR 318.7, 381.147; 21 CFR 173.310, 182.1810, 182.90, 182.6810
Tetrasodium pyrophosphate: 9 CFR 381.147; 21 CFR 133.169, 133.173, 133.179, 173.310, 181.29, 182.70, 182.6787, 182.6789

REFERENCES:

Budavari, S., O'Neill, M. J., Smith, A., and Heckelman, P. E. (1989) *The Merck Index.* 11th edn., Merck, Rahway, NJ.

Code of Federal Regulations (CFR) (1995) Volumes 9 (U. S. Department of Agriculture [USDA]), 21 (Food and Drug Administration [FDA]), and 27 (Bureau of Alcohol, Tobacco, and Firearms [BATF]). Office of the Federal Register, National Archives and Records Administration, U. S. Government Printing Office, Washington, DC.

Deshpande, S. S., Salunkhe, D. K., and Deshpande, U. S. (1995) Food Acidulants. In: Maga, J. A., and A. T. Tu (Eds.), *Food Additive Toxicology*. Marcel Dekker, Inc., New York. Ch. 2.

Doores, S. (1990) pH Control Agents and Acidulants. In: Branen, A. L., Davidson, P. M., and S. Salminen (Eds.), *Food Additives.* Marcel Dekker, Inc., New York. pp. 477–510.

Doores, S. (1993) Organic Acids. In: Davidson, P. M., and A. L. Branen (Eds.), *Antimicrobials in Foods*, 2nd edn., Marcel Dekker, Inc., New York. pp. 95–136.

Food Chemicals Codex (1981) 3rd edn., Committee on Food Chemicals Codex, Food and Nutrition Board, Institute of Medicine, National Academy of Sciences, National Academy Press, Washington, DC.

Food Chemicals Codex (1996) 4th edn., Committee on Food Chemicals Codex, Food and Nutrition Board, Institute of Medicine, National Academy of Sciences, National Academy Press, Washington, DC.

Lewis, R. J., Sr. (1989) *Food Additives Handbook.* Van Nostrand Reinhold, New York.

Lide, D. R. (Ed.) (1993) *Handbook of Chemistry and Physics.* 74th edn. CRC Press, Inc., Boca Raton, FL.

NAME:	Propionic acid
CATEGORY:	Antioxidants/ pH control agents/ Preservatives/ Flavour enhancers and modifiers
FOOD USE:	Baked goods/ Dairy products/ Meat, poultry and eggs and products/ Beverages/ Sugars, sugar preserves and confectionery
SYNONYMS:	Carboxyethane/ Ethane carboxylic acid/ Ethylformic acid/ Metacetonic acid/ Methylacetic acid/ Propanoic acid/ Prozoin/ Pseudoacetic acid/ CAS 79-09-4/ DOT 1848
FORMULA:	CH3-CH2-COOH
MOLECULAR MASS:	74.08
ALTERNATIVE FORMS:	Calcium propionate; dilauryl thiopropionic acid; potassium propionate; sodium propionate; thiopropionic acid
PROPERTIES AND APPEARANCE:	Oily liquid (propionic acid) White, free-flowing powder (salts)
BOILING POINT IN °C AT VARIOUS PRESSURES (INCLUDING 760 mm Hg):	4.6 (1 mm) 85.8 (100 mm) 122.0 (400 mm) 141.1 (760 mm)
MELTING POINT IN °C:	−21.5
FLASH POINT IN °C:	58
IONISATION CONSTANT AT 25°C:	1.34×10^{-5}
DENSITY AT 20°C (AND OTHER TEMPERATURES) IN g/l:	0.99336 @ 20°/4° 0.998 @ 15°/4°
HEAT OF COMBUSTION AT 25°C:	365.03 kg calories/gram molecular weight

VAPOUR PRESSURE AT VARIOUS TEMPERATURES IN mm Hg:	4.6 (1 mm) 39.7 (10 mm) 65.8 (40 mm) 85.8 (100 mm) 122.0 (400 mm) 141.1 (760 mm)	
PURITY %:	≥99.5	
WATER CONTENT MAXIMUM IN %:	0.15	
HEAVY METAL CONTENT MAXIMUM IN ppm:	10	
ARSENIC CONTENT MAXIMUM IN ppm:	3	
ASH MAXIMUM IN %:	0.01	
SOLUBILITY % AT VARIOUS TEMPERATURE/pH COMBINATIONS:		
in water:	**@ 100°C** Sodium salt, 150	**@ 100°C** Calcium salt, 58.8
in vegetable oil:	**@ 25°C** Soluble	
in ethanol solution:	**@ 95%** **@ 25°C** Sodium propionate, 4 **@ 25°C** Calcium propionate, insoluble	

FUNCTION IN FOODS:

Propionic acid is GRAS for miscellaneous and general-purpose usage in the United States with no limitation other than good manufacturing practices. Propionic acid can be used as a flavouring agent and as an antifungal agent to control mould growth on the surface of cheese and buffer and inhibits rope formation in baked goods. Maximum levels recommended are 0.3% in cheese products, 0.32% in flour and in white bread and rolls, 0.38% in whole wheat products.

Calcium propionate is GRAS for miscellaneous and general-purpose usage in the United States with no limitation other than good manufacturing practices. Calcium propionate can be used as an antimycotic agent when migrating from food packaging material for cold-pack cheese or cheese food, pasteurised process cheese or cheese food, and pasteurised process cheese spread at a level not to exceed 0.3% alone or in combination with sodium propionate. It is used as an antifungal agent to control mould growth on the surface of cheese and butter and inhibits rope formation in baked goods. It can be used in confections and frostings, gelatins, puddings fillings, jams and jellies. Maximum levels recommended for use as a preservative are 0.3% alone or in combination with sodium propionate based on the weight of the flour used in the fresh pie dough, cold pack cheese, pasteurised process cheese food, pasteurised process cheese spread, 0.32% alone or in combination with sodium propionate based on the weight of the flour in pizza crust, and 0.1% by weight of artificially sweetened fruit jelly, fruit preserves and jams. It is not limited in bread, rolls or buns. It can be used as a yeast nutrient.

Dilauryl thiopropionic acid is GRAS for miscellaneous and general-purpose usage in the United States with no limitation other than good manufacturing practices. Dilauryl thiopropionic acid is used as a chemical preservative when the total content of antioxidants does not exceed 0.02% of the fat or oil content.

Potassium propionate is GRAS for miscellaneous and general-purpose usage in the United States with no limitation other than good manufacturing practices. Potassium propionate is used as an antifungal agent to control mould growth on the surface of cheese and butter and inhibits rope formation in baked goods. It can be used in baked goods, non-alcoholic beverages, soft candy, cheese, frostings, confections, fresh pie, fillings, gelatins, jams, jellies, meat products, pizza crust, and puddings.

Sodium propionate is GRAS for miscellaneous and general-purpose usage in the United States with no limitation other than good manufacturing practices. Sodium propionate can be used as a flavouring agent and an antimycotic agent when migrating from food packaging material for cold-pack cheese or cheese food, pasteurised process cheese or cheese food, and pasteurised process cheese spread at a level not to exceed 0.3% alone or in combination with calcium propionate. It is used as an antifungal agent to control mould growth on the surface of cheese and butter and inhibits rope formation in baked goods. It can be used in baked goods, non-alcoholic beverages, cheeses, confections and frostings, gelatins, puddings, fillings, jams and jellies, meat products and soft candy. Maximum levels recommended for use as a preservative are 0.3% alone or in combination with calcium propionate based on the weight of the flour used in the fresh pie dough, cold pack cheese, pasteurised process cheese food, and pasteurised process cheese spread, 0.32% alone or in combination with calcium propionate based on the weight of the flour in pizza crust, 0.1% by weight of artificially sweetened fruit jelly, preserves and jams. It is not limited in bread, rolls or buns.

Thiodipropionic acid is GRAS for miscellaneous and general-purpose usage in the United States with no limitation other than good manufacturing practices. Thiodipropionic acid is used as a chemical preservative when the total content of antioxidants does not exceed 0.02% of the fat or oil content.

TECHNOLOGY OF USE IN FOODS:

Propionic acid $pK_1 = 4.87$

Propionic acid has a pungent, sour odour and taste, and is limited in use by its sensory properties.

The salts of propionic acid have a slight cheese-like flavour. The sodium salt is more soluble than the calcium form. The salts are readily incorporated into powdered mixes.

Propionates may be added to bread dough without interfering with leavening since there is little or no effect on yeast growth. Sodium propionate is recommended for use in chemically leavened products since calcium interferes in the leavening action. Calcium propionate is preferred for use in bread and rolls because calcium propionate contributes to the enrichment of the product.

FOOD SAFETY ISSUES:

LD_{50} (mg/kg body weight) of propionic acid. Mouse 625 intravenous route. Rat 2,600–3,500 oral route.

LD_{50} (mg/kg body weight) calcium propionate. Rat 3,340 oral route; 580–1,020 intravenous route.

LD_{50} (mg/kg body weight) sodium propionate. Mouse 2,100 subcutaneous route. Rat 5,100 oral route; 1,380–3,200 intravenous route. Rabbit 1,640 cutaneous route.

Acceptable daily intake for humans of propionic acid and its calcium, potassium and sodium salts is not limited.

LEGISLATION:

USA:

Propionic acid: 21 CFR 184.1081

Calcium propionate: 9 CFR 318.7, 381.147; 21 CFR 133.123, 133.124, 133.169, 133.173, 133.179, 136.110, 136.115, 136.130, 136.160, 136.180, 150.141, 150.161, 181.23, 184.1081, 184.1221

Dilauryl thiopropionic acid: 21 CFR 182.3109

Potassium propionate: 9 CFR 318.7, 381.147

Sodium propionate: 9 CFR 318.7; 21 CFR 133.123, 133.124, 133.169, 133.173, 133.179, 150.141, 150.161, 181.23, 184.1784

Thiodipropionic acid: 21 CFR 182.3109

Thiopropionic acid: 21 CFR 182.3280

REFERENCES:

Budavari, S., O'Neill, M. J., Smith, A., and Heckelman, P. E. (1989) *The Merck Index.* 11th edn., Merck, Rahway, NJ.

Code of Federal Regulations (CFR) (1995) Volumes 9 (U. S. Department of Agriculture [USDA]), 21 (Food and Drug Administration [FDA]), and 27 (Bureau of Alcohol, Tobacco, and Firearms [BATF]). Office of the Federal Register, National Archives and Records Administration, U. S. Government Printing Office, Washington, DC.

Deshpande, S. S., Salunkhe, D. K., and Deshpande, U. S. (1995) Food Acidulants. In: Maga, J. A., and A. T. Tu (Eds.), *Food Additive Toxicology*. Marcel Dekker, Inc., New York. Ch. 2.

Doores, S. (1990) pH Control Agents and Acidulants. In: Branen, A. L., Davidson, P. M., and S. Salminen (Eds.), *Food Additives.* Marcel Dekker, Inc., New York. pp. 477–510.

Doores, S. (1993) Organic Acids. In: Davidson, P. M., and A. L. Branen (Eds.), *Antimicrobials in Foods*, 2nd edn., Marcel Dekker, Inc., New York. pp. 95–136.

Food Chemicals Codex (1981) 3rd edn., Committee on Food Chemicals Codex, Food and Nutrition Board, Institute of Medicine, National Academy of Sciences, National Academy Press, Washington, DC.

Food Chemicals Codex (1996) 4th edn., Committee on Food Chemicals Codex, Food and Nutrition Board, Institute of Medicine, National Academy of Sciences, National Academy Press, Washington, DC.

Lewis, R. J., Sr. (1989) *Food Additives Handbook.* Van Nostrand Reinhold, New York.

Lide, D. R. (Ed.) (1993) *Handbook of Chemistry and Physics.* 74th edn. CRC Press, Inc., Boca Raton, FL.

NAME:	Sodium diacetate
CATEGORY:	Chelating agents/ pH control agents/ Preservatives/ Flavour enhancers and modifiers
FOOD USE:	Baked goods/ Dairy products/ Edible oils and fats/ Meat, poultry and eggs and products/ Sugars, sugar preserves and confectionery
SYNONYMS:	Dykon/ Sodium acid acetate/ Sodium hydrogen diacetate/ CAS 126-96-5
FORMULA:	CH3-COONa-CH3-COOH·xH2O
MOLECULAR MASS:	142.09, anhydrous
ALTERNATIVE FORMS:	Calcium diacetate
PROPERTIES AND APPEARANCE:	White, hygroscopic crystalline solid
BOILING POINT IN °C AT VARIOUS PRESSURES (INCLUDING 760 mm Hg):	235
MELTING POINT IN °C:	58
PURITY %:	≥39 to ≤41 acetic acid; ≥58 to ≤60 sodium acetate
WATER CONTENT MAXIMUM IN %:	≤2
HEAVY METAL CONTENT MAXIMUM IN ppm:	≤10
ARSENIC CONTENT MAXIMUM IN ppm:	≤3
SOLUBILITY % AT VARIOUS TEMPERATURE/pH COMBINATIONS:	
in water:	@ 25°C 100
FUNCTION IN FOODS:	Sodium diacetate is GRAS for miscellaneous and general-purpose usage in the United States with no limitation other than good manufacturing practices. Sodium diacetate can be used as a preservative in

butter and wrapping materials, pH control agent, flavouring agent and adjuvant. It inhibits bread mould and rope-forming bacteria in bakery products and has little effect on baker's yeast. It delays mould growth in cheese spreads at concentrations of 0.1–2% and in malt syrups at 0.5%. Maximum levels recommended are 0.05% in snack foods, soup, and soup mixes (commercial), 0.1% in fats, oils, soft candy, and meat products, 0.25% in gravies and sauces, 0.4% in baked goods and baking mixes.

Calcium diacetate is GRAS for miscellaneous and general-purpose usage in the United States with no limitation other than good manufacturing practices. It is used as a sequestrant, preservative, pH control agent, flavouring agent and adjuvant.

TECHNOLOGY OF USE IN FOODS:

Sodium diacetate $pK_1 = 4.75$

Effective at pH of 3.5 to 4.5 to control mould growth in animal feed and silage

FOOD SAFETY ISSUES:

Acceptable daily intake for humans of sodium diacetate is limited conditionally up to 15 mg/kg body weight

LEGISLATION:

USA:
Sodium diacetate: 21 CFR 184.1754
Calcium diacetate: 21 CFR 182.6197

REFERENCES:

Budavari, S., O'Neill, M. J., Smith, A., and Heckelman, P. E. (1989) *The Merck Index.* 11th edn., Merck, Rahway, NJ.

Code of Federal Regulations (CFR) (1995) Volumes 9 (U. S. Department of Agriculture [USDA]), 21 (Food and Drug Administration [FDA]), and 27 (Bureau of Alcohol, Tobacco, and Firearms [BATF]). Office of the Federal Register, National Archives and Records Administration, U. S. Government Printing Office, Washington, DC.

Deshpande, S. S., Salunkhe, D. K., and Deshpande, U. S. (1995) Food Acidulants. In: Maga, J. A., and A. T. Tu (Eds.), *Food Additive Toxicology*. Marcel Dekker, Inc., New York. Ch. 2.

Doores, S. (1990) pH Control Agents and Acidulants. In: Branen, A. L., Davidson, P. M., and S. Salminen (Eds.), *Food Additives.* Marcel Dekker, Inc., New York. pp. 477–510.

Doores, S. (1993) Organic Acids. In: Davidson, P. M., and A. L. Branen (Eds.), *Antimicrobials in Foods*, 2nd edn., Marcel Dekker, Inc., New York. pp. 95–136.

Food Chemicals Codex (1981) 3rd edn., Committee on Food Chemicals Codex, Food and Nutrition Board, Institute of Medicine, National Academy of Sciences, National Academy Press, Washington, DC.

Food Chemicals Codex (1996) 4th edn., Committee on Food Chemicals Codex, Food and Nutrition Board, Institute of Medicine, National Academy of Sciences, National Academy Press, Washington, DC.

Lewis, R. J., Sr. (1989) *Food Additives Handbook.* Van Nostrand Reinhold, New York.

Lide, D. R. (Ed.) (1993) *Handbook of Chemistry and Physics.* 74th edn. CRC Press, Inc., Boca Raton, FL.

NAME:	Succinic acid
CATEGORY:	pH control agents/ Preservatives/ Flavour enhancers and modifiers/ Flour and baking additives
FOOD USE:	Baked goods/ Dairy products/ Edible oils and fats/ Meat, poultry and eggs and products/ Beverages/ Sugars, sugar preserves and confectionery/ Vinegar, pickles and sauces
SYNONYMS:	Amber acid/ Butanedioic acid/ Ethylenesuccinic acid/ 1,2-Ethanedicarboxylic acid/ 1,4-Butanedioic acid/ CAS 110-15-6
FORMULA:	HOOC-(CH2)2-COOH
MOLECULAR MASS:	118.09
ALTERNATIVE FORMS:	Succinic acid, diethyl ester; succinic anhydride
PROPERTIES AND APPEARANCE:	Colourless or white crystals
BOILING POINT IN °C AT VARIOUS PRESSURES (INCLUDING 760 mm Hg):	235 (decomposition)
MELTING RANGE IN °C:	185–190
IONISATION CONSTANT AT 25°C:	6.89×10^{-5} 2.47×10^{-6}
DENSITY AT 20°C (AND OTHER TEMPERATURES) IN g/l:	1.564 @ 20°/4° 1.572 @ 25°/4°
HEAT OF COMBUSTION AT 25°C:	356.36 kg calories/gram molecular weight
PURITY %:	≥90.0
HEAVY METAL CONTENT MAXIMUM IN ppm:	10

ARSENIC CONTENT MAXIMUM IN ppm:	3
ASH MAXIMUM IN %:	0.025
SOLUBILITY % AT VARIOUS TEMPERATURE/pH COMBINATIONS:	
in water:	**@ 25°C** 7.7 **@ 100°C** 100
in ethanol solution:	1 g/100 ml
in propylene glycol:	1 g/18.5 ml
FUNCTION IN FOODS:	Succinic acid is GRAS for miscellaneous and general-purpose usage in the United States with no limitation other than good manufacturing practices. Succinic acid is a flavour enhancer and pH control agent. It can be used to alter the acidity of acidified milk, acidified low-fat milk, and acidified skim milk. It readily combines with proteins in modifying the plasticity of doughs and aids in the production of edible fats with desired thermal properties. It is used in gelatin desserts and cake flavourings. It is also used as a component of polymeric resins. Maximum levels recommended are 0.0061% in meat, 0.084% in condiments and relishes, and 2.6% in jams and jellies. Succinic acid, diethyl ester is used as a flavouring agent with no limitation other than good manufacturing practices. Succinic anhydride is GRAS for miscellaneous and general-purpose usage in the United States with no limitation other than good manufacturing practices. Succinic anhydride is used as a dehydrating agent for the removal of moisture from foods and it imparts a stability to dry mixes. It also aids in the controlled release of carbon dioxide during leavening. It is used in the esterification of food starch at a level not to exceed 4%.
TECHNOLOGY OF USE IN FOODS:	Succinic acid $pK_1 = 4.16$; $pK_2 = 5.61$ Succinic acid is odourless and has a slightly bitter and acid taste. It is non-hygroscopic with a slow taste build-up. It can extend the shelf life of dessert powders, without damaging flavour.
FOOD SAFETY ISSUES:	LD_{50} (mg/kg body weight) succinic acid. Rat 2,260 oral route. Acceptable daily intake for humans of succinic acid is not limited.

LEGISLATION:

USA:
Succinic acid: 21 CFR 131.111, 131.136, 131.144, 175.300, 184.1091
Succinic acid, diethyl ester: 21 CFR 172.515
Succinic anhydride: 21 CFR 172.894

REFERENCES:

Budavari, S., O'Neill, M. J., Smith, A., and Heckelman, P. E. (1989) *The Merck Index.* 11th edn., Merck, Rahway, NJ.

Code of Federal Regulations (CFR) (1995) Volumes 9 (U. S. Department of Agriculture [USDA]), 21 (Food and Drug Administration [FDA]), and 27 (Bureau of Alcohol, Tobacco, and Firearms [BATF]). Office of the Federal Register, National Archives and Records Administration, U. S. Government Printing Office, Washington, DC.

Deshpande, S. S., Salunkhe, D. K., and Deshpande, U. S. (1995) Food Acidulants. In: Maga, J. A., and A. T. Tu (Eds.), *Food Additive Toxicology*. Marcel Dekker, Inc., New York. Ch. 2.

Doores, S. (1990) pH Control Agents and Acidulants. In: Branen, A. L., Davidson, P. M., and S. Salminen (Eds.), *Food Additives.* Marcel Dekker, Inc., New York. pp. 477–510.

Doores, S. (1993) Organic Acids. In: Davidson, P. M., and A. L. Branen (Eds.), *Antimicrobials in Foods*, 2nd edn., Marcel Dekker, Inc., New York. pp. 95–136.

Food Chemicals Codex (1981) 3rd edn., Committee on Food Chemicals Codex, Food and Nutrition Board, Institute of Medicine, National Academy of Sciences, National Academy Press, Washington, DC.

Food Chemicals Codex (1996) 4th edn., Committee on Food Chemicals Codex, Food and Nutrition Board, Institute of Medicine, National Academy of Sciences, National Academy Press, Washington, DC.

Lewis, R. J., Sr. (1989) *Food Additives Handbook.* Van Nostrand Reinhold, New York.

Lide, D. R. (Ed.) (1993) *Handbook of Chemistry and Physics.* 74th edn. CRC Press, Inc., Boca Raton, FL.

NAME:	Tartaric acid
CATEGORY:	Emulsifiers/stabilisers/ Chelating agents/ pH control agents/ Preservatives/ Flavour enhancers and modifiers/ Flour and baking additives/ Anti-caking agents/ Firming agents/ Humectants
FOOD USE:	Baked goods/ Dairy products/ Edible oils and fats/ Fish and seafoods and products/ Meat, poultry and eggs and products/ Fruit, vegetables and nuts and products/ Beverages/ Soft drinks/ Sugars, sugar preserves and confectionery/ Alcoholic drinks
SYNONYMS:	Dihydroxybutanedioic acid/ 2,3-Dihydroxybutanedioic acid/ 2,3-Dihydroxysuccinic acid/ d-a, b-Dihydroxysuccinic acid/ L form CAS 87-69-4
FORMULA:	HOOC-(CHOH)2-COOH
MOLECULAR MASS:	150.09
ALTERNATIVE FORMS:	Diethyl tartrate; potassium acid tartrate; potassium bitartrate; sodium potassium tartrate; sodium tartrate
PROPERTIES AND APPEARANCE:	Colourless to translucent crystals, or a white, fine to granular crystalline powder
MELTING RANGE IN °C:	168–170
IONISATION CONSTANT AT 25°C:	1.04×10^{-3} 4.55×10^{-5}
DENSITY AT 20°C (AND OTHER TEMPERATURES) IN g/l:	1.7598 @ 20°/4°C
HEAT OF COMBUSTION AT 25°C:	275.1 kg carlories/gram molecular weight 274.7 kg carlories/gram molecular weight (L form) 272.6 kg carlories/gram molecular weight (D, L anhydrous form)
PURITY %:	≥99.7
WATER CONTENT MAXIMUM IN %:	0.5

HEAVY METAL CONTENT MAXIMUM IN ppm:	10
ARSENIC CONTENT MAXIMUM IN ppm:	3
ASH MAXIMUM IN %:	0.05
SOLUBILITY % AT VARIOUS TEMPERATURE/pH COMBINATIONS:	
in water:	**@ 0°C** 115 **@ 10°C** 126 **@ 20°C** 139 **@ 30°C** 156 **@ 40°C** 176 **@ 50°C** 195 **@ 60°C** 217 **@ 70°C** 244 **@ 80°C** 273 **@ 90°C** 307 **@ 100°C** 343
in ethanol solution:	1 g/3 ml
FUNCTION IN FOODS:	Tartaric acid is GRAS for miscellaneous and general-purpose usage in the United States with no limitation other than good manufacturing practices. Tartaric acid is used as a firming agent, flavour enhancer, flavouring agent, humectant, sequestrant, and pH control agent. It is used to alter the acidity of acidified milk, acidified low-fat milk, acidified skim milk, margarine, oleomargarine, meat and poultry products, artificially sweetened fruit jelly, fruit preserves and jams. It can be used in fruit jams, jellies, preserves, sherbets and grape-flavoured beverages and canned fruits. It can be used as a dip (1–3%) for meat carcasses to lower microbial populations. Tartaric acid salts are used to control the degree of acidity in soft drinks and as a flavour enhancer, particularly in lime and grape-flavoured beverages. Tartaric acid can be used to reduce the pH of wine or juice, but not below pH 3.0. It can be added prior to fermentation of grapes or in combination with other acids after fermentations to correct deficiencies at a level not to exceed 9 g/l finished wine. It can be used to correct natural acid deficiencies in grape juice/wine and to reduce the pH value where ameliorating material is used in the production of grape wine. Potassium acid tartrate (cream of tartar) is GRAS for miscellaneous and general-purpose usage in the United States with no limitation other than good manufacturing practices. It can be used as an anti-caking and leavening agent, antimicrobial or pH control agent, formulation and processing aid, humectant, stabiliser and thickener and surface active agent. It is used in baked goods, candy, crackers, confections and frostings, gelatins and puddings, jams, jellies, and margarine. Its use is limited in wine at 25 lb/1000 gallons (*ca.* 50 mg/kg). It is limited at 2 oz/100 lb (*ca.* 1 mg/kg) of finished product or artificial sweetened fruit jelly, preserves and jam.

Potassium bitartrate can be used to stabilise grape wine at a level not to exceed 35 lb/1000 gallons (*ca.* 70 mg/kg) of grape wine.

Sodium potassium tartrate is GRAS for miscellaneous and general-purpose usage in the United States with no limitation other than good manufacturing practices. It acts as a pH control agent, and sequestrant in cheese, jams, jellies and margarine. It is limited as an emulsifier at a level not to exceed 3.0% of pasteurised process cheese, cheese food and cheese spread. It is limited at 2 oz/100 lb (*ca.* 1 mg/kg) of finished product or artificial sweetened fruit jelly, preserves and jam.

Sodium tartrate is GRAS for miscellaneous and general-purpose usage in the United States with no limitation other than good manufacturing practices. It functions as a pH control agent in fats, oils, jams and jellies. It is limited as an emulsifier at a level not to exceed 3.0% of pasteurised process cheese, cheese food and cheese spread. It is limited at 2 oz/100 lb (*ca.* 1 mg/kg) of finished product of artificial sweetened fruit jelly, fruit preserves and jam.

TECHNOLOGY OF USE IN FOODS:

Tartaric acid $pK_1 = 2.98$; $pK_2 = 4.34$

Tartaric acid has a sharp or bitter tart taste and is the most soluble of all acidulants.

When blended with citric acid, tartaric acid contributes tartness to sour apple and wild cherry flavours.

SYNERGISTS:

Tartaric acid acts with antioxidants to prevent rancidity.

FOOD SAFETY ISSUES:

LD_{50} (mg/kg body weight) tartaric acid. Mouse 4,960 oral route; 485 intravenous route. Rat 3,310–3,530 oral route. Dog 5,000 oral route.

Acceptable daily intake for humans of L-tartaric acid and its potassium and sodium salt is limited conditionally up to 30 mg/kg body weight.

LEGISLATION:

USA:

Tartaric acid: 9 CFR 318.7, 381.147; 21 CFR 131.111, 131.136, 131.144, Part 145, 150.141, 150.161, 166.110, 184.1099; 27 CFR 24.182, 24.246, 240.364, 240.512, 240.1051

Potassium acid tartrate: 9 CFR 318.7; 21 CFR 150.141, 150.161, 184.1077; 27 CFR 240.1051

Potassium bitartrate: 21 CFR 184.1077; 27 CFR 24.246

Sodium potassium tartrate: 9 CFR 318.7; 21 CFR 133.169, 133.173, 133.179, 150.141, 150.161, 184.1804

Sodium tartrate: 21 CFR 133.169, 133.173, 133.179, 150.141, 150.161, 184.1801

REFERENCES:

Budavari, S., O'Neill, M. J., Smith, A., and Heckelman, P. E. (1989) *The Merck Index.* 11th edn., Merck, Rahway, NJ.

Code of Federal Regulations (CFR) (1995) Volumes 9 (U. S. Department of Agriculture [USDA]), 21 (Food and Drug Administration [FDA]), and 27 (Bureau of Alcohol, Tobacco, and Firearms [BATF]). Office of the Federal Register, National Archives and Records Administration, U. S. Government Printing Office, Washington, DC.

Deshpande, S. S., Salunkhe, D. K., and Deshpande, U. S. (1995) Food Acidulants. In: Maga, J. A., and A. T. Tu (Eds.), *Food Additive Toxicology*. Marcel Dekker, Inc., New York. Ch. 2.

Doores, S. (1990) pH Control Agents and Acidulants. In: Branen, A. L., Davidson, P. M., and S. Salminen (Eds.), *Food Additives.* Marcel Dekker, Inc., New York. pp. 477–510.

Doores, S. (1993) Organic Acids. In: Davidson, P. M., and A. L. Branen (Eds.), *Antimicrobials in Foods*, 2nd edn., Marcel Dekker, Inc., New York. pp. 95–136.

Food Chemicals Codex (1981) 3rd edn., Committee on Food Chemicals Codex, Food and Nutrition Board, Institute of Medicine, National Academy of Sciences, National Academy Press, Washington, DC.

Food Chemicals Codex (1996) 4th edn., Committee on Food Chemicals Codex, Food and Nutrition Board, Institute of Medicine, National Academy of Sciences, National Academy Press, Washington, DC.

Lewis, R. J., Sr. (1989) *Food Additives Handbook.* Van Nostrand Reinhold, New York.

Lide, D. R. (Ed.) (1993) *Handbook of Chemistry and Physics.* 74th edn. CRC Press, Inc., Boca Raton, FL.

Part 2

Antioxidants

NAME:	Ascorbic acid and isomers (L-ascorbic acid and erythorbic acid)	
CATEGORY:	Antioxidant/ Sequestrant/ Reducing agent	
FOOD USE:	Fruit juices/ Carbonated beverages/ Wine/ Beer/ Frozen fruits/ Canned fruits and vegetables/ Sausages/ Cured meats/ Milk powder/ Citrus oils	
SYNONYMS:	L-ascorbic acid: 1-ascorbic acid/ 2,3-Didehydro-L-threo-hexono-1,4-lactone/ 3-Keto-L-gulofuranolactone/ Vitamin C/ E300/ Cevitamic acid/ Ascorbo®-120/ Ascorbo®-C/ Cap-Shure®/ AS-125-50 Erythorbic acid: D-araboascorbic acid/ D-erythro-hex-2-enoic acid-γ-lactone/ isoascorbic acid/ Neo-Cebitat®/ Eribate® (sodium erythorbate)	
FORMULA:	C6 H8 O6	
MOLECULAR MASS:	176.13	
ALTERNATIVE FORMS:	L-ascorbic acid: sodium ascorbate; calcium ascorbate Erythorbic acid: sodium erythorbate	
PROPERTIES AND APPEARANCE:	Both are white or slightly yellow crystalline powders	
MELTING RANGE IN °C:	L-ascorbic acid, 190–192, with decomposition Erythorbic acid, 164–171, with decomposition	
	L-ascorbic acid	Erythorbic acid
PURITY %:	>99	>99
WATER CONTENT MAXIMUM IN %	0.4	0.4
HEAVY METAL CONTENT MAXIMUM IN ppm:	20	20
ARSENIC CONTENT MAXIMUM IN ppm:	3	3
ASH MAXIMUM IN %	0.1	0.3

SOLUBILITY % AT VARIOUS TEMPERATURE/pH COMBINATIONS:

		L-ascorbic acid	Erythorbic acid
in water:	**@ 20°C**	33	40
in vegetable oil:	**@ 20°C**	Insoluble	Insoluble
in ethanol solution:	**@ 100%**	2	—

FUNCTION IN FOODS: L-ascorbic acid: antioxidant activity is due to quenching of various forms of oxygen (singlet oxygen, hydroxyl radicals as well as superoxide), reduction of free radicals, thus terminating radical reaction. Acts as a synergist. Functions as vitamin C, colour fixing, flavouring, raising agent, oxidant and a reducing agent. Prevents enzymatic browning of fruits and vegetables.

Erythorbic acid: weaker antioxidant than L-ascorbic acid, enhances curing action of nitrites on meat pigments, stabilises colour of meat, meat products and fruits, inhibits growth of microorganisms and formation of *N*-nitrosamines in cured meats.

TECHNOLOGY OF USE IN FOODS: Apply to surfaces of cured meat cuts, easily oxidised on exposure to air in aqueous solutions

SYNERGISTS: α-Tocopherol/ Citric acid/ BHA/ BHT

FOOD SAFETY ISSUES: Avoid acrid smoke and irritating fumes when heated

LEGISLATION:

L-ascorbic acid:

USA:	**UK and EUROPE:**	**AUSTRALIA/PACIFIC RIM:**
FDA and USDA: approved	approved	Japan: approved

Erythorbic acid:

USA:	**AUSTRALIA/PACIFIC RIM:**
USDA and FDA: approved	Japan: restricted for purpose as antioxidant

REFERENCES:

Ash, M., and Ash, I. (1995) *Handbook of Food Additives*. Gower Publishing Ltd., Hampshire UK.

Schuler, P. (1990) In: Hudson, B. J. F. (Ed.), *Food Antioxidants*. Elsevier Science Publishers, Essex, UK.

FAO (1992) *Compendium of Food Additives Specifications*, Volume 1. Joint FAO/WHO Expert Committee on Food Additives: FAO Food and Nutrition Paper 52/1, FAO, Rome.

Food Chemical Codex (1981) 3rd edition. National Academy Press, Washington, DC.

NAME:	**Ascorbyl palmitate**
CATEGORY:	Antioxidant
FOOD USE:	Baked goods/ Soft drinks/ Margarine/ Sausages/ Vegetable oils/ Shortening
SYNONYMS:	6-*O*-Palmitoyl-L-ascorbate/ L-ascorbyl palmitate/ 2,3-Didehydro-L-threo-hexono-1,4-lactone-6-palmitate/ 6-Palmitoyl-3-keto-2-gulofurano-lactone/ E304/ Pristene®/ MT-AP/ Palmitoyl-L-ascorbic acid/ Ascorbyl palmitate NF/ Vitamin C-palmitate
FORMULA:	$C_{22}H_{38}O_7$
MOLECULAR MASS:	414.55
ALTERNATIVE FORMS:	Ascorbyl stearate
PROPERTIES AND APPEARANCE:	White powder with a soapy taste and citrus-like odour
MELTING RANGE IN °C	107–117 (113°C with decomposition in the presence of oxygen; 180°C with decomposition in the absence of oxygen)
PURITY %:	>95
WATER CONTENT MAXIMUM IN %:	2
HEAVY METAL CONTENT MAXIMUM IN ppm:	10
ARSENIC CONTENT MAXIMUM IN ppm:	3
ASH MAXIMUM %	0.1
SOLUBILITY % AT VARIOUS TEMPERATURE/pH COMBINATIONS:	
in water:	@ **25°C** pH = 7.0 0.002; @ **50°C** pH = 8.1 0.01; @ **60°C** pH = 8.1 60.0; @ **70°C** 2.0; @ **100°C** 10.0

in vegetable oil:						
Coconut oil	**@ 25°C**	1.2	**@ 100°C**	50.0		
Olive oil	**@ 25°C**	0.30				
Peanut oil	**@ 25°C**	0.30	**@ 70°C**	1.60	**@ 100°C**	9.00
Sunflower oil	**@ 25°C**	0.28				
in ethanol solution:	**@ 50%**	**@ 25°C**	0.4			
	@ 95%	**@ 25°C**	108.0			
	@ 100%	**@ 25°C**	125.0			
	@ 100%	**@ 0°C**	45.0			
in propylene glycol:	**@ 25°C**	48.0				

FUNCTION IN FOODS: Antioxidant that prevents oxidative rancidity development by quenching singlet oxygen. A sequestrant, emulsifier, stabiliser, source of vitamin C and a colour preservative

TECHNOLOGY OF USE IN FOODS: Decomposition starts at 113°C in the presence of oxygen; unstable in alkaline pH

SYNERGISTS: α-Tocopherol

LEGISLATION:

USA:	**UK:**	**AUSTRALIA/PACIFIC RIM:**
FDA and USDA: approved. 0.02% maximum in margarine	approved	Japan: approved

REFERENCES:
Ash, M., and Ash, I. (1995) *Handbook of Food Additives*. Gower Publishing Co. Ltd., Hampshire, UK.

NAME:	**Sodium ascorbate**
CATEGORY:	Antioxidant/ Colour preservative
FOOD USE:	Canned meat products/ Bottled baby food/ Frozen fish
SYNONYMS:	L(+)ascorbic acid sodium salt/ Vitamin C, sodium salt/ monosodium ascorbate/ sodium-L-ascorbate/ Descote®/ E301
FORMULA:	C6 H7 NaO6
MOLECULAR MASS:	198.11
ALTERNATIVE FORMS:	Calcium ascorbate/ L-ascorbic acid
PROPERTIES AND APPEARANCE:	White to yellowish crystalline solid, odourless, soapy taste
MELTING RANGE IN °C	200 (decomposes at 218)
PURITY %:	>99 (and not more than the equivalent of 101.0 of $C_6H_7NaO_6$ after drying)
WATER CONTENT MAXIMUM IN %:	0.25
HEAVY METAL CONTENT MAXIMUM IN ppm:	0.002
ARSENIC CONTENT MAXIMUM IN ppm:	3
SOLUBILITY % AT VARIOUS TEMPERATURE/pH COMBINATIONS:	
in water:	**@ 25°C** 89%
in vegetable oil:	**@ 25°C** Insoluble
in ethanol solution:	**@ 25°C** Insoluble
FUNCTION IN FOODS:	Functions as an antioxidant, colour preservative, dietary supplement of vitamin C, nutrient supplement in foods and curing of meat

ALTERNATIVES: L-ascorbic acid, calcium ascorbate

FOOD SAFETY ISSUES: When heated to decomposition, it emits toxic fumes of Na_2O that have been shown to be mutagenic in humans

LEGISLATION:

USA:	**UK:**	**AUSTRALIA/PACIFIC RIM:**
FDA and USDA approved: 87.5 oz/100 gal of pickle; 7/8 oz/100 lb of meat; 500 ppm in meat alone or in combination with ascorbic acid or erythorbic acid	approved	Japan: approved

REFERENCES:

Ash, M., and Ash, I. (1995) *Handbook of Food Additives.* Gower Publishers Ltd., Hampshire, UK.
Food Chemical Codex (1981) 3rd edition. National Academy Press, Washington, DC.

NAME:	Calcium ascorbate
CATEGORY:	Antioxidant
FOOD USE:	Meat products/ Concentrated dairy products
SYNONYMS:	E302
FORMULA:	$C_{12}H_{14}CaO_{12} \cdot 2H_2O$
PROPERTIES AND APPEARANCE:	White to slight yellow odourless crystalline powder
PURITY %:	Not less than 98
HEAVY METAL CONTENT MAXIMUM IN ppm:	10
ARSENIC METAL MAXIMUM IN ppm:	3
SOLUBILITY % AT VARIOUS TEMPERATURE/pH COMBINATIONS:	
in water:	@ 25°C 55%
in vegetable oil:	@ 25°C Insoluble
in ethanol solution:	@ 25°C Insoluble
FUNCTION IN FOODS:	Antioxidant, vitamin C source and functions as a meat colour preservative
FOOD SAFETY ISSUES:	When heated to decomposition, it emits acrid smoke and irritating fumes
LEGISLATION:	**USA:** FDA approved; **UK:** approved; **EUROPE:** listed
REFERENCES:	*Food Chemical Codex.* 1981. 3rd edition. National Academy Press, Washington, DC. Ash, M., and Ash, I. (1995) *Handbook of Food Additives.* Gower Publishing Inc., Hampshire, UK.

NAME:	Butylated hydroxyanisole (BHA)
CATEGORY:	Antioxidant
FOOD USE:	Bakery products/ Meat products/ Spices/ Cereals/ Dehydrated mashed potatoes/ Beverage mixes/ Dessert mixes/ Nuts/ Vitamins/ Yeast/ Vegetable oils/ Animal fats/ Processed cheeses/ Margarine/ Essential oils/ Chewing gum base
SYNONYMS:	Mixture of two isomers: 3-tertiary butyl-4-hydroxyanisole and 2-tertiary butyl-4-hydroxyanisole/ (1,1-dimethylethyl)-4-methoxyphenol/ E320/ Antracine® 12/ Embanox®/ Nipantiox®/ Sustane® BHA/ Sustane® 1-F/ Tenox® BHA/ Tenox® 4B/ Tenox® 5B
FORMULA:	(CH3)3 C C6H3 O CH3 OH
MOLECULAR MASS:	180.25
PROPERTIES AND APPEARANCE:	White waxy flakes or tablets
BOILING POINT IN °C AT VARIOUS PRESSURES (INCLUDING 760 mm Hg):	at 733 mmHg = 264–270; at 760 mm Hg = 270
MELTING RANGE IN °C:	48–55
FLASH POINT IN °C:	130
PURITY %:	Not less than 98.5 of 2-isomer and not less than 85 of 3-isomer
HEAVY METAL CONTENT MAXIMUM IN ppm:	10
ARSENIC CONTENT MAXIMUM IN ppm:	3
ASH MAXIMUM IN %:	0.05
SOLUBILITY % AT VARIOUS TEMPERATURE/pH COMBINATIONS:	
in water:	@ 20°C Insoluble

in vegetable oil:	**@ 25°C**	30% cottonseed oil 40% coconut, corn, peanut oils 50% soybean oil
in ethanol solution:	**100% @ 25°C**	>25%
in propylene glycol:	**@ 20°C**	70%

FUNCTION IN FOODS: Antioxidant preservative by terminating free radicals formed during autoxidation of unsaturated lipids. It also possesses antimicrobial activity as a phenolic compound

ALTERNATIVES: BHT; PG; TBHQ

TECHNOLOGY OF USE IN FOODS: Direct application: dissolves in lipid heated to 60°C, continue mild agitation for additional 20 min to ensure uniform distribution.

Concentration method: prepare an antioxidant concentrate solution by dissolving the antioxidant in a small quantity of heated lipid (93–121°C). This hot concentrate can be introduced into fat directly, or by metering.

Spray method: a dilute antioxidant solution can be directly sprayed onto food products such as nuts.

Other methods: antioxidants should be mixed with the seasonings for meat products such as sausages. For cereal products, antioxidants can be incorporated into the wax liner of the packages.

Usually applied with BHT for better efficiency than alone.

Should be protected from light and exposure to air.

SYNERGISTS: BHT; propyl gallate; methionine; lecithin; thiodipropionic acid; citric acid; phosphoric acid

FOOD SAFETY ISSUES: This antioxidant has not been subjected to great criticism of safety. However, suspected for tumour formation in animals with forestomach (Malaspina, 1987).

LEGISLATION:

USA:
Maximum usage level approved for general use; FDA 0.02% and USDA 0.01% of weight of fat.

Special applications include:
Chewing gum base – 0.01% by weight of chewing gum base
Active dry yeast or dry material

UK and EUROPE:
approved

CANADA:
approved

(conts)

(*USA: contd*)
Emulsion stabilisers – 0.02% by weight of emulsion, shortenings, stabiliser
Potato flakes, sweet potato flakes – 0.005% by weight, dry breakfast cereal, of food material, packaging material
Potato granules – 0.001% by weight of potato granules
Dry mixes for beverages – 0.009% of material and desserts
Beverages and desserts, prepared from dry mixes – 0.0002%
Dry diced glazed fruits – 0.0032%
Flavour substances – 0.5% of essential oil content

AUSTRALIA/PACIFIC RIM:
Japan: approved

REFERENCES:

Sustane® *Food-Grade Antioxidants*, Universal Oil Products Company, Food Antioxidants Department, Des Plaines, IL, 1994.

Tenox® *Food-Grade Antioxidants*, Eastman Chemical Company, Kingsport, TN. Publication ZG-262B, March 1995.

Malaspina, A. (1997) In: Miller, K. (Ed.), *Toxicological Aspects of Food*. Elsevier Applied Science, London, pp. 17–57.

Ash, M., and Ash, I. (1995) *Handbook of Food Additives*. Gower Publishing Ltd., Hampshire, UK.

Specchio, J. J. (1992) In: Hui, Y. H. (Ed.), *Encyclopedia of Food Science and Technology*, Volume 1, A–D, John Wiley & Sons, Inc., New York, pp. 73–78.

FAO (1992) *Compendium of Food Additives Specifications 1*. Joint FAO/WHO Expert Committee on Food Additives, FAO Food and Nutrition, Paper 52/1, FAO, Rome.

NAME:	Butylated hydroxytoluene (BHT)
CATEGORY:	Antioxidant
FOOD USE:	Breakfast cereals/ Baked goods/ Potato chips/ Vegetable oils/ Snack foods/ Butter/ Margarine/ Frozen seafoods/ Chewing gum base
SYNONYMS:	2,6-*bis* (1,1-dimethylethyl)-4-methylphenol/ 2,6-di-*tert*-butyl-*p*-cresol/ 2,6-di-*tert*-butyl-4-methylphenol/ E321/ Antracine® 8/ Ionol® CP/ Dalpac®/ Impruvol®/ Vianol®/ Tenox® BHT/ Tenox® 8/ Sustane® BHT
FORMULA:	[(CH3)3C]2 C6H2 CH3 OH
MOLECULAR MASS:	220.34
PROPERTIES AND APPEARANCE:	White granular crystals with slight odour
BOILING POINT IN °C AT VARIOUS PRESSURES (INCLUDING 760 mm Hg):	265
MELTING RANGE IN °C:	69–72
FLASH POINT IN °C:	118
PURITY %:	Not less than 99
HEAVY METAL CONTENT MAXIMUM IN ppm:	10
ARSENIC CONTENT MAXIMUM IN ppm:	3
ASH MAXIMUM IN %:	0.05
SOLUBILITY % AT VARIOUS TEMPERATURE/pH COMBINATIONS:	
in water:	**@ 20°C** Insoluble
in vegetable oil:	**@ 25°C** 30% in coconut, cottonseed, corn, peanut and soybean oils

in ethanol solution solution: **@ 100%** 25%

in propylene glycol: **@ 20°C** Insoluble

FUNCTION IN FOODS: Antioxidant preservative; prevents oxidative rancidity development in oil-containing foods by terminating free radicals formed during autoxidation of unsaturated lipids. It possesses antimicrobial activity as a phenolic compound

ALTERNATIVES: BHA; PG; TBHQ

TECHNOLOGY OF USE IN FOODS: Similar to BHA

SYNERGISTS: BHA

FOOD SAFETY ISSUES: This has not been subjected to great criticism over safety

LEGISLATION:

USA:
Maximum general usage level approved by USDA is 0.01% and FDA 0.02% of weight of lipids.

Special applications include:
- Enriched rice – 0.0033%
- Non-alcoholic beverages, frozen raw breaded shrimp, mixed nuts and margarine – 0.02% based on oil content
- Dry sausages – 0.003%
- Fresh pork sausages, brown-and-serve sausages, pre-grilled beef patties, pizza toppings, meatballs, dried meats – 0.01%
- Rendered animal fat or combination with vegetable fat, poultry fat or various poultry products – 0.01%
- Dry breakfast cereals – 0.005%
- Emulsion stabilised for shortening – 0.02%
- Potato granules – 0.001%
- Potato flakes, sweet potato flakes, dehydrated potato shreds – 0.005%

EUROPE and UK:
approved

CANADA:
approved

AUSTRALIA/PACIFIC RIM:
Japan: approved

REFERENCES:

Sustane® *Food-Grade Antioxidants*. Universal Oil Products Company, Food Antioxidants Department, Des Plaines, IL, 1994.

Tenox® *Food-Grade Antioxidants*. Eastman Chemical Company, Kingsport, TN. Publication ZG-262B, March 1995.

Ash, M., and Ash, I. (1995) *Handbook of Food Additives*. Gower Publishing Ltd., Hampshire, UK.

Specchio, J. J. (1992) In: Hui, Y. H., (Ed.), *Encyclopedia of Food Science and Technology*, Volume 1, A–D, John Wiley and Sons Inc., New York, pp. 73–78.

FAO (1992) *Compendium of Food Additives Specifications 1*. Joint FAO/WHO Expert Committee on Food Additives, FAO Food and Nutrition, Paper 52/1, FAO, Rome.

NAME:	Ethoxyquin
CATEGORY:	Antioxidant/ Stabiliser
FOOD USE:	Apples/ Pears/ Chili powder/ Paprika powder
SYNONYMS:	6-Ethoxy-1,2-dihydro-2,2,4-trimethylquinoline/ 1,2-Dihydro-6-ethoxy-2,2,4-trimethylquinoline/ Santoquine/ EMQ
FORMULA:	$C_{14}H_{19}NO$
MOLECULAR MASS:	217.34
PROPERTIES AND APPEARANCE:	Yellow-coloured liquid
BOILING POINT IN °C AT VARIOUS PRESSURES (INCLUDING 760 mm Hg):	125
FUNCTION IN FOODS:	Prevents brown spot formation (scald) in apples and pears during storage. Preserves the colour in ground chili and paprika. Used as antioxidant in animal feed
LEGISLATION:	**USA:** FDA approved: 100 ppm in chili powder; 100 ppm in paprika; 5 ppm in uncooked meat fat; 3 ppm in uncooked poultry fat; 0.5 ppm in eggs
REFERENCE:	Ash, M., and Ash, I. (1995) *Handbook of Food Additives.* Gower Publishing Ltd., Hampshire, UK.

NAME:	**Propyl gallate (PG)**
CATEGORY:	Antioxidant
FOOD USE:	Chewing gum base/ Non-alcoholic beverages/ Margarine/ Mixed nuts/ Fresh or dry sausages/ Pre-grilled beef patties/ Rendered animal fat/ Pizza toppings and meatballs
SYNONYMS:	*n*-Propyl-3,4,5-trihydroxybenzoate/ 3,4,5-Trihydroxybenzoic acid/ Gallic acid, propyl ester/ E310/ Nipa® 49/ Nipagallin® P/ Progallin® P/ Tenox® PG/ Sustane® PG
FORMULA:	(HO)3 C6H2 COO CH2 CH2 CH3
MOLECULAR MASS:	212.20
ALTERNATIVE FORMS:	Octyl gallate; dodecyl gallate
PROPERTIES AND APPEARANCE:	White crystalline powder with slight odour
BOILING POINT IN °C AT VARIOUS PRESSURES (INCLUDING 760 mm Hg):	Decomposes at about 148
MELTING RANGE IN °C:	146–150
FLASH POINT IN °C:	187
PURITY %:	Not less than 98% and not more than 102.5% on the dried basis
WATER CONTENT MAXIMUM IN %:	0.5
HEAVY METAL CONTENT MAXIMUM IN ppm:	10
ARSENIC CONTENT MAXIMUM IN ppm:	3
ASH MAXIMUM IN %:	0.1

SOLUBILITY % AT VARIOUS TEMPERATURE/pH COMBINATIONS:

in water:	**@ 20°C**	<1%
in vegetable oil:	**@ 20°C**	1% in cottonseed oil 2% in soybean oil insoluble in corn oil
in ethanol solution:	**@ 100%**	>60%

FUNCTION IN FOODS:
Prevents oxidative rancidity development in lipid-containing foods by terminating free radical formation during autoxidation of unsaturated lipids

ALTERNATIVES:
BHA; BHT; TBHQ; octyl gallate; dodecyl gallate

TECHNOLOGY OF USE IN FOODS:
Heat-sensitive and decomposes at 148°C; therefore, not good for foods subjected to heat treatment. Also poorly soluble in lipids and water; can form coloured complexes with metal ions, and affect the appearance of food

SYNERGISTS:
BHA; BHT

FOOD SAFETY ISSUES:
Not subjected to great criticism over safety

LEGISLATION:

USA:
Not allowed to use in combination with TBHQ.

For general used, FDA – 0.02% and USDA 0.01% alone or in combination with BHT or BHA by weight of lipid portion of food.

Special applications include:
- Chewing gum base, 0.1%
- Non-alcoholic beverages, 0.1%
- Margarine, 0.1%
- Mixed nuts, 0.02% based on oil content
- French beef or pork sausages
- Brown-and-serve sausages
- Pre-grilled beef patties

(conts)

UK:
approved

EUROPE:
listed

CANADA:
approved

AUSTRALIA/PACIFIC RIM:
Japan: approved

(*USA: contd*)

Pizza toppings and meatballs – for all these 0.01% based on weight of finished product

Rendered animal fats or combination of such fat with vegetable fat – 0.01% based on lipid content

REFERENCES:

Sustane® *Food-Grade Antioxidants*. Universal Oil Products Company, Food Antioxidants Department, Des Plaines, IL, 1994.

Tenox® *Food-Grade Antioxidants*. Eastman Chemical Company, Kingsport, TN. Publication ZG-262B, March 1995.

Ash, M., and Ash, I. (1995) *Handbook of Food Additives*. Gower Publishing Ltd., Hampshire, UK.

NAME:	**Tert-butyl hydroquinone (TBHQ)**
CATEGORY:	Antioxidant
FOOD USE:	Dry cereals/ Edible fats/ Margarine/ Pizza toppings/ Potato chips/ Poultry/ Dried meats/ Sausages/ Beef patties/ Vegetable oils
SYNONYMS:	2-(1,1-dimethylethyl)-1,4-benzenediol/ mono-*t*-butyl hydroquinone/ Sustane® TBHQ/ Tenox® TBHQ
FORMULA:	(CH3)3 C C6H3 (OH)2
MOLECULAR MASS:	166.22
PROPERTIES AND APPEARANCE:	White to tan colour solid crystals, having a characteristic odour
BOILING POINT IN °C AT VARIOUS PRESSURES (INCLUDING 760 mm Hg):	295
MELTING RANGE IN °C:	126.5–128.5
FLASH POINT IN °C:	171
PURITY %:	99
HEAVY METAL CONTENT MAXIMUM IN ppm:	10
ARSENIC CONTENT MAXIMUM IN ppm:	3
SOLUBILITY % AT VARIOUS TEMPERATURE/pH COMBINATIONS:	
in water:	**@ 20°C** <1% **@ 100°C** 5%
in vegetable oil:	**@ 20°C** 10% corn, cottonseed, and soybean oils
in ethanol solution:	**@ 100%** 25%
in propylene glycol:	**@ 20°C** 30%

FUNCTION IN FOODS: Prevents oxidative rancidity development in foods by terminating free radical formation

ALTERNATIVES: BHA; BHT

TECHNOLOGY OF USE IN FOODS: Similar to BHA. Some discoloration may occur in the presence of alkaline pH, certain proteins and sodium salts. It can be used as an antioxidant in frying oils

SYNERGISTS: BHA; citric acid

FOOD SAFETY ISSUES: Has shown mutagenicity *in vivo*; therefore, some countries consider that TBHQ does not meet current standards of toxicity testing

LEGISLATION:

USA:

Not allowed to use in combination with PG. For general usage,
- FDA – 0.02%, USDA – 0.01%, based on lipid content of food.

Specific food use:
- Non-alcoholic beverages
- Margarine, mixed nuts – 0.02% alone or in combination based on lipid content
- Dried meats
- Fresh pork or beef sausages
- Pre-grilled beef patties
- Pizza toppings
- Meatballs – 0.01% based on weight of finished product
- Rendered animal fats

EUROPE, UK, NORWAY, DENMARK, SWEDEN, SWITZERLAND:

not allowed for food use

CANADA:
not allowed for food use

AUSTRALIA/ PACIFIC RIM:
AUSTRALIA, NEW ZEALAND:
allowed for food use
Japan: not allowed for food use

REFERENCES:	Sustane® *Food-Grade Antioxidants*. Universal Oil Products Company, Food Antioxidants Department, Des Plaines, IL, 1994. Tenox® *Food-Grade Antioxidants.* Eastman Chemical Company, Kingsport, TN. Publication ZG-262B, March 1995. Ash, M., and Ash, I. (1995) *Handbook of Food Additives*. Gower Publishing Ltd., Hampshire, UK.

NAME:	**Tocopherols, mixed α (dl), γ and δ (synthetic)**
CATEGORY:	Antioxidant/ Vitamin supplement
FOOD USE:	Infant foods/ Milk fat/ Mayonnaise/ Salad dressings/ Vegetable oils/ Processed meats/ Fresh and frozen sausages/ Snacks and nuts/ Dehydrated potatoes
SYNONYMS:	dα-[2*R*,4′*R*,8′*R*]-2,5,7,8-tetramethyl-2-(4′,8′,12′-trimethyl-tridecyl)-6-chromanol; 5,7,8-trimethyltocol E307/ γ-7,8-dimethyltocol, E308/ δ-8-methyltocol, E309/ Pristene® 180, 184, 185, 186, 198/ Tenite®
FORMULA:	α = C29 H50 O2; γ = C28 H48 O2; δ = C27 H46 O2
MOLECULAR MASS:	α = 430.72; γ = 416.69; δ = 402.67
ALTERNATIVE FORMS:	Tocopheryl acetate
PROPERTIES AND APPEARANCE:	Yellow to brownish viscous oily liquid, odourless
BOILING POINT IN °C AT VARIOUS PRESSURES (INCLUDING 760 mm Hg):	α = 210; γ = 200–210; δ = 150
PURITY %:	α-Tocopherol content is not less than 96% and not more than 102%
SOLUBILITY % AT VARIOUS TEMPERATURE/pH COMBINATIONS:	
in water:	**@ 20°C** Insoluble
in vegetable oil:	Soluble
in ethanol solution:	Soluble
FUNCTION IN FOODS:	Functions as a free-radical terminator in autoxidation reactions, and is considered as an antioxidant. It supplies vitamin E as a nutrient to foods. Some plant foods, particularly vegetable oils, oilseeds and nuts, naturally contain tocopherols, but during processing they are partly destroyed
ALTERNATIVES:	Natural tocopherols
TECHNOLOGY OF USE IN FOODS:	Direct addition, spraying or dipping, can be used in frying oils

SYNERGISTS:	Citric acid; ascorbyl palmitate; lecithin	
FOOD SAFETY ISSUES:	Not subjected to safety criticism	
LEGISLATION:	**USA:**	**UK and EUROPE:**
	General manufacturing practices	Not to exceed 500 ppm

REFERENCES:

Tenox® *Food Grade Antioxidant*. Eastman Chemical Company, Publication ZG-262B, Kingsport, TN, March 1995.

Food Chemical Codex, 1981, 3rd edition. National Academy Press, Washington, DC.

Schuler, P. (1990) In: Hudson, B. J. F. (Ed.), *Food Antioxidants*. Elsevier Science Publishers, Essex, UK.

NAME:	**Beta-carotene**
CATEGORY:	Antioxidant/ Nutrient
FOOD USE:	Orange beverages/ Cheese/ Dairy products/ Butter/ Ice-cream/ Fats and oils/ Fruit juices/ Infant formula as vitamin A
SYNONYMS:	Provitamin A/ Food Orange 5/ Natural Yellow 26/ C140800/ C175130/ E160a
FORMULA:	$C_{40}H_{56}$
MOLECULAR MASS:	536.89
PROPERTIES AND APPEARANCE:	Purple hexagonal prisms or red leaflets
MELTING RANGE IN °C	178–179
FUNCTION IN FOODS:	Singlet oxygen quencher, thereby acts as an antioxidant. Colourant and function as provitamin A in foods.
REFERENCE:	*Food Chemical Codex* (1981) 3rd edition. National Academy Press, Washington, DC.

NAME:	Rosemary extract; natural spice extract
CATEGORY:	Antioxidant/ Natural
FOOD USE:	Edible fats and oils/ Processed meats and poultry/ Fresh and frozen sausages/ Salad dressings/ Mayonnaise/ Seasonings/ Snacks and nuts/ Soup bases/ Chewing gum/ Peanut butter/ Citrus oils
SYNONYMS:	Pristene® RO/ Pristene® RW/ Stabex®/ Herbalox®/ Freeze-Grad®-FP-15/ Flav-R-Keep® FP-51
PROPERTIES AND APPEARANCE:	Yellowish powder or oil- or water-dispersible liquid
DENSITY:	0.36 g/ml
PURITY %:	16–24.8% carnosic acid
WATER CONTENT MAXIMUM %:	5
HEAVY METAL CONTENT MAXIMUM IN ppm:	lead 20 cadmium 0.2 mercury 0.1
FUNCTION IN FOODS:	Retards development of oxidative rancidity by terminating free radicals. Used as a natural antioxidant
TECHNOLOGY OF USE IN FOODS:	Direct addition or metering into fat. Concentrated suspension can be dosed into bulk fat or food material. It can be used by dry-mixing in powdered food
SYNERGISTS:	Ascorbic acid; ascorbyl palmitate
LEGISLATION:	**USA:** 200–1000 ppm depending on the carnosic acid content and intended shelf-life and specific food product

REFERENCES:	Schuler, P. (1990) In: Hudson, B. J. F. (Ed.), *Food Antioxidants*. Elsevier Applied Science, Essex, UK. Stabex® SKW Chemicals. Product Data, Marietta, GA. Pristene® *Natural Based Food-Grade Antioxidants and Flavors*. Universal Oil Products Company, UOP2341-2, Food Antioxidant Department, Des Plaines, IL (1995).

NAME:	Tea extract
CATEGORY:	Antioxidant/ Natural
FOOD USE:	Salad oils/ Bulk vegetable oils/ Nuts and snacks (deep-fat fried)/ Extruded snack foods/ Seasoning mixes
SYNONYMS:	Teabalox® T-02/ Teabalox® O/ Green tea oleoresin
FORMULA:	Water-soluble catechins and gallates and oil-soluble oleoresin of green tea
PROPERTIES AND APPEARANCE:	Brown viscous liquid
FUNCTION IN FOODS:	Prevents oxidative rancidity development by terminating free radicals, and may chelate metal ions. Used as a natural antioxidant.
TECHNOLOGY OF USE IN FOODS:	Spray application to fried foods, extruded foods. At temperatures above 130°C, it can have reduced performance, but, thermal stable types (Herbalox O) are available. Migration of polyphenols from the oil phase to the water phase may occur in the presence of a significant quantity of unbound water, thereby reducing antioxidative activity
LEGISLATION:	**USA:** approved by FDA for use in foods, 0.025 to 0.15% of the fat
REFERENCE:	Kalsec Technical Data, Development Product Bulletin T-20. June 1993. Kalsec Inc., Kalamazoo, MI.

NAME:	Tocopherol, mixed natural concentrate	
CATEGORY:	Antioxidant/ Natural vitamin supplement	
FOOD USE:	Dairy products/ Cereal products/ Green vegetables (frozen)/ Margarine/ Sausages/ Vegetable oils/ Soft drinks/ Snacks and nuts/ Salad dressings/ Soup bases/ Seasonings/ Dehydrated potatoes/ Fresh and frozen sausages/ Processed meats and poultry/ Baked products	
SYNONYMS:	E306/ Pristene® MT70/ Prestene® MT-28/ Pristene® MT-50/ Vitamin E/ Eisai® Natural Vitamin E/ Tenox® GT-1/ Tenox® GT-2	
FORMULA:	Mixture of α, β, γ and δ tocopherol $\alpha = C_{29}H_{50}O_2$; β and $\gamma = C_{28}H_{48}O_2$; $\delta = C_{27}H_{46}O_2$	
PROPERTIES AND APPEARANCE:	Golden brown colour, slightly viscous liquid with a characteristic odour. It may show a slight separation of wax-like constituents in microcrystalline form. It oxidises and darkens slowly in air and on exposure to light, particularly when in alkaline media	
PURITY %:	28–70% tocopherol in vegetable oil. High α type – not less than 50% of total tocopherols of which not less than 50% consists of d-α-tocopherol and not less than 20% consists of d-β+d-γ+d-δ-tocopherols. Low α-type – not less than 50% of total tocopherols, of which not less than 80% consists of d-β+d-γ+d-δ tocopherols.	
SOLUBILITY % AT VARIOUS TEMPERATURE/pH COMBINATIONS:		
in water:	@ 20°C	Insoluble
in vegetable oil:		Soluble
in ethanol solution:		Soluble
in propylene glycol:		Soluble
FUNCTION IN FOODS:	Prevents oxidative rancidity development in lipid-containing foods by terminating free radicals. Dietary supplement of vitamin E, inhibits *N*-nitrosamine formation in pump-cured bacon	
ALTERNATIVES:	Synthetic tocopherols	

TECHNOLOGY OF USE IN FOODS:	Can be used in frying oils as tocopherols do not volatilise at 180°C. Spraying or dipping techniques for nuts (whole or broken) using 0.2% tocopherol solution in 96% ethanol. Should be stored in air-tight, light-proof containers.
SYNERGISTS:	Citric acid; ascorbic acid; ascorbyl palmitate; lecithin; amino acids; EDTA; BHA; BHT; PG
FOOD SAFETY ISSUES:	Has not been criticised over safety
LEGISLATION:	**USA:** GMP for general use. FDA: frozen raw breaded shrimp (amount required for intended technical effect), margarine, mixed nuts. USDA: rendered animal fat or combination of such fat with vegetable fat – 0.03% (30% concentration of tocopherol should be used). Dry sausages, semidry sausages, dried meats, uncooked or cooked sausages from fresh beef and/or pork, uncooked or cooked meatballs, Italian sausages, meat pizza toppings, brown-and-serve sausages, pre-grilled beef patties and restructured meats; 0.03% based on fat content, may not be used in combination with other antioxidants. Rendered poultry fat or various poultry products – 0.03% based on fat content; 0.02% in combination with BHA, BHT and PG.
REFERENCES:	Schuler, P. (1990) In: Hodson, B. J. F. (Ed.), *Food Antioxidants*. Elsevier Science Publishers, Essex, UK. *Food Chemical Codex* (1981) 3rd edition. National Academy Press, Washington, DC. Tenox® GT-1 and Tenox® GT-2. Eastman Chemical Company, Publication ZG-263D, Kingsport, TN, March 1996.

NAME:	Citric acid and its salts
CATEGORY:	Sequestrant/ Synergist for antioxidant/ Dispersing agent/ Acidifier/ Flavouring agent
FOOD USE:	Beverages/ Canned vegetables/ Baby foods/ Confectionery/ Ice-cream/ Baked products
SYNONYMS:	2-Hydroxy-1,2,3-propanetricarboxylic acid, β-Hydroxytricarballylic acid/ E330/ Cap-Shure® C-140E-75/ Citrostabil® NEU/ Citrostabil® – S/ Citrocoat® A1000HP/ Citrocoat® A2000HP/ Citrocoat® A4000TP/ Citrocoat® A4000TT/ Descote® Citric Acid 50%/ Liquinat®
FORMULA:	CH2(COOH)C(OH)(COOH)CH2 COOH
MOLECULAR MASS:	192.43
ALTERNATIVE FORMS:	Isopropylcitrate; triethylcitrate; ammonium citrate; mono-, di-, tri-sodium citrate; mono-, di-, tri-potassium citrate; calcium citrate; citric acid monohydrate
PROPERTIES AND APPEARANCE:	Colourless translucent crystals or powder with tart taste
MELTING RANGE IN °C:	152–154 (anhydrous); 135–153 (hydrated)
IONISATION CONSTANT AT 20°C:	$K_1 = 7.10 \times 10^{-4}$, $pK_a = 3.14$ $K_2 = 1.68 \times 10^{-5}$, $pK_a = 4.77$ $K_3 = 6.4 \times 10^{-5}$, $pK_a = 6.39$
PURITY %:	Not less than 99.5% of C6H8O7 calculated on the anhydrous basis
WATER CONTENT MAXIMUM IN %:	0.5% in the anhydrous form; 8.8% in hydrous form
HEAVY METAL CONTENT MAXIMUM IN ppm:	10
ARSENIC CONTENT MAXIMUM IN ppm:	3
ASH MAXIMUM IN %:	0.05

SOLUBILITY % AT VARIOUS TEMPERATURE/pH COMBINATIONS:

in water: **@ 20°C** 200 g/100 ml

in ethanol solution: 50 g/100 ml

FUNCTION IN FOODS: Control of pH in food, mainly an acidulant in beverages. Sequestrant of metal ions. Collects and deactivates metal ions, thus preventing the participation of metal ions in the initiation of autoxidation of unsaturated food lipids. Curing accelerator. Dispersant. As a flavour enhancer, it imparts tartness. Calcium citrate is a firming agent and a nutrient in baby foods

TECHNOLOGY OF USE IN FOODS: Readily soluble in water, and releases free acid. Therefore, it can be added directly to water-containing foods. Decomposes when heated

FOOD SAFETY ISSUES: Accelerates tooth decay. Moderate skin irritant and some allergic properties

LEGISLATION:

USA:
FDA and USDA approved.
2500 ppm – non-alcoholic beverages; 1600 ppm – ice-cream and ices;
4300 ppm – candy; 1200 ppm – baked products; 3600 ppm – chewing gum

UK:
approved

EUROPE:
listed

AUSTRALIA/PACIFIC RIM:
Japan: approved

REFERENCES:

Food Chemical Codex (1981) 3rd edition. National Academy Press, Washington, DC.

Ash, M., and Ash, I. (1995) *Handbook of Food Additives*. Gower Publishing Ltd., Hampshire, UK.

de la Teja, P. (1991) In: Smith, J. (Ed.), *Food Additive User's Handbook*. Van Nostrand Reinhold, New York.

Freydberg, N., and Gortner, W. A. (1982). *The Food Additive Book*. Bantam Books, Inc., New York.

NAME:	Ethylenediaminetetraacetic acid (EDTA)
CATEGORY:	Chelator/ Antioxidant
FOOD USE:	Margarine/ Mayonnaise/ Salad dressings/ Fish and fish products/ Beer/ Fruit drinks/ Vegetable juices/ Condiments
SYNONYMS:	Eidetic acid/ *N,N′*-1,2-ethanediyl*bis*[*N*-(carboxymethyl)glycine]/ edathamil, ethylenediamine-*N,N,N′,N′*-tetraacetic acid
FORMULA:	(HOOC CH2)2 NCH2 CH2N(CH2 COOH)2
MOLECULAR MASS:	292.28
ALTERNATIVE FORMS:	Calcium disodium EDTA; sodium dihydrogen EDTA; disodium EDTA
PROPERTIES AND APPEARANCE:	Colourless crystals
BOILING POINT IN °C AT VARIOUS PRESSURES (INCLUDING 760 mm Hg):	238
MELTING RANGE IN °C:	36.4; decomposes at 245°C
FLASH POINT IN °C:	160
HEAVY METAL CONTENT MAXIMUM IN ppm:	20
ARSENIC CONTENT MAXIMUM IN ppm:	3
SOLUBILITY % AT VARIOUS TEMPERATURE/pH COMBINATIONS:	
in water:	Freely soluble
in ethanol solution:	Insoluble
FUNCTION IN FOODS:	Chelation of transition metal ions such as copper and iron, thus preventing initiation of autoxidation of unsaturated lipids. EDTA salts improve the stability of foods

SYNERGISTS:	BHT; PG
LEGISLATION:	GMP
REFERENCES:	Ash, M., and Ash, I. (1995) *Handbook of Food Additives*. Gower Publishing Ltd., Hampshire, UK. Freydberg, N., and Gortner, W. A. (1982) *The Food Additives Book*. Bantam Books Inc., New York, NY. Dean, J. A. (1992) *Lange's Handbook of Chemistry*. 14th edition. McGraw-Hill, Inc., New York.

NAME:	L-Tartaric acid
CATEGORY:	Sequestrant/ Synergist of antioxidant
FOOD USE:	Non-alcoholic beverages/ Ice-cream confectionery/ Baked products/ Desserts/ Chewing gum/ Condiments/ Jam/ Jelly/ Canned vegetables and fruits
SYNONYMS:	L(+)-tartaric acid/ L-2,3-dihydroxybutane-dioic acid/ d-α-β-di-hydroxysuccinic acid/ 2,3-dihydroxysuccinic acid/ E334/ potassium acid tartrate (cream of tartar, potassium bitartrate)
FORMULA:	COOH·CHOH·CHOH·COOH
MOLECULAR MASS:	150.09
ALTERNATIVE FORMS:	Sodium tartrate (L-tartaric acid disodium salt)
PROPERTIES AND APPEARANCE:	Colourless or translucent crystals or white crystalline powder, odourless with acid taste
MELTING RANGE IN °C:	168–170
IONISATION CONSTANT AT 25°C:	$K_1 = 1.04 \times 10^{-3}$, $pK_a = 2.98$ $K_2 = 4.55 \times 10^{-3}$, $pK_a = 4.34$
PURITY %:	99
WATER CONTENT MAXIMUM IN %:	0.5
HEAVY METAL CONTENT MAXIMUM IN ppm:	10
ARSENIC CONTENT MAXIMUM IN ppm:	3
ASH MAXIMUM IN %:	0.05
SOLUBILITY % AT VARIOUS TEMPERATURE/pH COMBINATIONS:	
in water:	**@ 25°C** 139
in alcohol:	**@ 25°C** 33

FUNCTION IN FOODS:	Augments fruit flavours in beverages and candies; prevents discoloration or flavour changes from occurring during oxidative rancidity as a stabilising agent; prevents fermentation in baked products; firming agent, humectant, acidulant, pH control agent, sequestrant
TECHNOLOGY OF USE IN FOODS:	Direct addition
LEGISLATION:	**USA:** FDA and USDA approved: 960 ppm in non-alcoholic beverages; 570 ppm in ice-cream; 5400 ppm in candy; 130 ppm in baked products; 60 ppm in gelatins and puddings; 3700 ppm in chewing gum; 10,000 ppm in condiments
REFERENCES:	*Food Chemical Codex* (1981) 3rd edition. National Academy Press, Washington, DC. Ash, M., and Ash, I. (1995) *Handbook of Food Additives.* Gower Publishing Ltd., Hampshire, UK.

Part 3

Colourings

NAME:	Alkanet
CATEGORY:	Food colour
FOOD USE:	Alcoholic beverages/ Confectionery/ Ice-cream
SYNONYMS:	C.I. 75530 (Natural Red 20)/ Alkannin
FORMULA:	$C_{16}H_{16}O_5$
MOLECULAR MASS:	288.13
ALTERNATIVE FORMS:	None known
PROPERTIES AND APPEARANCE:	A dark red/brown powder or paste
BOILING POINT IN °C AT VARIOUS PRESSURES (INCLUDING 760 mm Hg):	N/A
MELTING RANGE IN °C:	Not known
FLASH POINT IN °C:	N/A
IONISATION CONSTANT AT 25°C:	N/A
DENSITY AT 20°C (AND OTHER TEMPERATURES) IN g/l:	N/A
HEAT OF COMBUSTION AT 25°C IN J/kg:	N/A
VAPOUR PRESSURE AT VARIOUS TEMPERATURES IN mm Hg:	N/A
PURITY %:	No data available
WATER CONTENT MAXIMUM IN %:	No data available

HEAVY METAL CONTENT MAXIMUM IN ppm:	50							
ARSENIC CONTENT MAXIMUM IN ppm:	1							
ASH MAXIMUM IN %:	2							
SOLUBILITY % AT VARIOUS TEMPERATURE/pH COMBINATIONS:								
in water:	**@ 20°C**	Slightly soluble	**@ 50°C**	Slightly soluble	**@ 100°C**	Slightly soluble		
in vegetable oil:	**@ 20°C**	Insoluble	**@ 50°C**	Insoluble	**@ 100°C**	Insoluble		
in sucrose solution:	**@ 10%**	Very slightly soluble	**@ 40%**	Very slightly soluble	**@ 60%**	Very slightly soluble		
in sodium chloride solution:	**@ 5%**	Insoluble	**@ 10%**	Insoluble	**@ 15%**	Insoluble		
in ethanol solution:	**@ 5%**	Very slightly soluble	**@ 20%**	Very slightly soluble	**@ 95%**	Soluble	**@ 100%**	Soluble
in propylene glycol:	**@ 20°C**	Not known	**@ 50°C**	Not known	**@ 100°C**	Not known		
FUNCTION IN FOODS:	Red to purple-red food colour. Soluble in alcohol.							
ALTERNATIVES:	Other purple/red food colours, although not of exactly the same hue, including: Orange; Beetroot; Anthocyanin; Allura red; Ponceau 4R; Carmoisine							
TECHNOLOGY OF USE IN FOODS:	Alcohol-soluble. Red to purple-red colour extracted from the roots of *Alkanna tinctoria* and *Anchusa tinctoria*. Red shade below pH 6, blue shade above. Used to colour alcoholic drinks and for confectionery. Only slightly soluble in water.							
SYNERGISTS:	None known							
ANTAGONISTS:	None known							
FOOD SAFETY ISSUES:	The Joint FAO/WHO Expert Committee on Food Additives has insufficient data to evaluate and determine an ADI.							

LEGISLATION:

USA:	UK and EUROPE:	CANADA:	AUSTRALIA:
Not permitted	Not permitted	Permitted	Permitted

REFERENCE:

Smith, J. (Ed.) (1991) *Food Additive User's Handbook.* Blackie Academic and Professional.

STRUCTURE OF ALKANET:

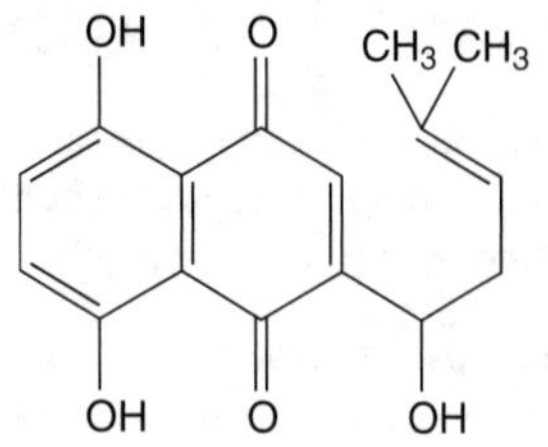

NAME:	Allura Red AC
CATEGORY:	Food colour
FOOD USE:	Baked goods/ Cereals and cereals products/ Dairy products/ Fish and seafood products/ Meat and poultry products/ Soft drinks/ Fruit, vegetable and nut products/ Beverages/ Alcoholic drinks/ Sugar and preserves/ Confectionery/ Edible ices/ Vinegar, pickles and sauces/ Decorations and coatings/ Soups/ Seasonings/ Snacks / Desserts
SYNONYMS:	FD & C Red 40/ E129/ C.I. 16035 (Food Red 17)/ Disodium 2-hydroxy-1-(2-methoxy-5-methyl-4-sulfonatophenylazo) naphthalene-6-sulfonate
FORMULA:	C18 H14 N2 Na2 O8 S2
MOLECULAR MASS:	496.42
ALTERNATIVE FORMS:	Calcium salt/ Potassium salt/ Aluminium lake (insoluble)
PROPERTIES AND APPEARANCE:	Dark red powder or granules
BOILING POINT IN °C AT VARIOUS PRESSURES (INCLUDING 760 mm Hg):	N/A
MELTING RANGE IN °C:	Not known
FLASH POINT IN °C:	N/A
IONISATION CONSTANT AT 25°C:	N/A
DENSITY AT 20°C (AND OTHER TEMPERATURES) IN g/l:	N/A
HEAT OF COMBUSTION AT 25°C IN J/kg:	N/A
VAPOUR PRESSURE AT VARIOUS TEMPERATURES IN mm Hg:	N/A

PURITY %:	Not less than 85% total colouring matter, calculated as the sodium salt					
WATER CONTENT MAXIMUM IN %:	15					
HEAVY METAL CONTENT MAXIMUM IN ppm:	<40					
ARSENIC CONTENT MAXIMUM IN ppm:	<3					
ASH MAXIMUM IN %:	N/A					
SOLUBILITY % AT VARIOUS TEMPERATURE/pH COMBINATIONS:						
in water:	**@ 20°C**	22%	**@ 50°C**	25%	**@ 100°C**	27%
in vegetable oil:	**@ 20°C**	Insoluble	**@ 50°C**	Insoluble	**@ 100°C**	Insoluble
in sucrose solution:	**@ 10%**	Soluble	**@ 40%**	Soluble	**@ 60%**	Soluble
in sodium chloride solution:	**@ 5%**	Soluble	**@ 10%**	Insoluble	**@ 15%**	Insoluble
in ethanol solution:	**@ 5%**	Soluble	**@ 20%**	**@ 25°C** 9.5% **@ 60°C** 22.0%	**@ 95%**	**@ 25°C** 0.001% **@ 60°C** 0.05%
	@ 100%	**@ 25°C** 0.001% **@ 60°C** 0.05%				
in propylene glycol:		**@ 25°C** 18%		**@ 60°C** 22%	**@ 100°C**	Not known
FUNCTION IN FOODS:	Water-soluble red food colour or insoluble red food colour as the aluminium lake.					
ALTERNATIVES:	Other red food colours, although not of exactly the same hue, include: Carmoisine; Ponceau 4R; Erythrosine; Carmine; Beet red; Anthocyanins (in acid media); Red 2G (where permitted)					
TECHNOLOGY OF USE IN FOODS:	A very heat-stable, yellowish-red colour with excellent light stability. Becomes bluer in alkaline media. Some fade with ascorbic acid.					
SYNERGISTS:	None known					
ANTAGONISTS:	Ascorbic acid					

FOOD SAFETY ISSUES: None known

LEGISLATION:

USA:
FD&E Red 40 certified food colour – Permitted

UK and EUROPE:
Permitted according to European Parliament and Council Directive 94/36/EC of June 30th 1994 on Colours for Use in Foodstuffs.
Refer to Annex V Part 2 for restrictions

CANADA:
Permitted when certified, but restrictions apply

AUSTRALIA:
Permitted, but restrictions apply

REFERENCE: Smith, J. (Ed.) (1991) *Food Additive User's Handbook.* Blackie Academic and Professional.

STRUCTURE OF ALLURA RED:

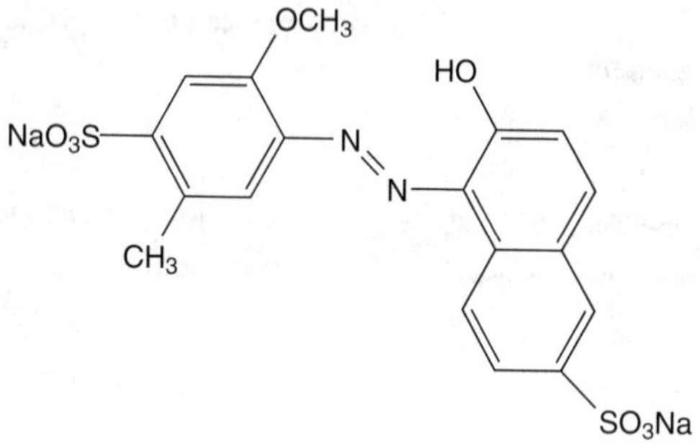

NAME:	Aluminium
CATEGORY:	Food colour
FOOD USE:	Surface coating of sugar/ Confectionery
SYNONYMS:	E173/ C.I. 77000 (pigment metal 1)
FORMULA:	Al
MOLECULAR MASS:	26.98
ALTERNATIVE FORMS:	N/A
PROPERTIES AND APPEARANCE:	Silver-grey powder or thin sheets (leaf)
BOILING POINT IN °C AT VARIOUS PRESSURES (INCLUDING 760 mm Hg):	1800
MELTING RANGE IN °C:	659.8
FLASH POINT IN °C:	N/A
IONISATION CONSTANT AT 25°C:	N/A
DENSITY AT 20°C (AND OTHER TEMPERATURES) IN g/l:	N/A
HEAT OF COMBUSTION AT 25°C IN J/kg:	N/A
VAPOUR PRESSURE AT VARIOUS TEMPERATURES IN mm Hg:	N/A
PURITY %:	Not less than 99%
WATER CONTENT MAXIMUM IN %:	N/A

HEAVY METAL CONTENT MAXIMUM IN ppm:	<40							
ARSENIC CONTENT MAXIMUM IN ppm:	<3							
ASH MAXIMUM IN %:	N/A							
SOLUBILITY % AT VARIOUS TEMPERATURE/pH COMBINATIONS:								
in water:	**@ 20°C**	Insoluble	**@ 50°C**	Insoluble	**@ 100°C**	Insoluble		
in vegetable oil:	**@ 20°C**	Insoluble	**@ 50°C**	Insoluble	**@ 100°C**	Insoluble		
in sucrose solution:	**@ 10%**	Insoluble	**@ 40%**	Insoluble	**@ 60%**	Insoluble		
in sodium chloride solution:	**@ 5%**	Insoluble	**@ 10%**	Insoluble	**@ 15%**	Insoluble		
in ethanol solution:	**@ 5%**	Insoluble	**@ 20%**	Insoluble	**@ 95%**	Insoluble	**@ 100%**	Insoluble
in propylene glycol:	**@ 20°C**	Insoluble	**@ 50°C**	Insoluble	**@ 100°C**	Insoluble		
FUNCTION IN FOODS:	Used as a dispersed powder or leaf to surface-colour sugar confectionery.							
ALTERNATIVES:	Silver powder or leaf							
TECHNOLOGY OF USE IN FOODS:	Applied as powder or leaf to the surface of hard sugar confectionery, then polished to a high gloss as a decoration. Other food colours can be mixed with the powder to achieve coloured coatings with a metallic sheen for hard sugar confectionery, known as dragees.							
SYNERGISTS:	None known							
ANTAGONISTS:	None known							
FOOD SAFETY ISSUES:	Media attention on the role of dietary aluminium in Alzheimer's disease. However, contributions to the diet from the use of aluminium as a food colour are negligible.							

LEGISLATION:	**USA:**	**UK and EUROPE:**	**CANADA:**	**AUSTRALIA:**
	Not permitted	Permitted according to European Parliament and Council Directive 94/36/EC of June 30th 1994 on Colours for Use in Foodstuffs. Refer to Annex IV for restrictions on use.	Permitted	Not permitted
REFERENCE:	Smith, J. (Ed.) (1991) *Food Additive User's Handbook.* Blackie Academic and Professional.			

NAME:	Amaranth
CATEGORY:	Food colour
FOOD USE:	Fish roe/ Confectionery/ Desserts/ Alcoholic beverages/ Edible ices/ Dairy products/ Canned products/ Decorations and coatings/ Soft drinks/ Sugar and preserves
SYNONYMS:	E123/ C.I. 16185 (Food Red 9)/ Trisodium 2-hydroxy-1-(4-sulfonato-1-naphthylazo) naphthalene-3-6-disulfonate
FORMULA:	$C_{20}H_{11}N_2Na_3O_{10}S_3$
MOLECULAR MASS:	604.48
ALTERNATIVE FORMS:	Calcium salt/ Potassium salt/ Aluminium lake (insoluble)
PROPERTIES AND APPEARANCE:	Reddish brown powder or granules
BOILING POINT IN °C AT VARIOUS PRESSURES (INCLUDING 760 mm Hg):	N/A
MELTING RANGE IN °C:	Not known
FLASH POINT IN °C:	N/A
IONISATION CONSTANT AT 25°C:	N/A
DENSITY AT 20°C (AND OTHER TEMPERATURES) IN g/l:	N/A
HEAT OF COMBUSTION AT 25°C IN J/kg:	N/A
VAPOUR PRESSURE AT VARIOUS TEMPERATURES IN mm Hg:	N/A
PURITY %:	Not less than 85% total colouring matter, calculated as the sodium salt
WATER CONTENT MAXIMUM IN %:	15

HEAVY METAL CONTENT MAXIMUM IN ppm:	40					
ARSENIC CONTENT MAXIMUM IN ppm:	3					
ASH MAXIMUM IN %:	N/A					
SOLUBILITY % AT VARIOUS TEMPERATURE/pH COMBINATIONS:						
in water:	**@ 20°C**	7%	**@ 50°C**	8%	**@ 100°C**	10%
in vegetable oil:	**@ 20°C**	Insoluble	**@ 50°C**	Insoluble	**@ 100°C**	Insoluble
in sucrose solution:	**@ 10%**	Soluble	**@ 40%**	Soluble	**@ 60%**	Soluble
in sodium chloride solution:	**@ 5%**	Soluble	**@ 10%**	Insoluble	**@ 15%**	Insoluble
in ethanol solution:	**@ 5%**	Soluble	**@ 20%**	Slightly soluble	**@ 95%**	<0.1%
	@ 100%	<0.1%				
in propylene glycol:	**@ 20°C**	0.4%	**@ 50°C**	0.4%	**@ 100°C**	0.5%
FUNCTION IN FOODS:	Water-soluble bluish-red food colour, or insoluble bluish-red food colour as the aluminium lake.					
ALTERNATIVES:	Other red food colours, although not of the same hue, include: Carmine; Carmoisine; Ponceau 4R; Allura Red; Erythrosine; Beet red; Anthocyanins (in acid media); Red 2G (where permitted)					
TECHNOLOGY OF USE IN FOODS:	A robust blue-red colour with good light stability and good heat stability to 105°C. Fades in the presence of sulphur dioxide and alkaline media. Alkaline media increases the blue hue.					
SYNERGISTS:	None known					
ANTAGONISTS:	Sulphur dioxide/ Alkaline media					
FOOD SAFETY ISSUES:	Possible increase in mammary tumours on long-term rat feeding studies. Delisted by USA and restricted use by European Union					

LEGISLATION:

USA:
Not permitted

UK and EUROPE:
Permitted according to European Parliament and Council Directive 94/36/EC of June 30th 1994 on Colours for Use in Foodstuffs.
Annex IV – only for aperitif wines, spirit drinks including products with less than 15% alcohol by volume – max. 30 mg/l.
fish roe max. 30 mg/l.

CANADA:
Permitted when certified, but restrictions apply

AUSTRALIA:
Permitted, but restrictions on use apply

REFERENCE:

Smith, J. (Ed.) (1991) *Food Additive Users Handbook*. Blackie Academic and Professional.

STRUCTURE OF AMARANTH:

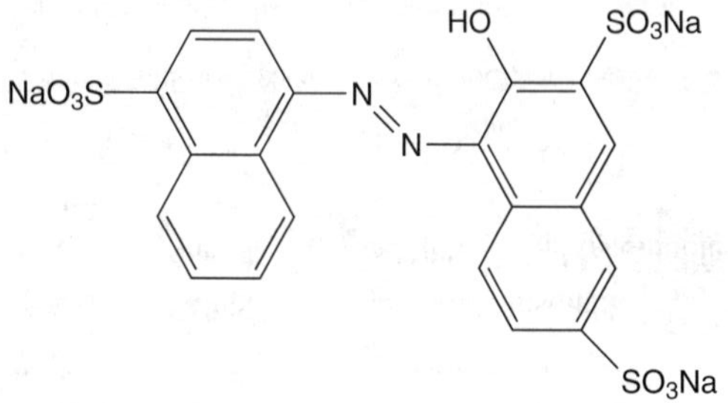

NAME:	Annatto
CATEGORY:	Food colour
FOOD USE:	Yellow fats/ Decorations/ Coatings/ Seasonings/ Edible ices/ Liqueurs/ Cheese/ Baked goods/ Dairy products/ Smoked fish/ Desserts/ Confectionery/ Vinegar, pickles and sauces/ Cereals and cereal products
SYNONYMS:	Bixin/ Norbixin/ E160b/ C.I. 75120 Natural Orange 4
FORMULA:	Bixin $C_{25}H_{30}O_4$; Norbixin $C_{25}H_{28}O_4$
MOLECULAR MASS:	Bixin 394.51; Norbixin 380.48
ALTERNATIVE FORMS:	*Cis* and *trans* bixin/ *Cis* and *trans* norbixin/ Alkali salts of norbixin
PROPERTIES AND APPEARANCE:	Reddish-brown powder, suspension or solution
BOILING POINT IN °C AT VARIOUS PRESSURES (INCLUDING 760 mm Hg):	N/A
MELTING RANGE IN °C:	Not known
FLASH POINT IN °C:	Not known
IONISATION CONSTANT AT 25°C:	Not known
DENSITY AT 20°C (AND OTHER TEMPERATURES) IN g/l:	N/A
HEAT OF COMBUSTION AT 25°C IN J/kg:	Not known
VAPOUR PRESSURE AT VARIOUS TEMPERATURES IN mm Hg:	N/A

PURITY %: Bixin powders 25%; norbixin powders 75%.
Alkali-extracted annatto 99.9%
Oil-extracted annatto 99.9%

WATER CONTENT MAXIMUM IN %: 5

HEAVY METAL CONTENT MAXIMUM IN ppm: 40

ARSENIC CONTENT MAXIMUM IN ppm: 3

ASH MAXIMUM IN %: Not known

SOLUBILITY % AT VARIOUS TEMPERATURE/pH COMBINATIONS:

in water:	**@ 20°C**	Insoluble in water. Soluble in alkaline solutions; pH12, 1% pigment concentration					
	@ 50°C	Soluble in alkaline solutions; pH12, 3% pigment concentration					
	@ 100°C	Soluble in alkaline solutions; pH12, 7% pigment concentration					
in vegetable oil:	**@ 20°C**	Norbixin insoluble; bixin soluble <0.05%					
	@ 50°C	Bixin soluble <0.1%					
	@ 100°C	Bixin soluble 0.25%					
in sucrose solution:	**@ 10%**	Insoluble	**@ 40%** Insoluble	**@ 60%**	Insoluble		
in sodium chloride solution:	**@ 5%**	Insoluble, unless in preparation containing emulsifier					
	@ 10%	Insoluble, unless in preparation containing emulsifier					
	@ 15%	Insoluble, unless in preparation containing emulsifier					
in ethanol solution:	**@ 5%**	Insoluble	**@ 20%** Insoluble	**@ 95%**	<0.1%	**@ 100%**	<0.1%
in propylene glycol:	**@ 20°C**	Insoluble	**@ 50°C** Insoluble	**@ 100°C**	Insoluble		

FUNCTION IN FOODS: An orange-yellow food colour suitable for a wide range of food applications

ALTERNATIVES: Other orange-yellow food colours, although not of exactly the same hue, include: Carminic acid; Sunset yellow; Paprika; beta-carotene/ mixed carotene

TECHNOLOGY OF USE IN FOODS:

Annatto can be used in the following forms:
(i) Dilute solution/ suspension; for oil-based applications.
(ii) Aqueous alkali solution; for aqueous application; for acidic environments, acid-proof forms are available.
Heat and light-stability are generally good. However, with annatto being a carotenoid, sensitivity to oxidation can be a problem.

SYNERGISTS:

Food-grade emulsifiers can be used to produce acid-stable forms. Antioxidants in oil-based forms can help stability.

An alkali such as KOH or NaOH renders norbixin water-soluble.

ANTAGONISTS:

Sulphur dioxide/ Divalent cations (e.g. Ca^{2+})/ Acids

FOOD SAFETY ISSUES:

None known

LEGISLATION:

USA:
Permitted as non-certified food colour according to 21 CFR 7330, 73.1030, 73.2030

UK and EUROPE:
Permitted according to European Parliament and Council Directive 94/36/EC of June 30th 1994 on Colours for Use in Foodstuffs.
Refer to Annexes III and IV for restrictions on use

CANADA:
Permitted

AUSTRALIA:
Permitted

OTHER COUNTRIES:
Generally permitted in most countries

REFERENCES:

Collins, P. (1992) The role of annatto in food colouring. *Food Ingredients & Processing International*, **February**, 23–27.

Smith, J. (Ed.) (1991) *Food Additive User's Handbook.* Blackie Academic and Professional.

STRUCTURE OF ANNATTO:

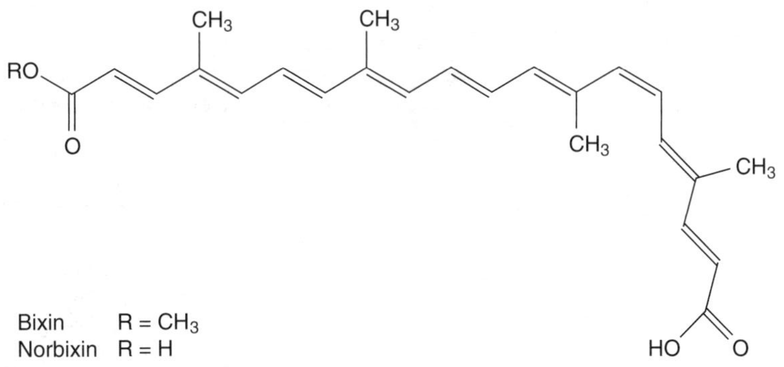

Bixin R = CH_3
Norbixin R = H

Note: also occurs as all-*trans* isomers.

NAME:	**Anthocyanin**
CATEGORY:	Food colour
FOOD USE:	Dairy products/ Fruit, vegetable and nut products/ Soft drinks/ Alcoholic drinks/ Sugars and preserves/ Confectionery/ Decorations and coatings/ Edible ices/ Desserts
SYNONYMS:	E163/ Grapeskin extract/ Enocianina/ Cyanidin (3,3′,4′,5,7-Pentahydroxy-flavylium chloride)/ Peonidin (3,4′,5,7-Tetrahydroxy-3′-methoxyflavylium chloride)/ Malvidin (3,4′,5,7-Tetrahydroxy-3′,5′-dimethoxyflavylium chloride)/ Delphinidin (3,5,7-Trihydroxy-2-(3,4,5-trihydroxyphenyl)-1-benzopyrilium chloride)/ Petunidin (3,3′,4′,5,7-Pentahydroxy-5′-methoxyflavylium chloride)/ Pelargonidin (3,5,7-Trihydroxy-2-(4-hydroxyphenyl)-1-benzopyrilium chloride)
FORMULA:	Cyanidin $C_{15}H_{11}O_6Cl$ Peonidin $C_{16}H_{13}O_6Cl$ Malvidin $C_{17}H_{15}O_7Cl$ Delphinidin $C_{15}H_{11}O_7Cl$ Petunidin $C_{16}H_{13}O_7Cl$ Pelargonidin $C_{15}H_{11}O_5Cl$
MOLECULAR MASS:	Cyanidin 322.6 Peonidin 336.7 Malvidin 366.7 Delphinidin 340.6 Petunidin 352.7 Pelargonidin 306.7
ALTERNATIVE FORMS:	Usually occur as glycosides (mono or di)
PROPERTIES AND APPEARANCE:	Not normally available in pure form. Generally available as an extract of the source fruit or vegetable. Extracts tend to be dark black/red liquids or powders with characteristic odour

BOILING POINT IN °C AT VARIOUS PRESSURES (INCLUDING 760 mm Hg):	N/A					
MELTING RANGE IN °C:	N/A					
FLASH POINT IN °C:	N/A					
IONISATION CONSTANT AT 25°C:	N/A					
DENSITY AT 20°C (AND OTHER TEMPERATURES) IN g/l:	N/A					
HEAT OF COMBUSTION AT 25°C IN J/kg:	N/A					
VAPOUR PRESSURE AT VARIOUS TEMPERATURES IN mm Hg:	N/A					
PURITY %:	A standard grapeskin extract contains around 1% total anthocyanin pigments.					
WATER CONTENT MAXIMUM IN %:	For a standard grapeskin extract, typically 70%					
HEAVY METAL CONTENT MAXIMUM IN ppm:	40					
ARSENIC CONTENT MAXIMUM IN ppm:	3					
ASH MAXIMUM In %:	2.5					
SOLUBILITY % AT VARIOUS TEMPERATURE/pH COMBINATIONS:						
in water:	@ **20°C**	Completely soluble	@ **50°C**	Completely soluble	@ **100°C**	Completely soluble, but pigments will degrade
in vegetable oil:	@ **20°C**	Insoluble	@ **50°C**	Insoluble	@ **100°C**	Insoluble
in sucrose solution:	@ **10%**	Completely soluble	@ **40%**	Completely soluble	@ **60%**	Completely soluble
in sodium chloride solution:	@ **5%**	N/A	@ **10%**	N/A	@ **15%**	N/A

in ethanol solution:	**@ 5%**	Completely soluble	**@ 20%**	Completely soluble	**@ 95%**	Slightly soluble
	@ 100%	Slightly soluble				
in propylene glycol:	**@ 20°C**	N/A	**@ 50°C**	N/A	**@ 100°C**	N/A

FUNCTION IN FOODS: A water-soluble red/purple food colour for use in acidic food products

ALTERNATIVES: Other red-purple food colours, although not of exactly the same hue, include: Beetroot red; Carmine; Erythrosine; Carmoisine; Allura red; Ponceau 4R; Amaranth; Red 2G

TECHNOLOGY OF USE IN FOODS: Anthocyanins are fully water-soluble, but require an acidic environment to exhibit their characteristic red colour.
At high temperatures (780°C), the anthocyanin pigments can degrade.

SYNERGISTS: Ascorbic acid (although can degrade anthocyanin at high levels); food acids (citric, etc.)

ANTAGONISTS: Sulphur dioxide (breaches monomeric anthocyanins, but has less effect on polymerised forms).
Metal ions (Sn, Fe, etc.)
Ascorbic acid at high levels.

FOOD SAFETY ISSUES: None known

LEGISLATION:

USA:	**UK and EUROPE:**	**CANADA:**	**AUSTRALIA:**	**OTHER COUNTRIES:**
Generally permitted. The 21 CFR permits anthocyanins in four ways: (i) 73.169 grape colour extract (ii) 73.170 grape skin extract (enocianina) (iii) 73.250 fruit juice (iv) 73.280 vegetable juice The actual category then depends on the anthocyanin in question	Permitted according to European Parliament and Council Directive 94/36/EC of June 30th 1994 on Colours for Use in Foodstuffs. Refer to Annex V Part 1 for restrictions	Generally permitted	Generally permitted	Permitted in most countries of the world

REFERENCES:

Jackman, R. L., Yada, R. Y. *et al.* (1987) Anthocyanins as food colorants – a review. *Journal of Food Biochemistry*, **11**, 201–247.

Smith, J. (Ed.) (1991) *Food Additive User's Handbook.* Blackie Academic & Professional.

Goto, T. (1987) Structure, stability and colour variation of natural anthocyanins. In: *Progress in the Chemistry of Organic Natural Products*, Springer-Verlag, p. 52.

STRUCTURE OF ANTHOCYANINS:

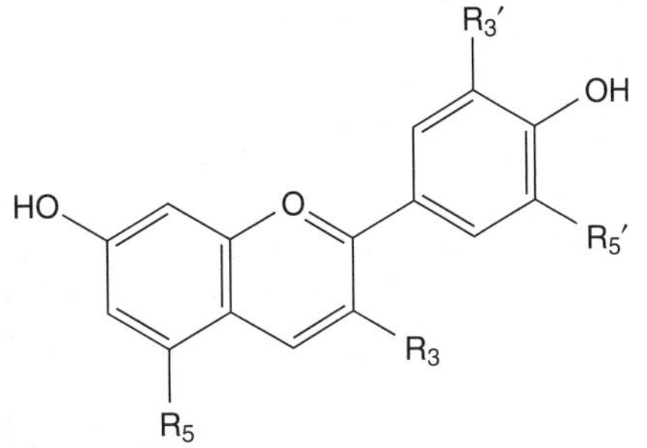

Anthocyanidins		R_3 = OH	R_5 = OH
		R_3'	R_5'
Pelargonidin	(Pg)	H	H
Cyanidin	(Cy)	OH	H
Peonidin	(Pn)	OCH_3	H
Delphinidin	(Dp)	OH	OH
Petunidin	(Pt)	OCH_3	OH
Malvidin	(Mv)	OCH_3	OCH3

Anthocyanins: Pg, Cy, Pn, Dp, Pt, Mv

with R_3 = O–sugar
R_5 = OH or O–glucose

NAME:	Beta-apo-8′-carotenal (C30)
CATEGORY:	Food colour
FOOD USE:	Soft drinks/ Confectionery/ Dairy products/ Margarine and oils/ Sauces/ Processed cheese/ Soups/ Salad dressings/ Ice-cream, edible ices/ Canned products/ Desserts
SYNONYMS:	E160e/ C.I. 40820 (Food Orange 6)/ Beta-apo-8′-carotenal/ *Trans*-beta-apo-8′-carotene-aldehyde
FORMULA:	$C_{30}H_{40}O$
MOLECULAR MASS:	416.65
ALTERNATIVE FORMS:	N/A
PROPERTIES AND APPEARANCE:	Dark violet crystals with metallic lustre or crystalline powder. Commercially available as suspensions of micronised crystals in vegetable oil or suspensions encapsulated to form fine granules
BOILING POINT IN °C AT VARIOUS PRESSURES (INCLUDING 760 mm Hg):	N/A
MELTING RANGE IN °C:	136–140
FLASH POINT IN °C:	N/A
IONISATION CONSTANT AT 25°C:	N/A
DENSITY AT 20°C (AND OTHER TEMPERATURES) IN g/l:	N/A
HEAT OF COMBUSTION AT 25°C IN J/kg:	N/A
VAPOUR PRESSURE AT VARIOUS TEMPERATURES IN mm Hg:	N/A
PURITY %:	Not less than 96% of total colouring matters
WATER CONTENT MAXIMUM IN %:	<0.5

HEAVY METAL CONTENT MAXIMUM IN ppm:	<40
ARSENIC CONTENT MAXIMUM IN ppm:	<3
ASH MAXIMUM IN %:	N/A

SOLUBILITY % AT VARIOUS TEMPERATURE/pH COMBINATIONS:

in water:	@ **25°C**	Insoluble	@ **50°C**	Insoluble	@ **100°C**	Insoluble		
in vegetable oil:	@ **20°C**	0.07 to 1.5%	@ **50°C**	0.07 to 1.5%	@ **100°C**	0.07 to 1.5%		
in sucrose solution:	@ **10%**	Insoluble	@ **40%**	Insoluble	@ **60%**	Insoluble		
in sodium chloride solution:	@ **5%**	Insoluble	@ **10%**	Insoluble	@ **15%**	Insoluble		
in ethanol solution:	@ **5%**	Insoluble	@ **20%**	Insoluble	@ **95%**	0.05%	@ **100%**	0.1%
in propylene glycol:	@ **20°C**	Insoluble	@ **50°C**	Insoluble	@ **100°C**	Insoluble		

FUNCTION IN FOODS:	Orange to orange-red food colour, depending on concentration
ALTERNATIVES:	Other orange food colours, although not the same hue: Paprika; Sunset Yellow; Annatto
TECHNOLOGY OF USE IN FOODS:	Available as a microcrystalline suspension in vegetable oil. To prepare a stock solution to colour margarine or oils, heat oil to 45° to 50°C and stir in about 30 g/l of suspension. If mixed with citrus oils, the oil may need to be heated to 70°C. If used as 10% water-dispersible fine granules, add granules slowly to 10–15 times its weight of water at about 40°C with continuous stirring until all the particles are completely dispersed. Reasonably stable to heat and light; stable to pH and acidic conditions. Stable to ascorbic acid. Can oxidise, losing colour; hence stabilisation with antioxidants such as tocopherols is recommended. When used to colour soft drinks it is recommended that the colour and flavour oils are homogenised before addition with the fruit juice and/or pulp to delay separation.
SYNERGISTS:	Antioxidants such as tocopherols/ascorbic acid

ANTAGONISTS: Oxidising agents/Peroxides

FOOD SAFETY ISSUES: Considered safe for food use. A vitamin A precursor.

LEGISLATION:

USA:
Permitted

UK and EUROPE:
Permitted according to European Parliament and Council Directive 94/36/EC of June 30th 1994 on Colours for Use in Foodstuffs.
Refer to Annex V Part 2 for restrictions

CANADA:
Permitted, but restricted in use

AUSTRALIA:
Permitted

REFERENCE: Megard, D. (1993) Stability of red beet pigments for use as food colorant: a review. *Food & Food Ingredients Journal*, **158**, 130–149.

STRUCTURE OF β-APO-8′-CAROTENAL:

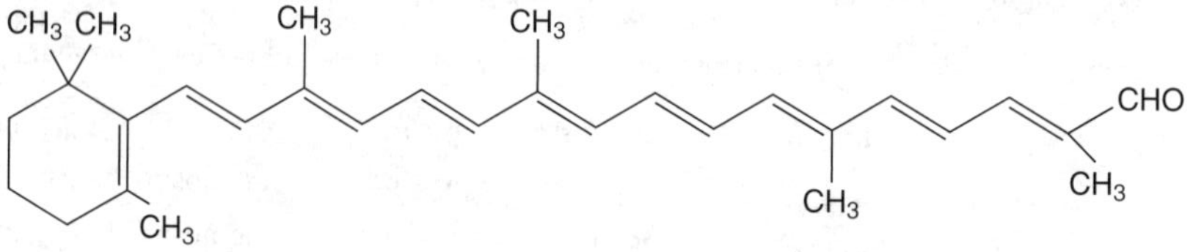

NAME:	**Brilliant Blue FCF**
CATEGORY:	Food colour
FOOD USE:	Baked goods/ Alcoholic drinks/ Sauces and pickles/ Confectionery/ Coatings and decorations/ Soups/ Canned products/ Dairy products/ Snacks/ Soft drinks/ Cereal and cereal products/ Meat and poultry products/ Desserts/ Edible ices/ Fish products
SYNONYMS:	E133/ FD&C Blue No. 1/ C.I. 42090 (Food Blue 2)/ disodium α-(4-(*N*-ethyl-3-sulfonatobenzyl amino) phenyl)-α-(4-*N*-ethyl-3-sulfonatobenzyl imino) cyclohexa-2,5-dienylidene) toluene-2-sulfonate
FORMULA:	$C_{37}H_{34}N_2Na_2O_9S_3$
MOLECULAR MASS:	792.84
ALTERNATIVE FORMS:	Calcium salt/ Potassium salt/ Aluminium lake (insoluble)
PROPERTIES AND APPEARANCE:	Reddish-blue powder or granules
BOILING POINT IN °C AT VARIOUS PRESSURES (INCLUDING 760 mm Hg):	N/A
MELTING RANGE IN °C:	Not known
FLASH POINT IN °C:	N/A
IONISATION CONSTANT AT 25°C:	N/A
DENSITY AT 20°C (AND OTHER TEMPERATURES) IN g/l:	N/A
HEAT OF COMBUSTION AT 25°C IN J/kg:	N/A
VAPOUR PRESSURE AT VARIOUS TEMPERATURES IN mm Hg:	N/A
PURITY %:	Content not less than 85% total colouring matter, calculated as the sodium salt

WATER CONTENT MAXIMUM IN %:	15
HEAVY METAL CONTENT MAXIMUM IN ppm:	40
ARSENIC CONTENT MAXIMUM IN ppm:	3
ASH MAXIMUM IN %:	N/A

SOLUBILITY % AT VARIOUS TEMPERATURE/pH COMBINATIONS:

in water:	@ **20°C** 20%	@ **50°C** 20%	@ **100°C** 20%	
in vegetable oil:	@ **20°C** Insoluble	@ **50°C** Insoluble	@ **100°C** Insoluble	
in sucrose solution:	@ **10%** Soluble	@ **40%** Soluble	@ **60%** Soluble	
in sodium chloride solution:	@ **5%** Soluble	@ **10%** Insoluble	@ **15%** Insoluble	
in ethanol solution:	@ **5%** 25°C 20% 60°C 20%	@ **20%** 25°C 20% 60°C 20%	@ **95%** 25°C 0.15% 60°C 0.25%	@ **100%** 25°C 0.15% 60°C 0.25%
in propylene glycol:	@ **20°C** 20%	@ **50°C** 20%	@ **100°C** 20%	

FUNCTION IN FOODS:	Water-soluble, bright-blue food colour; or insoluble bright-blue food colour as the aluminium lake
ALTERNATIVES:	Other blue food colours, although not the same hue: Patent Blue V; Green S; Indigotine
TECHNOLOGY OF USE IN FOODS:	A reasonably light-stable greenish-blue food colour with excellent heat stability. Stable in the presence of sodium benzoate, sulphur dioxide, fruit acids. Slight fading in the presence of 1% ascorbic acid. Fades in alkaline media.
SYNERGISTS:	None known
ANTAGONISTS:	Alkaline media/ Ascorbic acid
FOOD SAFETY ISSUES:	None known

LEGISLATION:

USA:
FD&C Blue No. 1 certified food colour

CANADA:
Permitted if certified

UK and EUROPE:
Permitted according to European Parliament and Council Directive 94/36/EC of June 30th 1994 on Colours for Use in Foodstuffs. Refer to Annex V Part 2 for restrictions

AUSTRALIA:
Permitted, but restrictions apply

REFERENCE: Smith, J. (Ed.) (1991) *Food Additive User's Handbook*. Blackie Academic & Professional.

STRUCTURE OF BRILLIANT BLUE FCF:

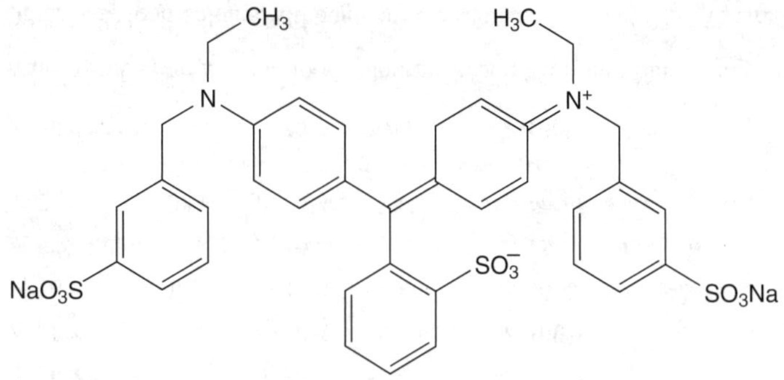

NAME:	Brown FK
CATEGORY:	Food colour
FOOD USE:	Smoked fish
SYNONYMS:	E154/ C.I. Food Brown 1/ A mixture of: 1. Sodium 4-(2,4-diaminophenylazo) benzene sulfonate 2. Sodium 4-(4,6-diamino-H-tolylazo) benzene sulfonate 3. Disodium 4,4′-(4,6-diamino-1,3-phenylene bisazo)di(benzene sulfonate) 4. Disodium 4,4′-(2,4-diamino-1,3-phenylene bisazo)di(benzene sulfonate) 5. Disodium 4,4′-(2,4-diamino-5-methyl-1,3-phenylene bisazo)di(benzene sulfonate) 6. Trisodium 4,4′,4″-(2,4-diaminobenzene-1,3,5-trisazo)tri-(benzene sulfonate)
FORMULA:	1. $C_{12}H_{11}N_4NaO_3S$ 2. $C_{13}H_{13}N_4NaO_3S$ 3. $C_{18}H_{14}N_6Na_2O_6S_2$ 4. $C_{18}H_{14}N_6Na_2O_6S_2$ 5. $C_{19}H_{16}N_6Na_2O_6S_2$ 6. $C_{24}H_{17}N_8Na_3O_9S_3$
MOLECULAR MASS:	1. 314.30; 2. 328.33; 3. 520.46; 4. 520.46; 5. 534.47; 6. 726.59
ALTERNATIVE FORMS:	Calcium salt/ Potassium salt/ Aluminium lake (insoluble)
PROPERTIES AND APPEARANCE:	Red-brown powder or granules
BOILING POINT IN °C AT VARIOUS PRESSURES (INCLUDING 760 mm Hg):	N/A
MELTING RANGE IN °C:	Not known
FLASH POINT IN °C:	N/A
IONISATION CONSTANT AT 25°C:	N/A
DENSITY AT 20°C (AND OTHER TEMPERATURES) IN g/l:	N/A
HEAT OF COMBUSTION AT 25°C IN J/kg:	N/A

VAPOUR PRESSURE AT VARIOUS TEMPERATURES IN mm Hg:	N/A					
PURITY %:	Not less than 70% total colouring matter calculated as the sodium salt					
WATER CONTENT MAXIMUM IN %:	30					
HEAVY METAL CONTENT MAXIMUM IN ppm:	40					
ARSENIC CONTENT MAXIMUM IN ppm:	3					
ASH MAXIMUM IN %:	N/A					
SOLUBILITY % AT VARIOUS TEMPERATURE/pH COMBINATIONS:						
in water:	@ **20°C**	20%	@ **50°C**	20%	@ **100°C**	20%
in vegetable oil:	@ **20°C**	Insoluble	@ **50°C**	Insoluble	@ **100°C**	Insoluble
in sucrose solution:	@ **10%**	20%	@ **40%**	20%	@ **60%**	20%
in sodium chloride solution:	@ **5%**	20%	@ **10%**	20%	@ **15%**	20%
in ethanol solution:	@ **5%**	20%	@ **20%**	20%	@ **95%**	Slightly soluble
	@ **100%**	Slightly soluble				
in propylene glycol:	@ **20°C**	Insoluble	@ **50°C**	Insoluble	@ **100°C**	Insoluble

FUNCTION IN FOODS: A water-soluble brown food colour

ALTERNATIVES: Other brown food colours, although not the same hue, include: Brown HT; Caramels; Blends of food colours to achieve a brown shade; For smoked fish, annatto can be used

TECHNOLOGY OF USE IN FOODS: A light-stable, reddish-brown food colour with high water solubility. Resistant to brine solutions and mainly used for colouring of smoked fish, but bonds onto protein and has been used to colour meat products.

Reasonably heat stable to 105°C. Fair resistance to fruit acids, alkaline media, benzoic acid. Fades in the presence of sulphur dioxide.

SYNERGISTS: None known

ANTAGONISTS: Sulphur dioxide

FOOD SAFETY ISSUES: Insufficient toxicological studies have been carried out to determine an allowable daily intake (ADI). Use restricted in Europe to kipper coloration (smoked cured herrings). Maximum level 20 mg/kg

LEGISLATION:

USA:
Not permitted

CANADA:
Not permitted

UK and EUROPE:
Permitted according to European Parliament and Council Directive 94/36/EC of June 30th 1994 on Colours for Use in Foodstuffs
Refer to Annex IV. Only permitted for colouring kippers (smoked cured herrings) at 20 mg/kg maximum level

AUSTRALIA:
Not permitted

REFERENCE: Smith, J. (Ed.) (1991) *Food Additive User's Handbook*. Blackie Academic & Professional.

STRUCTURE OF BROWN FK:

Essentially a mixture of:

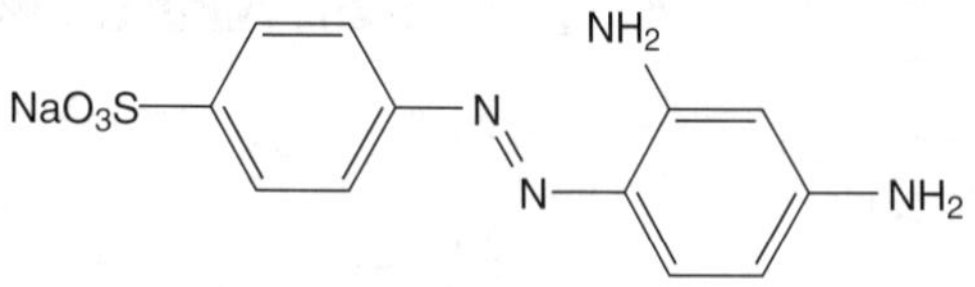

plus five related compounds

NAME:	Brown HT
CATEGORY:	Food colour
FOOD USE:	Baked goods/ Dairy products/ Confectionery/ Edible ices/ Desserts/ Soups/ Canned products/ Snacks/ Decorations and coatings/ Sauces, seasonings
SYNONYMS:	E155/ C.I. 20285 (Food Brown 3)/ Disodium 4,4′-(2,4-dihydroxy-5-hydroxymethyl-1,3-phenylene bisazo) di(naphthalene-l-sulfonate)/ Chocolate brown HT
FORMULA:	C27 H18 N4 Na2 O9 S2
MOLECULAR MASS:	652.57
ALTERNATIVE FORMS:	Potassium salt/ Calcium salt/ Aluminium lake (insoluble)
PROPERTIES AND APPEARANCE:	Reddish-brown powder or granules
BOILING POINT IN °C AT VARIOUS PRESSURES (INCLUDING 760 mm Hg):	N/A
MELTING RANGE IN °C:	N/A
FLASH POINT IN °C:	N/A
IONISATION CONSTANT AT 25°C:	N/A
DENSITY AT 20°C (AND OTHER TEMPERATURES) IN g/l:	N/A
HEAT OF COMBUSTION AT 25°C IN J/kg:	N/A
VAPOUR PRESSURE AT VARIOUS TEMPERATURES IN mm Hg:	N/A
PURITY %:	Not less than 70% total colouring matter calculated as the sodium salt

WATER CONTENT MAXIMUM IN %:	30							
HEAVY METAL CONTENT MAXIMUM IN ppm:	40							
ARSENIC CONTENT MAXIMUM IN ppm:	3							
ASH MAXIMUM IN %:	N/A							
SOLUBILITY % AT VARIOUS TEMPERATURE/pH COMBINATIONS:								
in water:	**@ 20°C**	20%	**@ 50°C**	20%	**@ 100°C**	20%		
in vegetable oil:	**@ 20°C**	Insoluble	**@ 50°C**	Insoluble	**@ 100°C**	Insoluble		
in sucrose solution:	**@ 10%**	20%	**@ 40%**	20%	**@ 60%**	Soluble		
in sodium chloride solution:	**@ 5%**	Soluble	**@ 10%**	Insoluble	**@ 15%**	Insoluble		
in ethanol solution:	**@ 5%**	Insoluble	**@ 20%**	Insoluble	**@ 95%**	Insoluble	**@ 100%**	Insoluble
in propylene glycol:	**@ 20°C**	15%	**@ 50°C**	15%	**@ 100°C**	20%		
FUNCTION IN FOODS:	Water-soluble brown food colour or insoluble brown food colour as the aluminium lake. A chocolate shade.							
ALTERNATIVES:	Other brown food colours, although not the same hue, include: Caramels; Blends of food colours to achieve a brown shade; Brown FK where permitted.							
TECHNOLOGY OF USE IN FOODS:	A heat- and light-stable food colour with high water solubility. Excellent for baked products, stable on baking to 200°C. Good stability in the presence of fruit acids. Good stability in alkaline media. Some fading in the presence of sulphur dioxide.							
SYNERGISTS:	None known							
ANTAGONISTS:	Sulphur dioxide							
FOOD SAFETY ISSUES:	None known							

LEGISLATION:

USA:
Not permitted

CANADA:
Not permitted

UK and EUROPE:
Permitted according to European Parliament and Council Directive 94/36/EC of June 30th 1994 on Colours for Use in Foodstuffs.
Refer to Annex V Part 2 for restrictions

AUSTRALIA:
Permitted, but restrictions apply

REFERENCE:

Smith, J. (Ed.) (1991) *Food Additive User's Handbook*. Blackie Academic & Professional.

STRUCTURE OF BROWN HT:

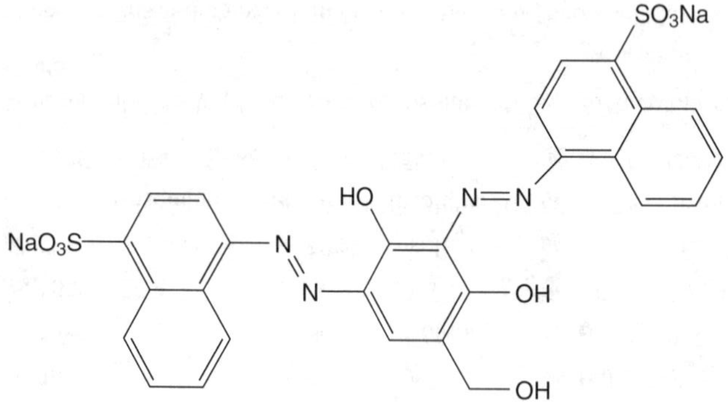

NAME:	Calcium carbonate
CATEGORY:	Food colour
FOOD USE:	Sugar confectionery/ Dairy creamers
SYNONYMS:	E170/ C.I. 77220 Pigment White 18/ Chalk
FORMULA:	$CaCO_3$
MOLECULAR MASS:	100.1
ALTERNATIVE FORMS:	N/A
PROPERTIES AND APPEARANCE:	White crystalline or amorphous powder
BOILING POINT IN °C AT VARIOUS PRESSURES (INCLUDING 760 mm Hg):	N/A
MELTING RANGE IN °C:	N/A
FLASH POINT IN °C:	N/A
IONISATION CONSTANT AT 25°C:	N/A
DENSITY AT 20°C (AND OTHER TEMPERATURES) IN g/l:	N/A
HEAT OF COMBUSTION AT 25°C IN J/kg:	N/A
VAPOUR PRESSURE AT VARIOUS TEMPERATURES IN mm Hg:	N/A
PURITY %:	Not less than 98 on anhydrous basis
WATER CONTENT MAXIMUM IN %:	2

HEAVY METAL CONTENT MAXIMUM IN ppm: 40

ARSENIC CONTENT MAXIMUM IN ppm: 3

ASH MAXIMUM IN %: N/A

SOLUBILITY % AT VARIOUS TEMPERATURE/pH COMBINATIONS:

	@ 20°C	@ 50°C	@ 100°C
in water:	Insoluble in pure water	Insoluble in pure water	Insoluble in pure water
in vegetable oil:	Insoluble	Insoluble	Insoluble

FUNCTION IN FOODS: Insoluble white food colour

ALTERNATIVES: Titanium dioxide

TECHNOLOGY OF USE IN FOODS: Insoluble white food colour used to impart a white opaque effect to food products such as sugar confectionery, non-dairy creamers and salad dressings.

Colour must be dispersed into a suitable medium since it is insoluble in solvents except acids. Dissolves readily in acidic media. Dissolves slowly in water containing carbon dioxide.

SYNERGISTS: None known

ANTAGONISTS: Food acids

FOOD SAFETY ISSUES: None known

LEGISLATION:

USA:	UK and EUROPE:	CANADA:	AUSTRALIA:
Not permitted as a food colour	Permitted according to European Parliament and Council Directive 94/36/EC of June 30th 1994 on Colours for Use in Foodstuffs. Refer to Annex V Part 1 for restrictions	Not permitted as a food colour	Not permitted as a food colour

REFERENCE:

Smith, J. (Ed.) (1991) *Food Additive User's Handbook*. Blackie Academic & Professional.

NAME:	**Canthaxanthin**
CATEGORY:	Food colour
FOOD USE:	Sausages/ Sauces/ Fish products/ Confectionery/ Soups/ Tomato-based products/ Processed cheese/ Ice-cream/ Soft drinks/ Salad dressings/ Desserts/ Meat products/ Dairy products
SYNONYMS:	E161g/C.I. 40850 (Food Orange 8)/ Beta-carotene-4,4′-dione/ 4,4′-Dioxo-beta-carotene/ 4,4′-Diketo-beta carotene
FORMULA:	$C_{40}H_{52}O_2$
MOLECULAR MASS:	564.86
ALTERNATIVE FORMS:	N/A
PROPERTIES AND APPEARANCE:	Deep violet crystals or crystalline powder
BOILING POINT IN °C AT VARIOUS PRESSURES (INCLUDING 760 mm Hg):	N/A
MELTING RANGE IN °C:	About 210 (decomposes)
FLASH POINT IN °C:	N/A
IONISATION CONSTANT AT 25°C:	N/A
DENSITY AT 20°C (AND OTHER TEMPERATURES) IN g/l:	N/A
HEAT OF COMBUSTION AT 25°C IN J/kg:	N/A
VAPOUR PRESSURE AT VARIOUS TEMPERATURES IN mm Hg:	N/A
PURITY %:	Not less than 96% of total colouring matters

WATER CONTENT MAXIMUM IN %:	5					
HEAVY METAL CONTENT MAXIMUM IN ppm:	40					
ARSENIC CONTENT MAXIMUM IN ppm:	3					
ASH MAXIMUM IN %:	N/A					
SOLUBILITY % AT VARIOUS TEMPERATURE/pH COMBINATIONS:						
in water:	**@ 20°C**	Insoluble	**@ 50°C**	Insoluble	**@ 100°C**	Insoluble
in vegetable oil:	**@ 20°C**	0.005%	**@ 50°C**	0.01%	**@ 100°C**	0.05%
in sucrose solution:	**@ 10%**	Insoluble	**@ 40%**	Insoluble	**@ 60%**	Insoluble
in sodium chloride solution:	**@ 5%**	Insoluble	**@ 10%**	Insoluble	**@ 15%**	Insoluble
in ethanol solution:	**@ 5%**	Slightly soluble	**@ 20%**	Very slightly soluble	**@ 95%**	Below 0.01%
	@ 100%	Below 0.01%				
in propylene glycol:	**@ 20°C**	Insoluble	**@ 50°C**	Insoluble	**@ 100°C**	Insoluble
FUNCTION IN FOODS:	Pink to red food colour, depending on concentration					
ALTERNATIVES:	Other pink to red food colours, although not the same hue: Carmine; Ponceau 4R; Erythrosine					
TECHNOLOGY OF USE IN FOODS:	Commercially available as colloidal canthaxanthin in a gelatin carbohydrate matrix containing antioxidants. Disperse beadlets in 10–15 times its weight of water at 30–40°C with continuous stirring until all the particles are completely dispersed. Good stability to heat, light, acidic and alkaline media. Stable to pH change and ascorbic acid. Can oxidise, losing colour; hence stabilisation with antioxidants such as tocopherols is recommended.					
SYNERGISTS:	Antioxidants such as tocopherols/ Ascorbic acid					

ANTAGONISTS: Oxidising agents/ Peroxides

FOOD SAFETY ISSUES: The poor solubility in the blood on ingestion has caused this colour to be retained by the body longer than some medical authorities like when taken in large quantities. Thus, its use has been restricted by some authorities.

LEGISLATION:

USA:
Permitted according to 21 CFR 73.75

CANADA:
Not permitted

UK and EUROPE:
Permitted according to European Parliament and Council Directive 94/36/EC of June 30th 1994 on Colours for Use in Foodstuffs.
Refer to Annex IV – Restricted for use in saucisses de Strasbourg at 15 mg/kg maximum.

REFERENCE: Smith, J. (Ed.) (1991) *Food Additive User's Handbook*. Blackie Academic & Professional.

STRUCTURE OF CANTHAXANTHIN:

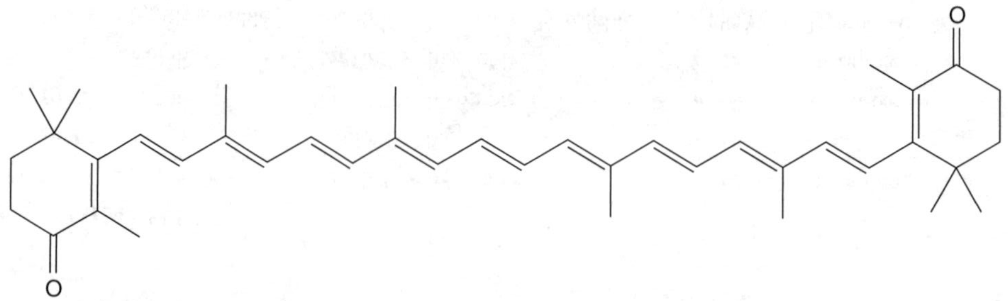

NAME:	Plain caramel
CATEGORY:	Food colour
FOOD USE:	Baked goods/ Cereal and cereal products/ Dairy products/ Meat and poultry products/ Soft drinks/ Alcoholic drinks/ Sugars and preserves/ Confectionery/ Vinegars, pickles and sauces/ Decorations and coatings/ Edible ices/ Desserts/ Seasonings
SYNONYMS:	E150a/ Spirit caramel
FORMULA:	Not known
MOLECULAR MASS:	Not known
ALTERNATIVE FORMS:	N/A
PROPERTIES AND APPEARANCE:	A dark-brown black viscous liquid or brown powder
BOILING POINT IN °C AT VARIOUS PRESSURES (INCLUDING 760 mm Hg):	N/A
MELTING RANGE IN °C:	N/A
FLASH POINT IN °C:	N/A
IONISATION CONSTANT AT 25°C:	N/A
DENSITY AT 20°C (AND OTHER TEMPERATURES) IN g/l:	N/A
HEAT OF COMBUSTION AT 25°C IN J/kg:	N/A
VAPOUR PRESSURE AT VARIOUS TEMPERATURES IN mm Hg:	N/A
PURITY %:	Colour intensity 0.01–0.12 based on absorbance of a 0.1% solution of caramel in water in a 1-cm cell at 610 nm.

WATER CONTENT MAXIMUM IN %:	Typically 40 for liquids					
HEAVY METAL CONTENT MAXIMUM IN ppm:	25					
ARSENIC CONTENT MAXIMUM IN ppm:	1					
ASH MAXIMUM IN %:	N/A					
SOLUBILITY % AT VARIOUS TEMPERATURE/pH COMBINATIONS:						
in water:	@ **20°C**	Completely soluble	@ **50°C**	Completely soluble	@ **100°C**	Completely soluble
in vegetable oil:	@ **20°C**	Insoluble	@ **50°C**	Insoluble	@ **100°C**	Insoluble
in sucrose solution:	@ **10%**	Completely soluble	@ **40%**	Completely soluble	@ **60%**	Completely soluble
in sodium chloride solution:	@ **5%**	Completely soluble	@ **10%**	Completely soluble	@ **15%**	Completely soluble
in ethanol solution:	@ **5%**	Completely soluble	@ **20%**	Completely soluble	@ **95%**	Completely soluble
	@ **100%**	Completely soluble				
in propylene glycol:	@ **20°C**	Dispersible	@ **50°C**	Dispersible	@ **100°C**	Dispersible
FUNCTION IN FOODS:	A water- or alcohol-soluble brown food colour					
ALTERNATIVES:	E150b–d in non-alcoholic applications; Brown HT; Brown FK					
TECHNOLOGY OF USE IN FOODS:	Fully water- and/or alcohol-soluble liquid and powdered forms available depending on the nature of the application.					
SYNERGISTS:	N/A					
ANTAGONISTS:	N/A					
FOOD SAFETY ISSUES:	There is concern over E150 caramels generally, although they have been fully toxicologically tested and found to be safe.					

LEGISLATION:

USA:
Permitted according to 21 CFR Part 73.85

CANADA:
Permitted

UK and EUROPE:
Permitted according to European Parliament and Council Directive 94/36/EC of June 30th 1994 on Colours for Use in Foodstuffs.
Refer to Annex V Part 1 for restrictions on use

AUSTRALIA:
Permitted

REFERENCE:

Smith, J. (Ed.) (1991) *Food Additive User's Handbook*. Blackie Academic & Professional.

NAME:	Ammonia caramel
CATEGORY:	Food colour
FOOD USE:	Baked goods/ Cereal and cereal products/ Dairy products/ Meat and poultry products/ Soft drinks/ Sugars and preserves/ Confectionery/ Vinegars, pickles and sauces/ Decorations and coatings/ Edible ices/ Desserts/ Seasonings/ Savoury dry mixes
SYNONYMS:	E150c
FORMULA:	Not known
MOLECULAR MASS:	Not known
ALTERNATIVE FORMS:	N/A
PROPERTIES AND APPEARANCE:	Dark brown to black liquids or powders
BOILING POINT IN °C AT VARIOUS PRESSURES (INCLUDING 760 mm Hg):	N/A
MELTING RANGE IN °C:	N/A
FLASH POINT IN °C:	N/A
IONISATION CONSTANT AT 25°C:	N/A
DENSITY AT 20°C (AND OTHER TEMPERATURES) IN g/l:	N/A
HEAT OF COMBUSTION AT 25°C IN J/kg:	N/A
VAPOUR PRESSURE AT VARIOUS TEMPERATURES IN mm Hg:	N/A
PURITY %:	Colour intensity 0.08–0.36 based on absorbance of a 0.1% solution of caramel in water in a 1-cm cell at 610 nm.

WATER CONTENT MAXIMUM IN %:	Typically 40 for liquids					
HEAVY METAL CONTENT MAXIMUM IN ppm:	25					
ARSENIC CONTENT MAXIMUM IN ppm:	1					
ASH MAXIMUM IN %:	N/A					
SOLUBILITY % AT VARIOUS TEMPERATURE/pH COMBINATIONS:						
in water:	**@ 20°C**	Completely soluble	**@ 50°C**	Completely soluble		
	@ 100°C	Completely soluble				
in vegetable oil:	**@ 20°C**	Insoluble	**@ 50°C**	Insoluble	**@ 100°C**	Insoluble
in sucrose solution:	**@ 10%**	Completely soluble	**@ 40%**	Completely soluble		
	@ 60%	Completely soluble				
in sodium chloride solution:	**@ 5%**	Completely soluble	**@ 10%**	Completely soluble		
	@ 15%	Completely soluble				
in ethanol solution:	**@ 5%**	Not stable (haze or precipitate)	**@ 20%**	Not stable (haze or precipitate)		
	@ 95%	Not stable (haze or precipitate)	**@ 100%**	Not stable (haze or precipitate)		
in propylene glycol:	**@ 20°C**	Dispersible	**@ 50°C**	Dispersible	**@ 100°C**	Dispersible
FUNCTION IN FOODS:	A water-soluble brown food colour					
ALTERNATIVES:	E150a, b & d; Brown HT; Brown FK					
TECHNOLOGY OF USE IN FOODS:	Fully water-soluble in both liquid and powder forms. To dissolve powders, use warm water. For dry mixes (e.g. instant desserts), the powders form can be used and dissolves when reconstituted by the consumer. Note: ammonia caramels are positively charged.					
SYNERGISTS:	N/A					
ANTAGONISTS:	N/A					

FOOD SAFETY ISSUES: There is media and public concern over E150 caramels. However, they have been fully toxicologically tested and found to be safe.

LEGISLATION:

USA:
Permitted according to 21 CFR Part 73.85

CANADA:
Permitted

UK and EUROPE:
Permitted according to European Parliament and Council Directive 94/36/EC of June 30th 1994 on Colours for Use in Foodstuffs. Refer to Annex V Part 1 for restrictions on use

AUSTRALIA:
Permitted

REFERENCE:

Smith, J. (Ed.) (1991) *Food Additive User's Handbook*. Blackie Academic & Professional.

NAME:	Caustic sulphite caramel
CATEGORY:	Food colour
FOOD USE:	Baked goods/ Cereal and cereal products/ Dairy products/ Meat and poultry products/ Soft drinks/ Sugars and preserves/ Confectionery/ Vinegars, pickles and sauces/ Decorations and coatings/ Edible ices/ Desserts/ Seasonings/ Savoury dry mixes
SYNONYMS:	E150b
FORMULA:	Not known
MOLECULAR MASS:	Not known
ALTERNATIVE FORMS:	N/A
PROPERTIES AND APPEARANCE:	Dark brown to black liquids or powders
BOILING POINT IN °C AT VARIOUS PRESSURES (INCLUDING 760 mm Hg):	N/A
MELTING RANGE IN °C:	N/A
FLASH POINT IN °C:	N/A
IONISATION CONSTANT AT 25°C:	N/A
DENSITY AT 20°C (AND OTHER TEMPERATURES) IN g/l:	N/A
HEAT OF COMBUSTION AT 25°C IN J/kg:	N/A
VAPOUR PRESSURE AT VARIOUS TEMPERATURES IN mm Hg:	N/A
PURITY %:	Colour intensity 0.05–0.13 based on absorbance of a 0.1% solution of caramel in water in a 1-cm cell at 610 nm.

WATER CONTENT MAXIMUM IN %: Typically 40 for liquids

HEAVY METAL CONTENT MAXIMUM IN ppm: 25

ARSENIC CONTENT MAXIMUM IN ppm: 1

ASH MAXIMUM IN %: N/A

SOLUBILITY % AT VARIOUS TEMPERATURE/pH COMBINATIONS:

in water:	**@ 20°C**	Completely soluble	**@ 50°C**	Completely soluble		
	@ 100°C	Completely soluble				
in vegetable oil:	**@ 20°C**	Insoluble	**@ 50°C**	Insoluble	**@ 100°C**	Insoluble
in sucrose solution:	**@ 10%**	Completely soluble	**@ 40%**	Completely soluble		
	@ 60%	Completely soluble				
in sodium chloride solution:	**@ 5%**	Completely soluble	**@ 10%**	Completely soluble		
	@ 15%	Completely soluble				
in ethanol solution:	**@ 5%**	Not stable (haze or precipitate)	**@ 20%**	Not stable (haze or precipitate)		
	@ 95%	Not stable (haze or precipitate)	**@ 100%**	Not stable (haze or precipitate)		
in propylene glycol:	**@ 20°C**	Dispersible	**@ 50°C**	Dispersible	**@ 100°C**	Dispersible

FUNCTION IN FOODS: A water-soluble brown food colour.

ALTERNATIVES: E150a, c and d; Brown HT; Brown FK

TECHNOLOGY OF USE IN FOODS: Fully water-soluble in both liquid and powder forms. To dissolve powders, use warm water.

For dry mixes (i.e. instant desserts), the powdered form is used and dissolves when reconstituted by the consumer.

Note: caustic sulphite caramels are negatively charged.

SYNERGISTS: N/A

ANTAGONISTS: N/A

FOOD SAFETY ISSUES:

There is media and public concern over E150 caramels. However, they have been fully toxicologically tested and found to be safe.

LEGISLATION:

USA:
Permitted according to 21 CFR Part 73.85

CANADA:
Permitted

UK and EUROPE:
Permitted according to European Parliament and Council Directive 94/36/EC of June 30th 1994 on Colours for Use in Foodstuffs.
Refer to Annex V Part 1 for restrictions on use

AUSTRALIA:
Permitted

REFERENCE:

Smith, J. (Ed.) (1991) *Food Additive User's Handbook.* Blackie Academic & Professional.

NAME:	Sulphite ammonia caramel
CATEGORY:	Food colour
FOOD USE:	Baked goods/ Cereals and cereal products/ Dairy products/ Meat and poultry products/ Soft drinks/ Sugars and preserves/ Confectionery/ Vinegars, pickles and sauces/ Decorations and coatings/ Edible ices/ Desserts/ Seasonings/ Savoury dry mixes
SYNONYMS:	E150d
FORMULA:	Not known
MOLECULAR MASS:	Not known
ALTERNATIVE FORMS:	N/A
PROPERTIES AND APPEARANCE:	Dark brown to black liquids or powders
BOILING POINT IN °C AT VARIOUS PRESSURES (INCLUDING 760 mm Hg):	N/A
MELTING RANGE IN °C:	N/A
FLASH POINT IN °C:	N/A
IONISATION CONSTANT AT 25°C:	N/A
DENSITY AT 20°C (AND OTHER TEMPERATURES) IN g/l:	N/A
HEAT OF COMBUSTION AT 25°C IN J/kg:	N/A
VAPOUR PRESSURE AT VARIOUS TEMPERATURES IN mm Hg:	N/A
PURITY %:	Colour intensity 0.10–0.60 based on absorbance of a 0.1% solution of caramel in water in a 1-cm cell at 610 nm

WATER CONTENT MAXIMUM IN %:	Typically 40% for liquids			
HEAVY METAL CONTENT MAXIMUM IN ppm:	25			
ARSENIC CONTENT MAXIMUM IN ppm:	1			
ASH MAXIMUM IN %:	N/A			
SOLUBILITY % AT VARIOUS TEMPERATURE/pH COMBINATIONS:				
in water:	**@ 20°C**	Completely soluble	**@ 50°C**	Completely soluble
	@ 100°C	Completely soluble		
in vegetable oil:	**@ 20°C**	Insoluble	**@ 50°C**	Insoluble
	@ 100°C	Insoluble		
in sucrose solution:	**@ 10%**	Completely soluble	**@ 40%**	Completely soluble
	@ 60%	Completely soluble		
in sodium chloride solution:	**@ 5%**	Completely soluble	**@ 10%**	Completely soluble
	@ 15%	Completely soluble		
in ethanol solution:	**@ 5%**	Not stable (haze or precipitate)	**@ 20%**	Not stable (haze or precipitate)
	@ 95%	Not stable (haze or precipitate)	**@ 100%**	Not stable (haze or precipitate)
in propylene glycol:	**@ 20°C**	Dispersible	**@ 50°C**	Dispersible
	@ 100°C	Dispersible		
FUNCTION IN FOODS:	A water-soluble brown food colour			
ALTERNATIVES:	E150a, b and c; Brown HT; Brown FK			
TECHNOLOGY OF USE IN FOODS:	Fully water-soluble in both liquid and powder forms. To dissolve powders, use warm water. For dry mixes (e.g. instant desserts), the powderer form can be used and dissolves when reconstituted by the consumer. Note: Caustic sulphite caramels are negatively charged.			
SYNERGISTS:	N/A			

ANTAGONISTS: N/A

FOOD SAFETY ISSUES: There is media and public concern over E150 caramels. However, they have been fully toxicologically tested and found to be safe.

LEGISLATION:

USA:	**UK and EUROPE:**	**CANADA:**	**OTHER COUNTRIES:**
Permitted according to 21 CFR Part 73.85	Permitted according to European Parliament and Council Directive 94/36/EC of June 30th 1994 on Colours for Use in Foodstuffs. Refer to Annex V Part 5 for restrictions on use	Permitted **AUSTRALIA:** Permitted	Generally permitted in most countries

REFERENCE: Smith, J. (Ed.) (1991) *Food Additive Users Handbook*. Blackie Academic & Professional.

NAME:	Beetroot red
CATEGORY:	Food colour
FOOD USE:	Baked goods/ Cereals and cereal products/ Dairy products/ Meat and poultry products/ Fruit, vegetable and nut products/ Sugars and preserves/ Confectionery/ Vinegar, pickles and sauces/ Decorations and coatings/ Edible ices/ Desserts/ Seasonings
SYNONYMS:	Beet red/ Betanin
FORMULA:	$C_{24}H_{26}N_2O_{13}$
MOLECULAR MASS:	550.48
ALTERNATIVE FORMS:	None known
PROPERTIES AND APPEARANCE:	Red or dark-red liquid, paste, powder or solid
BOILING POINT IN °C AT VARIOUS PRESSURES (INCLUDING 760 mm Hg):	Not applicable
MELTING RANGE IN °C:	Not known
FLASH POINT IN °C:	Not known
IONISATION CONSTANT AT 25°C:	Not known
DENSITY AT 20°C (AND OTHER TEMPERATURES) IN g/l:	Not applicable
HEAT OF COMBUSTION AT 25°C IN J/kg:	Not applicable
VAPOUR PRESSURE AT VARIOUS TEMPERATURES IN mm Hg:	Not known
PURITY %:	≥99.6% betanin
WATER CONTENT MAXIMUM IN %:	N/A

HEAVY METAL CONTENT MAXIMUM IN ppm:	40					
ARSENIC CONTENT MAXIMUM IN ppm:	3					
ASH MAXIMUM IN %:	Not known					
SOLUBILITY % AT VARIOUS TEMPERATURE/pH COMBINATIONS:						
in water:	**@ 20°C**	Readily soluble – 0.5%	**@ 50°C**	Readily soluble – 0.5%		
	@ 100°C	Readily soluble – 0.5%				
in vegetable oil:	**@ 20°C**	Insoluble	**@ 50°C**	Insoluble	**@ 100°C**	Insoluble
in sucrose solution:	**@ 10%**	Readily soluble – 0.5%	**@ 40%**	Readily soluble – 0.5%		
	@ 60%	Readily soluble – 0.5%				
in sodium chloride solution:	**@ 5%**	Readily soluble – 0.5%	**@ 10%**	Readily soluble – 0.5%		
	@ 15%	Readily soluble – 0.1%				
in ethanol solution:	**@ 5%**	Readily soluble – 0.5%	**@ 20%**	Readily soluble – 0.5%		
	@ 95%	Slightly soluble	**@ 100%**	Slightly soluble		
in propylene glycol:	**@ 20°C**	Readily soluble – 0.5%	**@ 50°C**	Readily soluble – 0.5%		
	@ 100°C	Readily soluble – 0.5%				
FUNCTION IN FOODS:	A red/blue food colour suited to applications with little or no heat processing					
ALTERNATIVES:	Other red food colours, although not of exactly the same hue, include: Anthocyanins; Carmine; Carminic acid; Ponceau 4R; Carmoisine; Allura red; Amaranth; Erythrosine					
TECHNOLOGY OF USE IN FOODS:	Beetroot red in either liquid or powder form is suitable for a wide range of food applications. However, exposure to heat will degrade the pigment.					
SYNERGISTS:	Ascorbic acid					
ANTAGONISTS:	Sulphur dioxide/ Metal ions					
FOOD SAFETY ISSUES:	None known					

LEGISLATION:

USA:
Permitted as non-certified food colour according to 21 CFR 73.260

UK and EUROPE:
Permitted according to European Parliament and Council Directive 94/36/EC of June 30th 1994 on Colours for Use in Foodstuffs.
Refer to Annex V Part 1 for restrictions on use

CANADA:
Permitted

AUSTRALIA:
Permitted

REFERENCE:

Smith, J. (Ed.) (1991) *Food Additive User's Handbook*. Blackie Academic & Professional.

STRUCTURE OF BETANIN:

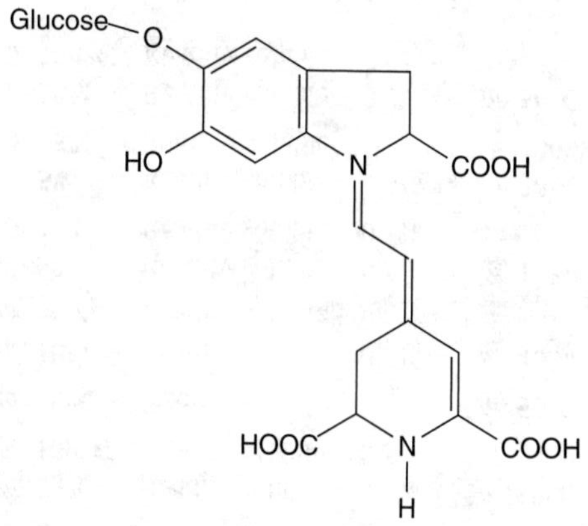

NAME:	Carmine
CATEGORY:	Food colour
FOOD USE:	Baked goods/ Cereals and cereal products/ Dairy products/ Fish, seafood products/ Meat and poultry products/ Fruit, vegetable and nut products/ Soft drinks/ Sugars and preserves/ Confectionery/ Vinegar, pickles and sauces/ Decorations and coatings/ Edible ices/ Desserts/ Seasonings
SYNONYMS:	Cochineal/ E120/ C.I. 75470/ Hydrated aluminium chelate (lake) of carminic acid
FORMULA:	See carminic acid
MOLECULAR MASS:	See carminic acid
ALTERNATIVE FORMS:	Calcium aluminium lake of carminic acid
PROPERTIES AND APPEARANCE:	Red to dark red, friable, solid or powder
BOILING POINT IN °C AT VARIOUS PRESSURES (INCLUDING 760 mm Hg):	Not applicable
MELTING RANGE IN °C:	Not applicable
FLASH POINT IN °C:	Not known
IONISATION CONSTANT AT 25°C:	Not known
DENSITY AT 20°C (AND OTHER TEMPERATURES) g/l:	Not applicable
HEAT OF COMBUSTION AT 25°C IN J/kg:	Not known
VAPOUR PRESSURE AT VARIOUS TEMPERATURES IN mm Hg:	Not applicable
PURITY %:	≥50% carminic acid

WATER CONTENT MAXIMUM IN %:	21.0%					
HEAVY METAL CONTENT MAXIMUM IN ppm:	40					
ARSENIC CONTENT MAXIMUM IN ppm:	3					
SOLUBILITY % AT VARIOUS TEMPERATURE/pH COMBINATIONS:						
in water:	**@ 20°C**	Insoluble in water Soluble in alkali – pH12–2%	**@ 50°C**	Soluble in alkali – pH12–10%	**@ 100°C**	Soluble in alkali – pH12–15%
in vegetable oil:	**@ 20°C**	Insoluble	**@ 50°C**	Insoluble	**@ 100°C**	Insoluble
in sucrose solution:	**@ 10%**	Insoluble	**@ 40%**	Insoluble	**@ 60%**	Insoluble
in sodium chloride solution:	**@ 5%**	Insoluble	**@ 10%**	Insoluble	**@ 15%**	Insoluble
in ethanol solution:	**@ 5%** **@ 100%**	Insoluble Insoluble	**@ 20%**	Insoluble	**@ 95%**	Insoluble
in propylene glycol:	**@ 20°C**	Insoluble	**@ 50°C**	Insoluble	**@ 100°C**	Insoluble
FUNCTION IN FOODS:	A red-pink food colour					
ALTERNATIVES:	Other red food colours, although not of exactly the same hue, include: Beetroot red; Anthocyanin; Carmoisine; Allura red; Red 2G; Ponceau 4R; Erythrosine					
TECHNOLOGY OF USE IN FOODS:	Carmine can be used in its insoluble lake form as a pigment to impart surface colour or colour to opaque foodstuffs. Alternatively, an alkaline solution can be used to provide a water-soluble form. Stability to heat, light and oxidation is excellent.					
SYNERGISTS:	Alkali salts can produce water-soluble solutions of carmine					
ANTAGONISTS:	The action of heat in an acidic environment can de-lake the carmine					
FOOD SAFETY ISSUES:	None known					

LEGISLATION:

USA:
Permitted as a non-certified food colour according to 21 CFR 73.100, 73.1100, 73.2087

UK and EUROPE:
Permitted according to European Parliament and Council Directive 94/36/EC of June 30th 1994 on Colours for Use in Foodstuffs. Refer to Annex V Part 2 for a list of restrictions

CANADA:
Permitted

AUSTRALIA:
Permitted

OTHER COUNTRIES:
Generally permitted in most countries

REFERENCES:

Smith, J. (Ed.) (1991) *Food Additive User's Handbook*. Blackie Academic & Professional.

Lloyd, A. G. (1980) Extraction and chemistry of cochineal. *Food Chemistry*, **5**, 91–107.

NAME:	Carminic acid
CATEGORY:	Food colour
FOOD USE:	Dairy products/ Meat and poultry products/ Soft drinks/ Sugars and preserves/ Confectionery/ Decorations and coatings/ Edible ices/ Desserts/ Seasonings
SYNONYMS:	Cochineal/ E120/ C.I. 75470/ 7-β-D-glucopyranosyl-3,5,6,8-tetrahydroxy-l-methyl-9,10-dioxoanthracene-2-carboxylic acid
FORMULA:	C22 H20 O13
MOLECULAR MASS:	492.39
ALTERNATIVE FORMS:	The aluminium lake is known as carmine
PROPERTIES AND APPEARANCE:	A red/orange powder, or a dark red liquid
BOILING POINT IN °C AT VARIOUS PRESSURES (INCLUDING 760 mm Hg):	Not applicable
MELTING RANGE IN °C:	Not known
FLASH POINT IN °C:	Not known
IONISATION CONSTANT AT 25°C:	Not known
DENSITY AT 20°C (AND OTHER TEMPERATURES) IN g/l:	Not applicable
HEAT OF COMBUSTION AT 25°C IN J/kg:	Not known
VAPOUR PRESSURE AT VARIOUS TEMPERATURES IN mm Hg:	Not applicable
PURITY %:	≥98% in aqueous products; powdered forms generally ≥80%

WATER CONTENT MAXIMUM IN %:	5					
HEAVY METAL CONTENT MAXIMUM IN ppm:	40					
ARSENIC CONTENT MAXIMUM IN ppm:	3					
ASH MAXIMUM IN %:	Not known					
SOLUBILITY % AT VARIOUS TEMPERATURE/pH COMBINATIONS:						
in water:	**@ 20°C**	Readily soluble – 3%	**@ 50°C**	Readily soluble – 5%	**@ 100°C**	Readily soluble – 10%
in vegetable oil:	**@ 20°C**	Insoluble	**@ 50°C**	Insoluble	**@ 100°C**	Insoluble
in sucrose solution:	**@ 10%**	Readily soluble – 5%	**@ 40%**	Readily soluble – 5%	**@ 60%**	Readily soluble – 5%
in sodium chloride solution:	**@ 5%**	Not known	**@ 10%**	Not known	**@ 15%**	Not known
in ethanol solution:	**@ 5%**	Readily soluble – 5%	**@ 20%**	Readily soluble – 5%	**@ 95%**	Soluble – 5%
	@ 100%	Soluble – 5%				
in propylene glycol:	**@ 20°C**	Not known	**@ 50°C**	Not known	**@ 100°C**	Not known
FUNCTION IN FOODS:	A bright yellow/orange to red/purple food colour dependent on pH. Acidic environments give yellow colours, becoming progressively redder as the pH approaches neutral					
ALTERNATIVES:	Sunset yellow; Carmine; Beetroot red; Riboflavin; Beta-carotene; Paprika					
TECHNOLOGY OF USE IN FOODS:	Carminic acid is generally used in an aqueous alcoholic solution and gives a colour that is very pH-dependent: pH3 → yellow pH7 → red/purple. Stability to heat, light and oxygen is excellent.					
SYNERGISTS:	None known					
ANTAGONISTS:	Sulphur dioxide/ Metal ions (e.g. Al^{3+})					
FOOD SAFETY ISSUES:	None known					

LEGISLATION:

USA:
Permitted as non-certified food colour according to 21 CFR 73.100, 73.1100

UK and EUROPE:
Permitted according to European Parliament and Council Directive 94/36/EC of June 30th 1994 on Colours for Use in Foodstuffs.
Refer to Annex V Part 2 for restrictions

CANADA:
Permitted

AUSTRALIA:
Permitted

OTHER COUNTRIES
Generally permitted in most countries

REFERENCES:

Lloyd, A. G. (1980) Extraction and chemistry of cochineal. *Food Chemistry*, **5**, 91–107.
Smith, J. (Ed.) (1991) *Food Additive Users Handbook.* Blackie Academic & Professional.

STRUCTURE OF CARMINIC ACID:

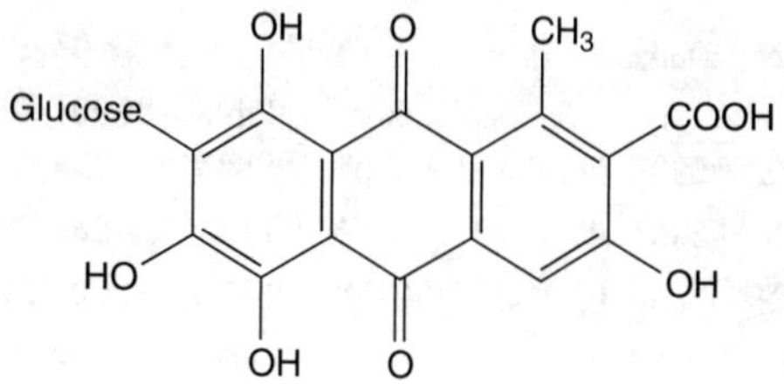

NAME:	Carmoisine
CATEGORY:	Food colour
FOOD USE:	Baked goods/ Meat and poultry products/ Soft drinks/ Fruit, vegetable and nut products/ Beverages/ Edible ices/ Sugars and preserves/ Confectionery/ Desserts/ Vinegar, pickles and sauces/ Decorations and coatings/ Seasonings/ Cereals and cereal products/ Dairy products/ Soups/ Snacks
SYNONYMS:	Azorubine/ E122/ C.I. 14720 (Food Red 3)/ Disodium 4-hydroxy-3-(4-sulfonato-1-naphthylazo) naphthalene-1-sulfonate.
FORMULA:	$C_{20}H_{12}N_2Na_2O_7S_2$
MOLECULAR MASS:	502.44
ALTERNATIVE FORMS:	Calcium salt/ Potassium salt/ Aluminium lake
PROPERTIES AND APPEARANCE:	Red to maroon powder or granules
BOILING POINT IN °C AT VARIOUS PRESSURES (INCLUDING 760 mm Hg):	N/A
MELTING RANGE IN °C:	N/A
FLASH POINT IN °C:	N/A
IONISATION CONSTANT AT 25°C:	N/A
DENSITY AT 20°C (AND OTHER TEMPERATURES) IN g/l:	N/A
HEAT OF COMBUSTION AT 25°C IN J/kg:	N/A
VAPOUR PRESSURE AT VARIOUS TEMPERATURES IN mm Hg:	N/A

PURITY %:	Not less than 85% dye content calculated as the sodium salt							
WATER CONTENT MAXIMUM IN %:	15							
HEAVY METAL CONTENT MAXIMUM IN ppm:	40							
ARSENIC CONTENT MAXIMUM IN ppm:	3							
ASH MAXIMUM IN %:	N/A							
SOLUBILITY % AT VARIOUS TEMPERATURE/pH COMBINATIONS:								
in water:	@ **20°C**	8%	@ **50°C**	8%	@ **100°C**	8%		
in vegetable oil:	@ **20°C**	Insoluble	@ **50°C**	Insoluble	@ **100°C**	Insoluble		
in sucrose solution:	@ **10%**	8%	@ **40%**	8%	@ **60%**	8%		
in sodium chloride solution:	@ **5%**	Soluble	@ **10%**	Insoluble	@ **15%**	Insoluble		
in ethanol solution:	@ **5%**	Soluble	@ **20%**	Slightly soluble	@ **95%**	<0.1%	@ **100%**	<0.1%
in propylene glycol:	@ **20°C**	1%	@ **50°C**	1%	@ **100°C**	2%		
FUNCTION IN FOODS:	Water-soluble red food colour or insoluble red food colour as the aluminium lake							
ALTERNATIVES:	Other red food colours, although not of exactly the same hue: Carmine; Ponceau 4R; Allura red; Erythrosine; Beet red; Anthocyanins (in acidic media); Red 2G (where permitted)							
TECHNOLOGY OF USE IN FOODS:	A robust red colour with good light stability. Fairly stable in use to 105°C. Fades in the presence of sulphur dioxide, and alkaline media. Will fade with heat.							
SYNERGISTS:	None known							
ANTAGONISTS:	Sulphur dioxide/ Alkaline media							
FOOD SAFETY ISSUES:	None known							

LEGISLATION:

USA:
Not permitted

CANADA:
Not permitted

UK and EUROPE:
Permitted according to European Parliament and Council Directive 94/36/EC of June 30th 1994 on Colours for Use in Foodstuffs.
Refer to Annex V Part 2 for list of permitted food categories and maximum levels

AUSTRALIA:
Permitted, but restrictions apply

REFERENCE:

Smith, J. (Ed.) (1991) *Food Additive User's Handbook*. Blackie Academic and Professional.

STRUCTURE FOR CARMOISINE:

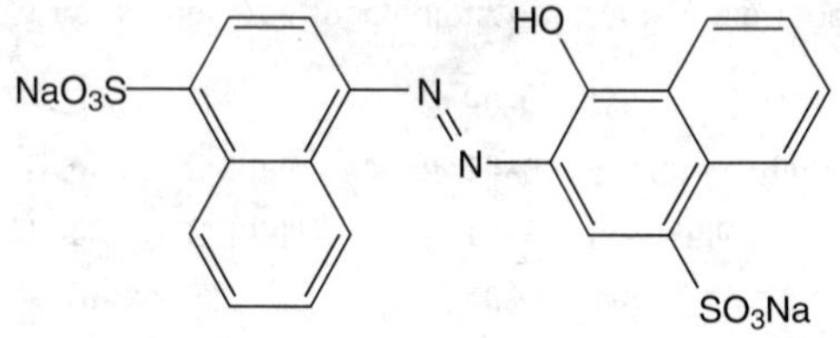

NAME:	Beta-carotene
CATEGORY:	Food colour
FOOD USE:	Baked goods/ Cereal and cereal products/ Dairy products/ Fish and seafood products/ Meat and poultry products/ Egg products/ Fruit, vegetable and nut products/ Soft drinks/ Sugars and preserves/ Confectionery/ Vinegar, pickles and sauces/ Decorations and coatings/ Edible ices/ Desserts/ Seasonings
SYNONYMS:	E160a (ii)/ C.I. 40800 Food Orange 5
FORMULA:	C40 H56
MOLECULAR MASS:	536.88
ALTERNATIVE FORMS:	*Cis* and *trans* forms
PROPERTIES AND APPEARANCE:	Red to brownish-red crystals or crystalline powder
BOILING POINT IN °C AT VARIOUS PRESSURES (INCLUDING 760 mm Hg):	Not applicable
MELTING RANGE IN °C:	176–182
FLASH POINT IN °C:	N/A
IONISATION CONSTANT AT 25°C:	N/A
DENSITY AT 20°C (AND OTHER TEMPERATURES) IN g/l:	Not applicable
HEAT OF COMBUSTION AT 25°C IN J/kg:	N/A
VAPOUR PRESSURE AT VARIOUS TEMPERATURES IN mm Hg:	Not applicable
PURITY %:	$\nless 96$

WATER CONTENT MAXIMUM IN %:	2							
HEAVY METAL CONTENT MAXIMUM IN ppm:	40							
ARSENIC CONTENT MAXIMUM IN ppm:	3							
ASH MAXIMUM IN %:	0.2							
SOLUBILITY % AT VARIOUS TEMPERATURE/pH COMBINATIONS:								
in water:	**@ 20°C**	Insoluble	**@ 50°C**	Insoluble	**@ 100°C**	Insoluble		
in vegetable oil:	**@ 20°C**	0.05%	**@ 50°C**	0.10%	**@ 100°C**	0.25%		
in sucrose solution:	**@ 10%**	Insoluble	**@ 40%**	Insoluble	**@ 60%**	Insoluble		
in sodium chloride solution:	**@ 5%**	Insoluble	**@ 10%**	Insoluble	**@ 15%**	Insoluble		
in ethanol solution:	**@ 5%**	Insoluble	**@ 20%**	Insoluble	**@ 95%**	<0.01%	**@ 100%**	<0.01%
in propylene glycol:	**@ 20°C**	Insoluble	**@ 50°C**	Insoluble	**@ 100°C**	Insoluble		
FUNCTION IN FOODS:	A bright yellow to orange food colour							
ALTERNATIVES:	Other yellow-orange food colours, although not of exactly the same hue, include: Mixed carotenes; Annatto; Carminic acid; Sunset yellow; Paprika							
TECHNOLOGY OF USE IN FOODS:	For oil-based applications (e.g. yellow fats) a suspension in oil can be applied directly or made into a stock solution in vegetable oil at 45–50°C. For aqueous food products water-dispersible forms are attained, generally with a food-grade emulsifier. Typical water-dispersible forms contain 1.5 or 10% beta-carotene, and in powdered form need to be slowly added to water at 40°C (approximately 1 part to 15 parts of water).							
SYNERGISTS:	Antioxidants/ Emulsifiers							
ANTAGONISTS:	Sulphur dioxide/ Oxygen/ metal ions							
FOOD SAFETY ISSUES:	None known							

LEGISLATION:

USA:
Permitted as a non-certified food colour according to 21 CFR 73.95

UK and EUROPE:
Permitted according to European Parliament and Council Directive 94/36/EC of June 30th 1994 on Colours for Use in Foodstuffs. Refer to Annex V, Part I for restrictions

CANADA:
Permitted

AUSTRALIA:
Permitted

OTHER COUNTRIES:
Generally permitted in most countries

REFERENCES:

Smith, J. (Ed.) (1991) *Food Additive User's Handbook.* Blackie Academic & Professional.
Killeit, U. (1991) Beta-carotene – more than a colorant. *Food marketing and technology*, **November**, 23–27.

STRUCTURE OF β-CAROTENE:

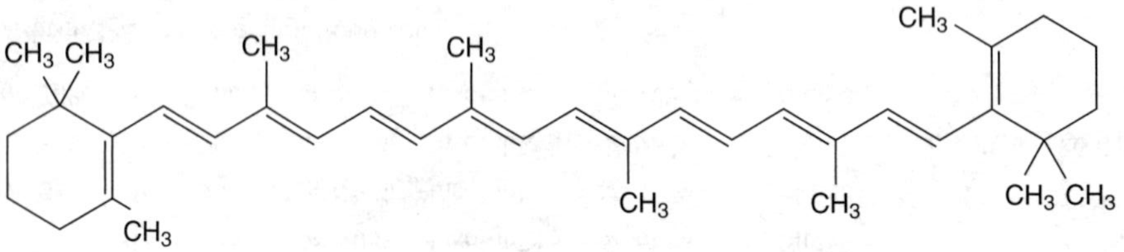

$C_{40}H_{56}$

NAME:	**Mixed carotenes**
CATEGORY:	Food colour
FOOD USE:	Baked goods/ Cereal and cereal products/ Dairy products/ Fish and seafood products/ Meat and poultry products/ Egg products/ Fruit, vegetable and nut products/ Soft drinks/ Sugars and preserves/ Confectionery/ Vinegar, pickles and sauces/ Decorations and coatings/ Edible ices/ Desserts/ Seasonings
SYNONYMS:	E160a(i)/ C.I. Food Orange 5/ C.I. No. 75130
FORMULA:	C40 H56
MOLECULAR MASS:	536.88
ALTERNATIVE FORMS:	*Cis* and *trans* forms/ contains α,β,γ-carotene
PROPERTIES AND APPEARANCE:	Red to orange-red crystals or suspension
BOILING POINT IN °C AT VARIOUS PRESSURES (INCLUDING 760 mm Hg):	N/A
MELTING RANGE IN °C:	176–182
FLASH POINT IN °C:	N/A
IONISATION CONSTANT AT 25°C:	N/A
DENSITY AT 20°C (AND OTHER TEMPERATURES) IN g/l:	N/A
HEAT OF COMBUSTION AT 25°C IN J/kg:	N/A
VAPOUR PRESSURE AT VARIOUS TEMPERATURES IN mm Hg:	N/A
PURITY %:	$\nless 5$

WATER CONTENT MAXIMUM IN %:	2							
HEAVY METAL CONTENT MAXIMUM IN ppm:	40							
ARSENIC CONTENT MAXIMUM IN ppm:	3							
ASH MAXIMUM IN %:	0.2							
SOLUBILITY % AT VARIOUS TEMPERATURE/pH COMBINATIONS:								
in water:	@ **20°C**	Insoluble	@ **50°C**	Insoluble	@ **100°C**	Insoluble		
in vegetable oil:	@ **20°C**	0.05%	@ **50°C**	0.10%	@ **100°C**	0.25%		
in sucrose solution:	@ **10%**	Insoluble	@ **40%**	Insoluble	@ **60%**	Insoluble		
in sodium chloride solution:	@ **5%**	Insoluble	@ **10%**	Insoluble	@ **15%**	Insoluble		
in ethanol solution:	@ **5%**	Insoluble	@ **20%**	Insoluble	@ **95%**	<0.01%	@ **100%**	<0.01%
in propylene glycol:	@ **20°C**	Insoluble	@ **50°C**	Insoluble	@ **100°C**	Insoluble		

FUNCTION IN FOODS: A bright-yellow to orange food colour

ALTERNATIVES: Other yellow-orange food colours, although not of exactly the same hue, include: Beta-carotene; Annatto; Carminic acid; Sunset yellow; Paprika; Crocin; Tartrazine; Safflower

TECHNOLOGY OF USE IN FOODS: For oil-based applications (e.g. yellow fats), a suspension in oil can be applied directly.

For aqueous food products, water-dispersible forms are attained, generally using a food-grade emulsifier.

SYNERGISTS: Antioxidants/ Emulsifiers

ANTAGONISTS: Sulphur dioxide/ Oxygen/ Metal ions

FOOD SAFETY ISSUES: None known

LEGISLATION:	**USA:** Not permitted	**UK and EUROPE:** Permitted according to European Parliament and Council Directive 94/36/EC of June 30th 1994 on Colours for Use in Foodstuffs. Refer to Annex V, Part 1 for restrictions.	**CANADA:** Not permitted **AUSTRALIA:** Permitted

REFERENCES:

Smith, J. (Ed.) (1991) *Food Additive User's Handbook*. Blackie Academic & Professional.

Tan, B. (1989) Palm carotenoids, tocopherols and tocotrienols. *Journal of the American Oil Chemists Society*, **66**(6), 770–776.

NAME:	Chlorophyll
CATEGORY:	Food colour
FOOD USE:	Dairy products/ Soft drinks/ Alcoholic drinks/ Sugars and preserves/ Confectionery/ Vinegars, pickles and sauces/ Decorations and coatings/ Edible ices/ Desserts
SYNONYMS:	E140(i)/ CI75810/ Natural Green 3/ Magnesium chlorophyll/ Magnesium phaeophytin
FORMULA:	Chlorophyll a $C_{55}H_{72}MgN_4O_5$; Chlorophyll b $C_{55}H_{70}MgN_4O_6$
MOLECULAR MASS:	Chlorophyll a 893.51; Chlorophyll b 907.49
ALTERNATIVE FORMS:	N/A
PROPERTIES AND APPEARANCE:	Waxy solid ranging in colour from olive green to dark green depending on the content of co-ordinated magnesium
BOILING POINT IN °C AT VARIOUS PRESSURES (INCLUDING 760 mm Hg):	N/A
MELTING RANGE IN °C:	Chlorophyll a 117–120; Chlorophyll b 120–130
FLASH POINT IN °C:	N/A
IONISATION CONSTANT AT 25°C:	N/A
DENSITY AT 20°C (AND OTHER TEMPERATURES) IN g/l:	N/A
HEAT OF COMBUSTION AT 25°C IN J/kg:	N/A
VAPOUR PRESSURE AT VARIOUS TEMPERATURES IN mm Hg:	N/A
PURITY %:	$\nless 10$

WATER CONTENT MAXIMUM IN %:	2					
HEAVY METAL CONTENT MAXIMUM IN ppm:	40					
ARSENIC CONTENT MAXIMUM IN ppm:	3					
ASH MAXIMUM IN %:	2					
SOLUBILITY % AT VARIOUS TEMPERATURE/pH COMBINATIONS:						
in water:	**@ 20°C**	Insoluble	**@ 50°C**	Insoluble	**@ 100°C**	Insoluble
in vegetable oil:	**@ 20°C**	Completely soluble	**@ 50°C**	Completely soluble	**@ 100°C**	Completely soluble
in sucrose solution:	**@ 10%**	Insoluble	**@ 40%**	Insoluble	**@ 60%**	Insoluble
in sodium chloride solution:	**@ 5%**	Insoluble	**@ 10%**	Insoluble	**@ 15%**	Insoluble
in ethanol solution:	**@ 5%**	Insoluble	**@ 20%**	Slightly soluble	**@ 95%**	5%
	@ 100%	5%				
in propylene glycol:	**@ 20°C**	Insoluble	**@ 50°C**	Insoluble	**@ 100°C**	Insoluble

FUNCTION IN FOODS: An oil-soluble olive-green food colour; can be rendered water-dispersible by use of a suitable emulsifier.

ALTERNATIVES: Other green colours, although not of the same hue (and not necessarily oil-soluble) include: Green S E142; Copper chlorophyll(in)

TECHNOLOGY OF USE IN FOODS: Fully oil-soluble, so can be used in concentrated or dilution in oil for direct addition to oil-based foodstuffs.

Exposure to meat (>65°C) – particularly if prolonged – will denature the pigment, with the colour becoming progressively more yellow.

For water-based foodstuffs, a water-dispersible form based on a suitable emulsifier can be used. Again, sensitivity to heat and oxidation will limit potential.

SYNERGISTS: Antioxidants/ Copper (but this forms the copper complex, E141)

ANTAGONISTS: Sulphur dioxide/ Metal ions

FOOD SAFETY ISSUES: None known

LEGISLATION:

USA:
Not permitted unless as a vegetable juice
21 CFR 73.260

UK and EUROPE:
Permitted according to European Parliament and Council Directive 94/36/EC of June 30th 1994 on Colours for Use in Foodstuffs.
Refer to Annex V, Part 1 for restrictions.

CANADA:
Permitted

AUSTRALIA:
Permitted

OTHER COUNTRIES:
Generally permitted in most countries of the world

REFERENCE:

Smith, J. (Ed.) (1991) *Food Additive User's Handbook*. Blackie Academic & Professional.

STRUCTURE OF CHLOROPHYLL:

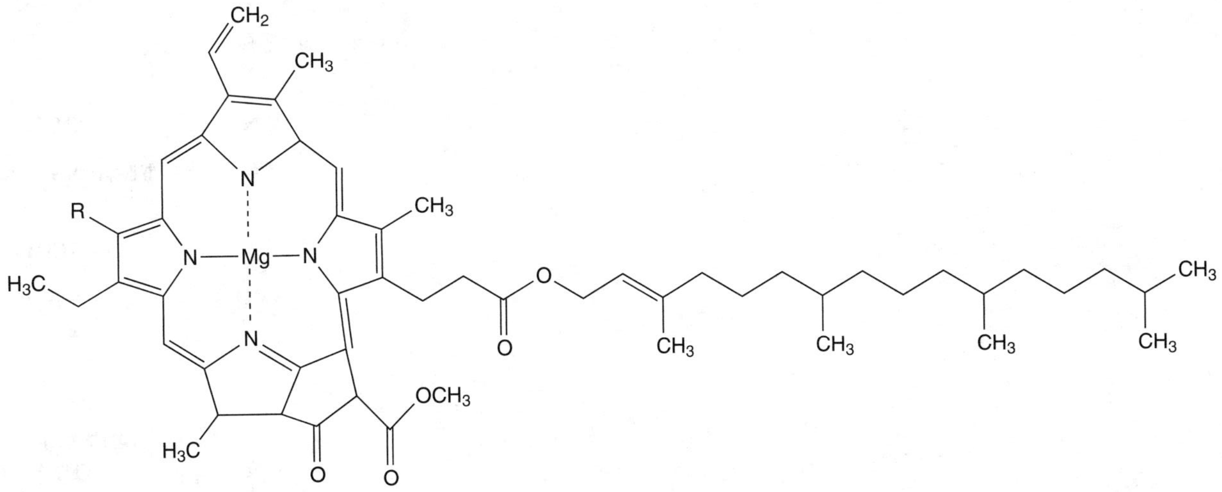

Chlorophyll a $R = CH_3$

Chlorophyll b $R = CHO$

NAME:	Chlorophyllin
CATEGORY:	Food colour
FOOD USE:	Dairy products/ Cereals/ Soft drinks/ Sugars and preserves/ Confectionery/ Vinegars, pickles and sauces/ Decorations and coatings/ Edible ices/ Desserts
SYNONYMS:	E140 (ii)/ C.I. 75815/ C.I. Natural Green 5
FORMULA:	Chlorophyllin a $C_{34}H_{34}N_4O_5$; Chlorophyllin b $C_{34}H_{32}N_4O_6$
MOLECULAR MASS:	Chlorophyllin a 578.68; Chlorophyllin b 592.66
ALTERNATIVE FORMS:	Sodium chlorophyllin/ Potassium chlorophyllin
PROPERTIES AND APPEARANCE:	Dark green to blue/black powder
BOILING POINT IN °C AT VARIOUS PRESSURES (INCLUDING 760 mm Hg):	N/A
MELTING RANGE IN °C:	Not known
FLASH POINT IN °C:	N/A
IONISATION CONSTANT AT 25°C:	N/A
DENSITY AT 25°C (AND OTHER TEMPERATURES) IN g/l:	N/A
HEAT OF COMBUSTION AT 25°C IN J/kg:	N/A
VAPOUR PRESSURE AT VARIOUS TEMPERATURES IN mm Hg:	N/A
PURITY %:	$\nless 95$ total chlorophyllins
WATER CONTENT MAXIMUM IN %:	5

HEAVY METAL CONTENT MAXIMUM IN ppm:	40					
ARSENIC CONTENT MAXIMUM IN ppm:	3					
ASH MAXIMUM IN %:	2					
SOLUBILITY % AT VARIOUS TEMPERATURE/pH COMBINATIONS:						
in water:	**@ 20°C**	Completely soluble	**@ 50°C**	Completely soluble	**@ 100°C**	Completely soluble
in vegetable oil:	**@ 20°C**	Insoluble	**@ 50°C**	Insoluble	**@ 100°C**	Insoluble
in sucrose solution:	**@ 10%**	Completely soluble	**@ 40%**	Completely soluble	**@ 60%**	Completely soluble
in sodium chloride solution:	**@ 5%**	Completely soluble	**@ 10%**	Soluble	**@ 15%**	Soluble
in ethanol solution:	**@ 5%**	Completely soluble	**@ 20%**	Completely soluble	**@ 95%**	Slightly soluble
	@ 100%	Slightly soluble				
in propylene glycol:	**@ 20°C**	Insoluble	**@ 50°C**	Insoluble	**@ 100°C**	Insoluble
FUNCTION IN FOODS:	A water-soluble green to olive-green food colour					
ALTERNATIVES:	Other green colours, although not of the same hue, include: Green S E142; Copper chlorophyll(in); Chlorophyll					
TECHNOLOGY OF USE IN FOODS:	Readily water-soluble, since in powdered form a predilution is helpful. Stability to heat and oxidation is relatively poor, so may only be suitable in certain applications.					
SYNERGISTS:	Antioxidants/ Copper (however, the copper complex is formed)					
ANTAGONISTS:	Sulphur dioxide/ Metal ions					
FOOD SAFETY ISSUES:	None known					

LEGISLATION:

USA:	UK and EUROPE:	CANADA:	AUSTRALIA:
Not permitted	Permitted according to European Parliament and Council Directive 94/36/EC of June 30th 1994 on Colours for Use in Foodstuffs. Refer to Annex V, Part 1 for restrictions.	Permitted	Permitted

REFERENCE: Smith, J. (Ed.) (1991) *Food Additive User's Handbook*. Blackie Academic & Professional.

STRUCTURE OF CHLOROPHYLLIN:

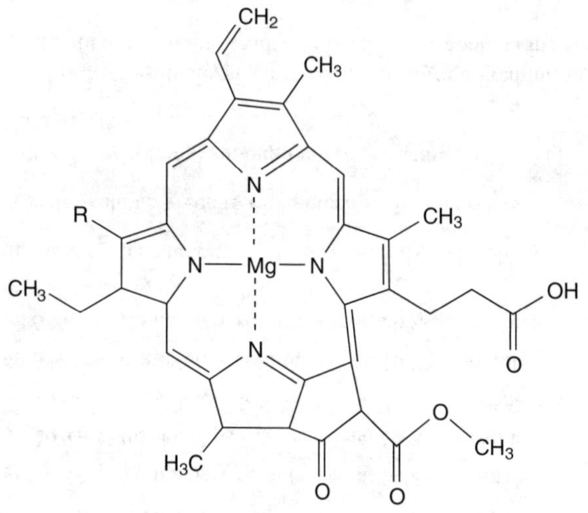

Chlorophyllin a R = CH_3

Chlorophyllin b R = CHO

NAME:	Copper chlorophyll
CATEGORY:	Food colour
FOOD USE:	Dairy products/ Soft drinks/ Alcoholic drinks/ Sugars and preserves/ Confectionery/ Vinegars, pickles and sauces/ Decorations and coatings/ Edible ices/ desserts
SYNONYMS:	E141 (i)/ C.I. 75815/ C.I. Natural Green 3/ Copper phaeophytin
FORMULA:	Copper chlorophyll a $C_{55}H_{72}CuN_4O_5$; Copper chlorophyll b $C_{55}H_{70}CuN_4O_6$
MOLECULAR MASS:	Copper chlorophyll a 932.75; Copper chlorophyll b 946.73
ALTERNATIVE FORMS:	N/A
PROPERTIES AND APPEARANCE:	Waxy solid ranging in colour from blue green to dark green.
BOILING POINT IN °C AT VARIOUS PRESSURES (INCLUDING 760 mm Hg):	N/A
MELTING RANGE IN °C:	Not known
FLASH POINT IN °C:	N/A
IONISATION CONSTANT AT 25°C:	N/A
DENSITY AT 20°C (AND OTHER TEMPERATURES) IN g/l:	N/A
HEAT OF COMBUSTION AT 25°C IN J/kg:	N/A
VAPOUR PRESSURE AT VARIOUS TEMPERATURES IN mm Hg:	N/A
PURITY %:	$\not<$10% total copper chlorophylls
WATER CONTENT MAXIMUM IN %:	2

HEAVY METAL CONTENT MAXIMUM IN ppm:	40					
ARSENIC CONTENT MAXIMUM IN ppm:	3					
ASH MAXIMUM IN %:	2					
SOLUBILITY % AT VARIOUS TEMPERATURE/pH COMBINATIONS:						
in water:	**@ 20°C**	Insoluble	**@ 50°C**	Insoluble	**@ 100°C**	Insoluble
in vegetable oil:	**@ 20°C**	Completely soluble	**@ 50°C**	Completely soluble	**@ 100°C**	Completely soluble
in sucrose solution:	**@ 10%**	Insoluble	**@ 40%**	Insoluble	**@ 60%**	Insoluble
in sodium chloride solution:	**@ 5%**	Insoluble	**@ 10%**	Insoluble	**@ 15%**	Insoluble
in ethanol solution:	**@ 5%**	Insoluble	**@ 20%**	Slightly soluble	**@ 95%**	5%
	@ 100%	5%				
in propylene glycol:	**@ 20°C**	Insoluble	**@ 50°C**	Insoluble	**@ 100°C**	Insoluble
FUNCTION IN FOODS:	An oil-soluble blue-green food colour; can be rendered water-dispersible by use of a suitable emulsifier					
ALTERNATIVES:	Other green colours, although not of time same hue (and not necessarily oil-soluble), include: Green S E142; Copper chlorophyllin; Chlorophyll					
TECHNOLOGY OF USE IN FOODS:	Fully oil-soluble, so predilute or direct addition to oil-based foodstuffs. Heat stability is far superior to that of the uncoppered form, as is stability to oxidation. Water-dispersible forms based on a suitable emulsifier are used for aqueous-based foodstuffs.					
SYNERGISTS:	None known					
ANTAGONISTS:	Sulphur dioxide					
FOOD SAFETY ISSUES:	None known; however, copper chlorophyll is not considered natural since the copper addition is performed artificially					

	USA:	UK and EUROPE:	CANADA:	AUSTRALIA:
LEGISLATION:	Not permitted	Permitted according to European Parliament and Council Directive 94/36/EC of June 30th 1994 on Colours for Use in Foodstuffs. Refer to Annex V, Part 1 for restrictions	Not Permitted	Permitted

REFERENCE:

Smith, J. (Ed.) (1991) *Food Additive User's Handbook.* Blackie Academic and Professional.

NAME:	Copper chlorophyllin
CATEGORY:	Food colour
FOOD USE:	Dairy products/ Cereals/ Soft drinks/ Sugars and preserves/ Confectionery/ Vinegars, pickles and sauces/ Decorations and coatings/ Edible ices/ Desserts
SYNONYMS:	E141 (ii)/ C.I. 75815/ C.I. Natural Green 5
FORMULA:	Copper chlorophyllin a $C_{34}H_{32}CuN_4O_5$; Copper chlorophyllin b $C_{34}H_{30}CuN_4O_5$
MOLECULAR MASS:	Copper chlorophyllin a 640.20; Copper chlorophyllin b 654.18
ALTERNATIVE FORMS:	Sodium copper chlorophyllin/ Potassium copper chlorophyllin
PROPERTIES AND APPEARANCE:	Dark green to blue/black powder
BOILING POINT IN °C AT VARIOUS PRESSURES (INCLUDING 760 mm Hg):	N/A
MELTING RANGE IN °C:	Not known
FLASH POINT IN °C:	N/A
IONISATION CONSTANT AT 25°C:	N/A
DENSITY AT 20°C (AND OTHER TEMPERATURES) IN g/l:	N/A
HEAT OF COMBUSTION AT 25°C IN J/kg:	N/A
VAPOUR PRESSURE AT VARIOUS TEMPERATURES IN mm Hg:	N/A
PURITY %:	$\nless$95% copper chlorophyllin
WATER CONTENT MAXIMUM IN %:	5

HEAVY METAL CONTENT MAXIMUM IN ppm:	40					
ARSENIC CONTENT MAXIMUM IN ppm:	3					
ASH MAXIMUM IN %:	2					
SOLUBILITY % AT VARIOUS TEMPERATURE/pH COMBINATIONS:						
in water:	**@ 20°C**	Completely soluble	**@ 50°C**	Completely soluble	**@ 100°C**	Completely soluble
in vegetable oil:	**@ 20°C**	Insoluble	**@ 50°C**	Insoluble	**@ 100°C**	Insoluble
in sucrose solution:	**@ 10%**	Completely soluble	**@ 40%**	Completely soluble	**@ 60%**	Completely soluble
in sodium chloride solution:	**@ 5%**	Completely soluble	**@ 10%**	Soluble	**@ 15%**	Soluble
in ethanol solution:	**@ 5%**	Completely soluble	**@ 20%**	Completely soluble	**@ 95%**	Slightly soluble
	@ 100%	Slightly soluble				
in propylene glycol:	**@ 20°C**	Insoluble	**@ 50°C**	Insoluble	**@ 100°C**	Insoluble

FUNCTION IN FOODS: A water-soluble blue-green food colour

ALTERNATIVES: Other green colours, although not of the same hue, include: Green S E142; Chlorophyllin; Copper chlorophyll; Chlorophyll

TECHNOLOGY OF USE IN FOODS: Readily water soluble, since in powdered form a predilution is helpful.

Stability to heat and oxidation is far superior to that of the uncoppered chlorophyllin.

SYNERGISTS: None known

ANTAGONISTS: Sulphur dioxide/ Metal ions

FOOD SAFETY ISSUES: None known; however, copper chlorophyllin is not considered natural since the copper addition is performed artificially

LEGISLATION:	**USA:** Not permitted	**UK and EUROPE:** Permitted according to European Parliament and Council Directive 94/36/EC of June 30th 1994 on Colours for Use in Foodstuffs. Refer to Annex V, Part 1 for restrictions	**CANADA:** Not permitted	**AUSTRALIA:** Permitted

REFERENCE:

Smith, J. (Ed.) (1991) *Food Additive User's Handbook.* Blackie Academic and Professional.

NAME:	Crocin
CATEGORY:	Food colour
FOOD USE:	Baked goods/ Dairy products/ Egg products/ Soft drinks/ Sugars and preserves/ Confectionery/ Vinegar, pickles and sauces/ Decorations and coatings/ Desserts/ Seasonings/ Flour confectionery/ Edible ices/ Fish and seafood products/ Cereal and cereal products
SYNONYMS:	Gardenia extract/ Saffron extract
FORMULA:	$C_{44}H_{64}O_{24}$
MOLECULAR MASS:	976.90
ALTERNATIVE FORMS:	Mono and di-glycosides of crocetin
PROPERTIES AND APPEARANCE:	A dark-brown/ orange liquid
BOILING POINT IN °C AT VARIOUS PRESSURES (INCLUDING 760 mm Hg):	N/A
MELTING RANGE IN °C:	Not known
FLASH POINT IN °C:	N/A
IONISATION CONSTANT AT 25°C:	N/A
DENSITY AT 20°C (AND OTHER TEMPERATURES) IN g/l:	N/A
HEAT OF COMBUSTION AT 25°C IN J/kg:	N/A
VAPOUR PRESSURE AT VARIOUS TEMPERATURES IN mm Hg:	N/A
PURITY %:	Typically contains 10% pigment

WATER CONTENT MAXIMUM IN %:	N/A					
HEAVY METAL CONTENT MAXIMUM IN ppm:	20					
ARSENIC CONTENT MAXIMUM IN ppm:	3					
ASH MAXIMUM IN %:	2					
SOLUBILITY % AT VARIOUS TEMPERATURE/pH COMBINATIONS:						
in water:	@ **20°C**	Completely soluble	@ **50°C**	Completely soluble	@ **100°C**	Completely soluble
in vegetable oil:	@ **20°C**	Insoluble	@ **50°C**	Insoluble	@ **100°C**	Insoluble
in sucrose solution:	@ **10%**	Completely soluble	@ **40%**	Completely soluble	@ **60%**	Completely soluble
in sodium chloride solution:	@ **5%**	Soluble	@ **10%**	Soluble	@ **15%**	Soluble
in ethanol solution:	@ **5%**	Completely soluble	@ **20%**	Completely soluble	@ **95%**	Soluble
	@ **100%**	Soluble				
in propylene glycol:	@ **20°C**	Soluble	@ **50°C**	Soluble	@ **100°C**	Soluble
FUNCTION IN FOODS:	A bright egg-yellow, water-soluble food colour					
ALTERNATIVES:	Other yellow colours, although not of exactly the same hue, include: Beta-carotene; Tartrazine; Sunset yellow; Turmeric/ curcumin; Safflower					
TECHNOLOGY OF USE IN FOODS:	Readily soluble in water. Can be applied directly into aqueous foodstuffs. Not suitable for oil-based products.					
SYNERGISTS:	Ascorbic acid/ Antioxidants					
ANTAGONISTS:	Sulphur dioxide/ Metal ions/ Oxygen					
FOOD SAFETY ISSUES:	None known					

LEGISLATION:

USA:
Not permitted

CANADA:
Permitted as saffron

UK and EUROPE:
Not permitted, although saffron extract is allowed as a spice/flavour

AUSTRALIA/PACIFIC RIM:
Australia: Permitted
Japan: Permitted for use as a natural colour

REFERENCES:

Timberlake, C. F., and Henry, B. S. (1986) Plant pigment as natural food colours. *Endeavour* (*new series*), **10**, 31–35.

Smith, J. (Ed.) (1991) *Food Additive User's Handbook*. Blackie Academic & Professional.

ANY OTHER RELEVANT INFORMATION:

Crocin was permitted in the UK according to the Colouring Matter in Foodstuff Regulations 1973 (as amended). However, the recently adopted (1996) EC directive on food colours does not list crocin as a permitted colour.

STRUCTURE OF CROCIN:

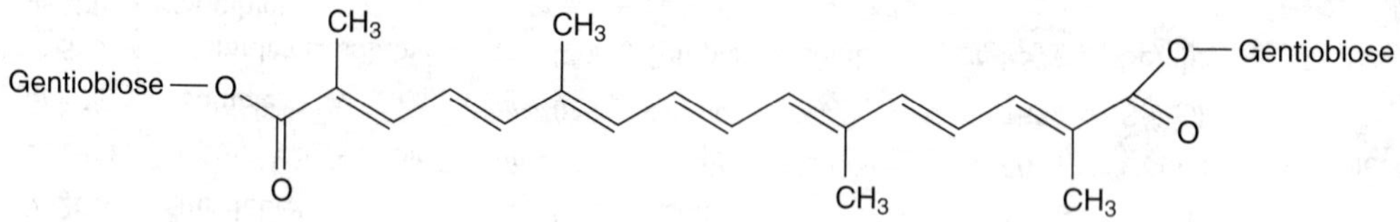

NAME:	Curcumin
CATEGORY:	Food colour
FOOD USE:	Baked goods/ Cereals and cereal products/ Dairy products/ Smoked fish/ Fish and seafood products/ Meat and poultry products/ Egg products/ Fruit, vegetable and nut products/ Canned soft drinks/ Snacks/ Soups/ Sugars and preserves/ Confectionery/ Vinegar, pickles and sauces/ Mustard/ Decorations and coatings/ Edible ices/ Desserts/ Seasonings
SYNONYMS:	Turmeric yellow/ Curcuma/ C.I. 75300 Natural Yellow 3/ Diferoyl methane/ (1,7-*bis*(4-hydroxy-3-methoxyphenyl)hepta-1,6-diene-3,5-dione/ E100
FORMULA:	$C_{21}H_{20}O_6$
MOLECULAR MASS:	368.39
ALTERNATIVE FORMS:	Demethoxy curcumin/ Bis-demethoxy curcumin/ Aluminium lake
PROPERTIES AND APPEARANCE:	Orange-yellow crystalline powder
BOILING POINT IN °C AT VARIOUS PRESSURES (INCLUDING 760 mm Hg):	Not applicable
MELTING RANGE IN °C:	179–182
FLASH POINT IN °C:	Not known
IONISATION CONSTANT AT 25°C:	N/A
DENSITY AT 20°C (AND OTHER TEMPERATURES) IN g/l:	Not applicable
HEAT OF COMBUSTION AT 25°C IN J/kg:	N/A
VAPOUR PRESSURE AT VARIOUS TEMPERATURES IN mm Hg:	N/A

PURITY %:	∢90% total colouring matters							
WATER CONTENT MAXIMUM IN %:	2							
HEAVY METAL CONTENT MAXIMUM IN ppm:	40							
ARSENIC CONTENT MAXIMUM IN ppm:	3							
ASH MAXIMUM IN %:	1							
SOLUBILITY % AT VARIOUS TEMPERATURE/pH COMBINATIONS:								
in water:	@ **20°C**	Insoluble	Readily soluble in alkali (but becomes orange (red colour))					
	@ **50°C**	Insoluble	Readily soluble in alkali (but becomes orange (red colour))					
	@ **100°C**	Insoluble	Readily soluble in alkali (but becomes orange (red colour))					
in vegetable oil:	@ **20°C**	<0.1%	@ **50°C**	0.1%	@ **100°C**	0.25%		
in sucrose solution:	@ **10%**	Insoluble	@ **40%**	Insoluble	@ **60%**	Insoluble		
in sodium chloride solution:	@ **5%**	Insoluble	@ **10%**	Insoluble	@ **15%**	Insoluble		
in ethanol solution:	@ **5%**	Insoluble	@ **20%**	Insoluble	@ **95%**	<0.1%	@ **100%**	0.1%
in propylene glycol:	@ **20°C**	Insoluble	@ **50°C**	<0.1%	@ **100°C**	0.7%		
FUNCTION IN FOODS:	A bright-yellow food colour, mainly suited to food applications requiring little or no light stability							
ALTERNATIVES:	Other yellow food colours, although not of exactly the same hue, include: Quinoline yellow; Riboflavin; Tartrazine; Crocin; Safflower; Beta-carotene							
TECHNOLOGY OF USE IN FOODS:	Curcumin is generally used in an emulsified solution, in which case it is water-dispersible. Contact with alkalis should be avoided. The greatest limitation on curcumin is its tendency to fade on exposure to light (photo-oxidation). One contradiction to this is in high boilings, where curcumin exhibits good light stability; this is due to the low water activity in these types of applications.							
SYNERGISTS:	A food-grade emulsifier is generally needed to render curcumin water-dispersible.							

ANTAGONISTS: Strong alkalis such as NaOH/ Sulphur dioxide/ Metal ions (e.g. Sn^{2+})/ Boric acid and/or borates.

FOOD SAFETY ISSUES: None known

LEGISLATION:

USA:	UK and EUROPE:	CANADA:
GRAS, permitted as non-certified food colour according to 21CFR 73.600	Permitted according to European Parliament and Council Directive 94/36/EC of June 30th 1994 on Colours for Use in Foodstuffs. Refer to Annex V, Part 2 for a substantial list of permitted food categories and maximum levels.	Permitted **AUSTRALIA:** Permitted **OTHER COUNTRIES:** Generally permitted in most countries

REFERENCE: Smith, J. (Ed.) (1991) *Food Additives Users Handbook*. Blackie Academic & Professional.

STRUCTURE OF CURCUMIN:

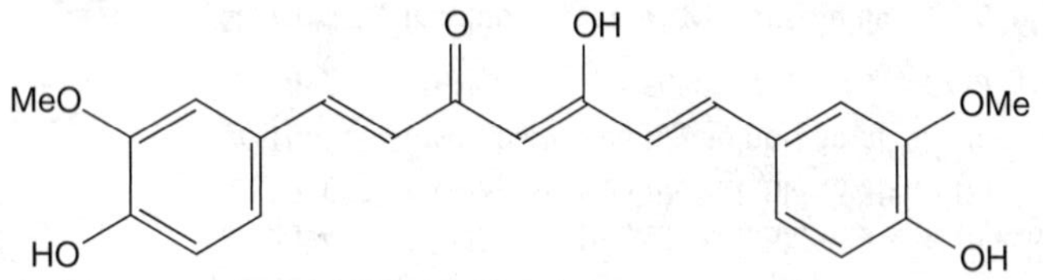

NAME:	**Erythrosine**
CATEGORY:	Food colour
FOOD USE:	Cocktail cherries/ Meat and poultry products/ Candied cherries/ Decorations and coatings/ Bigarreau cherries in syrup and in cocktails/ Canned products/ Confectionery
SYNONYMS:	E127/ C.I. 45430 (Food Red 14)/ FD & C Red No 3/ Disodium 2-(2,4,5,7-tetraiodo-3-oxido-6-oxoxanthen-9-yl)benzoate monohydrate/ Disodium 9(*O*-carboxy-phenyl)-6-hydroxy-2,4,5,7-tetraiodo-3*H*-xanthen-3-one
FORMULA:	$C_{20}H_6I_4Na_2O_5 \cdot H_2O$
MOLECULAR MASS:	897.88
ALTERNATIVE FORMS:	Calcium salt/ Potassium salt/ Aluminium lake (insoluble)
PROPERTIES AND APPEARANCE:	Bluish-red powder or granules
BOILING POINT IN °C AT VARIOUS PRESSURES (INCLUDING 760 mm Hg):	N/A
MELTING RANGE IN °C:	Not known
FLASH POINT IN °C:	N/A
IONISATION CONSTANT AT 25°C:	N/A
DENSITY AT 20°C (AND OTHER TEMPERATURES) IN g/l:	N/A
HEAT OF COMBUSTION AT 25°C IN J/kg:	N/A
VAPOUR PRESSURE AT VARIOUS TEMPERATURES IN mm Hg:	N/A
PURITY %:	Content not less than 87% total colouring matter, calculated as the anhydrous sodium salt

WATER CONTENT MAXIMUM IN %:	15							
HEAVY METAL CONTENT MAXIMUM IN ppm:	40							
ARSENIC CONTENT MAXIMUM IN ppm:	3							
ASH MAXIMUM IN %:	N/A							
SOLUBILITY % AT VARIOUS TEMPERATURE/pH COMBINATIONS:								
in water:	@ **20°C**	9% (pH 7)	@ **50°C**	17% (pH 7)	@ **100°C**	20% (pH 7)		
in vegetable oil:	@ **20°C**	Insoluble	@ **50°C**	Insoluble	@ **100°C**	Insoluble		
in sucrose solution:	@ **10%**	Soluble	@ **40%**	Soluble	@ **60%**	Soluble		
in sodium chloride solution:	@ **5%**	Slightly soluble	@ **10%**	Insoluble	@ **15%**	Insoluble		
in ethanol solution:	@ **5%**	@ **25°C** 10%	@ **20%**	@ **25°C** 8.0% @ **60°C** 10%	@ **95%**	Insoluble	@ **100%**	Insoluble
in propylene glycol:	@ **20°C**	20%	@ **50°C**	20%	@ **100°C**	20%		
FUNCTION IN FOODS:	Water-soluble red food colour, or insoluble red food colour as the aluminium lake.							
ALTERNATIVES:	Other red food colours, although not of exactly the same hue, include: Carmine; Anthocyanins (in acidic media); Carmoisine; Allura Red; Ponceau 4R; Beet Red; Red 2G (where permitted)							
TECHNOLOGY OF USE IN FOODS:	In solutions below pH 3–4, erythrosine forms erythrosinic acid, which is only slightly soluble. Used to colour cherries since the insoluble colour can be formed in the cherry and therefore the colour will not leach out to colour the syrup or other fruit. Poor light-stability. Precipitates in acid media. Faded by alkaline media.							
SYNERGISTS:	None known							
ANTAGONISTS:	Alkaline media/ Ascorbic acid (insoluble in 1%)/ Sulphur dioxide (insoluble in 25 ppm)/ Acid media – precipitates below pH 3–4.							

FOOD SAFETY ISSUES:	The colour contains iodine, and it has been suggested that iodine supplementation in the diet may be associated with an increased incidence of thyrotoxicosis. The use of this colour has been restricted.
LEGISLATION:	**USA:** FD&C Red No. 3 certified food colour **UK and EUROPE:** Permitted according to European Parliament and Council Directive 94/36/EC of 30th June 1994 on Colours for Use in Foodstuffs. Annex IV – only for: cocktail cherries and candied cherries max 200 mg/kg; bigarreau cherries in syrup and in cocktails max 150 mg/kg **CANADA:** Permitted when certified, but restrictions apply **AUSTRALIA:** Not permitted
REFERENCE:	Smith, J. (Ed.) (1991) *Food Additive User's Handbook*. Blackie Academic & Professional.
STRUCTURE OF ERYTHROSINE:	

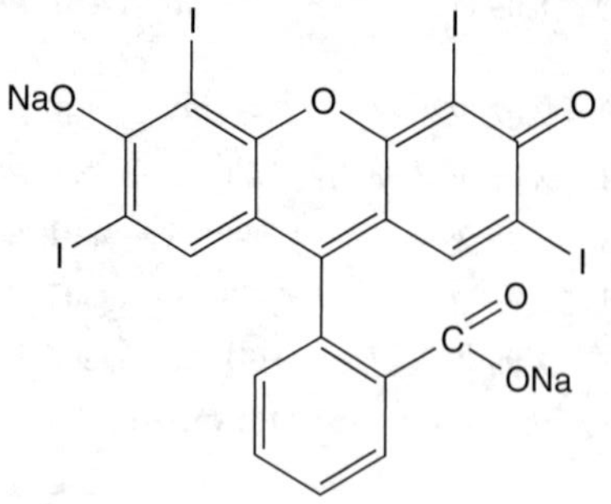

NAME:	Ethyl ester of beta-apo-8′-carotenoic acid (C30)
CATEGORY:	Food colour
FOOD USE:	Soft drinks/ Confectionery/ Processed cheese/ Sauces/ Margarine and oils/ Soups/ Salad dressings/ Ice-cream, edible ices/ Canned products/ Desserts
SYNONYMS:	E160f/ C.I. 40825 Food Orange 7/ Beta apo-8′-carotenoic ester/ Ethyl beta-apo-8′-carotenoate/ Beta-apo-8′-carotenoic acid ethyl ester/ Ethyl 8′-apo-beta-caroten-8′-oate
FORMULA:	C32 H44 O2
MOLECULAR MASS:	460.70
ALTERNATIVE FORMS:	Ethyl ester where permitted
PROPERTIES AND APPEARANCE:	Red to violet-red crystals or crystalline powder. Commercially available as a suspension of micronised crystals in vegetable oil
BOILING POINT IN °C AT VARIOUS PRESSURES (INCLUDING 760 mm Hg):	N/A
MELTING RANGE IN °C:	134–138
FLASH POINT IN °C:	N/A
IONISATION CONSTANT AT 25°C:	N/A
DENSITY AT 20°C (AND OTHER TEMPERATURES) IN g/l:	N/A
HEAT OF COMBUSTION AT 25°C IN J/kg:	N/A
VAPOUR PRESSURE AT VARIOUS TEMPERATURES IN mm Hg:	N/A

PURITY %:	Not less than 96% of total colouring matter					
WATER CONTENT MAXIMUM IN %:	N/A					
HEAVY METAL CONTENT MAXIMUM IN ppm:	40					
ARSENIC CONTENT MAXIMUM IN ppm:	3					
ASH MAXIMUM IN %:	N/A					
SOLUBILITY % AT VARIOUS TEMPERATURE/pH COMBINATIONS:						
in water:	**@ 20°C**	Insoluble	**@ 50°C**	Insoluble	**@ 100°C**	Insoluble
in vegetable oil:	**@ 20°C**	0.7–1%	**@ 50°C**	0.7–1%	**@ 100°C**	1–1.5%
in sucrose solution:	**@ 10%**	Insoluble	**@ 40%**	Insoluble	**@ 60%**	Insoluble
in sodium chloride solution:	**@ 5%**	Insoluble	**@ 10%**	Insoluble	**@ 15%**	Insoluble
in ethanol solution:	**@ 5%**	Insoluble	**@ 20%**	Insoluble	**@ 95%**	Slightly soluble
	@ 100%	Slightly soluble				
in propylene glycol:	**@ 20°C**	Insoluble	**@ 50°C**	Insoluble	**@ 100°C**	Insoluble
FUNCTION IN FOODS:	Yellow to orange food colour depending on concentration.					
ALTERNATIVES:	Other yellow to orange food colours, although not the same hue, include: Beta-carotene; Curcumin; Annatto; Tartrazine/ Sunset Yellow; Quinoline Yellow					
TECHNOLOGY OF USE IN FOODS:	Available as a microcrystalline suspension in vegetable oil. To prepare a stock solution to colour margarine or oils, heat oil to 45° to 50°C and stir in about 50 g per litre of 20% suspension. If mixed with citrus oils, heat to 70°C to disperse. Reasonably stable to heat and light; stable to pH change and acidic conditions; stable to ascorbic acid. Can oxidise, losing colour, and stabilisation with antioxidants such as tocopherols is recommended.					

When used to colour soft drink emulsions it is recommended that the product is homogenised to delay any separation of colour.

SYNERGISTS: Antioxidants (such as tocopherols)/ Ascorbic acid

ANTAGONISTS: Oxidising agents/ Peroxides

FOOD SAFETY ISSUES: Considered safe for food use.

LEGISLATION:

USA:	**UK and EUROPE:**	**CANADA:**
Not permitted	Permitted according to European Parliament and Council Directive 94/36/EC of June 30th 1994 on Colours for Use in Foodstuffs. Refer to Annex V, Part 2 for restrictions	Permitted, but restricted in use **AUSTRALIA:** Permitted

REFERENCE: Smith, J. (Ed.) (1991) *Food Additive User's Handbook*. Blackie Academic & Professional.

STRUCTURE OF ETHYL ESTER OF β-APO-8′-CAROTENOIC ACID:

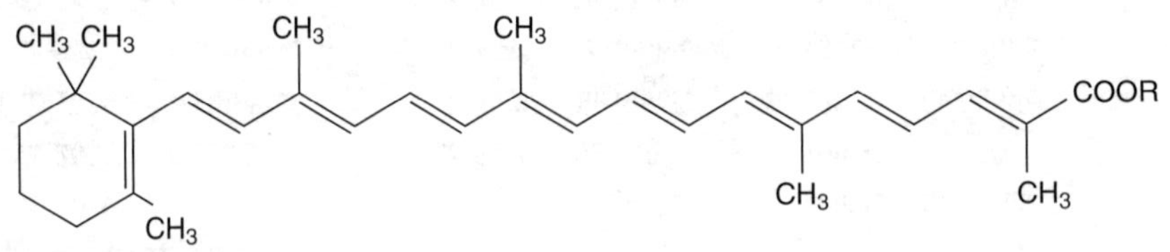

NAME:	Fast Green FCF
CATEGORY:	Food colour
FOOD USE:	Baked goods/ Fruit and vegetable and nut products/ Snacks/ Confectionery/ Cereal and cereal products/ Sauces/ Soft drinks/ Decorations and coatings/ Soups/ Desserts/ Dairy products/ Seasonings/ Canned products/ Edible ices/ Meat and poultry products
SYNONYMS:	FD&C Green No. 3/ C.I. 42053 Food Green 3/ *N*-ethyl-*N*-(4-(ethyl)((3-sulphophenyl)methyl)amino)phenyl)(4-hydroxy-2-sulphophenyl)methylene)-2,5-cyclohexadien-1-ylidene)-3-sulphobenzene methanaminium hydroxide, inner salt disodium salt
FORMULA:	C35 H30 N2 O10 S3 Na2
MOLECULAR MASS:	780.41
ALTERNATIVE FORMS:	Potassium salt/ Aluminium lake (insoluble)
PROPERTIES AND APPEARANCE:	Dark black/green granules or powder
BOILING POINT IN °C AT VARIOUS PRESSURES (INCLUDING 760 mm Hg):	N/A
MELTING RANGE IN °C:	N/A
FLASH POINT IN °C:	N/A
IONISATION CONSTANT AT 25°C:	N/A
DENSITY AT 20°C (AND OTHER TEMPERATURES) IN g/l:	N/A
HEAT OF COMBUSTION AT 25°C IN J/kg:	N/A
VAPOUR PRESSURE AT VARIOUS TEMPERATURES IN mm Hg:	N/A

PURITY %:	85% miniumum							
WATER CONTENT MAXIMUM IN %:	15							
HEAVY METAL CONTENT MAXIMUM IN ppm:	40							
ARSENIC CONTENT MAXIMUM IN ppm:	3							
ASH MAXIMUM IN %:	N/A							
SOLUBILITY % AT VARIOUS TEMPERATURE/pH COMBINATIONS:								
in water:	**@ 20°C**	20%	**@ 50°C**	20%	**@ 100°C**	20%		
in vegetable oil:	**@ 20°C**	Insoluble	**@ 50°C**	Insoluble	**@ 100°C**	Insoluble		
in sucrose solution:	**@ 10%**	20%	**@ 40%**	20%	**@ 60%**	20%		
in sodium chloride solution:	**@ 5%**	20%	**@ 10%**	10%	**@ 15%**	5%		
in ethanol solution:	**@ 5%**	20%	**@ 20%**	10%	**@ 95%**	0.02%	**@ 100%**	0.01%
in propylene glycol:	**@ 20°C**	20%	**@ 50°C**	20%	**@ 100°C**	Insoluble		
FUNCTION IN FOODS:	Water-soluble blue-green food colour, or insoluble blue-green food colour as the aluminium lake.							
ALTERNATIVES:	Other blue-green food colours, although not the same hue, include: Green S; Copper chlorophyllin							
TECHNOLOGY OF USE IN FOODS:	A fairly light- and heat-stable food colour. Stable to food acids and sulphur dioxide; turns bluer in alkaline media and fades. Sensitive to oxidation.							
SYNERGISTS:	None known							
ANTAGONISTS:	Alkaline media/ Ascorbic acid							
FOOD SAFETY ISSUES:	None known							

LEGISLATION:

USA: Permitted 21 CFR 74.203	**CANADA:** Permitted, but restrictions apply
UK and EUROPE: Not permitted	**AUSTRALIA:** Not permitted

REFERENCE:

Smith, J. (Ed.) (1991) *Food Additive User's Handbook.* Blackie Academic and Professional.

STRUCTURE OF FAST GREEN FCF:

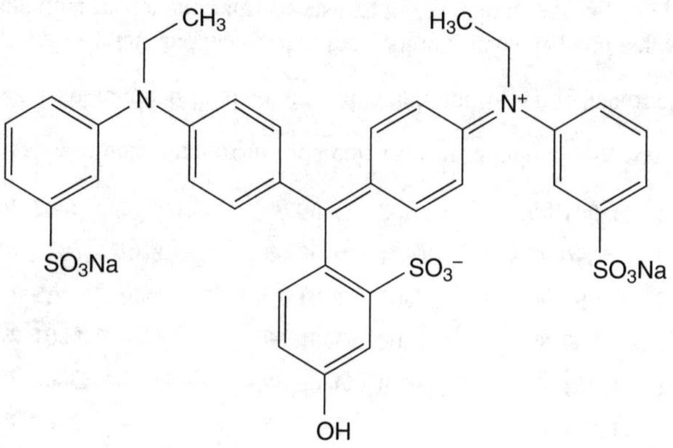

NAME:	Gold
CATEGORY:	food colour
FOOD USE:	Surface coating of confectionery/ Decoration of chocolates/ Liqueurs
SYNONYMS:	E175/ C.I. 77480 (Pigment Metal 3)/ Aurum
FORMULA:	Au
MOLECULAR MASS:	197.0
ALTERNATIVE FORMS:	N/A
PROPERTIES AND APPEARANCE:	Gold-coloured powder or thin sheets (leaf)
BOILING POINT IN °C AT VARIOUS PRESSURES (INCLUDING 760 mm Hg):	2610
MELTING RANGE IN °C:	1063
FLASH POINT IN °C:	N/A
IONISATION CONSTANT AT 25°C:	N/A
DENSITY AT 20°C (AND OTHER TEMPERATURES) IN g/l:	N/A
HEAT OF COMBUSTION AT 25°C IN J/kg:	N/A
VAPOUR PRESSURE AT VARIOUS TEMPERATURES IN mm Hg:	N/A
PURITY %:	Not less than 90%
WATER CONTENT MAXIMUM IN %:	N/A

HEAVY METAL CONTENT MAXIMUM IN ppm:	<40
ARSENIC CONTENT MAXIMUM IN ppm:	<3
ASH MAXIMUM IN %:	N/A

SOLUBILITY % AT VARIOUS TEMPERATURE/pH COMBINATIONS:

in water:	**@ 20°C**	Insoluble	**@ 50°C**	Insoluble	**@ 100°C**	Insoluble		
in vegetable oil:	**@ 20°C**	Insoluble	**@ 50°C**	Insoluble	**@ 100°C**	Insoluble		
in sucrose solution:	**@ 10%**	Insoluble	**@ 40%**	Insoluble	**@ 60%**	Insoluble		
in sodium chloride solution:	**@ 5%**	Insoluble	**@ 10%**	Insoluble	**@ 15%**	Insoluble		
in ethanol solution:	**@ 5%**	Insoluble	**@ 20%**	Insoluble	**@ 95%**	Insoluble	**@ 100%**	Insoluble
in propylene glycol:	**@ 20°C**	Insoluble	**@ 50°C**	Insoluble	**@ 100°C**	Insoluble		

FUNCTION IN FOODS:	Used as a dispersed powder or gold leaf to surface-colour sugar or chocolate confectionery, or as an additive to certain liqueurs.
ALTERNATIVES:	Aluminium powder pigmented with yellow/orange lake food colours and polished to a gloss after application.
TECHNOLOGY OF USE IN FOODS:	Applied as leaf or powder to the surface of hard sugar or chocolate confectionery, then polished to a high gloss as a decoration. Can be added as fine particles or leaf to certain liqueurs.
SYNERGISTS:	None known
ANTAGONISTS:	None known
FOOD SAFETY ISSUES:	None known

LEGISLATION:

USA:
Not permitted

CANADA:
Not permitted

UK and Europe:
Permitted according to European Parliament and Council Directive 94/36/EC of June 30th 1994 on Colours for Use in Foodstuffs.
Refer to Annex IV for restrictions on use.

AUSTRALIA:
Not permitted

REFERENCE:

Smith, J. (Ed.) (1991) *Food Additive User's Handbook*. Blackie Academic and Professional.

NAME:	Green S
CATEGORY:	Food colour
FOOD USE:	Baked goods/ Fruit and vegetable and nut products/ Edible ices/ Confectionery/ Cereal and cereal products/ Seasonings/ Soft drinks/ Decorations and coatings/ Soups/ Desserts/ Dairy products/ Snacks/ Canned products/ Meat and poultry products/ Sauces
SYNONYMS:	E142/ C.I. 44090 (Food Green 4)/ Brilliant Green BS/ Sodium *N*-[4-[[4-(dimethylamino)phenyl](2-hydroxy-3,6-disulfo-1-naphthalenyl)-methylene]2,5-cyclohexadien-1-ylidene]-*N*-methylmethanaminium/ Sodium 5-[4-dimethylamino-α-(4-dimethyliminocyclohexa-2,5-dienylidene)benzyl]-6-hydroxy-7-sulfonato-naphthalene-2-sulfonate
FORMULA:	$C_{27}H_{25}N_2NaO_7S_2$
MOLECULAR MASS:	576.63
ALTERNATIVE FORMS:	Calcium salt/ Potassium salt/ Aluminium lake (insoluble)
PROPERTIES AND APPEARANCE:	Dark blue or dark green powder or granules
BOILING POINT IN °C AT VARIOUS PRESSURES (INCLUDING 760 mm Hg):	N/A
MELTING RANGE IN °C:	Not known
FLASH POINT IN °C:	N/A
IONISATION CONSTANT AT 25°C:	N/A
DENSITY AT 20°C (AND OTHER TEMPERATURES) IN g/l:	N/A
HEAT OF COMBUSTION AT 25°C IN J/kg:	N/A
VAPOUR PRESSURE AT VARIOUS TEMPERATURES IN mm Hg:	N/A

PURITY %:	Not less than 80% total colouring matter calculated as the sodium salt
WATER CONTENT MAXIMUM IN %:	20
HEAVY METAL CONTENT MAXIMUM IN ppm:	40
ARSENIC CONTENT MAXIMUM IN ppm:	3
ASH MAXIMUM IN %:	N/A

SOLUBILITY % AT VARIOUS TEMPERATURE/pH COMBINATIONS:

in water:	**@ 20°C**	5%	**@ 50°C**	7%	**@ 100°C**	10%		
in vegetable oil:	**@ 20°C**	Insoluble	**@ 50°C**	Insoluble	**@ 100°C**	Insoluble		
in sucrose solution:	**@ 10%**	5%	**@ 40%**	5%	**@ 60%**	5%		
in sodium chloride solution:	**@ 5%**	Soluble	**@ 10%**	Insoluble	**@ 15%**	Insoluble		
in ethanol solution:	**@ 5%**	5%	**@ 20%**	5%	**@ 95%**	0.2%	**@ 100%**	0.2%
in propylene glycol:	**@ 20°C**	2%	**@ 50°C**	2%	**@ 100°C**	3%		

FUNCTION IN FOODS:	Water-soluble blue-green food colour, or insoluble blue-green food colour as the aluminium lake
ALTERNATIVES:	Other blue-green food colours, although not the same hue, include: Brilliant Blue FCF; Patent Blue V; Indigotine
TECHNOLOGY OF USE IN FOODS:	A heat-stable blue-green food colour useful as a component in green colour for canned peas and as a dulling agent in brown shades. The colour has moderate light stability, reasonable stability to acid conditions, and fair stability to alkaline conditions.
SYNERGISTS:	None known
ANTAGONISTS:	Sulphur dioxide/ Metal ions
FOOD SAFETY ISSUES:	None known

LEGISLATION:

USA:
Not permitted

CANADA:
Not permitted

UK and EUROPE:
Permitted according to European Parliament and Council Directive 94/36/EC of June 30th 1994 on Colours for Use in Foodstuffs.
Refer to Annex V, Part 2 for restrictions

AUSTRALIA:
Permitted, but restrictions apply

REFERENCE:

Smith, J. (Ed.) (1991) *Food Additive User's Handbook*. Blackie Academic and Professional.

STRUCTURE OF GREEN S:

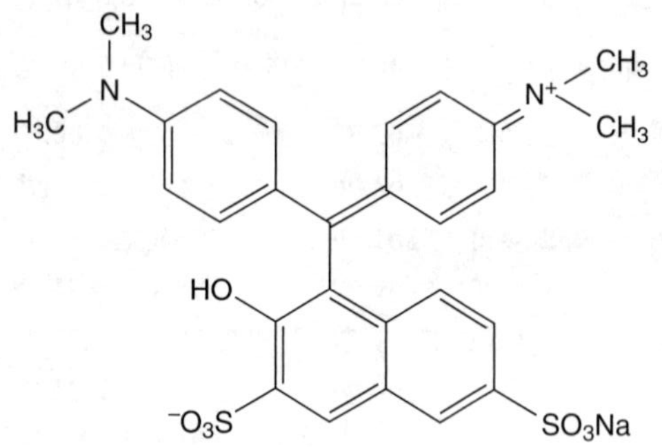

NAME:	Indigotine
CATEGORY:	Food colour
FOOD USE:	Confectionery/ Preserves/ Desserts
SYNONYMS:	Indigo Carmine/ E132/ FD&C Blue No. 2/ C.I. 73015 (Food Blue 1)/ Disodiom 3,3′-dioxo-2,2′-bis-indolylidene-5,5′-disulfonate
FORMULA:	$C_{16}H_8N_2Na_2O_8S_2$
MOLECULAR MASS:	466.36
ALTERNATIVE FORMS:	Calcium salt/ Potassium salt/ Aluminium lake (insoluble)
PROPERTIES AND APPEARANCE:	Dark blue powder or granules
BOILING POINT IN °C AT VARIOUS PRESSURES (INCLUDING 760 mm Hg):	N/A
MELTING RANGE IN °C:	Not known
FLASH POINT IN °C:	N/A
IONISATION CONSTANT AT 25°C:	N/A
DENSITY AT 20°C (AND OTHER TEMPERATURES) IN g/l:	N/A
HEAT OF COMBUSTION AT 25°C IN J/kg:	N/A
VAPOUR PRESSURE AT VARIOUS TEMPERATURES IN mm Hg:	N/A
PURITY %:	Content not less that 85% total colouring matter, calculated as the sodium salt
WATER CONTENT MAXIMUM IN %:	15

HEAVY METAL CONTENT MAXIMUM IN ppm:	40
ARSENIC CONTENT MAXIMUM IN ppm:	3

SOLUBILITY % AT VARIOUS TEMPERATURE/pH COMBINATIONS:

in water:	**@ 20°C**	1.6%	**@ 50°C**	2.2%	**@ 100°C**	2.5%						
in vegetable oil:	**@ 20°C**	Insoluble	**@ 50°C**	Insoluble	**@ 100°C**	Insoluble						
in sucrose solution:	**@ 10%**	1.5%	**@ 40%**	1.5%	**@ 60%**	1.5%						
in sodium chloride solution:	**@ 5%**	Soluble	**@ 10%**	Insoluble	**@ 15%**	Insoluble						
in ethanol solution:	**@ 5%**	1.0%	**@ 20%**	**@ 25°C**	0.5%	**@ 95%**	0.007%	**@ 100%**	**@ 60°C**	0.007%		
in propylene glycol:	**@ 20°C**	0.1%	**@ 50°C**	0.1%	**@ 100°C**	0.1%						

FUNCTION IN FOODS:	Water-soluble blue food colour, or insoluble blue food colour as the aluminium lake
ALTERNATIVES:	Other blue food colours, although not the same hue, include: Brilliant Blue FCF; Green S; Patent Blue V
TECHNOLOGY OF USE IN FOODS:	A blue food colour with poor solubility and poor light stability. Not stable to oxidation or pH change
SYNERGISTS:	None known
ANTAGONISTS:	Ascorbic acid/ Sulphur dioxide/ Citric acid/ Sodium bicarbonate/ Sodium carbonate/ Ammonium hydroxide/ Dextrose 10%
FOOD SAFETY ISSUES:	None known

LEGISLATION:

USA:
FD&C Blue No. 2 certified food colour

UK AND EUROPE:
Permitted according to European Parliament and Council Directive 94/36/EC of June 30th 1994 on Colours for Use in Foodstuffs.
Refer to Annex V, Part 2 for restrictions

CANADA:
Permitted when certified, but restrictions apply

AUSTRALIA:
Permitted, but restrictions apply

REFERENCE: Smith, J (Ed.) (1991) *Food Additive User's Handbook.* Blackie Academic & Professional.

STRUCTURE OF INDIGOTINE:

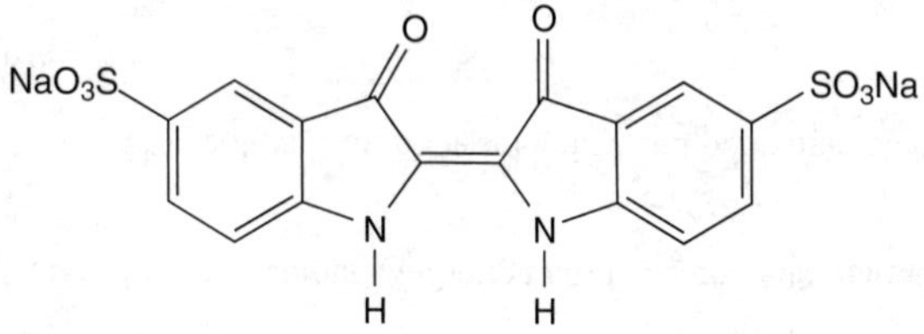

NAME:	**Black iron oxide**							
CATEGORY:	Food colour							
FOOD USE:	Confectionery coatings/ Pet foods/ Canned foods							
SYNONYMS:	E172/ C.I. 77499 (Pigment Black II)/ Iron oxide black/ Ferroso ferric oxide/ Iron (II, III) oxide							
FORMULA:	$FeO \cdot Fe_2O_3$							
MOLECULAR MASS:	231.55							
ALTERNATIVE FORMS:	N/A							
WATER CONTENT MAXIMUM IN %:	N/A							
HEAVY METAL CONTENT MAXIMUM IN ppm:	<40							
ARSENIC CONTENT MAXIMUM IN ppm:	<5							
SOLUBILITY % AT VARIOUS TEMPERATURE/pH COMBINATIONS:								
in water:	**@ 20°C**	Insoluble	**@ 50°C**	Insoluble	**@ 100°C**	Insoluble		
in vegetable oil:	**@ 20°C**	Insoluble	**@ 50°C**	Insoluble	**@ 100°C**	Insoluble		
in sucrose solution:	**@ 10%**	Insoluble	**@ 40%**	Insoluble	**@ 60%**	Insoluble		
in sodium chloride solution:	**@ 5%**	Insoluble	**@ 10%**	Insoluble	**@ 15%**	Insoluble		
in ethanol solution:	**@ 5%**	Insoluble	**@ 20%**	Insoluble	**@ 95%**	Insoluble	**@ 100%**	Insoluble
in propylene glycol:	**@ 20°C**	Insoluble	**@ 50°C**	Insoluble	**@ 100°C**	Insoluble		
ALTERNATIVES:	Vegetable carbon black powder							
TECHNOLOGY OF USE IN FOODS:	An insoluble heat- and light-stable black powder used mainly in canned or highly processed food and sugar-coated confectionery (dragees).							
SYNERGISTS:	None known							

ANTAGONISTS: None known

FOOD SAFETY ISSUES: None known

LEGISLATION:

USA:
Not permitted

UK and EUROPE:
Permitted according to European Parliament and Council Directive 94/36/EC of June 30th 1994 on Colours for Use in Foodstuffs

CANADA:
Permitted

AUSTRALIA:
Not permitted

REFERENCE:

Smith, J (Ed.) (1991) *Food Additive User's Handbook.* Blackie Academic & Professional.

NAME:	Red iron oxide
CATEGORY:	Food colour
FOOD USE:	Confectionery coatings/ Pet foods/ Canned foods
SYNONYMS:	E172/ C.I. 77491 (Pigment Red 101 and 102)/ Iron oxide red/ Anhydrous ferric oxide/ Anhydrous iron (III) oxide
FORMULA:	Fe2O3
MOLECULAR MASS:	159.70
ALTERNATIVE FORMS:	N/A
PROPERTIES AND APPEARANCE:	Red powder
BOILING POINT IN °C AT VARIOUS PRESSURES (INCLUDING 760 mm Hg):	N/A
MELTING RANGE IN °C:	N/A
FLASH POINT IN °C:	N/A
IONISATION CONSTANT AT 25°C:	N/A
DENSITY AT 20°C (AND OTHER TEMPERATURES) IN g/l:	N/A
HEAT OF COMBUSTION AT 25°C IN J/kg:	N/A
VAPOUR PRESSURE AT VARIOUS TEMPERATURES IN mm Hg:	N/A
PURITY %:	Not less than 60% total iron, expressed as iron
WATER CONTENT MAXIMUM IN %:	N/A

HEAVY METAL CONTENT MAXIMUM IN ppm:	40							
ARSENIC CONTENT MAXIMUM IN ppm:	5							
ASH MAXIMUM IN %:	N/A							
SOLUBILITY % AT VARIOUS TEMPERATURE/pH COMBINATIONS:								
in water:	**@ 20°C**	Insoluble	**@ 50°C**	Insoluble	**@ 100°C**	Insoluble		
in vegetable oil:	**@ 20°C**	Insoluble	**@ 50°C**	Insoluble	**@ 100°C**	Insoluble		
in sucrose solution:	**@ 10%**	Insoluble	**@ 40%**	Insoluble	**@ 60%**	Insoluble		
in sodium chloride solution:	**@ 5%**	Insoluble	**@ 10%**	Insoluble	**@ 15%**	Insoluble		
in ethanol solution:	**@ 5%**	Insoluble	**@ 20%**	Insoluble	**@ 95%**	Insoluble	**@ 100%**	Insoluble
in propylene glycol:	**@ 20°C**	Insoluble	**@ 50°C**	Insoluble	**@ 100°C**	Insoluble		
FUNCTION IN FOODS:	Insoluble red food colour							
ALTERNATIVES:	Other insoluble red colours or blends, but these are less heat-stable normally: Carmine; Ponceau 4R lake; Allura Red as lake							
TECHNOLOGY OF USE IN FOODS:	An insoluble orange red heat- and light-stable powder used mainly in canned or highly processed food and sugar-coated confectionery (dragees)							
SYNERGISTS:	None known							
ANTAGONISTS:	None known							
FOOD SAFETY ISSUES:	None known							

LEGISLATION:	**USA:** Not permitted	**UK and EUROPE:** Permitted according to European Parliament and Council Directive 94/36/EC of June 30th 1994 on Colours for Use in Foodstuffs	**CANADA:** Permitted	**AUSTRALIA:** Not permitted
REFERENCE:	Smith, J (Ed.) (1991) *Food Additive User's Handbook.* Blackie Academic & Professional.			

NAME:	Yellow iron oxide
CATEGORY:	Food colour
FOOD USE:	Confectionery coatings/ Pet foods/ Canned foods
SYNONYMS:	E172/ C.I. 77492 (Pigment Yellow 42 and 43)/ Iron oxide yellow/ Hydrated ferric oxide/ Hydrated iron (III) oxide
FORMULA:	FeO(OH) H2O
MOLECULAR MASS:	88.85
ALTERNATIVE FORMS:	N/A
PROPERTIES AND APPEARANCE:	Yellow powder
BOILING POINT IN °C AT VARIOUS PRESSURES (INCLUDING 760 mm Hg):	N/A
MELTING RANGE IN °C:	N/A
FLASH POINT IN °C:	N/A
IONISATION CONSTANT AT 25°C:	N/A
DENSITY AT 20°C (AND OTHER TEMPERATURES) IN g/l:	N/A
HEAT OF COMBUSTION AT 25°C IN J/kg:	N/A
VAPOUR PRESSURE AT VARIOUS TEMPERATURES IN mm Hg:	N/A
PURITY %:	Not less than 60% total iron, expressed as iron
WATER CONTENT MAXIMUM IN %:	N/A

HEAVY METAL CONTENT MAXIMUM IN ppm:	40							
ARSENIC CONTENT MAXIMUM IN ppm:	5							
ASH MAXIMUM IN %:	N/A							
SOLUBILITY % AT VARIOUS TEMPERATURE/pH COMBINATIONS:								
in water:	**@ 20°C**	Insoluble	**@ 50°C**	Insoluble	**@ 100°C**	Insoluble		
in vegetable oil:	**@ 20°C**	Insoluble	**@ 50°C**	Insoluble	**@ 100°C**	Insoluble		
in sucrose solution:	**@ 10%**	Insoluble	**@ 40%**	Insoluble	**@ 60%**	Insoluble		
in sodium chloride solution:	**@ 5%**	Insoluble	**@ 10%**	Insoluble	**@ 15%**	Insoluble		
in ethanol solution:	**@ 5%**	Insoluble	**@ 20%**	Insoluble	**@ 95%**	Insoluble	**@ 100%**	Insoluble
in propylene glycol:	**@ 20°C**	Insoluble	**@ 50°C**	Insoluble	**@ 100°C**	Insoluble		
FUNCTION IN FOODS:	Insoluble yellow food colour							
ALTERNATIVES:	Other insoluble yellow colours or blends, but these are normally less heat-stable: Tartrazine lake/ Quinoline Yellow lake							
TECHNOLOGY OF USE IN FOODS:	An insoluble yellow, heat- and light-stable powder used mainly in canned or highly processed food and sugar-coated confectionery (dragees)							
SYNERGISTS:	None known							
ANTAGONISTS:	None known							
FOOD SAFETY ISSUES:	None known							

LEGISLATION:

USA:	**UK and EUROPE:**	**CANADA:**	**AUSTRALIA:**
Not permitted	Permitted according to European Parliament and Council Directive 94/36/EC of June 30th 1994 on Colours for Use in Foodstuffs	Permitted	Not permitted

REFERENCE:

Smith, J. (Ed.) (1991) *Food Additive User's Handbook.* Blackie Academic & Professional.

NAME:	Litholrubine BK
CATEGORY:	Food colour
FOOD USE:	Edible cheese rind
SYNONYMS:	E180/ C.I. 15850 :1 (Pigment Red 57)/ Rubinpigment/ Carmine 6B/ Calcium 3-hydroxy-4-(4-methyl-2-sulfonatophenylazo)-2-naphthalene carboxylate
FORMULA:	$C_{18}H_{12}CaN_2O_6S$
MOLECULAR MASS:	424.45
PROPERTIES AND APPEARANCE:	Red powder
BOILING POINT IN °C AT VARIOUS PRESSURES (INCLUDING 760 mm Hg):	N/A
MELTING RANGE IN °C:	N/A
FLASH POINT IN °C:	N/A
IONISATION CONSTANT AT 25°C:	N/A
DENSITY AT 20°C (AND OTHER TEMPERATURES) IN g/l:	N/A
HEAT OF COMBUSTION AT 25°C IN J/kg:	N/A
VAPOUR PRESSURE AT VARIOUS TEMPERATURES IN mm Hg:	N/A
PURITY %:	Not less than 90% total colouring matters
WATER CONTENT MAXIMUM IN %:	10

HEAVY METAL CONTENT MAXIMUM IN ppm: 40

ARSENIC CONTENT MAXIMUM IN ppm: 3

SOLUBILITY % AT VARIOUS TEMPERATURE/pH COMBINATIONS:

in water:	**@ 20°C**	Insoluble	**@ 50°C**	Insoluble	**@ 100°C**	Insoluble		
in vegetable oil:	**@ 20°C**	Soluble	**@ 50°C**	Soluble	**@ 100°C**	Soluble		
in sucrose solution:	**@ 10%**	Insoluble	**@ 40%**	Insoluble	**@ 60%**	Insoluble		
in sodium chloride solution:	**@ 5%**	Insoluble	**@ 10%**	Insoluble	**@ 15%**	Insoluble		
in ethanol solution:	**@ 5%**	Insoluble	**@ 20%**	Insoluble	**@ 95%**	Insoluble	**@ 100%**	Insoluble
in propylene glycol:	**@ 20°C**	Insoluble	**@ 50°C**	Insoluble	**@ 100°C**	Insoluble		

FUNCTION IN FOODS: Water-insoluble bright-red food colour used for cheese rind and surface marking.

ALTERNATIVES: Carmine or other red insoluble lake colours can be dispersed into wax to give a similar coloration to Litholrubine BK

TECHNOLOGY OF USE IN FOODS: A red, water-insoluble food colour. Soluble in oil and hydrocarbon solvents. Used to colour cheese wax rind

SYNERGISTS: None known

ANTAGONISTS: Not known

FOOD SAFETY ISSUES: No ADI set due to insufficient toxicological data being available.

LEGISLATION:

USA:
Not permitted

UK and EUROPE:
Permitted according to European Parliament and Council Directive 94/36/EC of 30th June 1994 on Colours for Use in Foodstuffs.
Refer to Annex IV – restricted to colouring edible cheese rind.

CANADA:
Not permitted

AUSTRALIA:
Not permitted

REFERENCE: Smith, J. (Ed.) (1991) *Food Additive User's Handbook.* Blackie Academic & Professional.

STRUCTURE OF LITHOLRUBINE BK:

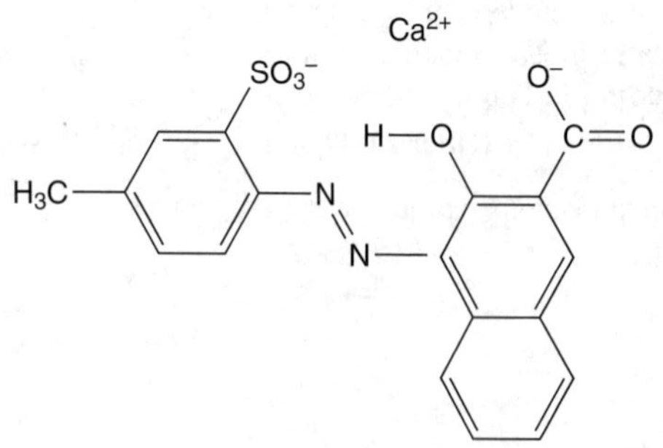

NAME:	Monascus
CATEGORY:	Food colour
FOOD USE:	Alcoholic beverages/ Bean curds/ Meat products/ Rice cakes/ Snacks/ Fish dishes/ Flour confectionery/ Sauces/ Sugar confectionery
SYNONYMS:	Monascus pigments: Rubrupunctatine (1); Monascorubine (2); Rubropunctatamine (3); Monascorubramine (4); Monascine (5); Ankaflavine (6)/ C.I. 1975 Natural Red 2
FORMULA:	(1) $C_{21}H_{22}O_5$; (2) $C_{23}H_{26}O_5$; (3) $C_{21}H_{23}O_4N$; (4) $C_{23}H_{27}O_4N$; (5) $C_{21}H_{24}O_5$; (6) $C_{23}H_{28}O_5$
MOLECULAR MASS:	(1) 354.2; (2) 382.2; (3) 341.2; (4) 368.2; (5) 356.2; (6) 384.2
ALTERNATIVE FORMS:	Available in various shades of yellow to red, depending on particular growth and extract
PROPERTIES AND APPEARANCE:	Yellow to orange red powder or liquid
BOILING POINT IN °C AT VARIOUS PRESSURES (INCLUDING 760 mm Hg):	N/A
MELTING RANGE IN °C:	N/A
FLASH POINT IN °C:	N/A
IONISATION CONSTANT AT 25°C:	N/A
DENSITY AT 20°C (AND OTHER TEMPERATURES) IN g/l:	N/A
HEAT OF COMBUSTION AT 25°C IN J/kg:	N/A
VAPOUR PRESSURE AT VARIOUS TEMPERATURES IN mm Hg:	N/A
PURITY %:	Not specified

WATER CONTENT MAXIMUM IN %:	N/A					
HEAVY METAL CONTENT MAXIMUM IN ppm:	40					
ARSENIC CONTENT MAXIMUM IN ppm:	3					
ASH MAXIMUM IN %:	N/A					
SOLUBILITY % AT VARIOUS TEMPERATURE/pH COMBINATIONS:						
in water:	@ **20°C**	Depends on extract	@ **50°C**	Depends on extract	@ **100°C**	Depends on extract
in vegetable oil:	@ **20°C**	Depends on extract	@ **50°C**	Depends on extract	@ **100°C**	Depends on extract
in sucrose solution:	@ **10%**	Depends on extract	@ **40%**	Depends on extract	@ **60%**	Depends on extract
in sodium chloride solution:	@ **5%**	Depends on extract	@ **10%**	Depends on extract	@ **15%**	Depends on extract
in ethanol solution:	@ **5%**	Depends on extract	@ **20%**	Depends on extract	@ **95%**	Depends on extract
	@ **100%**	Depends on extract				
in propylene glycol:	@ **20°C**	Depends on extract	@ **50°C**	Depends on extract	@ **100°C**	Depends on extract

FUNCTION IN FOODS: Food colour, food ingredient when red rice flour used, or red mould rice. Bacteriostat in some food systems. Various shades of colour can be produced and/or extracted from yellow through orange to reds.

TECHNOLOGY OF USE IN FOODS: Monascus colours are a group of colours produced by the moulds *Monascus purpureus*, *M. anka*, *M. ruber* grown on substrates, particularly rice. The dried red mass can be used as a coloured food ingredient, or extracts can be prepared. Depending on the substrate and growth conditions, various shades can be produced. The group of pigments can also be extracted and separated. Some extracts are water-soluble, and yellow to red colours are possible. The colours have reasonable heat- and light-stability. Traditionally, monascus colour has been used for years in the Far East.

SYNERGISTS: Ascorbic acid can improve light stability.

FOOD SAFETY ISSUES:

Inadequate toxicological data are available to give an ADI. Full characterisation of the various pigments is not available. French researchers claim that citrinin A toxin produced by various fungi is identical with a component of a red pigment produced by the monascus fungi, casting doubt on the safety in use, despite the long history of use in Asia.

LEGISLATION:

USA:	UK and EUROPE:	CANADA:	AUSTRALIA/PACIFIC RIM:
Not permitted	Not permitted	Not permitted	Australia: Not permitted Japan and Far East: Permitted as a natural food ingredient and colour

REFERENCES:

Blanc, P. J. *et al.* (1994) Pigments of monascus. *Journal of Food Science*, **59**, 862–865.

Blanc, P. J. (1994) Monascus colour contains toxin. *International Journal of Food Microbiology*, **27**, 201–203.

Fabre, C. E. *et al.* (1995) Production and food applications of red pigments of *Monascus ruber*. *Journal of Food Science*, **58**, 1099–1102.

ANY OTHER RELEVANT INFORMATION:

In China, red mould rice is also called 'Angkak'.

STRUCTURE OF MONASCUS PIGMENTS:

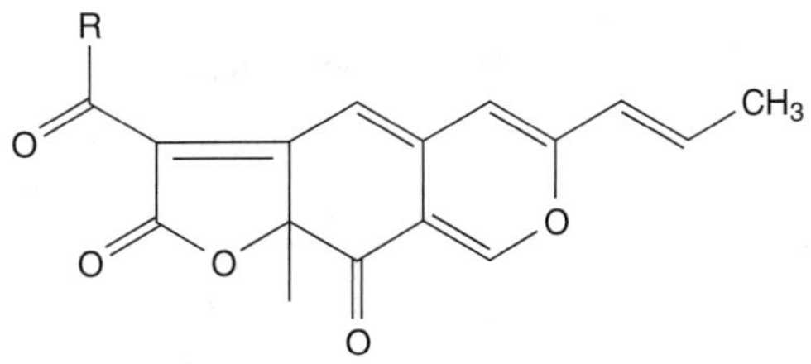

Monascin $R = C_5H_{11}$

Ankaflavin $R = C_7H_{15}$

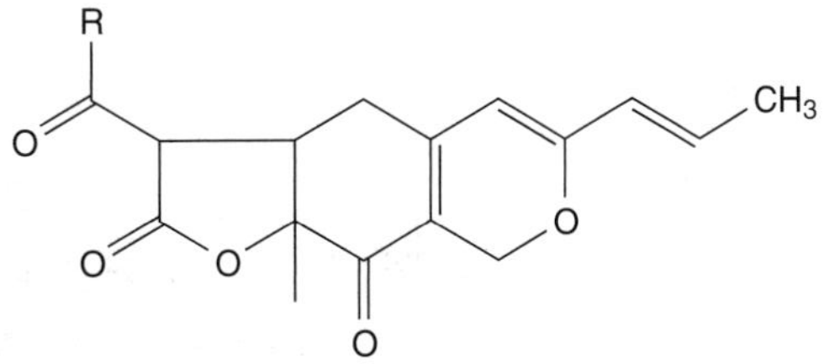

Rubropunctatin $R = C_5H_{11}$

Monascorubin $R = C_7H_{15}$

NAME:	Brilliant Black BN
CATEGORY:	Food colour
FOOD USE:	Sugar confectionery/ Jams and preserves/ Soft drinks
SYNONYMS:	Black PN/ E151/ C.I. 28440 (Food black 1)/ Tetra sodium 4-acetamido-5-hydroxy-6-[7-sulfonato-4-(4 sulfonato phenylazo)-1-naphthylazo] naphthalene-1,7-disulfonate
FORMULA:	$C_{28}H_{17}N_5Na_4O_{14}S_4$
MOLECULAR MASS:	867.69
ALTERNATIVE FORMS:	Calcium salt/ Potassium salt
PROPERTIES AND APPEARANCE:	Violet black powder or granules
BOILING POINT IN °C AT VARIOUS PRESSURES (INCLUDING 760 mm Hg):	N/A
MELTING RANGE IN °C:	N/A
FLASH POINT IN °C:	N/A
IONISATION CONSTANT AT 25°C:	N/A
DENSITY AT 20°C (AND OTHER TEMPERATURES) IN g/l:	N/A
HEAT OF COMBUSTION AT 25°C IN J/kg:	N/A
VAPOUR PRESSURE AT VARIOUS TEMPERATURES IN mm Hg:	N/A
PURITY %:	Not less than 80% total colouring matter
WATER CONTENT MAXIMUM IN %:	20

HEAVY METAL CONTENT MAXIMUM IN ppm:	40							
ARSENIC CONTENT MAXIMUM IN ppm:	3							
ASH MAXIMUM IN %:	N/A							
SOLUBILITY % AT VARIOUS TEMPERATURE/pH COMBINATIONS:								
in water:	**@ 20°C**	5%	**@ 50°C**	8%	**@ 100°C**	15%		
in vegetable oil:	**@ 20°C**	Insoluble	**@ 50°C**	Insoluble	**@ 100°C**	Insoluble		
in sucrose solution:	**@ 10%**	5%	**@ 40%**	4%	**@ 60%**	2%		
in sodium chloride solution:	**@ 5%**	Soluble	**@ 10%**	Insoluble	**@ 15%**	Insoluble		
in ethanol solution:	**@ 5%**	5%	**@ 20%**	4%	**@ 95%**	<0.1%	**@ 100%**	<0.1%
in propylene glycol:	**@ 20°C**	1%	**@ 50°C**	1.2%	**@ 100°C**	1.5%		
FUNCTION IN FOODS:	Water-soluble violet food colour							
ALTERNATIVES:	Blends of other food colours to achieve a violet hue							
TECHNOLOGY OF USE IN FOODS:	A water-soluble violet food colour with excellent light stability, but poor heat stability. It has reasonable resistance to fruit acids and alkaline conditions. Fades in the presence of sulphur dioxide.							
SYNERGISTS:	None known							
ANTAGONISTS:	Sulphur dioxide/ Ascorbic acid							
FOOD SAFETY ISSUES:	No known issues. Temporary ADI given. Additional toxicological data requested.							

LEGISLATION:

USA:
Not permitted

UK and EUROPE:
Permitted according to European Parliament and Council Directive 94/36/EC of June 30th 1994 on Colouring of Foodstuffs.
Refer to Annex V Part 2 for restrictions on use

CANADA:
Not permitted

AUSTRALIA:
Permitted, but restrictions apply

REFERENCE:

Smith, J. (Ed.) (1991) *Food Additive User's Handbook*. Blackie Academic & Professional.

STRUCTURE OF BRILLIANT BLACK BN:

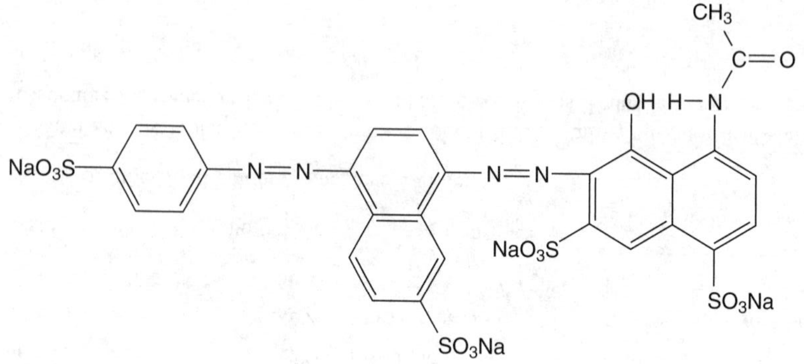

NAME:	Patent Blue V
CATEGORY:	Food colour
FOOD USE:	Baked goods/ Fruit and vegetable and nut products/ Edible ices/ Confectionery/ Cereal and cereal products/ Seasonings/ Canned products/ Decorations and coatings/ Soups/ Soft drinks/ Dairy products/ Snacks/ Desserts/ Meat and poultry products
SYNONYMS:	E131/ C.I. 42051 (Food Blue 5)/ Calcium or sodium compound of (4-(α(4-diethylaminophenyl)-5-hydroxy-2,4-disulfophenyl-methylidene)2,5-cyclohexadien-1-ylidene)diethyl-ammonium hydroxide inner salt
FORMULA:	$C_{27}H_{31}N_2O_7S_2Ca_{1/2}$; $C_{27}H_{31}N_2O_7S_2Na$
MOLECULAR MASS:	Calcium salt 579.72; Sodium salt 582.67
ALTERNATIVE FORMS:	Potassium salt/ Aluminium lake (insoluble)
PROPERTIES AND APPEARANCE:	Dark blue powder or granules
BOILING POINT IN °C AT VARIOUS PRESSURES (INCLUDING 760 mm Hg):	N/A
MELTING RANGE IN °C:	N/A
FLASH POINT IN °C:	N/A
IONISATION CONSTANT AT 25°C:	N/A
DENSITY AT 20°C (AND OTHER TEMPERATURES) IN g/l:	N/A
HEAT OF COMBUSTION AT 25°C IN J/kg:	N/A
VAPOUR PRESSURE AT VARIOUS TEMPERATURES IN mm Hg:	N/A

PURITY %: Content not less than 85% total colouring matter, calculated as the sodium salt

WATER CONTENT MAXIMUM IN %: 15

HEAVY METAL CONTENT MAXIMUM IN ppm: 40

ARSENIC CONTENT MAXIMUM IN ppm: 3

SOLUBILITY % AT VARIOUS TEMPERATURE/pH COMBINATIONS:

in water:	**@ 20°C**	4%	**@ 50°C**	5%	**@ 100°C**	6%		
in vegetable oil:	**@ 20°C**	Insoluble	**@ 50°C**	Insoluble	**@ 100°C**	Insoluble		
in sucrose solution:	**@ 10%**	4%	**@ 40%**	2%	**@ 60%**	1%		
in sodium chloride solution:	**@ 5%**	Soluble	**@ 10%**	Insoluble	**@ 15%**	Insoluble		
in ethanol solution:	**@ 5%**	4%	**@ 20%**	2%	**@ 95%**	<0.1%	**@ 100%**	<0.1%
in propylene glycol:	**@ 20°C**	2.0%	**@ 50°C**	2.5%	**@ 100°C**	3%		

FUNCTION IN FOODS: Water-soluble blue food colour, or insoluble blue food colour as the aluminium lake

ALTERNATIVES: Other blue food colours, although not the same hue, include: Brilliant Blue FCF; Green S; Indigotine

TECHNOLOGY OF USE IN FOODS: A light-stable blue food colour, stable to heat to 105°C. Fades in the presence of sulphur dioxide, ascorbic acid, fruit acids and alkaline media.

SYNERGISTS: None known

ANTAGONISTS: Sulphur dioxide/ Ascorbic acid/ Fruit acids/ Alkaline media

FOOD SAFETY ISSUES: None known

LEGISLATION:

USA:
Not permitted

UK and EUROPE:
Permitted according to European Parliament and Council Directive 94/36/EC of June 30th 1994 on Colours for use in Foodstuffs.
Refer to Annex V, Part 2 for restrictions on use

CANADA:
Not permitted

AUSTRALIA:
Not permitted

REFERENCE:

Smith, J. (Ed.) (1991) *Food Additive User's Handbook*. Blackie Academic & Professional.

STRUCTURE OF PATENT BLUE V:

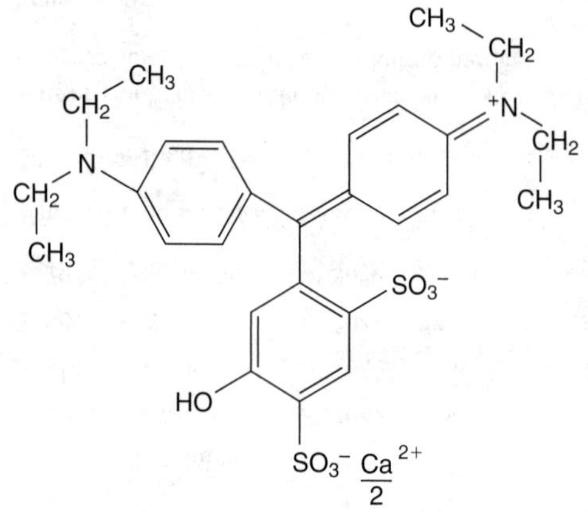

NAME:	Ponceau 4R
CATEGORY:	Food colour
FOOD USE:	Baked goods/ Cereals and cereal products/ Soft drinks/ Fruit, vegetable and nut products/ Meat and poultry products/ Edible ices/ Canned products/ Confectionery/ Desserts/ Sugars and preserves/ Decorations and coatings/ Seasonings/ Beverages/ Dairy products/ Snacks/ Fish and seafood products
SYNONYMS:	E124/ C.I. 16255 (Food Red 7)/ Cochineal red A/ New coccine/ Trisodium 2-hydroxy-1-(4-sulfonato-1-naphthylazo) naphthalene-6,7-disulfonate
FORMULA:	$C_{20}H_{14}N_2Na_3O_{10}S_3$
MOLECULAR MASS:	604.48
ALTERNATIVE FORMS:	Calcium salt/ Potassium salt/ Aluminium lake (insoluble)
PROPERTIES AND APPEARANCE:	Red powder or granules
BOILING POINT IN °C AT VARIOUS PRESSURES (INCLUDING 760 mm Hg):	N/A
MELTING RANGE IN °C:	N/A
FLASH POINT IN °C:	N/A
IONISATION CONSTANT AT 25°C:	N/A
DENSITY AT 20°C (AND OTHER TEMPERATURES) IN g/l:	N/A
HEAT OF COMBUSTION AT 25°C IN J/kg:	N/A
VAPOUR PRESSURE AT VARIOUS TEMPERATURES IN mm Hg:	N/A
PURITY %:	Not less than 80% total colouring matter, calculated as the sodium salt

WATER CONTENT MAXIMUM IN %:	20							
HEAVY METAL CONTENT MAXIMUM IN ppm:	40							
ARSENIC CONTENT MAXIMUM IN ppm:	3							
ASH MAXIMUM IN %:	N/A							
SOLUBILITY % AT VARIOUS TEMPERATURE/pH COMBINATIONS:								
in water:	**@ 20°C**	14%	**@ 50°C**	20%	**@ 100°C**	30%		
in vegetable oil:	**@ 20°C**	Insoluble	**@ 50°C**	Insoluble	**@ 100°C**	Insoluble		
in sucrose solution:	**@ 10%**	10%	**@ 40%**	5%	**@ 60%**	2%		
in sodium chloride solution:	**@ 5%**	Soluble	**@ 10%**	Insoluble	**@ 15%**	Insoluble		
in ethanol solution:	**@ 5%**	10%	**@ 20%**	5%	**@ 95%**	0.1%	**@ 100%**	<0.1%
in propylene glycol:	**@ 20°C**	4%	**@ 50°C**	5%	**@ 100°C**	6%		
FUNCTION IN FOODS:	Water-soluble red food colour, or insoluble red food colour as the aluminium lake							
ALTERNATIVES:	Other red food colours, although not of exactly the same hue, include: Carmine; Carmoisine; Allura Red; Erythrosine; Beet Red; Anthocyanins (in acidic media); Red 2G (where permitted)							
TECHNOLOGY OF USE IN FOODS:	A robust red colour with good light stability. Good heat stability to 105°C. Fades in alkaline media. Some fading with sulphur dioxide and ascorbic acid.							
SYNERGISTS:	None known							
ANTAGONISTS:	Sulphur dioxide/ Alkaline media/ Ascorbic acid							
FOOD SAFETY ISSUES:	None known							

LEGISLATION:

USA:	UK and EUROPE:	CANADA:
Not permitted	Permitted according to European Parliament and Council Directive 94/36/EC of June 30th 1994 on Colours for Use in Foodstuffs. Refer to Annex V, Part 2 for restrictions on use	Not permitted **AUSTRALIA:** Permitted, but restrictions apply

REFERENCE:

Smith, J. (Ed.) (1991) *Food Additive User's Handbook*. Blackie Academic & Professional.

STRUCTURE OF PONCEAU 4R:

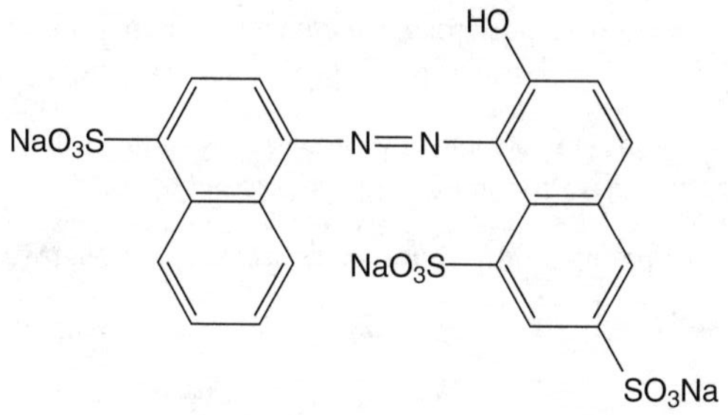

NAME:	Quinoline Yellow
CATEGORY:	Food colour
FOOD USE:	Baked goods/ Meat and poultry products/ Egg products/ Fruit, vegetable and nut products/ Beverages/ Soft drinks/ Sugars and preserves/ Confectionery/ Edible ices/ Vinegar, pickles and sauces/ Decorations and coatings/ Desserts/ Cereals and cereal products/ Dairy products/ Snacks/ Seasonings/ Soups
SYNONYMS:	E104/ C.I. 47005 (Food Yellow 13)/ The disodium salts of the disulfonates of 2-(2-quinolyl) indan-1, and 3-dione (principal components)
FORMULA:	$C_{18}H_{9}NNa_{2}O_{8}S_{2}$ (principal component)
MOLECULAR MASS:	477.38 (principal component)
ALTERNATIVE FORMS:	Calcium salt/ Potassium salt/ Aluminium lake (insoluble)
PROPERTIES AND APPEARANCE:	A green-yellow powder or granules
BOILING POINT IN °C AT VARIOUS PRESSURES (INCLUDING 760 mm Hg):	N/A
MELTING RANGE IN °C:	N/A
FLASH POINT IN °C:	N/A
IONISATION CONSTANT AT 25°C:	N/A
DENSITY AT 20°C (AND OTHER TEMPERATURES) IN g/l:	N/A
HEAT OF COMBUSTION AT 25°C IN J/kg:	N/A
VAPOUR PRESSURE AT VARIOUS TEMPERATURES IN mm Hg:	N/A
PURITY %:	Not less than 70% total dye content calculated as the sodium salt

WATER CONTENT MAXIMUM IN %:	30							
HEAVY METAL CONTENT MAXIMUM IN ppm:	40							
ARSENIC CONTENT MAXIMUM IN ppm:	3							
ASH MAXIMUM IN %:	N/A							
SOLUBILITY % AT VARIOUS TEMPERATURE/pH COMBINATIONS:								
in water:	**@ 20°C**	14%	**@ 50°C**	14%	**@ 100°C**	14%		
in vegetable oil:	**@ 20°C**	Insoluble	**@ 50°C**	Insoluble	**@ 100°C**	Insoluble		
in sucrose solution:	**@ 10%**	14%	**@ 40%**	10%	**@ 60%**	5%		
in sodium chloride solution:	**@ 5%**	Soluble	**@ 10%**	Insoluble	**@ 15%**	Insoluble		
in ethanol solution:	**@ 5%**	0.1%	**@ 20%**	<0.1%	**@ 95%**	Insoluble	**@ 100%**	Insoluble
in propylene glycol:	**@ 20°C**	<0.1%	**@ 50°C**	0.1%	**@ 100°C**	0.2%		
FUNCTION IN FOODS:	Water-soluble yellow food colour, or insoluble yellow food colour as the aluminium lake							
ALTERNATIVES:	Other yellow food colours, although not of exactly the same hue, include: Tartrazine; Riboflavin; Curcumin; Crocin; Safflower; Beta-carotene							
TECHNOLOGY OF USE IN FOODS:	Very robust in use, excellent stability up to 105°C. Less light-stable in alkaline medium. Some fade in the presence of benzoic and/or ascorbic acids.							
SYNERGISTS:	None known							
ANTAGONISTS:	Sodium hydroxide/ Benzoic acid/ Ascorbic acid							
FOOD SAFETY ISSUES:	None known							

LEGISLATION:	**USA:** Not permitted	**UK and EUROPE:** Permitted according to European Parliament and Council Directive 94/36/EC of June 30th 1994 on Colours for Use in Foodstuffs. Refer to Annex V, Part 2 for list of permitted food categories and maximum levels	**CANADA:** Not permitted **AUSTRALIA:** Permitted, but restrictions apply

REFERENCE:

Smith, J. (Ed.) (1991) *Food Additive User's Handbook*. Blackie Academic & Professional.

STRUCTURE OF QUINOLINE YELLOW:

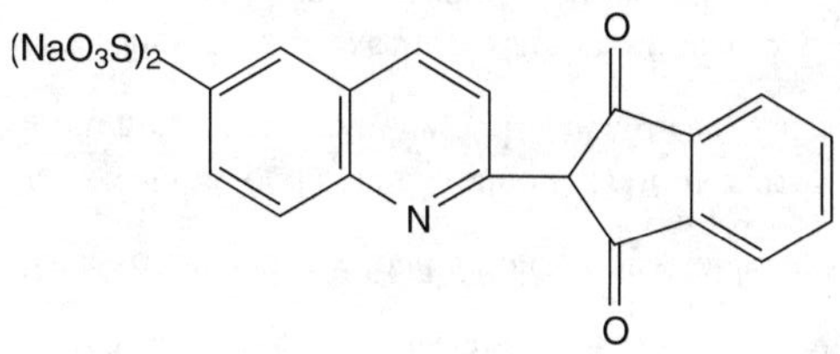

NAME:	Ponceau SX
CATEGORY:	Food colour
FOOD USE:	Colouring of cherries
SYNONYMS:	C.I. 14700 (Food Red 1)/ Formerly FD&C Red 4/ 2-(5-Sulpho-2,4-xylylazo)-1-naphthol-4-sulfonate disodium salt
FORMULA:	$C_{18}H_{14}N_2O_7S_2Na_2$
MOLECULAR MASS:	480.23
ALTERNATIVE FORMS:	Potassium salt/ Calcium salt
PROPERTIES AND APPEARANCE:	Dark red powder
BOILING POINT IN °C AT VARIOUS PRESSURES (INCLUDING 760 mm Hg):	N/A
MELTING RANGE IN °C:	N/A
FLASH POINT IN °C:	N/A
IONISATION CONSTANT AT 25°C:	N/A
DENSITY AT 20°C (AND OTHER TEMPERATURES) IN g/l:	N/A
HEAT OF COMBUSTION AT 25°C IN J/kg:	N/A
VAPOUR PRESSURE AT VARIOUS TEMPERATURES IN mm Hg:	N/A
PURITY %:	87
WATER CONTENT MAXIMUM IN %:	10

HEAVY METAL CONTENT MAXIMUM IN ppm:	40					
ARSENIC CONTENT MAXIMUM IN ppm:	3					
SOLUBILITY % AT VARIOUS TEMPERATURE/pH COMBINATIONS:						
in water:	@ **20°C**	Soluble	@ **50°C**	Soluble	@ **100°C**	Soluble
in vegetable oil:	@ **20°C**	Insoluble	@ **50°C**	Insoluble	@ **100°C**	Insoluble
in sucrose solution:	@ **10%**	Soluble	@ **40%**	Soluble	@ **60%**	Soluble
in sodium chloride solution:	@ **5%**	Soluble	@ **10%**	Partially soluble	@ **15%**	Slightly soluble
in ethanol solution:	@ **5%**	Soluble	@ **20%**	Soluble	@ **95%**	Slightly soluble
	@ **100%**	Slightly soluble				
in propylene glycol:	@ **20°C**	Soluble	@ **50°C**	Soluble	@ **100°C**	Soluble
FUNCTION IN FOODS:	Bright yellowish-red food colour					
ALTERNATIVES:	Erythrosine, but the shade is bluer					
TECHNOLOGY OF USE IN FOODS:	Water-soluble yellowish-red food colour, stable to heat to about 205°C. Excellent stability to fruit acids and benzoic acid. Poor stability to alkaline media. Good stability to sulphur dioxide. Major use for colouring of cherries, where permitted.					
SYNERGISTS:	None known					
ANTAGONISTS:	Alkaline media					
FOOD SAFETY ISSUES:	No ADI. Reproductive toxicology studies required					
LEGISLATION:	**USA:** Not now permitted	**CANADA:** Permitted (when certified) to a maximum of 150 ppm.				
	UK and EUROPE: Not permitted	**AUSTRALIA:** Not permitted				

REFERENCE: Walford, J. (Ed.) (1980) *Developments in Food Colours-1*. Applied Science Publishers Ltd.

STRUCTURE OF PONCEAU SX:

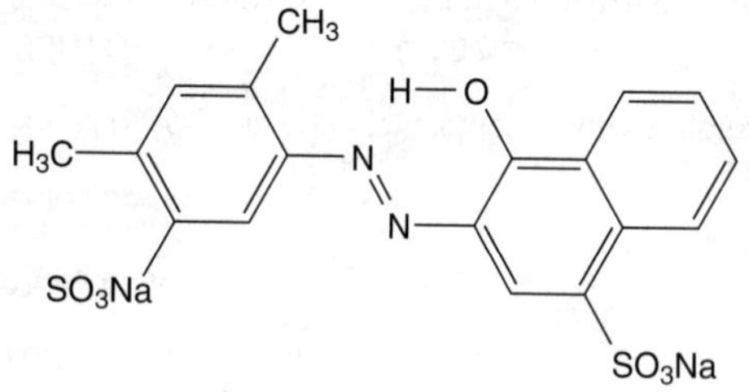

NAME:	**Silver**
CATEGORY:	Food colour
FOOD USE:	Surface coating of confectionery/ Decoration of chocolates/ Liqueurs
SYNONYMS:	E174/ Argentum/ C.I. 77820
FORMULA:	Ag
MOLECULAR MASS:	107.87
PROPERTIES AND APPEARANCE:	Silver powder or thin sheets (leaf)
BOILING POINT IN °C AT VARIOUS PRESSURES (INCLUDING 760 mm Hg):	1955
MELTING RANGE IN °C:	960.5
FLASH POINT IN °C:	N/A
IONISATION CONSTANT AT 25°C:	N/A
DENSITY AT 20°C (AND OTHER TEMPERATURES) IN g/l:	N/A
HEAT OF COMBUSTION AT 25°C IN J/kg:	N/A
VAPOUR PRESSURE AT VARIOUS TEMPERATURES IN mm Hg:	N/A
PURITY %:	Not less than 99.5
WATER CONTENT MAXIMUM IN %:	N/A
HEAVY METAL CONTENT MAXIMUM IN ppm:	40

ARSENIC CONTENT MAXIMUM IN ppm: 3

ASH MAXIMUM IN %: N/A

SOLUBILITY % AT VARIOUS TEMPERATURE/pH COMBINATIONS:

in water:	**@ 20°C**	Insoluble	**@ 50°C**	Insoluble	**@ 100°C**	Insoluble		
in vegetable oil:	**@ 20°C**	Insoluble	**@ 50°C**	Insoluble	**@ 100°C**	Insoluble		
in sucrose solution:	**@ 10%**	Insoluble	**@ 40%**	Insoluble	**@ 60%**	Insoluble		
in sodium chloride solution:	**@ 5%**	Insoluble	**@ 10%**	Insoluble	**@ 15%**	Insoluble		
in ethanol solution:	**@ 5%**	Insoluble	**@ 20%**	Insoluble	**@ 95%**	Insoluble	**@ 100%**	Insoluble
in propylene glycol:	**@ 20°C**	Insoluble	**@ 50°C**	Insoluble	**@ 100°C**	Insoluble		

FUNCTION IN FOODS: Used as a dispersed powder or silver leaf to surface-colour sugar or chocolate confectionery, or as an additive to certain liqueurs.

ALTERNATIVES: Aluminium powder or leaf

TECHNOLOGY OF USE IN FOODS: Applied as leaf or powder to the surface of hard sugar or chocolate confectionery, then polished to a high gloss as a decoration. Can be added as fine particles or leaf to certain liqueurs.

SYNERGISTS: None known

ANTAGONISTS: Moist air can cause oxidation and loss of gloss

FOOD SAFETY ISSUES: None known

LEGISLATION:

USA:
Not permitted

UK and EUROPE:
Permitted according to European Parliament and Council Directive 94/36/EC of 30th June 1994 on Colours for Use in Foodstuffs.
Refer to Annex V for restrictions of use

CANADA:
Permitted

AUSTRALIA:
Not permitted

REFERENCE:

Smith, J. (Ed.) (1991) *Food Additive User's Handbook*. Blackie Academic & Professional.

NAME:	Sunset Yellow
CATEGORY:	Food colour
FOOD USE:	Baked goods/ Cereals and cereal products/ Dairy products/ Fish and seafood products/ Meat and poultry products/ Egg products/ Fruit, vegetable and nut products/ Beverages/ Soft drinks/ Sugars and preserves/ Confectionery/ Alcoholic drinks/ Vinegar, pickles and sauces/ Decorations and coatings/ Edible ices/ Seasonings/ Desserts/ Snacks/ Soups
SYNONYMS:	FD&C Yellow No. 6/ C.I. 15985 Food Yellow 3/ Orange Yellow S/ Disodium 2-hydroxy-1-(4-sulfonatophenylazo) naphthalene-6-sulfonate
FORMULA:	$C_{16}H_{10}N_2Na_2O_7S_2$
MOLECULAR MASS:	452.37
ALTERNATIVE FORMS:	Potassium salt/ Calcium salt/ Aluminium lake
PROPERTIES AND APPEARANCE:	Bright orange-red powder or granules
BOILING POINT IN °C AT VARIOUS PRESSURES (INCLUDING 760 mm Hg):	Not applicable
MELTING RANGE IN °C:	N/A
FLASH POINT IN °C:	N/A
IONISATION CONSTANT AT 25°C:	N/A
DENSITY AT 20°C (AND OTHER TEMPERATURES) IN g/l:	Not applicable
HEAT OF COMBUSTION AT 25°C IN J/kg:	N/A
VAPOUR PRESSURE AT VARIOUS TEMPERATURES IN mm Hg:	N/A

PURITY %:	≮85% total colouring matter calculated as the sodium salt						
WATER CONTENT MAXIMUM IN %:	13						
HEAVY METAL CONTENT MAXIMUM IN ppm:	40						
ARSENIC CONTENT MAXIMUM IN ppm:	3						
ASH MAXIMUM IN %:	N/A						
SOLUBILITY % AT VARIOUS TEMPERATURE/pH COMBINATIONS:							
in water:	**@ 20°C**	18%	**@ 50°C**	19%	**@ 100°C**	19%	
in vegetable oil:	**@ 20°C**	Insoluble					
in sucrose solution:	**@ 10%**	18%	**@ 40%**	18%	**@ 60%**	18%	
in sodium chloride solution:	**@ 5%**	18%	**@ 10%**	18%	**@ 15%**	18%	
in ethanol solution:	**@ 5%**	18%	**@ 20%**	18%	**@ 100%**	**@ 60°C**	<0.1%
in propylene glycol:	**@ 20°C**	2%	**@ 50°C**	2%	**@ 100°C**	2%	
FUNCTION IN FOODS:	Water-soluble orange-yellow food colour, of insoluble pigment as the aluminium lake						
ALTERNATIVES:	Other orange-yellow colours, although not of exactly the same hue, include: Annatto; Carminic acid; Beta-carotene; Paprika						
TECHNOLOGY OF USE IN FOODS:	Very robust in use, excellent stability up to 205°C. Slight fade in 10% NaOH, considerable fade in 1% ascorbic acid and appreciable fade in SO_2. Calcium ions can lead to precipitation.						
SYNERGISTS:	None known						
ANTAGONISTS:	Ascorbic acid/ Sulphur dioxide/ Sodium hydroxide (and other strong alkalis)						
FOOD SAFETY ISSUES:	None known						

LEGISLATION:

USA:	UK and EUROPE:	CANADA:
GRAS; permitted as certified food colour according to 21 CFR 74.706	Permitted according to European Parliament and Council Directive 94/36/EC of June 30th 1994 on Colours for Use in Foodstuffs. Refer to Annex V, Part 2 for a substantial list of permitted food categories and maximum levels (50 ppm)	Permitted **AUSTRALIA:** Permitted **OTHER COUNTRIES:** Generally permitted in most countries

REFERENCE:

Smith, J. (Ed.) (1991) *Food Additive User's Handbook*. Blackie Academic & Professional.

STRUCTURE OF SUNSET YELLOW:

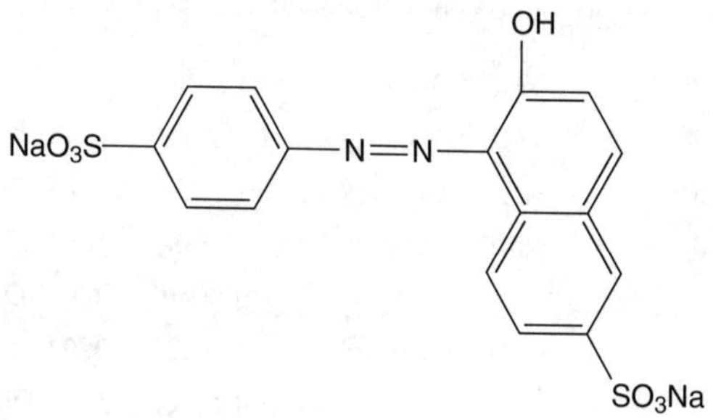

NAME:	Tartrazine
CATEGORY:	Food colour
FOOD USE:	Baked goods/ Cereals and cereal products/ Dairy products/ Fish and seafood products/ Meat and poultry products/ Egg products/ Fruit, vegetable and nut products/ Beverages/ Soft drinks/ Sugars and preserves/ Confectionery/ Alcoholic drinks/ Vinegar, pickles and sauces/ Decorations and coatings/ Edible ices/ Desserts including flavoured milk/ Seasonings/ Smoked fish/ Snacks/ Soups
SYNONYMS:	FD&C Yellow No. 5/ E102/ C.I. 19140 (Food Yellow 4)/ Trisodium 5-hydroxy-1-(4-sulfonatophenyl)-4-(4-sulfonatophenylazo)-*H*-pyrazole-3-carboxylate/ 5-Oxo-1-(*p*-sulfophenyl)-4-[(*p*-sulfophenyl)azo]-2-pyrazoline-3-carboxylic acid, trisodium salt
FORMULA:	C16 H9 N4 Na3 O9 S2
MOLECULAR MASS:	534.37
ALTERNATIVE FORMS:	Potassium salt/ Calcium salt/ Aluminium lake (insoluble)
PROPERTIES AND APPEARANCE:	A bright-yellow orange powder or granules
BOILING POINT IN °C AT VARIOUS PRESSURES (INCLUDING 760 mm Hg):	Not applicable
MELTING RANGE IN °C:	N/A
FLASH POINT IN °C:	N/A
IONISATION CONSTANT AT 25°C:	N/A
DENSITY AT 20°C (AND OTHER TEMPERATURES) IN g/l:	Not applicable
HEAT OF COMBUSTION AT 25°C IN J/kg:	N/A
VAPOUR PRESSURE AT VARIOUS TEMPERATURES IN mm Hg:	Not applicable

PURITY %:	Not less than 85% dye content total colouring matters calculated as the sodium salt
WATER CONTENT MAXIMUM IN %:	13
HEAVY METAL CONTENT MAXIMUM IN ppm:	40
ARSENIC CONTENT MAXIMUM IN ppm:	3
ASH MAXIMUM IN %:	N/A

SOLUBILITY % AT VARIOUS TEMPERATURE/pH COMBINATIONS:

in water:	**@ 20°C**	15%	**@ 50°C**	15%	**@ 100°C**	15%
in vegetable oil:	**@ 20°C**	Insoluble	**@ 50°C**	Insoluble	**@ 100°C**	Insoluble
in sucrose solution:	**@ 10%**	15%	**@ 40%**	15%	**@ 60%**	15%
in sodium chloride solution:	**@ 5%**	15%	**@ 10%**	15%	**@ 15%**	15%
in ethanol solution:	**@ 20%**	**@ 25°C** 10.7% **@ 60°C** 14.5%	**@ 100%**	**@ 25°C** Insoluble **@ 60°C** <0.1%		
in propylene glycol:	**@ 20°C**	6.5%	**@ 50°C**	6.5%	**@ 100°C**	6.5%

FUNCTION IN FOODS:	Water-soluble yellow food colour, or insoluble yellow food colour as the aluminium lake
ALTERNATIVES:	Other yellow food colours, although not of exactly the same hue, include: Quinoline Yellow; Riboflavin; Curcumin; Crocin; Safflower; Beta-carotene
TECHNOLOGY OF USE IN FOODS:	Very robust in use; excellent stability up to 105°C. Fades in presence of 10% sodium hydroxide. Appreciable fade with 1% ascorbic acid and over 25 ppm sulphur dioxide.
SYNERGISTS:	None known
ANTAGONISTS:	Ascorbic acid/ Sulphur dioxide/ Sodium hydroxide (and other strong alkalis)

FOOD SAFETY ISSUES: Media and public attention has been drawn to possible involvement in hyper-activity, particularly with respect to children. Scientific evidence in this respect is limited.

LEGISLATION:

USA:	**UK and EUROPE:**	**CANADA:**
GRAS; permitted as certified food colour according to 21 CFR 74.705	Permitted according to European Parliament and Council Directive 94/36/EC of June 30th 1994 on Colours for Use in Foodstuffs. Refer to Annex V, Part 2 for a substantial list of permitted food categories and maximum levels	Permitted **AUSTRALIA:** Permitted **OTHER COUNTRIES:** Generally permitted in most countries

REFERENCE: Smith, J. (Ed.) (1991) *Food Additive User's Handbook*. Blackie Academic & Professional.

FOOD USE CATEGORY: An important comment under this heading that applies to all colours is that although they can be used in the food categories listed the local legislation should be checked to ensure that the use of a particular colour is permitted is the application in any specified country.

STRUCTURE OF TARTRAZINE:

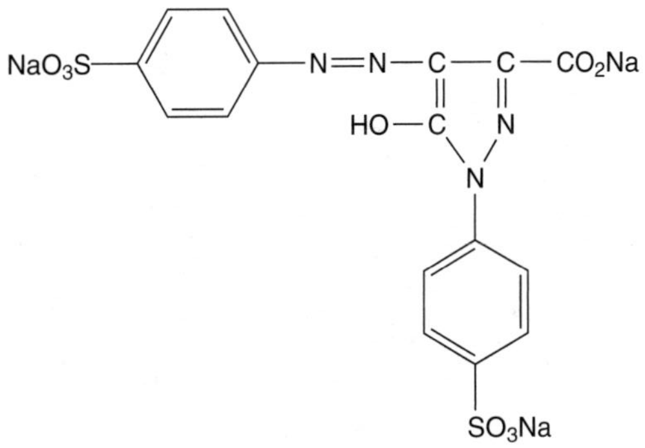

NAME:	Titanium dioxide
CATEGORY:	Food colour
FOOD USE:	Confectionery coatings/ Salad dressings/ Non-dairy creamers
SYNONYMS:	E171/ C.I. 77891 (Pigment White 6)
FORMULA:	TiO_2
MOLECULAR MASS:	79.88
PROPERTIES AND APPEARANCE:	Fine white powder
BOILING POINT IN °C AT VARIOUS PRESSURES (INCLUDING 760 mm Hg):	N/A
MELTING RANGE IN °C:	N/A
FLASH POINT IN °C:	N/A
IONISATION CONSTANT AT 25°C:	N/A
DENSITY AT 20°C (AND OTHER TEMPERATURES) IN g/l:	N/A
HEAT OF COMBUSTION AT 25°C IN J/kg:	N/A
VAPOUR PRESSURE AT VARIOUS TEMPERATURES IN mm Hg:	N/A
PURITY %:	99
WATER CONTENT MAXIMUM IN %:	0.5
HEAVY METAL CONTENT MAXIMUM IN ppm:	40

ARSENIC CONTENT MAXIMUM IN ppm:	3							
ASH MAXIMUM IN %:	N/A							
SOLUBILITY % AT VARIOUS TEMPERATURE/pH COMBINATIONS:								
in water:	**@ 20°C**	Insoluble	**@ 50°C**	Insoluble	**@ 100°C**	Insoluble		
in vegetable oil:	**@ 20°C**	Insoluble	**@ 50°C**	Insoluble	**@ 100°C**	Insoluble		
in sucrose solution:	**@ 10%**	Insoluble	**@ 40%**	Insoluble	**@ 60%**	Insoluble		
in sodium chloride solution:	**@ 5%**	Insoluble	**@ 10%**	Insoluble	**@ 15%**	Insoluble		
in ethanol solution:	**@ 5%**	Insoluble	**@ 20%**	Insoluble	**@ 95%**	Insoluble	**@ 100%**	Insoluble
in propylene glycol:	**@ 20°C**	Insoluble	**@ 50°C**	Insoluble	**@ 100°C**	Insoluble		

FUNCTION IN FOODS: Insoluble white food colour

ALTERNATIVES: Calcium carbonate

TECHNOLOGY OF USE IN FOODS: Insoluble white food colour used to impart a white opaque finish to sugar-panned confectionery, or as a background to added colours. Used to enhance the whiteness of non-dairy creamers, salad dressings or similar products.

Colour must be dispersed into a suitable medium since it is insoluble in solvents except acids.

SYNERGISTS: None known

ANTAGONISTS: None known

FOOD SAFETY ISSUES: None known

LEGISLATION:

USA:
Permitted GRAS according to 21 CFR 73.575

UK and EUROPE:
Permitted according to European Parliament and Council Directive 94/36/EC of 30th June 1994 on Colours for Use in Foodstuffs. Refer to Annex V, Part 1

CANADA:
Permitted

AUSTRALIA and PACIFIC RIM:
Australia: Not permitted
Japan: Permitted

REFERENCE:

Smith, J. (Ed.) (1991) *Food Additive User's Handbook*. Blackie Academic & Professional.

NAME:	Red 2G
CATEGORY:	Food colour
FOOD USE:	Sausages/ Preserves/ Confectionery/ Meat products
SYNONYMS:	E128/ C.I. 18050 (Food Red 10)/ Azogeranine/ Disodium 8-acetamido-1-hydroxy-2-phenylazonaphthalene 3,6-disulfonate
FORMULA:	C18 H13 N3 Na2 O8 S2
MOLECULAR MASS:	509.43 (sodium salt)
ALTERNATIVE FORMS:	Potassium salt/ Calcium salt/ Aluminium lake (insoluble)
PROPERTIES AND APPEARANCE:	Red powder or granules
BOILING POINT IN °C AT VARIOUS PRESSURES (INCLUDING 760 mm Hg):	N/A
MELTING RANGE IN °C:	N/A
FLASH POINT IN °C:	N/A
IONISATION CONSTANT AT 25°C:	N/A
DENSITY AT 20°C (AND OTHER TEMPERATURES) IN g/l:	N/A
HEAT OF COMBUSTION AT 25°C IN J/kg:	N/A
VAPOUR PRESSURE AT VARIOUS TEMPERATURES IN mm Hg:	N/A
PURITY %:	Content not less than 80% total colouring matters, calculated as the sodium salt
WATER CONTENT MAXIMUM IN %:	15

HEAVY METAL CONTENT MAXIMUM IN ppm:	40							
ARSENIC CONTENT MAXIMUM IN ppm:	3							
ASH MAXIMUM IN %:	N/A							
SOLUBILITY % AT VARIOUS TEMPERATURE/pH COMBINATIONS:								
in water:	**@ 20°C**	18%	**@ 50°C**	20%	**@ 100°C**	25%		
in vegetable oil:	**@ 20°C**	Insoluble	**@ 50°C**	Insoluble	**@ 100°C**	Insoluble		
in sucrose solution:	**@ 10%**	10%	**@ 40%**	6%	**@ 60%**	4%		
in sodium chloride solution:	**@ 5%**	Soluble	**@ 10%**	Insoluble	**@ 15%**	Insoluble		
in ethanol solution:	**@ 5%**	Insoluble	**@ 20%**	Insoluble	**@ 95%**	Insoluble	**@ 100%**	Insoluble
in propylene glycol:	**@ 20°C**	3%	**@ 50°C**	4%	**@ 100°C**	5%		
FUNCTION IN FOODS:	Water-soluble bright-red colour, or insoluble bright red food colour as the aluminium lake							
ALTERNATIVES:	Other red food colours, although not the same hue, include: Carmine; Carmoisine; Allura Red; Erythrosine; Beet Red; Anthocyanins (in acidic media); Ponceau 4R							
TECHNOLOGY OF USE IN FOODS:	A bright-red robust colour with good light and heat stability. Quite stable to acidic and alkaline conditions. Used extensively in sausages to give a characteristic pink/red shade which browns on cooking.							
SYNERGISTS:	None known							
ANTAGONISTS:	None known							
FOOD SAFETY ISSUES:	Red 2G has been shown to produce Heinz bodies in animals during metabolism. Its use has been restricted in Europe.							

LEGISLATION:

USA:
Not permitted

CANADA:
Not permitted

UK and EUROPE:
Permitted according to European Parliament and Council Directive 94/36/EC of June 30th 1994 on Colour for Use in Foodstuffs.
Refer to Annex IV – only permitted for breakfast sausages with a minimum cereal content of 6% and burger meat with a minimum vegetable and/or cereal content of 4% at a maximum level of 20 mg/kg

AUSTRALIA:
Not permitted

REFERENCE:

Smith, J. (Ed.) (1991) *Food Additive User's Handbook*. Blackie Academic & Professional.

STRUCTURE OF RED 2G:

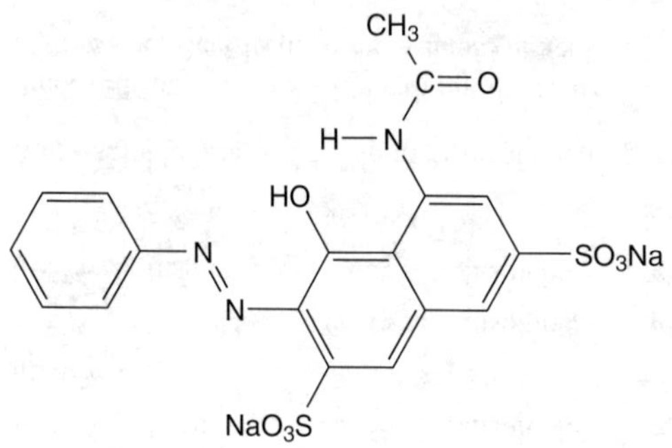

NAME:	Vegetable carbon
CATEGORY:	Food colour
FOOD USE:	Baked goods/ Cereal and cereal products/ Dairy products/ Meat and poultry products/ Sugars and preserves/ Confectionery/ Decorations and coatings/ Edible ices
SYNONYMS:	Carbon black/ E153
FORMULA:	C
MOLECULAR MASS:	12.01
ALTERNATIVE FORMS:	None known
PROPERTIES AND APPEARANCE:	Black powder, odourless and tasteless
BOILING POINT IN °C AT VARIOUS PRESSURES (INCLUDING 760 mm Hg):	Not applicable
MELTING RANGE IN °C:	Not applicable
FLASH POINT IN °C:	Not known
IONISATION CONSTANT AT 25°C:	Not known
DENSITY AT 20°C (AND OTHER TEMPERATURES) IN g/l:	Not applicable
HEAT OF COMBUSTION AT 25°C IN J/kg:	Not known
VAPOUR PRESSURE AT VARIOUS TEMPERATURES IN mm Hg:	Not applicable
PURITY %:	$\nless$95% carbon
WATER CONTENT MAXIMUM IN %:	2

HEAVY METAL CONTENT MAXIMUM IN ppm:	40							
ARSENIC CONTENT MAXIMUM IN ppm:	3							
ASH MAXIMUM IN %:	4.0							
SOLUBILITY % AT VARIOUS TEMPERATURE/pH COMBINATIONS:								
in water:	**@ 20°C**	Insoluble	**@ 50°C**	Insoluble	**@ 100°C**	Insoluble		
in vegetable oil:	**@ 20°C**	Insoluble	**@ 50°C**	Insoluble	**@ 100°C**	Insoluble		
in sucrose solution:	**@ 10%**	Insoluble	**@ 40%**	Insoluble	**@ 60%**	Insoluble		
in sodium chloride solution:	**@ 5%**	Insoluble	**@ 10%**	Insoluble	**@ 15%**	Insoluble		
in ethanol solution:	**@ 5%**	Insoluble	**@ 20%**	Insoluble	**@ 95%**	Insoluble	**@ 100%**	Insoluble
in propylene glycol:	**@ 20°C**	Insoluble	**@ 50°C**	Insoluble	**@ 100°C**	Insoluble		
FUNCTION IN FOODS:	A black food colour mainly used in sugar confectionery and commonly used as a shading agent							
ALTERNATIVES:	Other black food colours, although not of exactly the same hue, include: vegetable carbon.							
TECHNOLOGY OF USE IN FOODS:	Vegetable carbon is an insoluble pigment and as such is generally used in the form of a suspension.							
SYNERGISTS:	None known							
ANTAGONISTS:	None known							
FOOD SAFETY ISSUES:	None known							

LEGISLATION:

USA:
Not permitted

CANADA:
Not Permitted

UK and EUROPE:
Permitted according to European Parliament and Council Directive 94/36/EC of June 30th 1994 on Colours for Use in Foodstuffs.
Refer to Annex V, Part I for restrictions

AUSTRALIA:
Permitted

REFERENCE:

Smith, J. (Ed.) (1991) *Food Additive User's Handbook.* Blackie Academic & Professional.

NAME:	Safflower
CATEGORY:	Food colour (spice/flavour)
FOOD USE:	Baked goods/ Dairy products/ Egg products/ Soft drinks/ Sugars and preserves/ Confectionery/ Vinegars, pickles and sauces/ Desserts/ Seasonings/ Alcoholic drinks/ Edible ices
SYNONYMS:	Safflower Yellow/ Cartham/ Carthamin Yellow/ Carthamus/ C.I. Natural Yellow 5
FORMULA:	$C_{21}H_{22}O_{11}$
MOLECULAR MASS:	450.40
ALTERNATIVE FORMS:	N/A
PROPERTIES AND APPEARANCE:	Hygroscopic yellow to dark brown crystals, paste or powder with a faint characteristic odour
BOILING POINT IN °C AT VARIOUS PRESSURES (INCLUDING 760 mm Hg):	N/A
MELTING RANGE IN °C:	Unknown
FLASH POINT IN °C:	N/A
IONISATION CONSTANT AT 25°C:	N/A
DENSITY AT 20°C (AND OTHER TEMPERATURES) IN g/l:	N/A
HEAT OF COMBUSTION AT 25°C IN J/kg:	N/A
VAPOUR PRESSURE AT VARIOUS TEMPERATURES IN mm Hg:	N/A
PURITY %:	Typically extracts contain 0.5–1.0% pigment
WATER CONTENT MAXIMUM IN %:	N/A

HEAVY METAL CONTENT MAXIMUM IN ppm:	<40					
ARSENIC CONTENT MAXIMUM IN ppm:	<3					
ASH MAXIMUM IN %:	1					
SOLUBILITY % AT VARIOUS TEMPERATURE/pH COMBINATIONS:						
in water:	**@ 20°C**	Completely soluble	**@ 50°C**	Completely soluble	**@ 100°C**	Completely soluble
in vegetable oil:	**@ 20°C**	Insoluble	**@ 50°C**	Insoluble	**@ 100°C**	Insoluble
in sucrose solution:	**@ 10%**	Completely soluble	**@ 40%**	Completely soluble	**@ 60%**	Completely soluble
in sodium chloride solution:	**@ 5%**	Soluble	**@ 10%**	Soluble	**@ 15%**	Soluble
in ethanol solution:	**@ 5%**	Completely soluble	**@ 20%**	Soluble	**@ 95%**	Practically insoluble
	@ 100%	Practically insoluble				
in propylene glycol:	**@ 20°C**	Miscible	**@ 50°C**	Miscible	**@ 100°C**	Miscible
FUNCTION IN FOODS:	A bright-yellow water-soluble food colour that also imparts a characteristic flavour					
ALTERNATIVES:	Other yellow colours, although not of exactly the same hue, include: Beta-carotene; Tartrazine; Sunset Yellow; Turmeric/Curcumin; Crocin					
TECHNOLOGY OF USE IN FOODS:	Readily soluble in water and can be added directly to aqueous foodstuffs. For ease of addition, a pre-dilution in water can be used.					
SYNERGISTS:	None known					
ANTAGONISTS:	Sulphur dioxide					
FOOD SAFETY ISSUES:	None known					

LEGISLATION:

USA:
Not permitted as a food colour, but may be used as a characteristic flavouring

CANADA:
Not permitted, but may be used as a characteristic flavour

UK and EUROPE:
Not permitted as a food colour, but may be used as a characteristic flavour

AUSTRALIA:
Not permitted, but may be used as a characteristic flavour

REFERENCE:

Joint FAO/WHO Expert Committee on Food Additives (JECFA), FAO Food and Nutrition Paper 52/1 (1992).

NAME:	Santalin
CATEGORY:	Food colour (spice)
FOOD USE:	Meat and poultry products/ Alcoholic drinks/ Desserts/ Confectionery/ Fish and fish products/ Sugars and preserves/ Vinegars, pickles and sauces/ Decorations and coatings/ Seasonings/ Snack foods
SYNONYMS:	Sandalwood/ Red sandalwood/ Saunderswood/ Sanders Red
FORMULA:	Santalin A $C_{33}H_{26}O_{10}$; Santalin B $C_{34}H_{28}O_{10}$
MOLECULAR MASS:	Santalin A 582.52; Santalin B 596.54
ALTERNATIVE FORMS:	A much lesser pigment santalin 'C' is also reported in the literature
PROPERTIES AND APPEARANCE:	A dark red/brown oleoresin or crystalline powder
BOILING POINT IN °C AT VARIOUS PRESSURES (INCLUDING 760 mm Hg):	N/A
MELTING RANGE IN °C:	Santalin A 302–303; Santalin B 292–294
FLASH POINT IN °C:	N/A
IONISATION CONSTANT AT 25°C:	N/A
DENSITY AT 20°C (AND OTHER TEMPERATURES) IN g/l:	N/A
HEAT OF COMBUSTION AT 25°C IN J/kg:	N/A
VAPOUR PRESSURE AT VARIOUS TEMPERATURES IN mm Hg:	N/A
PURITY %:	Typically around 1% pigment content
WATER CONTENT MAXIMUM IN %:	N/A

HEAVY METAL CONTENT MAXIMUM IN ppm:	<20
ARSENIC CONTENT MAXIMUM IN ppm:	<1
ASH MAXIMUM IN %:	2

SOLUBILITY % AT VARIOUS TEMPERATURE/pH COMBINATIONS:

in water:	**@ 20°C**	Insoluble	**@ 50°C**	Insoluble	**@ 100°C**	Insoluble		
in vegetable oil:	**@ 20°C**	Insoluble	**@ 50°C**	Very slightly soluble	**@ 100°C**	Slightly soluble		
in sucrose solution:	**@ 10%**	Insoluble	**@ 40%**	Insoluble	**@ 60%**	Insoluble		
in sodium chloride solution:	**@ 5%**	Insoluble	**@ 10%**	Insoluble	**@ 15%**	Insoluble		
in ethanol solution:	**@ 5%**	Insoluble	**@ 20%**	Slightly soluble	**@ 95%**	Soluble	**@ 100%**	Soluble
in propylene glycol:	**@ 20°C**	Insoluble	**@ 50°C**	Miscible	**@ 100°C**	Miscible		

FUNCTION IN FOODS:	A spice or oleoresin that imparts a red/brown colour and characteristic flavour
ALTERNATIVES:	Other red/brown food colours, or combinations, although not of exactly the same hue, include: Caramel; Beetroot; Carmine; Paprika
TECHNOLOGY OF USE IN FOODS:	Directly soluble in alcohol, so an alcoholic solution can be used to colour/flavour alcoholic drinks. An emulsified formulation, usually based on polysorbate, can be used to add to aqueous foodstuffs. Such a formulation does not normally provide a clear solution in water, but rather a hazy red brown colour.
SYNERGISTS:	Emulsifiers, particularly polysorbates
ANTAGONISTS:	Sulphur dioxide
FOOD SAFETY ISSUES:	None known

LEGISLATION:

USA:
Not permitted as food colour specifically, but may be used as a spice/flavouring

UK and EUROPE:
Not permitted as a food colour, but may be used as a spice/flavour

CANADA:
Permitted as saunderswood

AUSTRALIA:
Not permitted as a food colour, but may be used as a spice/flavour

REFERENCES:

Verghese, J. (1986) Santalin – a peerless natural colourant. *Cosmetics & Toiletries*, **101**, 69–74.

Arnone, A., Camarda, L., *et al.* (1975) Structures of the red sandalwood pigments, santalins A and B. *J.C.S. Perkin's Trans*. I, 186–194.

ANY OTHER RELEVANT INFORMATION:

Santalin was permitted in the UK according to the Colouring Matter in Foodstuff Regulations 1973 (as amended). However, the 1996 EC directive on food colours does not list santalin as a permitted colour.

NAME:	Lutein
CATEGORY:	Food colour
FOOD USE:	Baked goods/ Dairy products/ Egg products/ Soft drinks/ Sugars and preserves/ Confectionery/ Decorations and coatings/ Edible ices/ Desserts
SYNONYMS:	E161b/ Mixed carotenoids/ Xanthophylls/ Tagetes
FORMULA:	C40 H51 O2
MOLECULAR MASS:	568.88
ALTERNATIVE FORMS:	Fatty acid esters (mono and di)
PROPERTIES AND APPEARANCE:	An orange-brown oleoresin
BOILING POINT IN °C AT VARIOUS PRESSURES (INCLUDING 760 mm Hg):	N/A
FLASH POINT IN °C:	N/A
IONISATION CONSTANT AT 25°C:	N/A
DENSITY AT 20°C (AND OTHER TEMPERATURES) IN g/l:	N/A
HEAT OF COMBUSTION AT 25°C IN J/kg:	N/A
VAPOUR PRESSURE AT VARIOUS TEMPERATURES IN mm Hg:	N/A
PURITY %:	≮4% calculated as lutein
WATER CONTENT MAXIMUM IN %:	2

HEAVY METAL CONTENT MAXIMUM IN ppm: 40

ARSENIC CONTENT MAXIMUM IN ppm: 3

ASH MAXIMUM IN %: 1

SOLUBILITY % AT VARIOUS TEMPERATURE/pH COMBINATIONS:

in water:	**@ 20°C**	Insoluble	**@ 50°C**	Insoluble	**@ 100°C**	Insoluble		
in vegetable oil:	**@ 20°C**	Very slightly soluble	**@ 50°C**	0.5%	**@ 100°C**	2%		
in sucrose solution:	**@ 10%**	Insoluble	**@ 40%**	Insoluble	**@ 60%**	Insoluble		
in sodium chloride solution:	**@ 5%**	Insoluble	**@ 10%**	Insoluble	**@ 15%**	Insoluble		
in ethanol solution:	**@ 5%**	Insoluble	**@ 20%**	Very slightly soluble	**@ 95%**	0.5%	**@ 100%**	0.5%
in propylene glycol:	**@ 20°C**	Insoluble	**@ 50°C**	Insoluble	**@ 100°C**	Insoluble		

FUNCTION IN FOODS: A bright egg-yellow food colour

ALTERNATIVES: Other yellow food colours, although not of exactly the same hue, include: Mixed carotenes; Beta-carotene; Carminic acid; Sunset yellow; Quinoline yellow; Crocin

TECHNOLOGY OF USE IN FOODS: For oil-based applications, a solution of the oleoresin in vegetable oil can be directly applied.

For aqueous-based food products, water-dispersible forms are required. These are generally oil-in-water emulsions with the pigment dissolved in the oil phase and using a food-grade emulsifier such as polysorbate.

SYNERGISTS: Antioxidants/ Emulsifiers (e.g. polysorbates)/ Vegetable oil

ANTAGONISTS: Sulphur dioxide/ Oxygen/ Metal ions

FOOD SAFETY ISSUES: None known

LEGISLATION:

USA:	UK and EUROPE:	CANADA:
Not permitted as food colour	Permitted according to European Parliament and Council Directive 94/36/EC of June 30th 1994 on Colours for Use in Foodstuffs. Refer to Annex V, Part 2 for restrictions	Permitted as xanthophylls **AUSTRALIA:** Permitted (GMP basis)

REFERENCES:

Smith, J. (Ed.) (1991) *Food Additive User's Handbook*. Blackie Academic & Professional.

Philip, T., Berry, J. W. (1976) A process for the purification of lutein-fatty acid esters from marigold petals. *Journal of Food Science*, **41**, 163–164.

STRUCTURE OF LUTEIN:

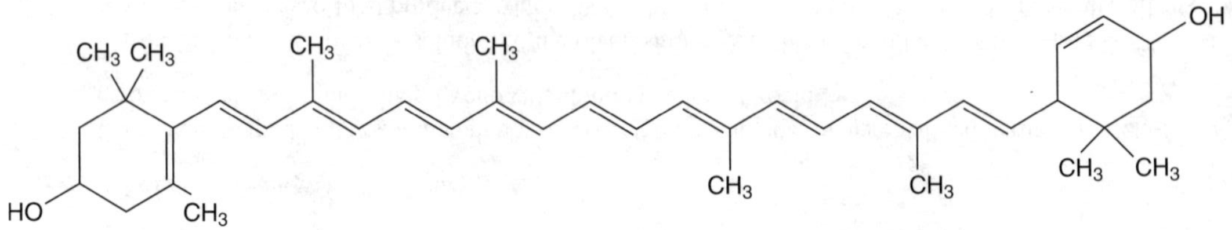

NAME:	Lycopene
CATEGORY:	Food colour
FOOD USE:	Baked goods/ Dairy products/ Egg products/ Soft drinks/ Sugars and preserves/ Confectionery/ Vinegar, pickles and sauces/ Decorations and coatings/ Desserts/ Seasonings
SYNONYMS:	Natural Yellow 27/ C.I. 75125/ ψ,ψ-carotene
FORMULA:	$C_{40}H_{56}$
MOLECULAR MASS:	536.85
PROPERTIES AND APPEARANCE:	An orange/red suspension in oil
BOILING POINT IN °C AT VARIOUS PRESSURES (INCLUDING 760 mm Hg):	N/A
MELTING RANGE IN °C:	169–171
FLASH POINT IN °C:	N/A
IONISATION CONSTANT AT 25°C:	N/A
DENSITY AT 20°C (AND OTHER TEMPERATURES) IN g/l:	N/A
HEAT OF COMBUSTION AT 25°C IN J/kg:	N/A
VAPOUR PRESSURE AT VARIOUS TEMPERATURES IN mm Hg:	N/A
PURITY %:	≮5% of total colouring matters
WATER CONTENT MAXIMUM IN %:	2

HEAVY METAL CONTENT MAXIMUM IN ppm:	40							
ARSENIC CONTENT MAXIMUM IN ppm:	3							
ASH MAXIMUM IN %:	0.1							
SOLUBILITY % AT VARIOUS TEMPERATURE/pH COMBINATIONS:								
in water:	**@ 20°C**	Insoluble	**@ 50°C**	Insoluble	**@ 100°C**	Insoluble		
in vegetable oil:	**@ 20°C**	Very slightly soluble	**@ 50°C**	0.05%	**@ 100°C**	0.10%		
in sucrose solution:	**@ 10%**	Insoluble	**@ 40%**	Insoluble	**@ 60%**	Insoluble		
in sodium chloride solution:	**@ 5%**	Insoluble	**@ 10%**	Insoluble	**@ 15%**	Insoluble		
in ethanol solution:	**@ 5%**	Insoluble	**@ 20%**	Insoluble	**@ 95%**	0.05%	**@ 100%**	0.05%
in propylene glycol:	**@ 20°C**	Insoluble	**@ 50°C**	Insoluble	**@ 100°C**	Insoluble		
FUNCTION IN FOODS:	An orange/red food colour							
ALTERNATIVES:	Paprika; Beta-carotene; Mixed carotenes; Annatto; Sunset yellow							
TECHNOLOGY OF USE IN FOODS:	For oil-based applications, a suspension in oil can be directly applied. For aqueous food products, water-dispersible forms are required, using food-grade emulsifiers.							
SYNERGISTS:	Antioxidants/ Vegetable oil/ Emulsifiers							
ANTAGONISTS:	Sulphur dioxide/ Oxygen/ Metal ions							
FOOD SAFETY ISSUES:	None known							

LEGISLATION:

USA:
Not permitted specifically, but if used in form of concentrated tomato extract, then it would be permitted as a non-certified colour according to 21 CFR 73 (vegetable juice)

UK and EUROPE:
Permitted according to European Parliament and Council Directive 94/36/EC of June 30th 1994 on Colours for Use in Foodstuffs.
Refer to Annex V, Part 2 for restrictions on use

CANADA:
Not permitted

AUSTRALIA:
Permitted

REFERENCES:

Smith, J. (Ed.) (1991) *Food Additive User's Handbook*. Blackie Academic & Professional.
Nir, Z., Hartal, D., Raveh, Y. (1993) Lycopene from tomatoes. *International Food Ingredients*, **6**.

STRUCTURE OF LYCOPENE:

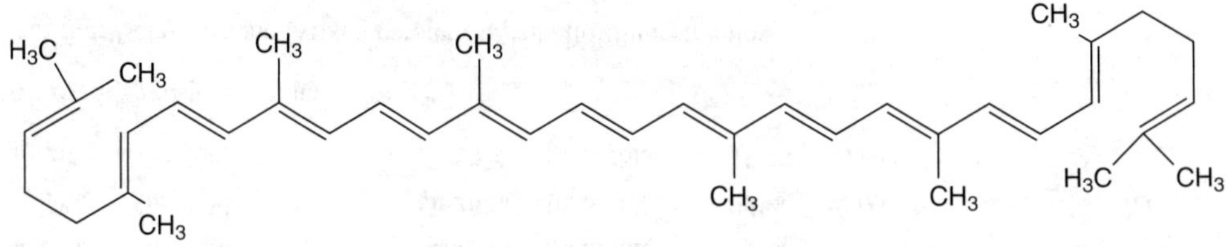

NAME:	Paprika extract
CATEGORY:	Food colour
FOOD USE:	Baked goods/ Dairy products/ Fish and seafood products/ Meat and poultry products/ Egg products/ Soft drinks/ Sugars and preserves/ Confectionery/ Vinegar, pickles and sauces/ Decorations and coatings/ Edible ices/ Desserts/ Seasonings
SYNONYMS:	Paprika oleoresin/ Capsaicin/ Capsorubin/ E160c
FORMULA:	Capsanthin, $C_{40}H_{56}O_3$; Capsorubin, $C_{40}H_{56}O_4$
MOLECULAR MASS:	Capsanthin, 584.85; Capsorubin, 600.85
ALTERNATIVE FORMS:	Fatty acid esters
PROPERTIES AND APPEARANCE:	A dark orange/brown oleoresin
BOILING POINT IN °C AT VARIOUS PRESSURES (INCLUDING 760 mm Hg):	N/A
MELTING RANGE IN °C:	N/A
FLASH POINT IN °C:	N/A
IONISATION CONSTANT AT 25°C:	N/A
DENSITY AT 20°C (AND OTHER TEMPERATURES) IN g/l:	N/A
HEAT OF COMBUSTION AT 25°C IN J/kg:	N/A
VAPOUR PRESSURE AT VARIOUS TEMPERATURES IN mm Hg:	N/A
PURITY IN %:	≥93 carotenoids, of which capsanthin/capsorubin not less than 30% of total carotenoids

WATER CONTENT MAXIMUM IN %: 2

HEAVY METAL CONTENT MAXIMUM IN ppm: 40

ARSENIC CONTENT MAXIMUM IN ppm: 3

ASH MAXIMUM IN %: 0.2

SOLUBILITY % AT VARIOUS TEMPERATURE/pH COMBINATIONS:

in water:	@ **20°C**	Insoluble	@ **50°C**	Insoluble	@ **100°C**	Insoluble		
in vegetable oil:	@ **20°C**	0.5%	@ **50°C**	1.0%	@ **100°C**	10.0%		
in sucrose solution:	@ **10%**	Insoluble	@ **40%**	Insoluble	@ **60%**	Insoluble		
in sodium chloride solution:	@ **5%**	Insoluble	@ **10%**	Insoluble	@ **15%**	Insoluble		
in ethanol solution:	@ **5%**	Insoluble	@ **20%**	Insoluble	@ **95%**	0.10%	@ **100%**	0.10%
in propylene glycol:	@ **20°C**	Insoluble	@ **50°C**	Insoluble	@ **100°C**	Insoluble		

FUNCTION IN FOODS: An orange/red food colour

ALTERNATIVES: Beta-carotene; Mixed carotenes; Annatto; Lycopene; Sunset Yellow

TECHNOLOGY OF USE IN FOODS: Paprika oleoresin can be dispersed in vegetable oil for direct application into oil-based food products.

For aqueous-based food products, water-dispersible forms are required and food-grade emulsifiers are used.

Care needs to be taken to use deodorised forms for sweet or dairy applications.

SYNERGISTS: Vegetable oil/ Antioxidants/ Emulsifiers

ANTAGONISTS: Sulphur dioxide/ Oxygen/ Metal ions

FOOD SAFETY ISSUES: None known

LEGISLATION:

USA:
Permitted as non-certified food colour according to 21 CFR 73

CANADA:
Permitted

UK and EUROPE:
Permitted according to European Parliament and Council Directive 94/36/EC of June 30th 1994 on Colour for Use in Foodstuffs. Refer to Annex V, Part 1 for restrictions on use

AUSTRALIA:
Permitted

REFERENCE: Smith, J. (ed.) (1991) *Food Additive Users Handbook.* Blackie Academic & Professional.

STRUCTURE OF PAPRIKA CAROTENOIDS:

Capsanthin

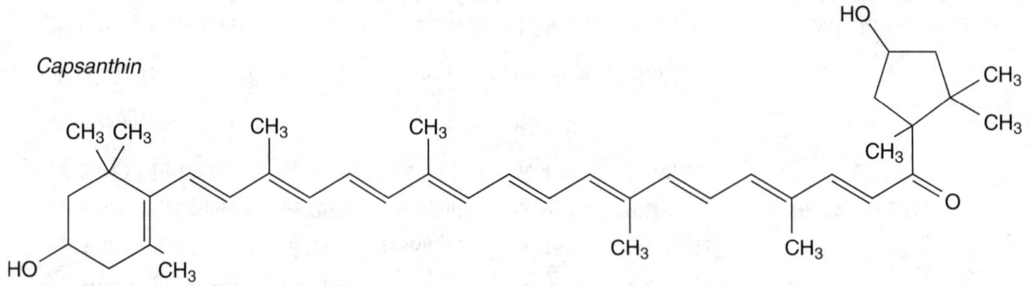

Capsorubin

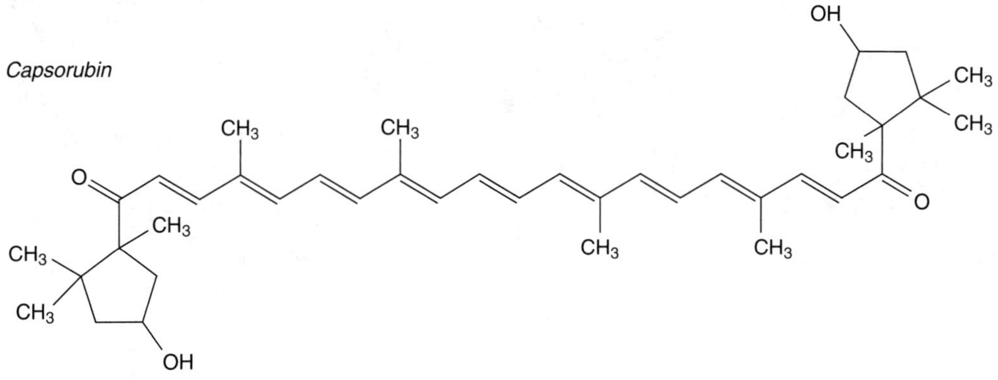

NAME:	**Riboflavin**
CATEGORY:	Food colour
FOOD USE:	Confectionery/ Cereal products/ Sauces/ Dairy products
SYNONYMS:	E101 (i)/ Lactoflavin/ Vitamin B2/ 7,8-Dimethyl-10-(D-ribo-2,3,4,5-tetra hydroxypentyl) benzo(g)pteridine-2,4 (3*H*,10*H*)-dione/ 7,8-Dimethyl-10-(1′-D-ribityl)isoalloxazine
FORMULA:	$C_{17}H_{20}N_4O_6$
MOLECULAR MASS:	376.37
PROPERTIES AND APPEARANCE:	Yellow to orange-yellow crystalline powder
BOILING POINT IN °C AT VARIOUS PRESSURES (INCLUDING 760 mm Hg):	N/A
MELTING RANGE IN °C:	278–282
FLASH POINT IN °C:	N/A
IONISATION CONSTANT AT 25°C:	N/A
DENSITY AT 20°C (AND OTHER TEMPERATURES) IN g/l:	N/A
HEAT OF COMBUSTION AT 25°C IN J/kg:	N/A
VAPOUR PRESSURE AT VARIOUS TEMPERATURES IN mm Hg:	N/A
PURITY %:	Not less than 98% on the anhydrous basis
WATER CONTENT MAXIMUM IN %:	1.5

HEAVY METAL CONTENT MAXIMUM IN ppm:	40					
ARSENIC CONTENT MAXIMUM IN ppm:	3					
SOLUBILITY % AT VARIOUS TEMPERATURE/pH COMBINATIONS:						
in water:	**@ 20°C**	Slightly soluble; readily soluble in alkaline solutions pH > 8.0	**@ 50°C**	Slightly soluble	**@ 100°C**	Soluble (0.03%)
in vegetable oil:	**@ 20°C**	Insoluble	**@ 50°C**	Insoluble	**@ 100°C**	Insoluble
in sucrose solution:	**@ 10%**	Slightly soluble	**@ 40%**	Slightly soluble	**@ 60%**	Slightly soluble
in sodium chloride solution:	**@ 5%**	Slightly soluble (more so than in water)	**@ 10%**	Slightly soluble (more so than in water)	**@ 15%**	Slightly soluble (more so than in water)
in ethanol solution:	**@ 5%**	Slightly soluble	**@ 20%**	Slightly soluble		
	@ 95%	Sparingly soluble (0.005%)	**@ 100%**	Sparingly soluble (0.005%)		
in propylene glycol:	**@ 20°C**	Slightly soluble	**@ 50°C**	Slightly soluble	**@ 100°C**	Slightly soluble
FUNCTION IN FOODS:	Greenish-yellow food colour (vitamin B2)					
ALTERNATIVES:	Other yellow food colours, although not the same hue, include: Curcumin; Tartrazine; Quinoline Yellow; Beta-carotene; Riboflavin-5′-phosphate					
TECHNOLOGY OF USE IN FOODS:	Can be added direct to foodstuffs to act as a pigment, or dissolved if at low concentrations. A stock solution in dilute alkali can be prepared to facilitate coloration of the foodstuff.					
SYNERGISTS:	Antioxidants					
ANTAGONISTS:	Strong alkalis/ Sulphur dioxide/ Metal ions					
FOOD SAFETY ISSUES:	None known					

LEGISLATION:

USA:
Not permitted as food colouring

UK and EUROPE:
Permitted according to European Parliament and Council Directive 94/36/EC of June 30th 1994 on Colours for Use in Foodstuffs. Refer to various Annexes for restrictions

CANADA:
Permitted

AUSTRALIA:
Permitted

REFERENCE:

Smith, J. (ed.) (1991) *Food Additive User's Handbook.* Blackie Academic & Professional.

STRUCTURE OF RIBOFLAVIN:

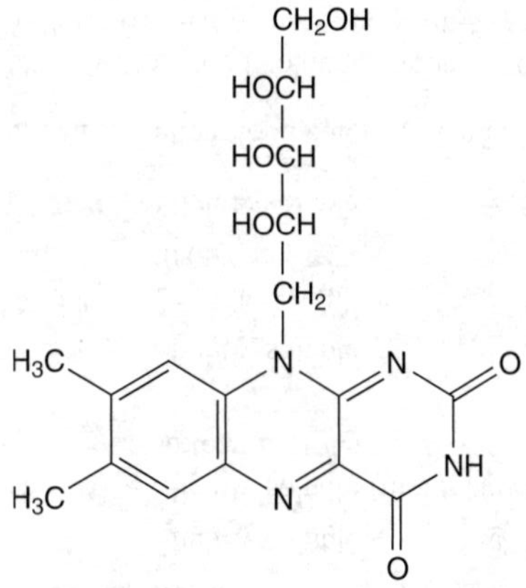

NAME:	**Riboflavin-5′-phosphate**
CATEGORY:	Food colour
FOOD USE:	Confectionery/ Cereal products/ Sauces/ Dairy products
SYNONYMS:	E101 (ii)/ Riboflavin-5′-phosphate sodium salt/ (2*R*,3*R*,4*S*)-5-(3′)10′-dihydro-7′,8′-dimethyl-2′,4′-dioxo-10′-benzo[y]pteridinyl)-2,3,4-trihydroxypentyl phosphate/ Monosodium salt of 5′-monophosphate ester of riboflavin
FORMULA:	C17 H20 N4 Na O9 P ·2H2O
MOLECULAR MASS:	541.36
PROPERTIES AND APPEARANCE:	Yellow to orange crystalline hygroscopic powder
BOILING POINT IN °C AT VARIOUS PRESSURES (INCLUDING 760 mm Hg):	N/A
FLASH POINT IN °C:	N/A
IONISATION CONSTANT AT 25°C:	N/A
DENSITY AT 20°C (AND OTHER TEMPERATURES) IN g/l:	N/A
HEAT OF COMBUSTION AT 25°C IN J/kg:	N/A
VAPOUR PRESSURE AT VARIOUS TEMPERATURES IN mm Hg:	N/A
PURITY %:	Not less than 95% total colouring matter calculated as the dihydrate
WATER CONTENT MAXIMUM IN %:	8
HEAVY METAL CONTENT MAXIMUM IN ppm:	≮40

ARSENIC CONTENT MAXIMUM IN ppm:	3							
SOLUBILITY % AT VARIOUS TEMPERATURE/pH COMBINATIONS:								
in water:	**@ 20°C**	Soluble (10%)	**@ 50°C**	Soluble (10%)	**@ 100°C**	Soluble (12%)		
in vegetable oil:	**@ 20°C**	Insoluble	**@ 50°C**	Insoluble	**@ 100°C**	Insoluble		
in sucrose solution:	**@ 10%**	Soluble	**@ 40%**	Soluble	**@ 60%**	Soluble		
in sodium chloride solution:	**@ 5%**	Slightly soluble (more soluble than in water)	**@ 10%**	Slightly soluble (more soluble than in water)	**@ 15%**	Slightly soluble (more soluble than in water)		
in ethanol solution:	**@ 5%**	Slightly soluble	**@ 20%**	Slightly soluble	**@ 95%**	Sparingly soluble (0.005%)	**@ 100%**	Sparingly soluble (0.005%)
in propylene glycol:	**@ 20°C**	Soluble	**@ 50°C**	Soluble	**@ 100°C**	Soluble		
FUNCTION IN FOODS:	Greenish-yellow food colour (vitamin B2)							
ALTERNATIVES:	Other yellow food colours, although not the same hue, include: Riboflavin; Curcumin; Tartrazine; Quinoline Yellow; Beta-carotene							
TECHNOLOGY OF USE IN FOODS:	Can be added directly to foodstuffs. A stock solution in water can be prepared to facilitate the colouration of the foodstuff.							
SYNERGISTS:	Antioxidants							
ANTAGONISTS:	Strong alkalis/ Sulphur dioxide/ Metal ions							
FOOD SAFETY ISSUES:	None known							

LEGISLATION:

USA:	UK and EUROPE:	CANADA:
Not permitted as food colour	Permitted according to European Parliament and Council Directive 94/36/EC of June 30th 1994 on Colours for Use in Foodstuffs. Refer to various Annexes for restrictions	Generally permitted **AUSTRALIA:** Permitted

REFERENCE:

Smith, J. (ed.) (1991) *Food Additive User's Handbook.* Blackie Academic & Professional.

Part 4

Emulsifiers

NAME:	Mono- and diglycerides of fatty acids – saturated
CATEGORY:	Emulsifier
FOOD USE:	Baked goods/ Cereals and cereal products/ Edible oils and fats/ Cake batters/ Ice-cream/ Margarine/ Table spreads/ Shortening/ Bread/ Toffees and caramels/ Potato products/ Extruded snacks
SYNONYMS:	Glyceryl monostearate/ E471/ GMS/ Glyceryl stearate/ Glyceryl monopalmitate/ Mono-Di
FORMULA:	CH2OH-CHOH-CH2OOC-(CH2)16·CH3 (for monostearate)
MOLECULAR MASS:	358
ALTERNATIVE FORMS:	Varies according to fatty acid composition. Varies according to mono-ester content (30%–90+%)
PROPERTIES AND APPEARANCE:	White to cream beads, flakes or powder
MELTING RANGE IN °C:	60–72
PURITY %:	Mono-ester varying from min 30% to 90% +
WATER CONTENT MAXIMUM IN %:	2
HEAVY METAL CONTENT MAXIMUM IN ppm:	10 (as Pb)
ARSENIC CONTENT MAXIMUM IN ppm:	3
ASH MAXIMUM IN %:	0.5

SOLUBILITY % AT VARIOUS TEMPERATURE/pH COMBINATIONS:

in water:	**@ 20°C**	Insoluble	**@ 50°C**	Dispersible	**@ 100°C**	Dispersible but may form cubic mesophases
in vegetable oil:	**@ 50°C**	Dispersible				
in sucrose solution:	**@ 60%**	Dispersible				

in ethanol solution:	**@ 95%**	Soluble
in propylene glycol:	**@ 50°C**	Soluble

FUNCTION IN FOODS:

Aeration in cakes and sponges. Amylose-complexing in cereals and potato products. Crumb softening in bread. Emulsifying in cream liqueurs, margarine. Emulsion stabilising in bakery compounds. Improvement of texture consistency in ice-cream, toffees. Creaming in cake margarine. Reduced stickiness in pasta and potato products, caramels. Extrusion aid in snacks. Lubrication in chewing gum base

Amylose complexing index – min 90% monoester: 87–92 (excellent)
min 40% monoester: 40–45 (poor)

HLB – 3–4 (low polarity)

ALTERNATIVES:

Mono- and diglycerides of fatty acids-unsaturated. Other emulsifying agents in varying respects at different dosages

TECHNOLOGY OF USE IN FOODS:

When producing an aqueous dispersion for cake aeration it is essential to control temperature to max 60–65°C, depending on emulsifier type, to ensure correct crystal formation. (see also ref. 1)

For ice-cream, minimum levels of 0.21% 1-monoglyceride are recommended (see also ref. 2)

For crumb-softening (anti-staling, anti-firming) levels of min 0.3–0.5% monoglyceride on flour weight are recommended added in the form of a hydrate (see also ref. 3)

Improves fat distribution and chewing properties/ reduces stickiness in toffees and caramels

SYNERGISTS:

Alpha-tending emulsifiers (acetic and lactic acid ester and propylene glycol esters) for maintaining crystal form for aeration

ANTAGONISTS:

In aeration dispersions – acids and alkalis

FOOD SAFETY ISSUES:

No acute toxic effects at practicable dosage levels. No harmful effects specifically associated with mono- and diglycerides. ADI not limited.

LEGISLATION:

USA:
US FDA 21 CFR §184.1505. GRAS: Dough strengthener, emulsifier, formulation aid, lubricant, surface-active agent, etc.

AUSTRALIA/PACIFIC RIM:
Generally permitted in all countries

UK and EUROPE:
E471 – ADI no limit
EC Directive 95/2 (OJ No.L61, 18.3.95) Miscellanous Additives (Enacted in UK as Statutory Instrument 1995 No.3187) Schedule 1 (Generally permitted for use in food) & Schedules 4, 7 & 8 Purity Criteria laid down in EC Directive 78/663 (OJ No.L223, 14.7.78)

REFERENCES:

Krog, N., Rusom, T., and Larsson, K. (1988) Applications in the food industry, In: Becher, P. (Ed.), *Encyclopedia of emulsion technology, vol 2.* Marcel Dekker, Inc., New York, Basel, p. 321.

Flack, E. (1996) The role of emulsifiers in low-fat food products. In: Roller, S., and S. A. Jones (Eds.), *Handbook of fat replacers*, CRC Press, Inc., Boca Raton, New York, London, Tokyo, p. 213.

Flack, E., and Krog, N. (1998) *Influence of monoglycerides on staling in wheat bread*. In: Turner, A. (Ed.), Food Technology International Europe Sterling Publications Ltd, London, p. 199.

FAO/WHO (1973) *Specifications for identity & purity of food additives, part v. emulsifiers*. Rome, pp. 17–18.

ANY OTHER RELEVANT INFORMATION:

Amylose complexing index (see Krog, N. (1970) Interaction of monoglyerides in different physical states with amylose and their anti-firming effects in bread. *J. Food. Tech.*, **5**, 77.

Acid value max 3; Iodine value max 3

NAME:	**Mono- and diglycerides of fatty acids – unsaturated**					
CATEGORY:	Emulsifier					
FOOD USE:	Margarine and table spreads/ Bakery compounds/ Cake margarine and shortening/ Whipped toppings/ Coffee whiteners					
SYNONYMS:	Glyceryl monostearate/ GMS/ Mono-Di/ E471					
FORMULA:	CH2OH-CHOH-CH2OOC-(CH2)7·CH=CH(CH2)7·CH3 (for monooleate)					
MOLECULAR MASS:	356					
ALTERNATIVE FORMS:	Varies according to fatty acid composition. Varies according to mono-ester content (30%–90+%)					
PROPERTIES AND APPEARANCE:	Cream to yellow viscous liquids to pastes					
MELTING RANGE IN °C:	<20–60					
PURITY %:	Mono-ester varying from min 30% to 90+%					
WATER CONTENT MAXIMUM IN %:	2					
HEAVY METAL CONTENT MAXIMUM IN ppm:	10 (as Pb)					
ARSENIC CONTENT MAXIMUM IN ppm:	3					
ASH MAXIMUM IN %:	0.5					
SOLUBILITY % AT VARIOUS TEMPERATURE/pH COMBINATIONS:						
in water:	**@ 20°C**	Insoluble	**@ 50°C**	Dispersible	**@ 100°C**	Dispersible
in vegetable oil:	**@ 20°C**	Dispersible				
in ethanol solution:	**@ 100%**	Soluble				
in propylene glycol:	**@ 100°C**	Soluble				

FUNCTION IN FOODS:	Emulsion-stabilising in margarine and spreads, bakery compounds and coffee whiteners. Aeration/creaming in cake margarine and shortening. Foam stabilising in whipped toppings Amylose-complexing index – min 90% monoester – 26–35 (poor) HLB – 3–4 (low polarity)	
ALTERNATIVES:	Mono- and diglycerides of fatty acids – saturated. Other emulsifying agents in varying respects at different dosages	
TECHNOLOGY OF USE IN FOODS:	For low-fat spreads, unsaturated mono-diglycerides (IV 80–105) are preferred for producing a fine stable dispersion of water droplets. Emulsifier dosage 0.5–1%, possibly in combination with 0.2–0.5% polyglycerol polyricinoleate (IV approx. 85) and hydrocolloids. For bakery compounds (40–80% soft fat), unsaturated mono-diglycerides (IV 60–80) are recommended at dosages of 1–2% to produce a fine stable dispersion with good shelf-life	
FOOD SAFETY ISSUES:	No acute toxic effects at practicable dosage levels. No harmful effects specifically associated with mono- and diglycerides. ADI – not limited.	
LEGISLATION:	**USA:** US FDA 21 CFR 184.1505. GRAS – Dough strengthener, emulsifiers, formulation aids, lubricants, surface-active agents, etc.	**UK and EUROPE:** E471 – ADI no limit EC Directive 95/2 (OJ No. L61, 18.3.95) Schedule 1 (enacted in UK as statutory instrument 1995 No. 3187) Miscellaneous additives generally permitted for use. Also schedules 4, 7 & 8 (parts 1, 2, 3 & 4) Purity criteria laid down in EC Directive 78/663 (OJ L223, 14/8/78)
REFERENCES:	Madsen, J. (1989) Low-calorie spread and melange production in Europe. In: *World Conference on Edible Oils and Fats* AOCS, Maastricht, October. Flack, E. Butter, margarine, spreads and baking fats. In: Gunstone, F. D., and F. B. Padley (Eds.), *Handbook of lipid technology*, Marcel Dekker, Inc., New York (in preparation). FAO/WHO (1973) *Specifications for identity and purity of food additives part v. emulsifiers*. Rome, pp. 17–18.	
ANY OTHER RELEVANT INFORMATION:	Acid value max 3; Iodine value 40–105	

NAME:	Acetic acid esters of mono- and diglycerides of fatty acids
CATEGORY:	Emulsifier
FOOD USE:	Chewing gum base/ Cake batters/ Topping powders/ Aerated desserts/ Coatings
SYNONYMS:	Acetylated monoglycerides/ Acetem/ E472a/ Acetylated mono- and diglycerides/ Acetoglycerides
FORMULA:	CH2·O·COCH3-CHOH-CH2·O·CO·(CH2)16·CH3 (for stearate)
MOLECULAR MASS:	400
ALTERNATIVE FORMS:	Varies according to degree of acetylation (50–96%) and fatty acid composition
PROPERTIES AND APPEARANCE:	Clear liquid to white paste
MELTING RANGE IN °C:	5–50
PURITY %:	Total acetic acid min 9%, max 32%
WATER CONTENT MAXIMUM IN %:	2
HEAVY METAL CONTENT MAXIMUM IN ppm:	10 (as Pb)
ARSENIC CONTENT MAXIMUM IN ppm:	3
ASH MAXIMUM IN %:	0.5
SOLUBILITY % AT VARIOUS TEMPERATURE/pH COMBINATIONS:	
in water:	**@ 20°C** Insoluble
FUNCTION IN FOODS:	Fat crystal modification – alpha tending. Foam stiffening and stabilising. Release agent for high sugar goods. Coating agent (mono-molecular film-forming) for nuts, fruit pieces, meat joints. Amylose complexing index HLB 2 (low polarity)

ALTERNATIVES: Lactic acid esters/ propylene glycol monostearate and palmitate

TECHNOLOGY OF USE IN FOODS: Stabilises monoglycerides in alpha crystal form in emulsions used to facilitate aeration of cake and sponge batters.

Maintains fats in alpha crystal form to stabilise aeration of whipped desserts, imitation creams and toppings.

Forms protective coating against loss of moisture and fat oxidation for nuts, raisins, etc.

FOOD SAFETY ISSUES: No acute toxic effects at practicable dosage levels. ADI not limited

LEGISLATION:

USA:	UK and EUROPE:
US FDA 21 CFR 172.828 – food additive	E472a – ADI no limit EC Directive 95/2 (OJ No.L61, 18.3.95) Miscellanous Additives (Enacted in UK as Statutory Instrument 1995 No. 3187) Schedule 1 (Generally permitted for use in food) & Schedules 4, 7 & 8 Purity Criteria laid down in EC Directive 78/663 (OJ No.L223, 14.7.78)

REFERENCES:

Andreasen, J. (1973) The efficiency of emulsifiers in whipped topping, at CIFST Annual Conference, Vancouver, Canada, April.

Martin, J. B., and Lutton, E. S. (1972) Preparation and phase behavior of acetyl monoglycerides. *J. Am. Oil Chem. Soc.* **49**, 683.

ANY OTHER RELEVANT INFORMATION: Acid value max 3; Iodine value max 5 or 40–50; Saponification value 280–385

NAME:	Citric acid esters of mono- and diglycerides of fatty acids
CATEGORY:	Emulsifier
FOOD USE:	Frying margarine/ Meat products/ Spreads/ Whipping cream/ Beverage emulsions
SYNONYMS:	Citroglycerides/ E472c/ Citrem/ Monoglyceride citrate
FORMULA:	CH2·O·C=O·CH2·COH·COOH·CH2·COOH-CHOH-CH2·O·CO(CH2)16·CH3 (for stearate)
MOLECULAR MASS:	532
ALTERNATIVE FORMS:	Varies according to degree of citration and fatty acid composition
PROPERTIES AND APPEARANCE:	White to off-white powder or flakes
MELTING RANGE IN °C:	58–64
FLASH POINT IN °C:	>100
PURITY %:	Total citric acid min 13%, max 50%
WATER CONTENT MAXIMUM IN %:	2
HEAVY METAL CONTENT MAXIMUM IN ppm:	10 (as Pb)
ARSENIC CONTENT MAXIMUM IN ppm:	3
ASH MAXIMUM IN %:	0.5
FUNCTION IN FOODS:	Provides emulsion stability through fine distribution of water droplets. Synergist/solubiliser for antioxidants. Amylose complexing index 36 (poor) HLB 10–12 (medium/high polarity)

TECHNOLOGY OF USE IN FOODS:	Crystallises in α crystal form. Excellent anti-spattering agent	
LEGISLATION:	**USA:**	**UK and EUROPE:**
	US FDA 21 CFR 172.832	E472c ADI – no limit. EC Directive 0.5/2 (OJ No. L61, 18.3.95) Schedule 1 (Enacted in UK as statutory instrument 1995 No. 3187) Miscellaneous Additives Generally Permitted For Use in Food. Also schedules 4 & 8 (Part 3) Purity Criteria Laid Down in EC Directive 78/663 (OJ No. L223, 14.7.78)
REFERENCE:	Flack, E. (1985) Foam stabilisation of dairy whipping cream. *Dairy industries international*, **50**(6).	
ANY OTHER RELEVANT INFORMATION:	Acid value 10–40; Saponification value 220–255.	

NAME:	Diacetyl tartaric acid esters of mono- and diglycerides of fatty acids
CATEGORY:	Emulsifier
FOOD USE:	Baked goods/ Bread/ Cereal products/ Coffee whiteners/ Biscuits/ Extruded snacks/ Sauces/ Soups/ Colour concentrates
SYNONYMS:	DATEM/ Mono- and diacetyl tartaric acid/ E472e/ Esters of mono- and diglycerides of fatty acids
FORMULA:	CH2·O·C=O·CH·O·COCH3·CH·O·COCH3·COOH-CHOH-CH2·O·CO·(CH2)16·CH3 (for stearate)
MOLECULAR MASS:	574
ALTERNATIVE FORMS:	Varies according to level of diacetyl tartaric acid and fatty acid composition
PROPERTIES AND APPEARANCE:	Pale liquid to paste to powder/ flakes
MELTING RANGE IN °C:	Liquid – 45
PURITY %:	Total tartaric acid content; min 10%, max 40%
WATER CONTENT MAXIMUM IN %:	2
HEAVY METAL CONTENT MAXIMUM IN ppm:	10 (as Pb)
ARSENIC CONTENT MAXIMUM IN ppm:	3
ASH MAXIMUM IN %:	0.5
SOLUBILITY % AT VARIOUS TEMPERATURE/pH COMBINATIONS:	
in water:	@ 20°C Dispersible
in vegetable oil:	@ 20°C Partially soluble

FUNCTION IN FOODS:

Used in bread and yeast-raised products at levels of 0.2–0.6% of flour weight. DATEM possess excellent dough-strengthening properties due to their ability to interact with gluten. They thus enhance gas (CO_2) retention and improve tolerance to mechanical handling. HLB 8–10 improves fat particle distribution in coffee whiteners, thereby improving whitening effect. Effective at levels of 0.125–0.5% in reduction of fat in biscuits.

Amylose complexing index 49 (fair)
HLB 7–8 (medium polarity)

ALTERNATIVES:

E472f/ SSL/ Succinic acid esters

TECHNOLOGY OF USE IN FOODS:

Invariably used in form of improvers. Fine-powdered forms are blended with up to 20% anti-caking agents such as calcium carbonate or tricalcium orthophosphate and with soya flour, etc. Liquid and paste forms are blended with fats (m.p. 35–38°C) and sugars.

Improvers are added to the dough to provide 0.2–0.6% DATEM on flour weight.

SYNERGISTS:

Ascorbic acid/ Potassium or calcium bromate/ Enzymes (alpha-amylase, hemicellulase)

LEGISLATION:

USA:
US FDA 21 CFR §182.1101. GRAS – emulsifier, flavouring agent and adjuvant. Narrower compositional range than EC, i.e. 17–20% esterified tartaric acid, 14–17% esterified acetic acid, min 12% glycerol – *Food Chemicals Codex*, 2nd edition (1972)

UK and EUROPE:
E472e; ADI 0–50 mg/kg body weight
EC Directive 95/2 (OJ No.L61, 18.3.95) Miscellaneous Additives (Enacted in UK as Statutory Instrument 1995 No.3187) Schedule 1 (Generally permitted for use in Food) Schedules 4 and 7 Purity Criteria laid down in EC Directive 78/663 (OJ, No.L223, 14.7.78)

CANADA:
Permitted and used at up to 0.6% on flour weight

REFERENCES:

Schuster, G. and Adams, W. (1981) Gibt es gemischte Wein-, Essig- und FettsaureEster des Glycerins (E472f): Seifen, Öle, Fette, Wachse, (heft 3), 61.

Tamstorf, S. Emulsifiers for bakery and starch products. Danisco Technical Paper TP 1001.

ANY OTHER RELEVANT INFORMATION:

Total tartaric acid min 10%, max 40%; Total glycerol min 11%, max 28%; Total acetic acid min 8%, max 32%; Acid value 60–110

NAME:	**Lactic acid esters of mono- and diglycerides of fatty acids**
CATEGORY:	Emulsifier
FOOD USE:	Shortening/ Cake and Sponge Batters/ Aerated desserts and toppings/ Imitation cream/ Cake margarine/ Whipping cream
SYNONYMS:	Lactylated monoglycerides/ LACTEM/ E472b/ Lactylated mono- and diglycerides/ Lactoglycerides
FORMULA:	CH2·O·CO-CHOH-CH3-CHOH-CH2·O·CO·(CH2)16·CH3 (for stearate)
MOLECULAR MASS:	430
ALTERNATIVE FORMS:	Varies according to lactic acid content (12–30%) and fatty acid composition
PROPERTIES AND APPEARANCE:	White to cream powder, flakes or paste
MELTING RANGE IN °C	45–50
WATER CONTENT MAXIMUM IN %:	2
HEAVY METAL CONTENT MAXIMUM IN ppm:	10 (as Pb)
ARSENIC CONTENT MAXIMUM IN ppm:	3
ASH MAXIMUM IN %	0.5
SOLUBILITY % AT VARIOUS TEMPERATURE/pH COMBINATIONS:	
in water:	**@ 20°C** Insoluble
FUNCTION IN FOODS:	Alpha-tending emulsifier used in combination with saturated monoglycerides to stabilise their α-crystalline form. Amylose complexing index 22 (poor) HLB 5–8 (medium polarity)

TECHNOLOGY OF USE IN FOODS:	Can modify the crystallisation behaviour of fats and improves their whipping effect in batters and creams	
LEGISLATION:	**USA:** US FDA 21 CFR §172.852 – emulsifier, plasticiser-surface active agent	**UK and EUROPE:** E472b – ADI no limit. EC Directive 95/2 (OJ No.L61, 18.3.95) Schedule 1 (enacted in UK as Statutory Instrument 1995 No.3187) Miscellaneous Additives Generally Permitted for Use in Food, also Schedule 8 (parts 3 and 4) Purity Criteria laid down in EC Directive 78/663 (OJ No.L223, 14.7.78)
REFERENCE:	Qi Si, J. The technology of imitation whipping cream. Danisco Technical Paper TP 2502.	
ANY OTHER RELEVANT INFORMATION:	Acid value max 5; Saponification value 245–320	

NAME:	**Mixed acetic and tartaric acid esters of mono- and diglycerides of fatty acids**
CATEGORY:	Emulsifier
FOOD USE:	Baked goods/ Bread
SYNONYMS:	E472
FORMULA:	CH2·OC=O-CH·O·COCH3·CHOCOCH3·COOH·CHOH-CH2OCO (CH2)16·CH3 (for stearate)
MOLECULAR MASS:	574
ALTERNATIVE FORMS:	Varies according to levels of esterified acetic and tartaric acids and fatty acid composition
PROPERTIES AND APPEARANCE:	Clear liquid to white powder flakes
FUNCTION IN FOODS:	Functionality in bread as DATEM E472e. Dough-strengthening properties are excellent. Amylose complexing fair
ALTERNATIVES:	DATEM E472e
LEGISLATION:	**USA:** Not listed **UK and EUROPE:** E472f. EC Directive 95/2 (OJ No.L61, 18.3.95) Miscellaneous Additives (Enacted in UK as Statutory Instrument 1995 No.3187) Schedule 1 (Generally permitted for use in food) & Schedule 7 Purity Criteria laid down in EC Directive 78/663 (OJ No.L223, 14.7.78)
ANY OTHER RELEVANT INFORMATION:	Total tartaric acid min 20%, max 40%; Total glycerol min 12%, max 27%; Total acetic acid min 10%, max 20%

NAME:	Succinic acid esters of mono- and diglycerides of fatty acids
CATEGORY:	Emulsifier
FOOD USE:	Bread
SYNONYMS:	Succinylated monoglycerides/ SMG
FORMULA:	CH2·O·C=O-CH2-CH2-COOH-CHOH-CH2·O·CO·(CH2)16·CH3
MOLECULAR MASS:	458 (for monostearate)
PROPERTIES AND APPEARANCE:	White beads
MELTING RANGE IN °C:	55–60
PURITY %:	Monoester content 12–20%; SMG min 55%
WATER CONTENT MAXIMUM IN %:	2
HEAVY METAL CONTENT MAXIMUM IN ppm:	10
ARSENIC CONTENT MAXIMUM IN ppm:	3
FUNCTION IN FOODS:	Dough conditioner and crumb softener in bread and fermented doughs Amylose complexing index 63 (good) HLB 5–7 (medium polarity)
ALTERNATIVES:	DATEM stearoyl lactylates/ Sucrose esters
TECHNOLOGY OF USE IN FOODS:	Dough conditioner in bread. Melt into fat (shortening) before addition to dough at 0.125–0.25% of flour weight.

LEGISLATION:	**USA:** US FDA 21 CFR §172.830 regulated emulsifier in shortenings, dough conditioner. Limitation 0.5–3.0% in bread. – max 0.5% on flour weight alone or in combination with stearoyl lactylates, Polysorbate 60 or ethoxylated monoglycerides	**UK and EUROPE:** Not listed; not permitted.
ANY OTHER RELEVANT INFORMATION:	Acid value 70–120; Iodine value max 3; Hydroxyl value 138–152	

NAME:	**Tartaric acid esters of mono- and diglycerides of fatty acids**
CATEGORY:	Emulsifier
FOOD USE:	Bread
SYNONYMS:	E472d
ALTERNATIVE FORMS:	Varies according to level of tartaric acid (15–50%) and fatty acid composition
PROPERTIES AND APPEARANCE:	Yellowish viscous liquid to hard wax
PURITY %:	Total glycerol min. 12, max 29
WATER CONTENT MAXIMUM IN %:	2
HEAVY METAL CONTENT MAXIMUM IN ppm:	10 (as Pb)
ARSENIC CONTENT MAXIMUM IN ppm:	3
ASH MAXIMUM IN %:	0.5
FUNCTION IN FOODS:	Dough conditioning effect in fermented doughs, but DATEM preferred. Has EC listing, but rarely used
ALTERNATIVES:	DATEM
LEGISLATION:	**USA:** Not listed **UK and EUROPE:** E472d – ADI no limit EC Directive 95/2 (OJ No.L61, 18.3.95) Miscellaneous Additives (Enacted in UK as Statutory Instrument 1995 No.3187) Schedule 1 (Generally permitted for use in food) & Schedule 7 Purity Criteria laid down in EC Directive 78/663 (OJ No.L223, 14.7.78)
ANY OTHER RELEVANT INFORMATION:	Acid value max 30

NAME:	**Ethoxylated mono- and diglycerides of fatty acids**	
CATEGORY:	Emulsifier	
FOOD USE:	Baked goods/ Bread/ Tin-grease emulsions/ Pan release/ Non-dairy creamers	
SYNONYMS:	Polyoxyethylene monoglycerides/ PEG 20 mono- and diglycerides	
FORMULA:	CH3(CH2)16·CO·O·CH2·COH·HCH2(OCH2CH2)·OH	
ALTERNATIVE FORMS:	Varies with fatty acid composition and degree of ethoxylation	
PROPERTIES AND APPEARANCE:	Cream to pale yellow paste	
MELTING RANGE IN °C:	Semi-liquid at room temperature	
FLASH POINT IN °C:	>300	
FUNCTION IN FOODS:	Dough conditioner: bread 0.5%. Emulsifier for cake batter: 0.45%. Coffee whitener 0.4%; pan release 0.5%	
ALTERNATIVES:	Mono- and diglycerides/ DATEM	
TECHNOLOGY OF USE IN FOODS:	Dough-strengthening very good; crumb-softening (amylose complexing) poor. HLB 10–12 (medium polarity)	
LEGISLATION:	**USA:** US FDA 21 CFR §172.834. Regulated emulsifier, dough conditioner. Limitation 0.2–0.5%	**UK and EUROPE:** Not listed
ANY OTHER RELEVANT INFORMATION:	Saponification value 65–75; Hydroxyl value 65–80	

NAME:	**Lecithins**
CATEGORY:	Emulsifier/ Antioxidant
FOOD USE:	Baked goods/ Chocolate/ Cake fillings/ Margarine/ Spreads/ Dairy products/ Processed cheese/ Instant foods/ Beverage powders
SYNONYMS:	Phosphatidylcholine/ 1,2-Diacyl-*sn*-glycero-3-phosphorylcholine/ E322
FORMULA:	Complex mixture of phosphatidylcholine, phosphatidylethanolamine, phosphatidylinosital, phosphatidic acid, glycolipids, etc.
ALTERNATIVE FORMS:	Depending upon source (soya, corn, sunflower, rapeseed, egg) and degree of refining and/or modification (fractionation, hydroxylation, etc.)
PROPERTIES AND APPEARANCE:	Cream to brown viscous liquid or paste to fawn granules
PURITY %:	Acetone insoluble min 62
WATER CONTENT MAXIMUM IN %:	1
HEAVY METAL CONTENT MAXIMUM IN ppm:	10
ARSENIC CONTENT MAXIMUM IN ppm:	3
SOLUBILITY % AT VARIOUS TEMPERATURE/pH COMBINATIONS:	
in water:	**@ 20°C** Dispersible
FUNCTION IN FOODS:	Antispattering agent in margarine. Emulsifying agent in baked goods, margarine, spreads, processed cheese, salad dressings. Release agent in baked goods. Wetting agent in instant foods, beverage powders. Antioxidant in margarine, edible oils and fats. Viscosity reduction in chocolate and coatings. Amylose complexing poor. HLB lecithin 3–4 (low polarity); modified lecithin 10–12 (medium polarity)

ALTERNATIVES:	Mono- and diglycerides. Polysorbates and sucrose esters for high HLB
TECHNOLOGY OF USE IN FOODS:	Variety of uses due to wide range of pH values through modification.
FOOD SAFETY ISSUES:	No known safety problems
LEGISLATION:	**USA:** US FDA 21 CFR §184.1400. GRAS **UK and EUROPE:** E322 – ADI not limited EC Directive 95/2 (OJ No.L61, 18.3.95) Miscellanous Additives (Enacted in UK as Statutory Instrument 1995 No.3187) Schedule 1 (Generally permitted for use in food) Schedules 4, 7, & 8 Purity Criteria laid down in EC Directive 78/663 (OJ No.L223, 14.7.78)
REFERENCES:	Bonekamp-Nasner, A. (1992) *Emulsifiers – Lecithin and Lecithin Derivatives in Chocolate, Confectionery Production.* London, pp. 66–68. Minifie, B. W. (1980) *Manufacturing confectioner*, London, April, pp. 47–50. Gunstone, F. D., Harwood, J. L., and Padley, F. B. (1994) *The Lipid Handbook*, 2nd edition. Chapman & Hall, London. Szuhaj, B. F. and List, G. R. (1985) *Lecithins*. American Oil Chemists Society.

NAME:	Polyglycerol esters of fatty acids
CATEGORY:	Emulsifier
FOOD USE:	Baked goods/ Cake batters/ Cake fillings/ Cake margarine/ Spreads/ Shortenings/ Synthetic cream/ Toppings
SYNONYMS:	E475/ Polyglyceryl esters of fatty acids/ Polyglycerolesters of non-polymerised fatty acids/ PGE
FORMULA:	CH2OH-CHOH-CH2-O-CH2·CHOH·CH2·O·CH2·CHOH·CH2·O·CO·(CH2)16·CH3 (for tri-glycerol stearate)
MOLECULAR MASS:	506
ALTERNATIVE FORMS:	Varies according to degree of polymerisation and fatty acid composition
PROPERTIES AND APPEARANCE:	Cream to light brown pastes, flakes, beads or powder
MELTING RANGE IN °C	30–58
PURITY %:	Total fatty acid ester content min 90
WATER CONTENT MAXIMUM IN %:	2
HEAVY METAL CONTENT MAXIMUM IN ppm:	10 (as Pb)
ARSENIC CONTENT MAXIMUM IN ppm:	3
ASH MAXIMUM IN %:	0.5
FUNCTION IN FOODS:	Generally more hydrophilic than monoglycerides. Improves cake batter performance, *viz.* crumb structure and cake volume. Often used together with distilled monoglycerides. Amylose complexing index 30–34 (poor) HLB 5–13 (medium polarity)
ALTERNATIVES:	Mono-diglycerides/ polysorbates

LEGISLATION:	**USA:** US FDA 21 CFR §172.854 – emulsifier, cloud inhibitor in salad oils GMP	**UK and EUROPE:** E475 – ADI 0–25 mg/kg body weight EC Directive 95/2 (OJ No.L61, 18.3.95) Miscellaneous Additives (Enacted in UK as Statutory Instrument 1995 No.3187) Schedule 3 (Other permitted Miscellaneous Additives) & Schedule 4 Purity Criteria laid down in EC Directive 78/663 (OJ No.L223, 14.7.78)
ANY OTHER RELEVANT INFORMATION:	Acid value max 3; Iodine value 0–80; Total glycerol + polyglycerol min 18%, max 60%	

NAME:	**Polyglycerol polyricinoleate**
CATEGORY:	Emulsifier
FOOD USE:	Chocolate/ Coatings/ Tin-greasing emulsions/ Couverture/ Pan release
SYNONYMS:	E476/ Polyglycerol esters of polycondensed fatty acids of castor oil/ Partial polyglycerol esters of polyricinoleic acid/ PGPR
FORMULA:	CH2OH-CHOH-CH2·O·CH2·CHOH·CH2·O·CH2·CHOH·CH2·O·CO·(CH2)7CH=CHCH2·CHOH(CH2)5·CH3
MOLECULAR MASS:	520
PROPERTIES AND APPEARANCE:	Brown viscous liquid
MELTING RANGE IN °C:	Liquid at room temperature
PURITY %:	Polyglycerols equal to or higher than heptaglycerol max 10%. Total fatty acid ester content min 90%
WATER CONTENT MAXIMUM IN %:	2
HEAVY METAL CONTENT MAXIMUM IN ppm:	10 (as Pb)
ARSENIC CONTENT MAXIMUM IN ppm:	3
FUNCTION IN FOODS:	Reduces casson plastic viscosity and yield value in chocolate when used at 0.1–0.5%. Creates very stable water-in-oil emulsions when used at 2–5%
ALTERNATIVES:	Lecithin and emulsifier YN in chocolate. Thermally oxidised soybean oil for tin greases
TECHNOLOGY OF USE IN FOODS:	Water-in-oil emulsions stable at high temperatures; thus, valuable as baking release agents

LEGISLATION:	**USA:** Not listed	**UK and EUROPE:** E476 ADI: 0–7.5 mg/kg body weight EC Directive 95/2 (OJ No.L61, 18.3.95) Miscellaneous Additives (Enacted in UK as Statutory Instrument 1995 No.3187) Schedule 3 (Other permitted Miscellaneous Additives) Includes specific purity criteria (schedule 5)
REFERENCE:	Chevalley, J. (1994) Chocolate flow properties. In: Beckett, S. T. (Ed.), *Industrial Chocolate Manufacture and Use*, 2nd edition. Blackie, Glasgow.	
ANY OTHER RELEVANT INFORMATION:	Acid value max 6; Iodine value 72–103; Hydroxyl value 80–100	

NAME:	**Propylene glycol esters of fatty acids**
CATEGORY:	Emulsifier
FOOD USE:	Cake batters/ Margarine/ Shortening/ Powdered desserts/ Toppings/ Synthetic cream
SYNONYMS:	Propane-1,2-diolesters of fatty acids/ PGMS/ E477/ PGME
FORMULA:	CH3-CHOH-CH2O·CO·(CH2)16·CH3 (for stearate)
MOLECULAR MASS:	330
ALTERNATIVE FORMS:	Varies according to fatty acid composition and monoester content
PROPERTIES AND APPEARANCE:	White waxy paste, flakes or beads
MELTING RANGE IN °C:	36–45
PURITY %:	Total fatty acid ester, min 85%; dimer + trimer of propane 1,2-diol, max 0.4%; Total propane 1,2-diol min 11%, max 31%
HEAVY METAL CONTENT MAXIMUM IN ppm:	10 (as Pb)
ARSENIC CONTENT MAXIMUM IN ppm:	3
ASH MAXIMUM IN %:	0.5
SOLUBILITY % AT VARIOUS TEMPERATURE/pH COMBINATIONS:	
in water:	**@ 20°C** Insoluble **@ 50°C** Dispersible
in vegetable oil:	**@ 20°C** Soluble
FUNCTION IN FOODS:	Alpha-tending. Excellent aerating and foam-stabilising properties in whipped dessert and topping powders. Effective aerating agent in baked goods, especially in combination with distilled monoglycerides and in shortenings. Component and co-emulsifier with distilled monoglycerides in cake improver gels.

	HLB 3–4 (low polarity). Amylose complexing <20 (poor)
ALTERNATIVES:	Acetylated monoglycerides/ Lactic acid esters
TECHNOLOGY OF USE IN FOODS:	Modifies the crystallisation behaviour of fats and improves their aeration effect in batters and creams. Stabilises monoglycerides in α crystal form in cake improver gels
LEGISLATION:	**USA:** US FDA 21 CFR §172.856 GMP **UK and EUROPE:** E477 – ADI 0–25 mg/kg body weight EC Directive 95/2 (OJ No.L61, 18.3.95) Miscellaneous Additives (Enacted in UK as Statutory Instrument 1995 No.3187) Schedule 3 (Other permitted Miscellaneous Additives) Purity Criteria laid down in EC Directive 78/663 (OJ No.L223, 14.7.78)
REFERENCES:	Buchheim, W., Barfod, N. and Krog, N. (1985) Relation between microstructure destabilisation phenomena and rheological properties of whippable emulsions. *Food Microstructure*, **4**. Westerbeek, J. M. M., and Prins, A. (1990) Function of α-tending emulsifiers and proteins in whippable emulsions. In: Dickensen, E. (Ed.), *Food Polymers, Gels & Colloids.* Royal Society of Chemistry, London, Special Publication, No. 82.

NAME:	Sodium stearoyl lactylate
CATEGORY:	Emulsifier
FOOD USE:	Baked goods/ Bread/ Shortening/ Coffee whiteners/ Biscuits
SYNONYMS:	E481/ SSL/ Sodium stearyl-2-lactylate
FORMULA:	CH3·CHOCO[CHOCO(CH2)16·CH3)]CH3·COONa
MOLECULAR MASS:	459
PROPERTIES AND APPEARANCE:	White powder, flakes or beads
MELTING RANGE IN °C:	42–52
PURITY %:	Total lactic acid (free + combined) min 15%, max 40%; sodium min 2.5%, max 5%
HEAVY METAL CONTENT MAXIMUM IN ppm:	10 (as Pb)
ARSENIC CONTENT MAXIMUM IN ppm:	3
FUNCTION IN FOODS:	Excellent dough-strengthener and crumb-softener in bread. (See also references to DATEM and distilled monoglycerides.) Improves fat particle distribution in coffee whiteners and improves whitening effect. Effective in reduction of fat in biscuits. Amylose complexing index 72 (very good). HLB value 10–12 (medium polarity).
ALTERNATIVES:	Combination of DATEM and distilled monoglycerides/ sucrose esters
TECHNOLOGY OF USE IN FOODS:	In bread: dough-strengthener and crumb-softener when used at levels of 0.3–0.5%. Fat-reducing agent in biscuits (use at 0.125–0.5%).

LEGISLATION:	**USA:** US FDA 21 CFR §173.846 – dough strengthener, emulsifier, processing aid, surface active agent, stabiliser, formulation aid, texturiser max 0.5% on flour weight in bread. GMP in non-standardised products **CANADA:** Max 0.375% on flour weight in bread	**UK and EUROPE:** E481 ADI: 0–20 mg/kg body weight EC Directive 95/2 (OJ No.L61, 18.3.95) Miscellaneous Additives (Enacted in UK as Statutory Instrument 1995 No.3187) Schedule 3 (Other permitted Miscellaneous Additives) Purity Criteria laid down in EC Directive 78/663 (OJ No.L223, 14.7.78)
REFERENCE:	Flack, E. (1996) The role of emulsifiers in low-fat food products. In: Roller, S. and S. A. Jones (Eds.), *Handbook of Fat Replacers*. CRC Press, Boca Raton, pp. 213–234.	
ANY OTHER RELEVANT INFORMATION:	Acid value 60–130	

NAME:	Calcium stearoyl lactylate
CATEGORY:	Emulsifier
FOOD USE:	Bread/ Baked goods
SYNONYMS:	CSL/ E482
FORMULA:	CH3-CHOCO[CHOCO(CH2)16-CH3]CH3-COOCaOH
MOLECULAR MASS:	468
PROPERTIES AND APPEARANCE:	White/cream powder, flakes or beads
MELTING RANGE IN °C:	45–55
PURITY %:	Total lactic acid (free + combined) min 15%, max 40%; calcium min 1%, max 5.2%
HEAVY METAL CONTENT MAXIMUM IN ppm:	10 (as Pb)
ARSENIC CONTENT MAXIMUM IN ppm:	3
FUNCTION IN FOODS:	Excellent dough-strengthener and crumb-softener in bread. Used at levels of 0.3–0.5%. (See also references to DATEM and distilled monoglycerides.) Amylose complexing index 65 (good) HLB value 5–6 (medium polarity)
ALTERNATIVES:	Combination of DATEM and distilled monoglycerides
TECHNOLOGY OF USE IN FOODS:	Dough-strengthener and crumb-softener in bread – use at levels of 0.3–0.5%.

LEGISLATION:

USA:
US FDA 21 CFR §172.844 – regulated dough conditioner, whipping agent, conditioning agent

AUSTRALIA/PACIFIC RIM:
Japan: approved with restrictions: bread 5.5 mg/kg max; cake 8 mg/kg max; confectionery 5 g/kg max; pasta 4.5 g/kg max

UK and EUROPE:
E482 – ADI: 0–20 mg/kg body weight
EC Directive 95/2 (OJ No.L61, 18.3.95) Miscellaneous Additives (Enacted in UK as Statutory Instrument 1995 No. 3187) Schedule 3 (Other permitted Miscellaneous Additives) Purity Criteria laid down in EC Directive 78/663 (OJ No.L223, 14.7.78)

REFERENCE:

Kamel, B. S. (1991) Emulsifiers. In: Smith, J. (Ed.), *Food Additive User's Handbook*. Blackie, Glasgow, London, pp. 169–201.

ANY OTHER RELEVANT INFORMATION:

Acid value 50–130

NAME:	Sorbitan esters of fatty acids
CATEGORY:	Emulsifier
FOOD USE:	Cakes/ Chocolate/ Toppings/ Dry yeast/ Margarine/ Coatings/ Coffee whiteners
SYNONYMS:	Anhydrosorbitol esters/ SPANS
FORMULA:	CH2-CHOH-CHOHC[CHOH-CH2-OOC(CH2)16-CH3]O
MOLECULAR MASS:	430
ALTERNATIVE FORMS:	Varies according to fatty acid composition
PROPERTIES AND APPEARANCE:	Amber liquid to tan beads/flakes
MELTING RANGE IN °C:	Liquid to 57°C, depending on fatty acid
WATER CONTENT MAXIMUM IN %:	1.5
HEAVY METAL CONTENT MAXIMUM IN ppm:	10
ARSENIC CONTENT MAXIMUM IN ppm:	3
SOLUBILITY % AT VARIOUS TEMPERATURE/pH COMBINATIONS:	
in water:	Insoluble in cold water, but dispersible in hot water
in vegetable oil:	Soluble above melting point of ester
FUNCTION IN FOODS:	Emulsifier (water-in-oil emulsions). Crystal modifier in margarine, chocolate and cooking oils Amylose complexing index <20 (poor) HLB 3–5 (Low polarity)
SYNERGISTS:	In blends with polysorbates to achieve specific HLB values

LEGISLATION:	**USA:** US FDA 21 CFR § 172.842 (monostearate) § 175.320	**UK and EUROPE:** E491 (sorbitan monostearate) E492 (tristearate) E493 (monolaurate) E494 (monooleate) E495 (monopalmitate) ADI: 0–25 mg/kg body weight EC Directive 95/2 (OJ No.L61, 18.3.95) Miscellaneous Additives (Enacted in UK as Statutory Instrument 1995 No.3187) Schedule 3 (Other permitted Miscellaneous Additives) & Schedule 4 Purity Criteria laid down in SI 1995 No.3187 refers to :-E491:FCC 1981 page 307 – E492/E495:FAO Food & Nutrition Paper No.4 pp293/7 – E493/E494:BPC 1973 pp465/6	**AUSTRALIA/PACIFIC RIM:** Japan: listed for use in chewing gum base, emulsifier, plastifier for chewing gum

REFERENCES:

Weyland, M. (1994) Functional effects of emulsifiers in chocolate. *Manufacturing Confectioner*, **May**, 111–117.

Chislett, L. R. and Walford, J. (1976) Sorbitan and polyoxyethylene sorbitan esters in food products. *Flavours*, (**March/April**), 61.

Nielsen, M. (1995) Sorbitan tristearate anticrystalliser in palm olein. At 21st ISF World Congress, The Hague. Danisco Technical Paper TP1502-le.

ANY OTHER RELEVANT INFORMATION: Acid value 5–15; Saponification value 140–188; Hydroxyl value 66–358

NAME:	Polysorbates
CATEGORY:	Emulsifier
FOOD USE:	Baked goods/ Bread/ Cake batters/ Cake fillings/ Ice-cream/ Margarine/ Salad dressings
SYNONYMS:	Polyoxyetheyene (20) sorbitan esters/ PEG20 sorbitan esters/ POE20 sorbitan esters/ Tweens
FORMULA:	$C_{64}\cdot H_{126}\cdot O_{26}$ (empirical) complex (for monostearate)
MOLECULAR MASS:	1310 (for monostearate)
ALTERNATIVE FORMS:	Varies according to fatty acid composition
PROPERTIES AND APPEARANCE:	Yellow to orange viscous liquid, or soft gel to waxy paste
MELTING RANGE IN °C:	Liquid – 45°C
WATER CONTENT MAXIMUM IN %:	3
HEAVY METAL CONTENT MAXIMUM IN ppm:	10
ARSENIC CONTENT MAXIMUM IN ppm:	3
SOLUBILITY % AT VARIOUS TEMPERATURE/pH COMBINATIONS:	
in water:	Soluble in cold and hot water
in vegetable oil:	Insoluble
FUNCTION IN FOODS:	Emulsifier: HLB 10–16 (high polarity) Solubiliser and wetting agent; solvent/diluent Amylose complexing index 28–32 (poor)
ALTERNATIVES:	Sucrose esters
FOOD SAFETY ISSUES:	Extensive toxicological testing has established no untoward effects – FAO/WHO report series no 53a 1974

LEGISLATION:	**USA:** US FDA 21 CFR § 172.515 (polysorbate 20-monolaurate), § 172.836 (60-tristearate), § 172.838 (65-tristearate), § 172.840 (80-monooleate)	**UK and EUROPE:** E432 (polysorbate 20-monolaurate) E433 (80-monooleate) E434 (40-monopalmitate) E435 (60-monostearate) E436 (65-tristearate) ADI: 0–25 mg/kg body weight Directive 95/2 (OJ No.L61, 18.3.95) Miscellaneous Additives (Enacted in UK as Statutory Instrument 1995 No.3187) Schedule 3 (Other permitted Miscellaneous Additives) & Schedule 4 Purity Criteria for Polysorbates 20,80,60 & 65 refer to *Food Chemicals Codex* 1981 pages 234–236 with the exception (for 20,80,60 & 65) of the descriptions relating to conforming to FDA specifications for fats and fatty acids. For Purity Criteria for Polysorbate 40 refer to Food and Nutrition Paper No 4 (1978) of United Nations' FAO at page 278
REFERENCE:	Del Vecchio, A. J. (1975) Emulsifiers and their use in soft wheat products. *Bakers Digest*, **49**(4), 28.	
ANY OTHER RELEVANT INFORMATION:	Acid value max 2; Saponification value 40–98; Hydroxyl value 44–108	

NAME:	Sucrose esters of fatty acids
CATEGORY:	Emulsifier
FOOD USE:	Bread/ Cakes/ Coffee milk drinks/ Ice-cream/ Spreads/ Yoghurt/ Toppings/ Noodles/ Colour solvent
SYNONYMS:	Sucrose fatty acid esters/ Saccharose esters
FORMULA:	Widely varying (Empirical $C_{30}H_{56}O_{12}$ for sucrose monostearate)
MOLECULAR MASS:	508 (for monostearate)
ALTERNATIVES FORMS:	Varies according to fatty acid composition. Varies according to degree of esterification from mono- to octa-esters
PROPERTIES AND APPEARANCE:	White to cream powders (saturated); yellowish pastes to waxes (unsaturated)
PURITY %:	Total sucrose fatty acid ester content min 80%; total glyceride content max 20% (Dimethyl formamide content 1 mg/kg max)
HEAVY METAL CONTENT MAXIMUM IN ppm:	10 (as Pb)
ARSENIC CONTENT MAXIMUM IN ppm:	3
ASH MAXIMUM IN %:	2
FUNCTION IN FOODS:	Wide range of usages in Japan which are mostly fulfilled by other types of emulsifiers in other countries, often at lower cost. Strongly hydrophilic types produce very stable oil-in-water emulsions Amylose complexing index <25 (poor) HLB 7–16 (high polarity)

LEGISLATION:

USA:
US FDA 21 CFR § 172.859 GMP-emulsifier, stabiliser, texturiser, coating for fruits

AUSTRALIA/PACIFIC RIM:
Japan: widely accepted

UK and EUROPE:
E473 ADI: 0–20 mg/kg body weight
EC Directive 95/2 (OJ No.L61, 18.3.95) Miscellaneous Additives (Enacted in UK as Statutory Instrument 1995 No.3187) Schedule 3 (Other permitted Miscellaneous Additives) & Schedule 4 Purity Criteria laid down in EC Directive 78/663 (OJ No.L223, 14.7.78)

REFERENCES:

Yin, Y., Walker, C. E., and Deffenbaugh, L. B. (1994) Emulsification properties of sugar esters. In: Akoh, C. C., and Swanson, B. G. (Eds.), *Carbohydrate Polyesters as Fat Substitutes.* Marcel Dekker, New York, pp. 111–136.

Pomeranz, Y. (1994) Sucrose esters in baked goods. In: Akoh, C. C., and Swanson, B. G. (Eds.), *Carbohydate Polyesters as Fat Substitutes*. Marcel Dekker, New York.

ANY OTHER RELEVANT INFORMATION:

Acid value max 6

NAME:	Ammonium phosphatides	
CATEGORY:	Emulsifier	
FOOD USE:	Chocolate/ Coatings	
SYNONYMS:	Emulsifier YN	
FORMULA:	CH2·O·R_1-CH2·O·R_2-CH2-O-P=O·OH-O-NH4 where R_1 and R_2 represent a fatty acid moiety or hydrogen	
PROPERTIES AND APPEARANCE:	Viscous semi-solid	
PURITY %:	Phosphorus 3.0–3.4%; nitrogen 1.2–1.5%	
HEAVY METAL CONTENT MAXIMUM IN ppm:	40	
ARSENIC CONTENT MAXIMUM IN ppm:	5	
ASH MAXIMUM IN %:	2.5	
SOLUBILITY % AT VARIOUS TEMPERATURE/pH COMBINATIONS:		
in water:	@ 20°C	Insoluble
in vegetable oil:	@ 20°C	Soluble
in ethanol solution:	@ 5%	Partially soluble
FUNCTION IN FOODS:	Reduces casson plastic viscosity and yield value at levels of 0.2–0.5%. Synthetic lecithin produced to overcome the flavour problems with natural lecithin when used in chocolate.	
ALTERNATIVES:	Soya lecithin	
TECHNOLOGY OF USE IN FOODS:	Controls viscosity of chocolate and coatings	
FOOD SAFETY ISSUES:	Long-term studies show no untoward effects up to 6% of the diet	

LEGISLATION:	**USA:** Not listed	**UK and EUROPE:** E 442 ADI: 0–30 mg/kg body weight EC Directive 95/2 (OJ No.L61, 18.3.95) Miscellaneous Additives (Enacted in UK as Statutory Instrument 1995 No.3187) Schedule 3 (Other permitted Miscellaneous Additives) & Schedule 4 includes specific Purity Criteria (Schedule 5) Maximum level in chocolate products, 10 g/kg
REFERENCES:	Chevalley, J. (1994) Chocolate flow properties. In: Beckett, S. T. (Ed.), *Industrial Chocolate Manufacture and Use*, 2nd edition. Blackie, Glasgow. Anonymous (1991) *Food Emulsifiers Confectionery Production, London*, **February**, pp. 136–140.	
ANY OTHER RELEVANT INFORMATION:	pH of aqueous extract 6.0–8.0	

NAME:	Di-octyl sodium sulphosuccinate	
CATEGORY:	Emulsifier	
FOOD USE:	Fruit drinks/ Sugar/ Molasses/ Chocolate drinks	
SYNONYMS:	Sodium dioctyl sulfosuccinate/ Dicusate sodium/ DSS	
FORMULA:	C8 H17 OOC CH2 CH(SO3Na)COO C8 H17	
MOLECULAR MASS:	445	
PROPERTIES AND APPEARANCE:	White waxy solid	
MELTING RANGE IN °C:	173–179	
SOLUBILITY % AT VARIOUS TEMPERATURE/pH COMBINATIONS:		
in water:	**@ 20°C** Slowly soluble	
in ethanol solution:	**@ 5%** Soluble	
FOOD SAFETY ISSUES:	LD_{50} (oral, rat) 1900 mg/kg. Moderately toxic by ingestion. Gives off toxic fumes of SO_2+Na_2O	
LEGISLATION:	**USA:** US FDA 21 CFR §73.1 Diluent for colour additives, mixtures for food additives, processing and in sugar industry. Limitations 9 ppm finished food; 25 ppm molasses	**UK and EUROPE:** Not listed

NAME:	Stearyl tartrate
CATEGORY:	Emulsifier
FOOD USE:	Baked goods/ Bread
SYNONYMS:	E483
PROPERTIES AND APPEARANCE:	Cream to pale yellow paste
MELTING RANGE IN °C:	67–77
PURITY %:	Total tartaric acid min 18%, max 35%
HEAVY METAL CONTENT MAXIMUM IN ppm:	10 (as Pb)
ASH MAXIMUM IN %:	5
FUNCTION IN FOODS:	Dough conditioning function very good; crumb softening function fair.
ALTERNATIVES:	DATEM/ SSL/ CSL
LEGISLATION:	**USA:** Not listed **UK and EUROPE:** E483 EC Directive 95/2 (OJ No.L61, 18.3.95) Miscellaneous Additives (Enacted in UK as Statutory Instrument 1995 No.3187) Schedule 3 (Other permitted Miscellaneous Additives) Purity Criteria laid down in EC Directive 78/663 (OJ No.L223, 14.7.78)
ANY OTHER RELEVANT INFORMATION:	Acid value max 6; Unsaponifiable matter 77–83%; Iodine value max 4 Listed in EC, but rarely used. DATEM preferred

Part 5

Enzymes

NAME:	**Alpha-amylase**
CATEGORY:	Enzymes
FOOD USE:	Cereals and starches/ Alcoholic beverages/ Non-alcoholic beverages/ Fruit and vegetable juices
SYNONYMS:	1,4-alpha-D-glucan glucanhydrolase/ EC 3.2.1.1/ CAS 9000-90-2/ EINECS 232-565-6
PROPERTIES AND APPEARANCE:	Off-white powder or suspension
SOLUBILITY (%) AT VARIOUS TEMPERATURE/pH COMBINATIONS:	
in water:	Soluble
in sucrose solution:	Soluble
in sodium chloride solution:	Soluble
pH OPTIMUM:	Cereals 5–6; pancreas 6.5; *Aspergillus niger* 5
TEMPERATURE OPTIMUM °C:	Cereals 50; pancreas 40; *A. niger* 55
SIDE ACTIVITIES:	Cereals: beta-amylase, beta-glucanase, neutral acid protease. Pancreas: esterase, lipase, protease. *A. niger*: cellulase, hemicellulase, acid protease, xylanase
FUNCTION IN FOODS:	Conversion of starch to glucose sugar in syrups (especially corn syrups), baking (to improve crumb softness and shelf life); brewing, distilling
TECHNOLOGY OF USE IN FOODS:	Baking: acceleration of fermentation. Improve bread flour to yield loaves of increased volume, improve crust colour and crumb structure – 0.002–0.006% of the flour. Starch liquefaction: reduction of maltose – as liquid for jet cooking, 0.05–0.07% DS; as enzyme/enzyme liquid, 0.05–0.1% DS. Alcoholic beverages-brewing: reduce viscosity of mash, as liquid, 0.025%, conversion of starch to sugars for fermentation, as liquid, 0.003%. Fruit and vegetable juices: remove starch to improve appearance and extraction, as liquid or powder, 0.0005–0.002% w/v, preparation of purées and tenderisation, mostly as liquid
SYNERGISTS:	Cereals: activated by calcium ions. Pancreas: heat stability increased by calcium salts. *A niger*: activated by calcium ions

ANTAGONISTS: Cereals – inhibited by oxidising agents

LEGISLATION:

USA:
GRAS sources: *A. niger*, *A. oryzae*, *Bacillus licheniformis*, *Bacillus subtilis*, *Bacillus stearothermophilus*, *Rhizopus oryzae*, barley malt.
FDA 21 CFR § 184.1027 (from *B. licheniformis*)

UK and EUROPE:
U.K. Permitted for use –
porcine or bovine pancreatic tissue, *A. niger*, *A. oryzae*, *B. licheniformis*, *B. subtilis*

CANADA:
Permitted for use in or upon ale, beer, light beer, malt liquor, porter, stout from *A. niger*, *A. oryzae*, *B.subtilis*

AUSTRALIA/PACIFIC RIM:
Japan: approved

REFERENCES:

Ash, M., and Ash, I. (1995) *Food Additives*. Gower Publishing Co., Brookfield, VT, USA.

Owusu-Ansah, Y. J. (1991) Enzymes. In: Smith, J. (Ed.), *Food Additive User's Handbook*. Blackie A & P, London, pp. 120–150.

NAME:	**Alpha-galactosidase**
CATEGORY:	Enzymes
FOOD USE:	Sugar beet
SYNONYMS:	Alpha-D-galactoside galactohydrolase/ EC 3.2.1.22/ EINECS 232-792-0
pH OPTIMUM:	*Aspergillus niger* 4.5; *Saccharomyces* spp. 5
TEMPERATURE OPTIMUM °C:	*Aspergillus niger* 65; *Saccharomyces* spp. 50
SIDE ACTIVITIES:	*Aspergillus niger* glucosidase, hemicellulase; *Saccharomyces* spp. glucosidase, invertase
FUNCTION IN FOODS:	Exo-hydrolysis of terminal non-reducing alpha-D-galactoside residues of polysaccharides, oligosaccharides, galactomannans, galactolipids. Enzyme used in production of sugar from sugar beets; production aid
TECHNOLOGY OF USE IN FOODS:	Enzyme used in production of sugar from sugar beets; production aid
SYNERGISTS:	*Aspergillus niger*; *Saccharomyces* spp.
ANTAGONISTS:	*Aspergillus niger*; *Saccharomyces* spp.
LEGISLATION:	**USA:** GRAS from *Morteirella vinaceae* var. *raffinoseutilizer*, FDA 21CFR § 173.145 **UK and EUROPE:** Approved **CANADA:** Approved **AUSTRALIA/PACIFIC RIM:** Japan: approved
REFERENCES:	Ash, M., and Ash, I. (1995) *Food Additives*. Gower Publishing Co., Brookfield, VT, USA. Owusu-Ansah, Y. J. (1991) Enzymes. In: Smith, J. (Ed.), *Food Additive User's Handbook*. Blackie A & P, London, pp. 120–150.
ANY OTHER RELEVANT INFORMATION:	*Aspergillus niger* acts on many galactosides; *Saccharomyces* spp. low galactomannase activity

NAME:	Amyloglucosidase
CATEGORY:	Enzymes
FOOD USE:	Cereals and starches/ Alcoholic beverages/ Non-alcoholic beverages/ Fruit and vegetable juices
SYNONYMS:	14-alpha-D-glucanohydrolase/ EC 3.2.1.3/ CAS 9032-08-0
PROPERTIES AND APPEARANCE:	Powder
pH OPTIMUM:	*Aspergillus oryzae* 4–5; *Aspergillus niger* 4–5; *Rhizopus* spp. 4–5; *Trichderma viridue* -; *Aspergillus oryzae* 5; *Bacillus subtilis* 6–7; *Bacillus licheniformis* 7–8
TEMPERATURE OPTIMUM °C:	*Aspergillus oryzae* 55; *Aspergillus niger* 55; *Rhizopus* spp. 55; *Trichderma viridue* -; *Aspergillus oryzae* 55; *Bacillus subtilis* 70–80; *Bacillus licheniformis* 90–95
SIDE ACTIVITIES:	*Aspergillus oryzae* also hydrolyses alpha-1,6 bonds in starch; *Aspergillus niger* -; *Rhizopus* spp. -; *Trichderma viridue* -; *Aspergillus oryzae* glucoamylase, acid protease; *Bacillus subtilis* beta-glucanase, acid and neutral protease; *Bacillus licheniformis* beta-glucanase, acid and neutral protease – both of these are thermolabile and rapidly inactivated at temperature of amylase use
FUNCTION IN FOODS:	Enzyme, degrading agent; degrades gelatinised starch into sugars in production of distilled spirits and vinegar
TECHNOLOGY OF USE IN FOODS:	Baking: acceleration of fermentation. Improve bread flour to yield loaves of increased volume, improve crust colour and crumb structure, 0.002–0.006% of the flour. Starch liquefaction: reduction of maltose, as liquid for jet cooking, 0.05–0.07% DS; as enzyme/enzyme liquid, 0.05–0.1% DS; production of glucose, liquid with syrup or without other enzymes 0.06–0.131% DS. Alcoholic beverages-brewing: reduce viscosity of mash, as liquid, 0.025%; conversion of starch to sugars for fermentation, as liquid, 0.003%; wine – remove haze and improve filtration, liquid or powder 0.002% w/v. Fruit and vegetable juices: remove starch to improve appearance and extraction, as liquid or powder, 0.0005–0.002% w/v, preparation of purées and tenderisation – mostly as liquid

SYNERGISTS:	*Aspergillus oryzae* -; *Aspergillus niger* -; *Rhizopus* spp. -; *Trichderma viridue* -; *Aspergillus oryzae* activated by calcium ions; *Bacillus subtilis* activated by calcium ions; *Bacillus licheniformis* low calcium dependence, especially in presence of high substrate
ANTAGONISTS:	*Aspergillus oryzae* -; *Aspergillus niger* -; *Rhizopus* spp. -; *Trichderma viridue* -; *Aspergillus oryzae* -; *Bacillus subtilis* inhibited by chelating agents; *Bacillus licheniformis* -
LEGISLATION:	**USA:** GRAS – *Aspergillus oryzae*; *Aspergillus niger*; *Rhizopus* spp.; *R. niveus.* Limitation 0.1% (of gelatinised starch) FDA 21 CFR § 173.110 **UK and EUROPE:** UK – *Aspergillus niger* **CANADA:** *Aspergillus oryzae*, *Aspergillus niger* – ale, beer, light beer, malt liquor, porter, stout, bread, flour, wholewheat flour, chocolate syrup, distiller's mash, precooked (instant) cereals, starch used in the production of dextrins, maltose, dextrose, glucose (glucose syrup) or glucose solids, unstandardised bakery products; *Rhizopus delemar* mash destined for vinegar manufacture; *R. niveus* distillers' mash; Multiplicisporus – brewers' mash, distiller's mash, mash for vinegar manufacture, starch used in the production of dextrins, maltose, dextrose, glucose (glucose syrup) or glucose solids (dried glucose syrup)
REFERENCES:	Ash, M., and Ash, I. (1995) *Food Additives.* Gower Publishing Co., Brookfield, VT. Owusu-Ansah, Y. J. (1991) Enzymes. In: Smith, J. (Ed.), *Food Additive User's Handbook.* Blackie A & P, London, pp. 120–150.
ANY OTHER RELEVANT INFORMATION:	Derived from *Rhizopus niveus* with diatomaceous earth as carrier

NAME:	**Anthocyanase**
CATEGORY:	Enzymes
FOOD USE:	Alcoholic beverages/ Wine
SYNONYMS:	Anthocyanin-beta-glucosidase
pH OPTIMUM:	*Aspergillus niger* 3–9
TEMPERATURE OPTIMUM °C:	*Aspergillus niger* 50
SIDE ACTIVITIES:	*Aspergillus niger* – beta-glucosidase
FUNCTION IN FOODS:	Decolorise wines
TECHNOLOGY OF USE IN FOODS:	Alcoholic beverages. Wine – decolorise wines as powder or liquid at 0.1–0.3%
REFERENCE:	Owusu-Ansah, Y. J. (1991) Enzymes. In: Smith, J. (Ed.), *Food Additive User's Handbook*. Blackie A & P, London, pp. 120–150.

NAME:	Beta-amylase
CATEGORY:	Enzymes
FOOD USE:	Cereals and starches/ Alcoholic beverages/ Non-alcoholic beverages/ Fruit and vegetable juices
SYNONYMS:	1,4-alpha-D-glucan maltohydrolase/ EC 3.2.1.2/ CAS 9001-91-3/ EINECS 232-566-1
pH OPTIMUM:	Cereals 5.5; soya bean 4–7; *Bacillus* spp. 5–7
TEMPERATURE OPTIMUM °C:	Cereals 55; soya bean 55; *Bacillus* spp. 60
SIDE ACTIVITIES:	Cereals: alpha-amylase. Soya bean: alpha-amylase, lipoxygenase. *Bacillus* spp.: alpha-amylase, beta-glucanase, neutral protease
FUNCTION IN FOODS:	Baking: acceleration of fermentation. Improve bread flour to yield loaves of increased volume, improve crust colour and crumb structure. Starch liquefaction: reduction of maltose – as liquid for jet cooking. Alcoholic beverages-brewing: reduce viscosity of mash, conversion of starch to sugars for fermentation. Fruit and vegetable juices: remove starch to improve appearance and extraction. Preparation of purees and tenderisation – mostly as liquid
TECHNOLOGY OF USE IN FOODS:	Baking: acceleration of fermentation. Improve bread flour to yield loaves of increased volume, improve crust colour and crumb structure, 0.002–0.006% of the flour. Starch liquefaction: reduction of maltose; as liquid for jet cooking, 0.05–0.07% DS; as enzyme/enzyme liquid, 0.05–0.1% DS. Alcoholic beverages-brewing: reduce viscosity of mash, as liquid, 0.025%; conversion of starch to sugars for fermentation, as liquid, 0.003%. Fruit and vegetable juices: remove starch to improve appearance and extraction; as liquid or powder, 0.0005–0.002% w/v. Preparation of purées and tenderisation – mostly as liquid
SYNERGISTS:	Cereals – activated by reducing agents. Soya bean – acid tolerant. *Bacillus* spp.
ANTAGONISTS:	Cereals; soya bean; *Bacillus* spp.

LEGISLATION:

USA:
GRAS status – barley malt

CANADA:
Permitted for use in or upon ale, beer, light beer, malt liquor, porter, stout from *A. niger, A. oryzae, B. subtilis*

UK and EUROPE:
Permitted

REFERENCES:

Ash, M., and Ash, I. (1995) *Food Additives*. Gower Publishing Co., Brookfield, VT.

Owusu-Ansah, Y. J. (1991) Enzymes. In: Smith, J. (Ed.), *Food Additive User's Handbook*. Blackie A & P, London, pp. 120–150.

NAME:	Beta-galactosidase
CATEGORY:	Enzymes
FOOD USE:	Milk and milk products
SYNONYMS:	Lactase
pH OPTIMUM:	*Aspergillus niger* 4.5; *Aspergillus oryzae* 4.5; *Bacillus* spp. 7.3; *Kluyveromyces* spp. 6.5; *Saccharomyces* spp. 6.5
TEMPERATURE OPTIMUM °C:	*Aspergillus niger* 55; *Aspergillus oryzae* 55; *Bacillus spp.* 60; *Kluyveromyces* spp. 45; *Saccharomyces* spp. 40
SIDE ACTIVITIES:	*Aspergillus niger* alpha-L-arabinase, glucanase, glucosidase, transferase, invertase, acid protease; *Aspergillus oryzae* alpha-L-arabinase, glucanase, glucosidase, transferase, invertase, acid protease; *Bacillus* spp. amylase, glucosidase, protease, transferase; *Kluyveromyces* spp. glucosidase, invertase, protease, transferase; *Saccharomyces* spp. glucosidase, invertase, protease, transferase
FUNCTION IN FOODS:	Hydrolysis of terminal non-reducing beta-D-galactose residues
TECHNOLOGY OF USE IN FOODS:	Milk and milk products – prevents grainy texture, stabilisation of proteins during freezing, removal of lactase for lactose-free products – soluble or immobilised
SYNERGISTS:	*Aspergillus niger*; *Aspergillus oryzae*; *Bacillus* spp.; *Kluyveromyces* spp.; *Saccharomyces* spp.
ANTAGONISTS:	*Aspergillus niger* glucose, galactose; *Aspergillus oryzae* glucose, galactose; *Bacillus* spp. glucose, galactose; *Kluyveromyces* spp. glucose, galactose; *Saccharomyces* spp. glucose, galactose

LEGISLATION:	**USA:** GRAS status – *Aspergillus niger*, *A. oryzae*, *Saccharomyces fragilis*, *Candida pseudotropicalis*	**UK and EUROPE:** *Aspergillus niger* **CANADA:** *Aspergillus niger*, *A. oryzae*, *Saccharomyces* spp. permitted in or on lactose-reducing enzyme preparations, milk destined for use in ice-cream mix
REFERENCE:	Owusu-Ansah, Y. J. (1991) Enzymes. In: Smith, J. (Ed.), *Food Additive User's Handbook*. Blackie A & P, London, pp. 120–150.	

NAME:	Beta-glucanase			
CATEGORY:	Enzymes			
FOOD USE:	Hydrolyses cellulose in brewing worts			
SYNONYMS:	Endo-1,3(4)-beta-D-glucanase/ EC 3.2.1.6/ CAS 9074-99-1 (b- from *Aspergillus niger*), 9012-54-8 (from *Bacillus subtilis*)/ laminarinase/ EINECS 232-980-2, 232-734-4 resp.			
pH OPTIMUM:	*Aspergillus niger* 5; *Bacillus subtilis* 7; *Penicillium emersonii* 4			
TEMPERATURE OPTIMUM °C:	*Aspergillus niger* 60; *Bacillus subtilis* 50–60; *Penicillium emersonii* 70			
SIDE ACTIVITIES:	*Aspergillus niger* amylase, glucoamylase, glucosidase; *Bacillus subtilis* amylase, glucosidase, protease; *Penicillium emersonii* amylase, dextranase, protease			
FUNCTION IN FOODS:	Endo-hydrolysis of terminal 1,4-alpha-D-glucose residues from non-reducing end of polyglucoside chains. Enzyme for beer filtration			
TECHNOLOGY OF USE IN FOODS:	Hydrolyses cellulose in brewing worts			
LEGISLATION:	**USA:**	**UK and EUROPE:**	**CANADA:**	**AUSTRALIA/ PACIFIC RIM:**
	GRAS from *Aspergillus niger*	Approved	Ale, beer, light beer, malt liquor, porter, stout, corn for degerming distiller's mash, mash destined for vinegar manufacture, unstandardised bakery products	Japan: approved
REFERENCES:	Ash, M., and Ash, I. (1995) *Food Additives*. Gower Publishing Co., Brookfield, VT. Owusu-Ansah, Y. J. (1991) Enzymes. In: Smith, J. (Ed.), *Food Additive User's Handbook*. Blackie A & P, London, pp. 120–150.			
ANY OTHER RELEVANT INFORMATION:	*Aspergillus niger* broad range with low specificity; *Bacillus subtilis* narrow range with higher specificity; *Penicillium emersonii* tolerant of low pH in some variants			

NAME:	**Beta-glucosidase**
CATEGORY:	Enzymes
FOOD USE:	Sugar beet/ Beer
SYNONYMS:	Beta-D-glucoside hydrolase/ EC 3.2.1.21
pH OPTIMUM:	*Aspergillus niger* 5, *Aspergillus oryzae* 5; *Bacillus* spp. 7; *Clostridium thermocellum* 9; *Saccharomyces* spp. 7; Sweet almond 7; *Trichoderma viride* 5
TEMPERATURE OPTIMUM °C:	*Aspergillus niger* 60; *Aspergillus oryzae* 65; *Bacillus* spp. 70; *Clostridium thermocellum* 60; *Saccharomyces* spp. 45; Sweet almond 50; *Trichoderma viride* 65
SIDE ACTIVITIES:	*Aspergillus niger* amylase, glucoamylase, protease; *Aspergillus oryzae* amylase, glucoamylase, protease; *Bacillus* spp. amylase, glucanase, protease; *Clostridium thermocellum* protease; *Saccharomyces* spp. glucanase, protease; Sweet almond: usually very pure; *Trichoderma viride* hemicellulase
FUNCTION IN FOODS:	Used in the final stages of starch and cellulose hydrolysis
TECHNOLOGY OF USE IN FOODS:	Used with other saccharolytic enzymes to release glucose from dextrans
LEGISLATION:	**UK and EUROPE:** Approved **AUSTRALIA/PACIFIC RIM:** Japan: approved
REFERENCES:	Ash, M., and Ash, I. (1995) *Food Additives*. Gower Publishing Co., Brookfield, VT. Owusu-Ansah, Y. J. (1991) Enzymes. In: Smith, J. (Ed.), *Food Additive User's Handbook*. Blackie A & P, London, pp. 120–150.

ANY OTHER RELEVANT INFORMATION:	*Aspergillus niger* used to reduce cellobiose inhibition on cellulase; *Aspergillus oryzae* broad specificity, unusually high thermo-tolerance; *Bacillus* spp. narrow specificity; *Clostridium thermocellum* very active on beta-1,3-bonds; *Saccharomyces* spp. broad specificity; Sweet almond: broad specificity; *Trichoderma viride* broad specificity

NAME:	Bromelain
CATEGORY:	Enzymes
FOOD USE:	Meat and other proteinaceous foods/ Meat and fish/ Eggs and egg products
SYNONYMS:	CAS 37189-34-7/ EINECS 253-387-5/ Bromelin
pH OPTIMUM:	5.5–7.0
TEMPERATURE OPTIMUM °C:	50
FUNCTION IN FOODS:	Chillproofing of beer. Natural enzyme, meat tenderising. Preparation of pre-cooked cereals, processing aid, tissue softening agent. Used in beer, bread, cereals, meat, poultry and wine
TECHNOLOGY OF USE IN FOODS:	Meat and other proteinaceous foods. Meat and fish: tenderisation of meat as liquid; produce fish hydrolysates as liquid at 2% of protein; enhance fish stick-water effluent treatment as liquid at 0.2%; removal of oil from tissues as liquid. Eggs and egg products – improve drying properties as liquid/powder
LEGISLATION:	**USA:** GRAS from pineapples **UK and EUROPE:** *Ananas bracteatus, Ananas comosus* **CANADA:** Source – pineapples. Permitted in or upon ale, beer, light beer, malt liquor, porter, stout, bread, flour, wholewheat flour, edible collagen sausage casings, hydrolysed animal, milk and vegetable proteins, meat cuts meat-tenderising preparations, pumping pickle for curing meat cuts **AUSTRALIA/PACIFIC RIM:** Japan: approved
REFERENCES:	Ash, M., and Ash, I. (1995) *Food Additives*. Gower Publishing Co., Brookfield, VT. Owusu-Ansah, Y. J. (1991) Enzymes. In: Smith, J. (Ed.), *Food Additive User's Handbook*. Blackie A & P, London, pp. 120–150.

NAME:	Catalase			
CATEGORY:	Enzymes			
FOOD USE:	Non-alcoholic beverages/ Soft drinks/ Milk			
SYNONYMS:	Hydrogen peroxide oxidoreductase/ EC 1.11.1.6/ CAS 9001-05-2/ EINECS 232-577-1			
pH OPTIMUM:	*Aspergillus niger* 5–8; bovine liver 7			
TEMPERATURE OPTIMUM °C:	*Aspergillus niger* 35; bovine liver 45			
SIDE ACTIVITIES:	*Aspergillus niger* – usually very pure; bovine liver – usually very pure			
FUNCTION IN FOODS:	Enzyme; in food preservation; production aid in cheese; in decomposing residual hydrogen peroxide in cheese manufacture, bleaching and oxidising processes			
TECHNOLOGY OF USE IN FOODS:	Non-alcoholic beverages – soft drinks, stabilisation of citrus terpenes as powder or liquid – combined with glucose oxidase. Milk – removal of hydrogen peroxide as liquid or powder			
SYNERGISTS:	*Aspergillus niger* stable at low pH; bovine liver -			
ANTAGONISTS:	*Aspergillus niger* -; bovine liver – inactivated by alkali			
LEGISLATION:	**USA:**	**UK and EUROPE:**	**CANADA:**	**AUSTRALIA/ PACIFIC RIM:**
	FDA 21 CFR § 173.135	Sources – bovine liver, *Aspergillus niger*	Sources – *Aspergillus niger, Micrococcus lysodeikticus*, bovine liver. Permitted in or upon soft drinks, egg albumin	Japan: approved
REFERENCES:	Ash, M., and Ash, I. (1995) *Food Additives*. Gower Publishing Co., Brookfield, VT. Owusu-Ansah, Y. J. (1991) Enzymes. In: Smith, J. (Ed.), *Food Additive User's Handbook*. Blackie A & P, London, pp. 120–150.			
ANY OTHER RELEVANT INFORMATION:	Molecular weight 240,000 Da			

NAME:	**Cellobiase (*Aspergillus niger*)**
CATEGORY:	Enzymes
FOOD USE:	Sugar manufacture
SYNONYMS:	Beta-D-glucoside glucohydrolase/ EC 3.2.1.21
pH OPTIMUM:	5
TEMPERATURE OPTIMUM °C:	60
SIDE ACTIVITIES:	Amylase, glucoamylase, protease, hemicellulase
FUNCTION IN FOODS:	Exo-hydrolysis of terminal non-reducing 1,4-alpha-D-glucose residues. Used to reduce product inhibition of cellobiose when cellulases used
TECHNOLOGY OF USE IN FOODS:	Cleaves cellobiose units from the non-reducing ends of cellulose polymers
LEGISLATION:	**USA:** GRAS **UK and EUROPE:** Approved **CANADA:** Distiller's mash, liquid coffee concentrate, spice extracts, natural flavour and colour extractives
REFERENCE:	Owusu-Ansah, Y. J. (1991) Enzymes. In: Smith, J. (Ed.), *Food Additive User's Handbook*. Blackie A & P, London, pp. 120–150.

NAME:	**Cellulase**
CATEGORY:	Enzymes
FOOD USE:	Alcoholic beverages/ Non-alcoholic beverages/ Fruit and vegetable juices/ Fats and oils
SYNONYMS:	1,4-(1,3; 1,4)-beta-D-glucan 4-glucanohydrolase/ EC 3.2.1.4/ CAS 9012-54-81/ EINECS 232-734-4
PROPERTIES AND APPEARANCE:	Off-white powder
pH OPTIMUM:	*Aspergillus niger* 5; *Basidiomycetes* spp. 4; *Penicillium funiculosum* 5; *Rhizopus* spp. 4; *Trichoderma* spp. 5
TEMPERATURE OPTIMUM °C:	*Aspergillus niger* 45; *Basidiomycetes* spp. 50; *Penicillium funiculosum* 65; *Rhizopus* spp. 45; *Trichoderma* spp. 55
SIDE ACTIVITIES:	*Aspergillus niger* amylase, cellobiase, glucosidase, glucoamylase; *Basidiomycetes* spp. hemicellulase, protease; *Penicillium funiculosum* amylase, glucoamylase, cellobiase; *Rhizopus* spp. amylase, glucoamylase, protease; *Trichoderma* spp. hemicellulase
FUNCTION IN FOODS:	Endo-hydrolysis of 1,4-beta-glucosidic links of cereal glucans, cellulose, lichenin. Enzyme; digestive aid in medicine and brewing industry; aids bacteria in the hydrolysis of cellulose; aids in removal of visceral masses during clam processing and of shells in shrimp processing
TECHNOLOGY OF USE IN FOODS:	Fruit and vegetable juices – remove starch to improve appearance and extraction as liquid or powder at 0.0005–0.002% w/v; fats and oils – vegetable oil extractions – hydrolyse cell wall materials, as liquid or powder at 0.5–2% DS; alcoholic beverages – assist in filtration by hydrolysing complex cell wall materials as liquid or powder at 0.1% DS; non-alcoholic beverages – coffee – cellulose breakdown during drying, as liquid or powder; tea – cellulose breakdown during fermentation
SYNERGISTS:	*Aspergillus niger*; *Basidiomycetes* spp.; *Penicillium funiculosum*; *Rhizopus* spp.; *Trichoderma* spp.
ANTAGONISTS:	*Aspergillus niger*; *Basidiomycetes* spp.; *Penicillium funiculosum*; *Rhizopus* spp.; *Trichoderma* spp.

LEGISLATION:

USA:
GRAS sources *Aspergillus niger* and *Trichoderma reesei*. Regulatory: FDA 21CFR § 173.120, GRAS

CANADA:
Permitted source *Aspergillus niger* for use in distiller's mash, liquid coffee concentrate, spice extracts, natural flavour and colour

UK and EUROPE:
UK: Permitted sources – bovine liver, *Aspergillus niger*

AUSTRALIA/PACIFIC RIM:
Japan: approved

REFERENCES:

Ash, M., and Ash, I. (1995) *Food Additives*. Gower Publishing Co., Brookfield, VT.

Owusu-Ansah, Y. J. (1991) Enzymes. In: Smith, J. (Ed.), *Food Additive User's Handbook*. Blackie A & P, London, pp. 120–150.

ANY OTHER RELEVANT INFORMATION:

Aspergillus niger generally low in C1-type activity; *Basidiomycetes* spp. good C1-type with broad specificity; *Penicillium funiculosum* product inhibition is usually low; *Rhizopus* spp. broad specificity; *Trichoderma* spp. high C1 activity. Molecular weight 31,000 Da

NAME:	Chymotrypsin (pancreatic)
CATEGORY:	Enzymes
FOOD USE:	Baked goods/ Alcoholic beverages, beer
SYNONYMS:	EC 3.4.21.1
pH OPTIMUM:	8–9
TEMPERATURE OPTIMUM °C:	35
SIDE ACTIVITIES:	Amylase, lipase, esterase
FUNCTION IN FOODS:	Preferential cleavage of tyrosine, tryptophan, phenylalanine, leucine residues
TECHNOLOGY OF USE IN FOODS:	Baked goods – modification of gluten in baking of biscuits as powder at up to 0.25% of flour; to reduce mixing time of dough as tablets at 75 HU per 100 g flour. Alcoholic beverages (beer), to provide nitrogen for yeast growth and aid in filtration and chillproofing as liquid or powder at 0.3% DS
ANTAGONISTS:	Inhibited by compounds in cereals, beans, potato, egg
REFERENCE:	Owusu-Ansah, Y. J. (1991) Enzymes. In: Smith, J. (Ed.), *Food Additive User's Handbook*. Blackie A & P, London, pp. 120–150.
ANY OTHER RELEVANT INFORMATION:	Bovine chymotrypsin is more thermotolerant

NAME:	**Dextranase (*Penicillium* spp.)**
CATEGORY:	Enzymes
FOOD USE:	Used to hydrolyse dextrans from starch
SYNONYMS:	1,6-alpha-D-glucan 6-glucanohydrolase/ EC 3.2.1.11/ CAS 9025-70-1/ EINECS 232-803-9
pH OPTIMUM:	5
TEMPERATURE OPTIMUM:	55
SIDE ACTIVITIES:	Cellulase, hemicellulase
FUNCTION IN FOODS:	Endo-hydrolysis of dextrans
TECHNOLOGY OF USE IN FOODS:	Used to hydrolyse dextrans from starch
LEGISLATION:	**USA:** GRAS **UK and EUROPE:** *Penicillium funiculosum*, *P. lilacinum* **CANADA:** Approved **AUSTRALIA/PACIFIC RIM:** Japan: approved
REFERENCES:	Ash, M., and Ash, I. (1995) *Food Additives*. Gower Publishing Co., Brookfield, VT. Owusu-Ansah, Y. J. (1991) Enzymes. In: Smith, J. (Ed.), *Food Additive User's Handbook*. Blackie A & P, London, pp. 120–150.
ANY OTHER RELEVANT INFORMATION:	Products are isomaltose and isomaltotriose

NAME:	**Diacetyl reductase (*Aerobacter aerogenes*)**
CATEGORY:	Enzymes
FOOD USE:	Alcoholic beverages
SYNONYMS:	Anthocyanin-beta-glycosidase
pH OPTIMUM:	6–8
TEMPERATURE OPTIMUM °C:	30
FUNCTION IN FOODS:	Removal of diacetyls in beer as liquid
TECHNOLOGY OF USE IN FOODS:	Removal of diacetyls in beer as liquid
SYNERGISTS:	Activated by NADH
ANTAGONISTS:	Inhibited by ethanol
REFERENCE:	Owusu-Ansah, Y. J. (1991) Enzymes. In: Smith, J. (Ed.), *Food Additive User's Handbook*. Blackie A & P, London, pp. 120–150.

NAME:	Ficin (*Ficus* spp.)
CATEGORY:	Enzymes
FOOD USE:	Meat and other proteinaceous foods/ Meat and fish/ Eggs and egg products
SYNONYMS:	EC 3.4.22.3/ CAS 9001-33-6/ EINECS 232-599-1/ Debricin/ Ficus protease/ Ficus proteinase
PROPERTIES AND APPEARANCE:	White powder
SOLUBILITY % AT VARIOUS TEMPERATURE/pH COMBINATIONS:	
in water:	Very soluble
pH OPTIMUM:	5–7
TEMPERATURE OPTIMUM °C:	65
SIDE ACTIVITIES:	Lysozyme, esterase, peroxidase
FUNCTION IN FOODS:	Enzyme for chillproofing of beer. Meat tenderising. Preparation of pre-cooked cereals. Processing aid; tenderising agent; tissue softening agent. Preferential cleavage of lysine, alanine, tyrosine, glycine, asparagine, leucine, valine
TECHNOLOGY OF USE IN FOODS:	Meat and other proteinaceous foods. Meat and fish – tenderisation of meat as liquid, produce fish hydrolysates as liquid at 2% of protein, enhance fish stick-water effluent treatment as liquid at 0.2%, removal of oil from tissues as liquid. Eggs and egg products – improve drying properties as liquid/powder
SYNERGISTS:	Reducing compounds
ANTAGONISTS:	Oxidising agents

LEGISLATION:

USA:
GRAS from figs. USDA 9CFR § 318.7, 381.147; BATF 27CFR § 240.1051

UK and EUROPE:
Approved

CANADA:
Source – fig tree latex. Permitted in or upon ale, beer, light beer, malt liquor, porter, stout, edible collagen sausage casings, hydrolysed animal, milk and vegetable proteins, meat cuts meat-tenderising preparations

REFERENCES:

Ash, M., and Ash, I. (1995) *Food Additives*. Gower Publishing Co., Brookfield, VT.

Owusu-Ansah, Y. J. (1991) Enzymes. In: Smith, J. (Ed.), *Food Additive User's Handbook*. Blackie A & P, London, pp. 120–150.

NAME:	**Glucoamylase**
CATEGORY:	Enzymes
FOOD USE:	Cereals and starches/ Alcoholic beverages/ Non-alcoholic beverages/ Fruit and vegetable juices
SYNONYMS:	1,4-alpha-D-glucan glucanohydrolase/ EC 3.2.1.3/ CAS 9032-08-0/ EINECS 232-877-2
PROPERTIES AND APPEARANCE:	Powder
pH OPTIMUM:	*Aspergillus awamori* 4–5; *Aspergillus niger* 3–5; *Aspergillus oryzae* 4.5; *Rhizopus* spp. 2.5–5
TEMPERATURE OPTIMUM °C:	*Aspergillus awamori* 60; *Aspergillus niger* 65; *Aspergillus oryzae* 60; *Rhizopus* spp. 55
SIDE ACTIVITIES:	*Aspergillus awamori* amylase, glucanase, cellulase, hemicellulase, protease; *Aspergillus niger* amylase, glucanase, cellulase, hemicellulase, protease; *Aspergillus oryzae* amylase, glucanase, cellulase, hemicellulase, protease; *Rhizopus* spp. amylase, glucanase, cellulase, hemicellulase, protease
FUNCTION IN FOODS:	Hydrolysis of starch dextrins to glucose. Food processing, low-carbohydrate beer. Exo-hydrolysis of terminal 1,4-alpha-D-glucose residues from non-reducing end of polyglucoside chains
TECHNOLOGY OF USE IN FOODS:	Baking: acceleration of fermentation. Improve bread flour to yield loaves of increased volume, improve crust colour and crumb structure, 0.002–0.006% of the flour. Starch liquefaction: reduction of maltose – as liquid for jet cooking, 0.05–0.07% DS; as enzyme/enzyme liquid, 0.05–0.1% DS; production of glucose – liquid with syrup or without other enzymes 0.06–0.131% DS. Alcoholic beverages-brewing: reduce viscosity of mash, as liquid, 0.025%; conversion of starch to sugars for fermentation, as liquid, 0.003%; wine – remove haze and improve filtration, liquid or powder 0.002% w/v. Fruit and vegetable juices: remove starch to improve appearance and extraction – as liquid or powder, 0.0005–0.002% w/v. Preparation of purées and tenderisation – mostly as liquid
SYNERGISTS:	*Aspergillus awamori*; *Aspergillus niger*; *Aspergillus oryzae*; *Rhizopus* spp.
ANTAGONISTS:	*Aspergillus awamori*; *Aspergillus niger*; *Aspergillus oryzae*; *Rhizopus* spp.

LEGISLATION:

USA:
GRAS – *Aspergillus oryzae*; *Aspergillus niger*; *Rhizopus* spp.; *R. niveus*

CANADA:
Aspergillus oryzae, *Aspergillus niger* – stout, bread, flour, wholewheat flour, chocolate ale, beer, light beer, malt liquor, porter, syrup, distiller's mash, pre-cooked (instant) cereals, starch used in the production of dextrins, maltose, dextrose, glucose (glucose syrup) or glucose solids, unstandardised bakery products; *Rhizopus delemar* mash destined for vinegar manufacture; *R. niveus* distillers' mash; Multiplici sporus – brewers' mash, distiller's mash, mash for vinegar manufacture, starch used in the production of dextrins, maltose, dextrose, glucose (glucose syrup) or glucose solids (dried glucose syrup)

UK and EUROPE:
Aspergillus niger

AUSTRALIA/PACIFIC RIM:
Japan: approved

REFERENCES:

Ash, M., and Ash, I. (1995) *Food Additives*. Gower Publishing Co., Brookfield, VT.

Owusu-Ansah, Y. J. (1991) Enzymes. In: Smith, J. (Ed.), *Food Additive User's Handbook*. Blackie A & P, London, pp. 120–150.

ANY OTHER RELEVANT INFORMATION:

Acid-tolerance and thermotolerance vary between sources. Molecular weight 97,000 Da

NAME:	**Glucose isomerase**
CATEGORY:	Enzymes
FOOD USE:	Conversion of glucose to fructose
SYNONYMS:	D-Xylose ketol isomerase/ EC 5.3.1.5
pH OPTIMUM:	*Actinoplanes missouriensis* 7.5; *Bacillus coagulans* 8; *Streptomyces* spp. 8; *Streptomyces albus* 6–7
TEMPERATURE OPTIMUM °C:	*Actinoplanes missouriensis* 60; *Bacillus coagulans* 60; *Streptomyces* spp. 63; *Streptomyces albus* 60–75
SIDE ACTIVITIES:	*Actinoplanes missouriensis* – usually none; *Bacillus coagulans* – usually none; *Streptomyces* spp. – usually none; *Streptomyces albus* – usually none.
FUNCTION IN FOODS:	A true xylose isomerase acting on glucose at high substrate concentration. Enzyme which converts glucose to fructose; used in production of high-fructose corn syrup
TECHNOLOGY OF USE IN FOODS:	Immobilised 0.0015–0.03% DS fixed bed 0.16 DS batch
SYNERGISTS:	In the immobilised form activated by magnesium and cobalt (the need for cobalt varies with preparation), magnesium is competed for by calcium and must therefore be in excess
LEGISLATION:	**USA:** GRAS status – 184.1372 (immobilised preparations) – *Actinoplanes missouriensis*; *Bacillus coagulans*; *Streptomyces olivaceous*; *Streptomyces murinus*, *Streptomyces rubiginosus*, *Streptomyces olivochromogenes*, *Arthrobacter globiformus* **CANADA:** *Actinoplanes missouriensis*; *Bacillus coagulans*; *Streptomyces* spp.; *Olivochromogenes* spp.; *S. olivaceous.* Glucose, glucose syrup partially or completely isomerised to fructose **UK and EUROPE:** Soluble – *Bacillus coagulans.* Immobilised – *Bacillus coagulans*, *Streptomyces olivaceous* **AUSTRALIA/PACIFIC RIM:** Japan: approved

REFERENCES:

Ash, M., and Ash, I. (1995) *Food Additives*. Gower Publishing Co., Brookfield, VT.

Owusu-Ansah, Y. J. (1991) Enzymes. In: Smith, J. (Ed.), *Food Additive User's Handbook*. Blackie A & P, London, pp. 120–150.

NAME:	**Glucose oxidase**
CATEGORY:	Enzymes
FOOD USE:	Alcoholic beverages/ Non-alcoholic beverages/ Fruits and vegetables/ Meat and other proteinaceous foods
SYNONYMS:	Beta-D-glucose, oxygen 1-oxidoreductase/ EC 1.1.3.4
pH OPTIMUM:	*Aspergillus niger* 4.5; *Aspergillus* spp. 2.5–8; *Penicillium notatum* 3–7
TEMPERATURE OPTIMUM °C:	*Aspergillus niger* 50; *Aspergillus* spp. 15–70; *Penicillium notatum* 50
SIDE ACTIVITIES:	*Aspergillus niger* catalase; *Aspergillus* spp. catalase; *Penicillium notatum* catalase
FUNCTION IN FOODS:	Enzyme which converts glucose to fructose; used in production of high-fructose corn syrup
TECHNOLOGY OF USE IN FOODS:	Alcoholic beverages: wine – remove oxygen as powder or liquid at 10–70 GOU/l. Non-alcoholic beverages: soft drinks – stabilisation of citrus terpenes as powder or liquid at 20–90 GOU/l. Fruits and vegetables – juices – remove oxygen as powder/liquid at 20–200 GOU/l. Meat and other proteinaceous foods – eggs and egg products – glucose removal from dried eggs as powder or liquid at 150–225 GOU/l white, 300–375 GOU/l whole
SYNERGISTS:	*Aspergillus niger* -; *Aspergillus* spp. acid tolerant, thermotolerant; *Penicillium notatum* -
ANTAGONISTS:	*Aspergillus niger*; *Aspergillus* spp.; *Penicillium notatum*
LEGISLATION:	**USA:** GRAS – *Aspergillus niger*. FDA 21CFR § 184.1372 **UK and EUROPE:** UK: *Aspergillus niger* **CANADA:** Source permitted – *Aspergillus niger* – permitted in or upon soft drinks, liquid whole egg, egg white and liquid egg yolk destined for drying **AUSTRALIA/PACIFIC RIM:** Japan: approved

REFERENCES:

Ash, M., and Ash, I. (1995) *Food Additives*. Gower Publishing Co., Brookfield, VT.

Owusu-Ansah, Y. J. (1991) Enzymes. In: Smith, J. (Ed.), *Food Additive User's Handbook*. Blackie A & P, London, pp. 120–150.

NAME:	**Hemicellulase (*Aspergillus* spp.)**
CATEGORY:	Enzymes
SYNONYMS:	Endo-1,4-beta-D-mannan hydrolase (EC 3.2.1.78)/ Exo-alpha-L-arabinofuran hydrolase (EC 3.2.1.55)/ Exo-1,3-beta-D-xylan hydrolase (EC 3.2.1.72)
pH OPTIMUM:	3–6
TEMPERATURE OPTIMUM °C:	70
SIDE ACTIVITIES:	Cellulase, glucosidase, pectinase, pentosanase
FUNCTION IN FOODS:	Enzyme which converts glucose to fructose; used in production of high-fructose corn syrup. Hydrolyses coffee gums; used in the extraction of essential oils and plant extracts
TECHNOLOGY OF USE IN FOODS:	Hydrolyses coffee gums; used in the extraction of essential oils and plant extracts
LEGISLATION:	**USA:** FDA 21CFR § 184.1372 **CANADA:** *Bacillus subtilis* – permitted in or upon distiller's mash, liquid coffee concentrate, mash destined for vinegar manufacture **AUSTRALIA/PACIFIC RIM:** Japan: approved
REFERENCES:	Ash, M., and Ash, I. (1995) *Food Additives*. Gower Publishing Co., Brookfield, VT. Owusu-Ansah, Y. J. (1991) Enzymes. In: Smith, J. (Ed.), *Food Additive User's Handbook*. Blackie A & P, London, pp. 120–150.
ANY OTHER RELEVANT INFORMATION:	These are complex enzyme systems that require specific substrates to identify and distinguish them

NAME:	**Inulinase**
CATEGORY:	Enzymes
FOOD USE:	Jerusalem artichoke/ Manufacture of fructose
SYNONYMS:	2,1-beta-D-fructan fructanhydrolase/ EC 3.2.1.7
pH OPTIMUM:	*Aspergillus* spp. 4.5; *Candida* spp. 5
TEMPERATURE OPTIMUM °C:	*Aspergillus* spp. 60; *Candida* spp. 40
SIDE ACTIVITIES:	*Aspergillus* spp. amylase, invertase, glucoamylase, protease; *Candida* spp. amylase, invertase, glucoamylase, protease
FUNCTION IN FOODS:	Endo-hydrolysis of inulin
TECHNOLOGY OF USE IN FOODS:	Endo-hydrolysis of inulin in Jerusalem artichoke for production of fructose
SYNERGISTS:	*Aspergillus* spp. -; *Candida* spp. -
ANTAGONISTS:	*Aspergillus* spp. -; *Candida* spp. -
REFERENCE:	Owusu-Ansah, Y. J. (1991) Enzymes. In: Smith, J. (Ed.), *Food Additive User's Handbook*. Blackie A & P, London, pp. 120–150.

NAME:	Invertase
CATEGORY:	Enzymes
FOOD USE:	Confectionery
SYNONYMS:	Beta-D-fructofuranoside fructohydrolase/ EC 3.2.1.26/ Sucrase/ Invertin
pH OPTIMUM:	*Candida* spp. 4.5; *Saccharomyces* spp. 4.5
TEMPERATURE OPTIMUM °C:	*Candida* spp. 50; *Saccharomyces* spp. 55
SIDE ACTIVITIES:	*Candida* spp. proteases; *Saccharomyces* spp. proteases
FUNCTION IN FOODS:	Hydrolysis of sucrose to glucose and fructose
TECHNOLOGY OF USE IN FOODS:	Used to hydrolyse sucrose to glucose and fructose, to prevent crystallisation in confectionery and to increase sweetness and liquidity in confectionery soft centres
SYNERGISTS:	*Candida* spp.; *Saccharomyces* spp.
ANTAGONISTS:	*Candida* spp.; *Saccharomyces* spp.

LEGISLATION:

USA:	UK and EUROPE:	CANADA:
GRAS – *Saccharomyces cerevisiae*	*Saccharomyces cerevisiae*	*Saccharomyces cerevisiae* – permitted in or upon soft-centred and liquid-centred confections, unstandardised baking goods

REFERENCES:

Ash, M., and Ash, I. (1995) *Food Additives*. Gower Publishing Co., Brookfield, VT.

Owusu-Ansah, Y. J. (1991) Enzymes. In: Smith, J. (Ed.), *Food Additive User's Handbook*. Blackie A & P, London, pp. 120–150.

ANY OTHER RELEVANT INFORMATION: *Candida* spp. maximum activity is shown at low substrate levels; *Saccharomyces* spp. all contain a bound mannan

NAME:	**Lipase**
CATEGORY:	Enzymes
FOOD USE:	Cheese/ Eggs and egg products/ Oils
SYNONYMS:	Triacylglycerol acylhydrolase, EC 3.1.1.3 (*Aspergillus niger*, *Candida cylindraceae*, *Mucor miehei*); fatty acid esterase, carboxylic-ester hydrolase, EC 3.1.1.1 (pancreatic lipase); aryl ester hydrolase, EC 3.1.1.2 (pregastric esterase, *Rhizopus* spp.)/ CAS 9001-62-1/ EINECS 232-619-9
pH OPTIMUM:	*Aspergillus niger* 5–7; *Candida cylindraceae* 8; *Mucor miehei* 7.5; pancreatic lipase 7.5–8; pregastric esterase 5.5–7; *Rhizopus* spp. 5–8
TEMPERATURE OPTIMUM °C:	*Aspergillus niger* 40; *Candida cylindraceae* 50; *Mucor miehei* 50; pancreatic lipase 40; pregastric esterase 30–60; *Rhizopus* spp. 40
SIDE ACTIVITIES:	*Aspergillus niger* amylase, cellulase, esterase, hemicellulase, pectinase, protease; *Candida cylindraceae* esterase, protease; *Mucor miehei* esterase, protease; pancreatic lipase – amylase, protease; pregastric esterase – amylase; *Rhizopus* spp. amylase, cellulase, esterase, protease
FUNCTION IN FOODS:	Hydrolyses fat to glycerol and fatty acid
TECHNOLOGY OF USE IN FOODS:	Cheese – flavour development as liquid or powder at 1% DS. Eggs and egg products – improve emulsification and whipping properties as powder or immobilised. Oils – oil hydrolysis – to produce free fatty acids at 2% DS, and interesterification. Production of value-added triacylglycerols from less-valued feed stock – mostly immobilised systems at 1–5% E/S
SYNERGISTS:	*Aspergillus niger*; *Candida cylindraceae*; *Mucor miehei*; pancreatic lipase; pregastric esterase; *Rhizopus* spp.
ANTAGONISTS:	*Aspergillus niger*; *Candida cylindraceae*; *Mucor miehei*; pancreatic lipase; pregastric esterase; *Rhizopus* spp.

LEGISLATION:

USA:
GRAS from calf, kid or lamb pancreatic tissue, *Aspergillus niger*, *Aspergillus oryzae. Mucor miehei*, *Bacillus licheniformis*, *Bacillus subtilis*

UK and EUROPE:
Edible oral and forestomach tissue of the calf, kid or lamb; porcine or bovine pancreatic tissue

CANADA:
Source – *Aspergillus niger*, *Aspergillus oryzae*, edible forestomach tissue of calves, kids or lambs; animal pancreatic tissue. Permitted in or on dairy-based flavouring preparations – liquid and dried egg white, romano cheese

REFERENCES:

Ash, M., and Ash, I. (1995) *Food Additives*. Gower Publishing Co., Brookfield, VT.

Owusu-Ansah, Y. J. (1991) Enzymes. In: Smith, J. (Ed.), *Food Additive User's Handbook*. Blackie A & P, London, pp. 120–150.

ANY OTHER RELEVANT INFORMATION:

Aspergillus niger high in esterase; *Candida cylindraceae* active on higher fats and oils; *Mucor miehei* high in true lipase activity; pancreatic lipase – preferential action on triacylglycerols; pregastric esterase – high esterase:lipase ratios; *Rhizopus* spp. – very varied specificities

NAME:	**Metallo-neutral proteases**
CATEGORY:	Enzymes
FOOD USE:	Fermented beverages/ Flour/ Hydrolysed proteins/ Meat
SYNONYMS:	EC 3.4.24.4
pH OPTIMUM:	*Aspergillus oryzae* 7; *Bacillus thermoproteolyticus* 8; *Bacillus* spp. 7
TEMPERATURE OPTIMUM °C:	*Aspergillus oryzae* 50; *Bacillus thermoproteolyticus* 65; *Bacillus* spp. 50
SIDE ACTIVITIES:	*Aspergillus oryzae* other proteases; *Bacillus thermoproteolyticus* other proteases; *Bacillus* spp. other proteases
FUNCTION IN FOODS:	Hydrolysis of proteins
TECHNOLOGY OF USE IN FOODS:	Preferential cleavage of bonds with hydrophobic residues
SYNERGISTS:	*Aspergillus oryzae*; *Bacillus thermoproteolyticus*; *Bacillus* spp.
ANTAGONISTS:	*Aspergillus oryzae* – reducing agents, chelating agents, halogens; *Bacillus thermoproteolyticus* – reducing agents, chelating agents, halogens; *Bacillus* spp. – reducing agents, chelating agents, halogens
REFERENCE:	Owusu-Ansah, Y. J. (1991) Enzymes. In: Smith, J. (Ed.), *Food Additive User's Handbook*. Blackie A & P, London, pp. 120–150.

NAME:	**Microbial rennet**
CATEGORY:	Enzymes
FOOD USE:	Cheese
SYNONYMS:	Microbial chymosin/ EC 3.4.23.4
pH OPTIMUM:	4.8–6
TEMPERATURE OPTIMUM °C:	30–40
FUNCTION IN FOODS:	Specific for one bond of kappa-casein
TECHNOLOGY OF USE IN FOODS:	Cheese – coagulation of casein as powder or solution at 0.01–0.15%
LEGISLATION:	**USA:** GRAS source FDA 21CFR 173.150 *Mucor miehei*, *Mucor pucillus*, *Endothia parasitica*, *Bacillus cereus* **UK and EUROPE:** Approved **CANADA:** Source – *Mucor miehei*, *Mucor pucillus*, *Endothia parasitica* – permitted in or upon cheese, cottage cheese, Emmentaler (Swiss) cheese, sour cream
REFERENCES:	Ash, M., and Ash, I. (1995) *Food Additives*. Gower Publishing Co., Brookfield, VT. Owusu-Ansah, Y. J. (1991) Enzymes. In: Smith, J. (Ed.), *Food Additive User's Handbook*. Blackie A & P, London, pp. 120–150.

NAME:	**Naringinase (*Penicillium* spp.)**
CATEGORY:	Enzymes
FOOD USE:	Fruit and vegetable juices
SYNONYMS:	Beta-L-l-rhamnosidase/ EC 3.2.1.40
pH OPTIMUM:	3–5
TEMPERATURE OPTIMUM °C:	40
SIDE ACTIVITIES:	Beta-glucosidase (EC 3.1.1.21)
FUNCTION IN FOODS:	Debittering of citrus juice as powder
TECHNOLOGY OF USE IN FOODS:	Fruit and vegetable juices – debittering of citrus juice as powder
REFERENCE:	Owusu-Ansah, Y. J. (1991) Enzymes. In: Smith, J. (Ed.), *Food Additive User's Handbook*. Blackie A & P, London, pp. 120–150.

NAME:	Papain (papaya species)
CATEGORY:	Enzymes
FOOD USE:	Meat and other proteinaceous foods/ Meat and fish/ Eggs and egg products
SYNONYMS:	EC 3.4.22.2/ CAS 9001-73-4/ EINECS 232-627-2/ Vegetable pepsin/ Papayotin
PROPERTIES AND APPEARANCE:	White to grey powder, slightly hygroscopic
pH OPTIMUM:	5–7
TEMPERATURE OPTIMUM °C:	65
SIDE ACTIVITIES:	Lysozyme, glucanase, glucosidase, cellulase
FUNCTION IN FOODS:	Direct food additive, enzyme, processing aid, texturiser; meat tenderiser, tissue softening agent. Chillproofing, antihazing agent for beer. Preferential cleavage of arginine, lysine
TECHNOLOGY OF USE IN FOODS:	Meat and other proteinaceous foods. Meat and fish – tenderisation of meat as liquid, produce fish hydrolysates as liquid at 2% of protein, enhance fish stick-water effluent treatment as liquid at 0.2%. Removal of oil from tissues as liquid. Eggs and egg products – improve drying properties as liquid/powder
SYNERGISTS:	Reducing compounds
ANTAGONISTS:	Oxidising agents

LEGISLATION:	**USA:** GRAS from papaya. FDA 21CFR § 184.1585, GRAS; USDA 9CFR § 318.7, 381.147; BATF 27CFR § 240.1051 **UK and EUROPE:** *Carica papaya*	**CANADA:** Source: papaya – permitted in or upon ale, beer, light beer, malt liquor, porter, stout, bread, flour, wholewheat flour, edible collagen sausage casings, hydrolysed animal, milk and vegetable proteins, meat cuts meat-tenderising preparations, pumping pickle for curing meat cuts. Also beef before slaughter and pre-cooked instant cereals **AUSTRALIA/PACIFIC RIM:** Japan: approved
REFERENCES:	Ash, M., and Ash, I. (1995) *Food Additives*. Gower Publishing Co., Brookfield, VT. Owusu-Ansah, Y. J. (1991) Enzymes. In: Smith, J. (Ed.), *Food Additive User's Handbook*. Blackie A & P, London, pp. 120–150.	
ANY OTHER RELEVANT INFORMATION:	Proteolytic enzyme derived from latex of the green fruit and leaves of *Carica papaya*. Molecular weight 21,000 Da	

NAME:	Pectinase
CATEGORY:	Enzymes
FOOD USE:	Alcoholic beverages/ Non-alcoholic beverages/ Coffee, cocoa/ Fruits and vegetables/ Fats and oils
SYNONYMS:	poly(1,4-alpha-D-galacturonide) glycano-hydrolase EC 3.2.1.15/ CAS 9032-75-1/ EINECS 232-885-6
pH OPTIMUM:	*Aspergillus* spp. 2.5–6; *Rhizopus* spp. 2.5–6
TEMPERATURE OPTIMUM °C:	*Aspergillus* spp. 40–60; *Rhizopus* spp. 30–50
SIDE ACTIVITIES:	*Aspergillus* spp. pectinesterase, pectin lyase, etc.; *Rhizopus* spp. pectinesterase, pectin lyase, etc.
FUNCTION IN FOODS:	Enzyme for wine, cider, fruit juice, natural flavour/colour extracts, citrus fruit skins for jams, vegetable stock for soup manufacture
TECHNOLOGY OF USE IN FOODS:	Alcoholic beverages: wine – clarification of wine – decreases pressing time and increases extraction yield as liquid complex at 0.01–0.02%. Non-alcoholic beverages – coffee – removal of gelatinous coating during drying – as liquid or powder at 20–50 ppm. Non-alcoholic beverages – cocoa – hydrolysis of pulp from beans during fermentation as liquid or powder. Fruits and vegetables – juices – improve extraction, 0.003–0.03%; aid clarification, 0.01–0.02% mostly liquid. Vegetables – production of hydrolysates as liquid at 11–20 ppm. Fats and oils – extractions – degrade pectin substances to release oil as liquid/powder at 0.5–3% DS
SYNERGISTS:	*Aspergillus* spp.; *Rhizopus* spp.
ANTAGONISTS:	*Aspergillus* spp.; *Rhizopus* spp.

LEGISLATION:

USA:
GRAS source *Aspergillus niger*, *Rhizopus oryzae*; FDA 173.130

UK and EUROPE:
UK permitted source for pectin esterase and pectin lyase – *Aspergillus niger*

CANADA:
Source: *Aspergillus niger*, *Rhizopus oryzae* – permitted in or upon cider, wine, distiller's mash, juice of named fruits, natural flavour and colour extractives, skin of citrus fruits destined for jam, marmalade and candied fruit production, vegetable stock for use in soup manufacture

REFERENCES:

Ash, M., and Ash, I. (1995) *Food Additives*. Gower Publishing Co., Brookfield, VT.

Owusu-Ansah, Y. J. (1991) Enzymes. In: Smith, J. (Ed.), *Food Additive User's Handbook*. Blackie A & P, London, pp. 120–150.

ANY OTHER RELEVANT INFORMATION: Wide variety of component activities

NAME:	Penicillin amidase
CATEGORY:	Enzymes
SYNONYMS:	Penicillin amidohydrolase/ EC 3.5.1.11
pH OPTIMUM:	*Bacillus* spp. 7–8; *Basidiomycetes* spp. 4–6
TEMPERATURE OPTIMUM °C:	*Bacillus* spp. 37; *Basidiomycetes* spp. 50
SIDE ACTIVITIES:	*Bacillus* spp. – usually very pure; *Basidiomycetes* spp. – usually very pure
REFERENCE:	Owusu-Ansah, Y. J. (1991) Enzymes. In: Smith, J. (Ed.), *Food Additive User's Handbook*. Blackie A & P, London, pp. 120–150.

NAME:	Pepsin (porcine mucosa)			
CATEGORY:	Enzymes			
FOOD USE:	Cheese			
SYNONYMS:	EC 3.4.23.1/ CAS 9001-75-6/ EINECS 232-629-3/ Pepsinum			
PROPERTIES AND APPEARANCE:	White or yellowish-white powder or lustrous transparent or translucent scales, odourless			
pH OPTIMUM:	1.8–2.2			
TEMPERATURE OPTIMUM °C:	40–60			
SIDE ACTIVITIES:	Usually very pure			
FUNCTION IN FOODS:	Preferential cleavage of phenylalanine, leucine			
TECHNOLOGY OF USE IN FOODS:	Cheese – coagulation of casein as powder or solution at 0.01–0.15%			
SYNERGISTS:	Reducing agents			
ANTAGONISTS:	Oxidising agents, aliphatic alcohols			
LEGISLATION:	**USA:**	**UK and EUROPE:**	**CANADA:**	**AUSTRALIA/ PACIFIC RIM:**
	GRAS from porcine and bovine stomachs	UK adult bovine abomasum	Source – glandular layer of porcine stomach. Permitted in or upon ale, beer, light beer, malt liquor, porter, stout, cheese, cottage cheese, cream cheese, cream cheese with named ingredients, cream cheese spread, cream cheese spread with named ingredients, defatted soy flour, pre-cooked instant cereals	Japan: approved

REFERENCES:	Ash, M., and Ash, I. (1995) *Food Additives*. Gower Publishing Co., Brookfield, VT. Owusu-Ansah, Y. J. (1991) Enzymes. In: Smith, J. (Ed.), *Food Additive User's Handbook*. Blackie A & P, London, pp. 120–150.
ANY OTHER RELEVANT INFORMATION:	A digestive enzyme of gastric juice which hydrolyses certain linkages of proteins to produce peptones. Molecular weight 36,000 Da

NAME:	**Peroxidase (horseradish)**
CATEGORY:	Enzymes
FOOD USE:	Dairy products
SYNONYMS:	Hydrogen peroxide oxidoreductase/ EC 1.11.1.7
pH OPTIMUM:	5–7
TEMPERATURE OPTIMUM °C:	45
SIDE ACTIVITIES:	Catalase
FUNCTION IN FOODS:	Elimination of hydrogen peroxide from treated milk
TECHNOLOGY OF USE IN FOODS:	Elimination of hydrogen peroxide from treated milk
REFERENCE:	Owusu-Ansah, Y. J. (1991) Enzymes. In: Smith, J. (Ed.), *Food Additive User's Handbook*. Blackie A & P, London, pp. 120–150.
ANY OTHER RELEVANT INFORMATION:	A specific haem protein enzyme

NAME:	**Pullulanase**
CATEGORY:	Enzymes
FOOD USE:	Sugar manufacture
SYNONYMS:	EC 3.2.1.4/ CAS 9075-68-7/ EINECS 232-983-9
PROPERTIES AND APPEARANCE:	White powder
pH OPTIMUM:	*Bacillus* spp. 4.5; *Klebsiella aerogenes* 5
TEMPERATURE OPTIMUM °C:	*Bacillus* spp. 60; *Klebsiella aerogenes* 50
SIDE ACTIVITIES:	*Bacillus* spp. protease, amylase; *Klebsiella aerogenes* protease, amylase
FUNCTION IN FOODS:	Food enzyme in production of maltose. Hydrolysis of 1,6-alpha-D-glucosidic link in pullulan, amylopectin, glycogen and limit dextrans
TECHNOLOGY OF USE IN FOODS:	Food enzyme in production of maltose. Hydrolysis of 1,6-alpha-D-glucosidic link in pullulan, amylopectin, glycogen and limit dextrans
LEGISLATION:	**UK and EUROPE:** *Klebsiella aerogenes*
REFERENCES:	Ash, M., and Ash, I. (1995) *Food Additives*. Gower Publishing Co., Brookfield, VT. Owusu-Ansah, Y. J. (1991) Enzymes. In: Smith, J. (Ed.), *Food Additive User's Handbook*. Blackie A & P, London, pp. 120–150.

NAME:	**Rennet (bovine abomasum)**			
CATEGORY:	Enzymes			
FOOD USE:	Cheese			
SYNONYMS:	Chymosin/ EC 3.4.23.4/ CAS 9001-98-3/ Bovine rennet/ Rennin			
PROPERTIES AND APPEARANCE:	Yellowish-white powder, peculiar odour, slight salty taste			
SOLUBILITY % AT VARIOUS TEMPERATURE/pH COMBINATIONS:				
in water:	Slightly soluble			
pH OPTIMUM:	4.8–6			
TEMPERATURE OPTIMUM °C:	30–40			
SIDE ACTIVITIES:	May contain pepsin			
FUNCTION IN FOODS:	Specific for one bond of kappa-casein			
TECHNOLOGY OF USE IN FOODS:	Cheese – coagulation of casein as powder or solution at 0.01–0.15%			
LEGISLATION:	**USA:**	**UK and EUROPE:**	**CANADA:**	**AUSTRALIA/ PACIFIC RIM:**
	GRAS source – ruminant fourth stomach (abomasum). FDA 21CFR § 184.1685, GRAS; USDA 9CFR § 318.7 (rennet-treated calcium reduced dried milk and calcium lactate, limitation 3.5% in sausages), 381.147	Sources – abomasum of calf, kid or lamb	Source – aqueous extracts from fourth stomach of calves, kids or lambs. Permitted in or upon unstandardised milk-based dessert preparations, cheese, cottage cheese, cream cheese, cream cheese with named additives, cream cheese spread, cream cheese spread with named additives	Japan: approved

REFERENCES:

Ash, M., and Ash, I. (1995) *Food Additives*. Gower Publishing Co., Brookfield, VT.

Owusu-Ansah, Y. J. (1991) Enzymes. In: Smith, J. (Ed.), *Food Additive User's Handbook*. Blackie A & P, London, pp. 120–150.

NAME:	Subtilisin
CATEGORY:	Enzymes
FOOD USE:	Baked goods/ Alcoholic beverages (beer)
SYNONYMS:	EC 3.4.21.14
pH OPTIMUM:	*Bacillus amyloliquifaciens* 9–11; *Bacillus licheniformis* 9–11; *Bacillus subtilis* 9–10; *Aspergillus oryzae* 8–10
TEMPERATURE OPTIMUM °C:	*Bacillus amyloliquifaciens* 60–70; *Bacillus licheniformis* 60–70; *Bacillus subtilis* 55; *Aspergillus oryzae* 60–70
SIDE ACTIVITIES:	*Bacillus amyloliquifaciens* amylase, glucanase, other protease; *Bacillus licheniformis* amylase, glucanase, other protease; *Bacillus subtilis* amylase, glucanase, other protease; *Aspergillus oryzae* amylase, glucanase, other protease
FUNCTION IN FOODS:	Baked goods – modification of gluten in baking of biscuits as powder at up to 0.25% of flour; to reduce mixing time of dough as tablets at 75 HU per 100 g flour. Alcoholic beverages (beer) to provide nitrogen for yeast growth and aid in filtration and chillproofing as liquid or powder at 0.3% DS
TECHNOLOGY OF USE IN FOODS:	Baked goods – modification of gluten in baking of biscuits as powder at up to 0.25% of flour; to reduce mixing time of dough as tablets at 75 HU per 100 g flour. Alcoholic beverages (beer) to provide nitrogen for yeast growth and aid in filtration and chillproofing as liquid or powder at 0.3% DS
ANTAGONISTS:	*Bacillus amyloliquifaciens* organophosphorus compounds; *Bacillus licheniformis* organophosphorus compounds; *Bacillus subtilis* organophosphorus compounds; *Aspergillus oryzae* organophosphorus compounds

NAME:	**Tannase**
CATEGORY:	Enzymes
FOOD USE:	Non-alcoholic beverages
SYNONYMS:	Tannic acid acylhydrolase/ EC 3.1.1.20
pH OPTIMUM:	*Aspergillus niger* 4.5; *Aspergillus oryzae* 3–5
TEMPERATURE OPTIMUM °C:	*Aspergillus niger* 55; *Aspergillus oryzae* 45
SIDE ACTIVITIES:	*Aspergillus niger* amylase, glucoamylase; *Aspergillus oryzae* amylase, glucoamylase, protease, cellulase
FUNCTION IN FOODS:	Removal of polyphenolics as liquid or powder 0.03%
TECHNOLOGY OF USE IN FOODS:	Removal of polyphenolics as liquid or powder 0.03%
LEGISLATION:	**AUSTRALIA/PACIFIC RIM:** Japan: approved
REFERENCES:	Ash, M., and Ash, I. (1995) *Food Additives*. Gower Publishing Co., Brookfield, VT. Owusu-Ansah, Y. J. (1991) Enzymes. In: Smith, J. (Ed.), *Food Additive User's Handbook*. Blackie A & P, London, pp. 120–150.
ANY OTHER RELEVANT INFORMATION:	*Aspergillus niger*; *Aspergillus oryzae*

NAME:	Trypsin (pancreatic)
CATEGORY:	Enzymes
FOOD USE:	Baked goods/ Alcoholic beverages (beer)
SYNONYMS:	EC 3.4.21.4/ CAS 9002-07-7/ EINECS 232-650-8
pH OPTIMUM:	8–9
TEMPERATURE OPTIMUM °C:	45
SIDE ACTIVITIES:	Amylase, lipase, esterase
FUNCTION IN FOODS:	Preferential cleavage of arginine, lysine
TECHNOLOGY OF USE IN FOODS:	Baked goods – modification of gluten in baking of biscuits as powder at up to 0.25% of flour; to reduce mixing time of dough as tablets at 75 HU per 100 g flour. Alcoholic beverages (beer) to provide nitrogen for yeast growth and aid in filtration and chillproofing as liquid or powder at 0.3% DS
ANTAGONISTS:	Compounds in cereals, beans, potato, egg
LEGISLATION:	**USA:** GRAS from porcine, bovine pancreas **UK and EUROPE:** Porcine, bovine pancreatic tissue **AUSTRALIA/PACIFIC RIM:** Japan: approved
REFERENCES:	Ash, M., and Ash, I. (1995) *Food Additives*. Gower Publishing Co., Brookfield, VT. Owusu-Ansah, Y. J. (1991) Enzymes. In: Smith, J. (Ed.), *Food Additive User's Handbook*. Blackie A & P, London, pp. 120–150.
ANY OTHER RELEVANT INFORMATION:	Bovine more thermotolerant

NAME:	Xylanase
CATEGORY:	Enzymes
FOOD USE:	Fruit juice
SYNONYMS:	1,3-beta-D-xylanhydrolase/ EC 3.2.1.32
pH OPTIMUM:	*Aspergillus niger* 3–5; *Aspergillus oryzae* 4; *Bacillus* spp. 7–9
TEMPERATURE OPTIMUM °C:	*Aspergillus niger* 45–55; *Aspergillus oryzae* 45; *Bacillus* spp. 55–65
SIDE ACTIVITIES:	*Aspergillus niger* amylase, glucoamylase, glucosidase, cellulase; *Aspergillus oryzae* amylase, glucanase, glucoamylase, cellulase; *Bacillus* spp. amylase, glucanase, protease
FUNCTION IN FOODS:	Endo-hydrolysis of 1,3-beta-D-xylose units of xylans
TECHNOLOGY OF USE IN FOODS:	Endo-hydrolysis of 1,3-beta-D-xylose units of xylans
SYNERGISTS:	*Aspergillus niger*; *Aspergillus oryzae*; *Bacillus* spp.
ANTAGONISTS:	*Aspergillus niger*; *Aspergillus oryzae*; *Bacillus* spp.
REFERENCE:	Owusu-Ansah, Y. J. (1991) Enzymes. In: Smith, J. (Ed.), *Food Additive User's Handbook*. Blackie A & P, London, pp. 120–150.

Part 6

Flavour Enhancers

NAME:	**Acetic acid**
CATEGORY:	Flavour enhancer
FOOD USE:	Baked goods/ Catsup/ Cheese/ Chewing gum/ Condiments/ Dairy products/ Fats/ Fats (rendered)/ Gravies/ Mayonnaise/ Meat products/ Oils/ Pickles/ Relishes/ Salad dressings/ Sauces
SYNONYMS:	Acetic acid (aqueous solution)/ Acetic acid, glacial/ Acide acetique (French)/ Acido acetico (Italian)/ Azijnzuur (Dutch)/ Essigsaeure (German)/ Octowy kwas (Polish)/ CAS 64-19-7/ EINECS 200-580-7/ Ethanoic acid/ Ethylic acid/ Methanecarboxylic acid/ Pyroligneus acid/ Vinegar acid/ Ry-So/ Sour Dough Base/ Vanease/ BFP white sour/ FEMA No. 2006/ E260
FORMULA:	CH3 COOH
MOLECULAR MASS:	60.05
PROPERTIES AND APPEARANCE:	Clear, colourless liquid; pungent odour. Miscible with water, alcohol, glycerol, carbon tetrachloride; practically insoluble in carbon disulphide
BOILING POINT IN °C AT VARIOUS PRESSURES (INCLUDING 760 mm Hg):	117–118
MELTING RANGE IN °C:	16.7
FLASH POINT IN °C:	43
DENSITY AT 20°C (AND OTHER TEMPERATURES) IN g/l:	1.049
FUNCTION IN FOODS:	Acidifier; boiler water additive; colour diluent; curing/ pickling agent; flavour enhancer; flavouring agent; pH control agent; solvent/vehicle
ANTAGONISTS:	Incompatible with: chromic acid, nitric acid, 2-amino-ethanol, NH_4NO_3, ClF_3, chlorosulphonic acid, (O_3 + diallyl methyl carbinol), ethylenediamine, ethylene imine, (HNO_3 + acetone), oleum, $HClO_4$, permanganates, $P(OCN)_3$, KOH, NaOH, *n*-xylene, carbonates and many oxides and phosphates

FOOD SAFETY ISSUES:

Human poison by unspecified route. Moderately toxic by various routes; severe eye and skin irritant; caustic – can cause severe burns, lacrimation, and conjunctivitis.

Human systemic effects by ingestion: changes in oesophagus, ulceration or bleeding from the small and large intestines. Human systemic irritant effects and mucous membrane irritant.

Experimental reprotoxicity effects; mutagenic data.

Common air contaminant. Combustible liquid; moderate fire and explosive hazard when exposed to heat or flame; can react vigorously with oxidising materials.

To fight fire – use CO_2, dry chemical, alcohol foam, foam and mist. When heated to decomposition will emit irritating fumes. Potentially explosive with 5-azido-tetrazole, hydrogen peroxide, potassium permanganate, sodium peroxide, phosphorus trichloride. Potentially violent reactions with acetaldehyde and acetic anhydride. Ignites on contact with potassium-tert-butoxide

LEGISLATION:

USA:	**UK and EUROPE:**	**CANADA:**	**AUSTRALIA/ PACIFIC RIM:**
FDA – 21CFR 182.1005. Approved for use in foods as flavour enhancer, flavouring agent/adjuvant, curing/ pickling agent, pH control agent, solvent/vehicle at GRAS quantities. GRAS – limitations of: – 0.26% in baked goods – 0.8% in cheese and dairy products – 0.5% in chewing gum – 9.0% in condiments and relishes – 0.5% in fats and oils, – 0.3% in gravies and sauces – 0.6% in meat products (*conts*)	EEC Regulations (E260). Used as antibacterial preservative; acidity stabiliser; diluent for colours; flavouring agent. No limits on ADI UK: approved for use Europe: listed	18-10-79 Permitted for use in cream cheese spread, canned asparagus, and gelatin as pH-adjusting agents, acid-reacting materials, and water-correcting agents at GMP levels. Permitted for use as class 1 preservatives in preserved fish, meat, meat by-products, poultry, and pickles at food manufacturing practice levels. Approved for use in	Japan: approved as acetic acid glacial; used as acidity regulator and food acid

	(*USA: contd*) – 0.15% in all other food categories when used in accordance with good manufacturing practices USDA – 9CFR 318.7 – sufficient for purpose. 21CFR 182.70, 172.814, 184.1005, 73.85, 178.1010	unstandardised foods at GMP levels

REFERENCES:

Lewis, R. J., Sr. (1989) *Food Additives Handbook*. Van Nostrand Reinhold, New York.

Smoley, C. K. (1993) *Everything added to Food in the United States*. U.S. Food and Drug Administration.

NAME:	**Brown algae**
CATEGORY:	Flavour enhancer
FOOD USE:	Flavouring/ Seasonings/ Spices
SYNONYMS:	*Analipus japonicus*/ *Eisenia bicyclis*/ *Kjellmaniella gyrate*/ *Hizikia fusiforme*/ *Laminaria angustata*/ *L. japonica*/ *L. longicruris*/ *L. claustonia*/ *L. digitata*/ *Macrocystis pyrifera*/ CAS 977026928
FUNCTION IN FOODS:	Flavour enhancer; flavour adjuvant; flavouring agent
ALTERNATIVES:	Red algae
FOOD SAFETY ISSUES:	Heated to decomposition will emit acrid smoke and irritating fumes
LEGISLATION:	**USA:** FDA – 21CFR 184.1120: approved for use in foods as flavour enhancer, flavouring adjuvant for spices and seasonings at GRAS quantities. GRAS use at level not in excess of the amount reasonably required to accomplish the intended effect FDA – 21CFR 172.365
REFERENCE:	Lewis, R. J., Sr. (1989) *Food Additives Handbook*. Van Nostrand Reinhold. New York.

NAME:	Red algae
CATEGORY:	Flavour enhancer
FOOD USE:	Flavourings/ Seasonings/ Spices
SYNONYMS:	*Porphyra crispata*/ *P. deutata*/ *P. perforata*/ *P. suborbiculata*/ *P. tenera*/ *Rhodymenia palmata*/ *Gloiopeltis furcata*/ CAS 977090042
FUNCTION IN FOODS:	Flavour enhancer; flavour adjuvant
ALTERNATIVES:	Brown algae
LEGISLATION:	**USA:** FDA – 21CFR 184.1121: approved for use in foods as flavour enhancer, flavouring adjuvant for spices and seasonings at GRAS quantities. GRAS use at level not in excess of the amount reasonably required to accomplish the intended effect
REFERENCE:	Lewis, R. J., Sr. (1989) *Food Additives Handbook*. Van Nostrand Reinhold. New York.

NAME:	Ammoniated glycyrrhizin
CATEGORY:	Flavour enhancer
FOOD USE:	Baked goods/ Beverages (alcoholic)/ Beverages (non-alcoholic)/ Candy (hard)/ Candy (soft)/ Chewing gum/ Herbs/ Plant protein products/ Seasonings/ Vitamin or mineral dietary supplements/ Pharmaceuticals
SYNONYMS:	CAS 053956040/ Ammonium glycyrrhizinate, pentahydrate/ Monoammonium glycyrrhizinate/ Ammonium glycyrrhizinate/ Magnasweet®/ MAG
FORMULA:	$C42 H65 N O16 \cdot 5H2O$
MOLECULAR MASS:	839.91
ALTERNATIVE FORMS:	Ammonium glycyrrhizinate, pentahydrate/ Ammonium glycyrrhizinate/ Monoammonium glycyrrhizinate
PROPERTIES AND APPEARANCE:	Obtained by extraction from ammoniated glycyrrhizin; derived from roots of *Glycyrrhiza glabra*, family Leguminosae. White powder, sweet taste; insoluble in glacial acetic acid; soluble in ammonia. About 100 times as sweet as sucrose; hygroscopic
FUNCTION IN FOODS:	Flavour enhancer; flavouring agent; surface-active agent; foaming agent; aromatisation of food; sweetening agent for taste correction of food and drugs and production of confectioneries; sweetness potentiators; masking agents; reduce metallic aftertaste from high-intensity sweeteners
ALTERNATIVES:	For Licorice taste: Licorice/Licorice root extract; Licorice; Thaumatin. For high sweetness intensity: aspartame; ethyl maltol; glycine; Sucralose
TECHNOLOGY OF USE IN FOODS:	Some relative sweetness measured against 20 mg/100 ml of sucrose solution (70 times sweeter) or 10 mg/100 ml of sucrose solution (93 times sweeter). Slow onset of sweetness, but taste is long. Effectively masks bitter, harsh, and astringent aftertastes common to pharmaceutical active ingredients. Enhances and magnifies natural and artifical flavours.

	Pre-solubilised: shelf-stable solution of glycerin or propylene glycol ideal for use when rapid dispersion is desired or when powder is unsuitable for processing. Powder forms: good dispersion and compression properties designed for dry mixes, aqueous systems and tableting soluble in aqueous, alcoholic and hydro-alcoholic type formulations can be incorporated into products containing propylene glycol and glycerin oil-based systems will require use of suitable emulsifiers; most favoured solubility pH 3.5–8.0. Dissolving of powdered form will require moderate agitation and elevated temperature of solvent to 140–160°C (liquid form could be used in absence of these conditions)
SYNERGISTS:	Shows synergism when mixed with sucrose. Works synergistically with other natural and artificial sweeteners to provide better overall sweetness and in proper combination can create systems that simulate sucrose. Works synergistically with primary sweeteners (i.e. sucrose, fructose, and high-intensity sweeteners)
ANTAGONISTS:	Elevated saccharin levels may intensify metallic aftertaste
FOOD SAFETY ISSUES:	When heated to decomposition will emit acrid smoke and irritating fumes
LEGISLATION:	**USA:** FDA – 21 CFR 184.1408: approved for use in foods as flavour enhancer, flavouring agent, surface active agent at GRAS quantities. GRAS – with limitations of: – 0.05% in baked goods – 0.1% in alcoholic beverages – 0.15% in non-alcoholic beverages – 1.1% in chewing gum – 16.0% in hard candy – 0.15% in herbs and seasonings – 0.15% in plant protein products – 3.1% in soft candy – 0.5% in vitamin or mineral dietary supplements – 0.1% in all other foods except sugar substitutes when used in accordance with food manufacturing practices Not permitted for use as a non-nutritive sweetener in sugar substitutes

REFERENCES:	Lewis, R. J., Sr. (1989) *Food Additives Handbook*. Van Nostrand Reinhold, New York. Krutosikova, A., and Uher, M. (1992) *Natural and Synthetic Sweet Substances*. Ellis Horwood Series, New York.
ANY OTHER RELEVANT INFORMATION:	Manufacturer/Distributors: MacAndrews & Forbes Company, Third Street and Jefferson Avenue, Camden, New Jersey. 08104, USA. Tel: (609) 964-8840 Telfax: (609) 964-6029 Telex: 84-5337

NAME:	Esterase-lipase
CATEGORY:	Flavour enhancer
FOOD USE:	Cheese/ Fats/ Milk products/ Oil
PROPERTIES AND APPEARANCE:	Derived from *Mucor miehei*
TECHNOLOGY OF USE IN FOODS:	Enzyme used as flavour enhancer in cheeses, fats, and oils
FOOD SAFETY ISSUES:	When heated to decomposition will emit acrid smoke and irritating fumes
LEGISLATION:	**USA:** FDA – 21CFR 173.140: use at a level not in excess of the amount reasonably required to accomplish the intended effect
REFERENCE:	Lewis, R. J., Sr. (1989) *Food Additives Handbook.* Van Nostrand Reinhold, New York.

NAME:	Disodium guanylate			
CATEGORY:	Flavour enhancer			
FOOD USE:	Canned foods/ Poultry/ Sauces/ Snack items/ Soups			
SYNONYMS:	CAS 5550-12-9/ E627/ Disodium GMP/ Disodium guanylate (FCC)/ Disodium-5′-GMP/ Disodium-5′-guanylate/ GMP disodium salt/ 5′-GMP-disodium salt/ GMP sodium salt/ Sodium GMP/ Sodium guanosine-5′-monophosphate/ Sodium guanylate/ Sodium-5′-guanylate/ GMP/ Guanylic acid sodium salt/ Guanine riboside-5-phosphoric acid/ Guanosine 5′-disodium phosphate/ I+G/ Luxor 1576, 1626, 1639, EB-400			
FORMULA:	C10H14N5O8P·2Na or Na2C10H12N5O8P·2HOH			
MOLECULAR MASS:	409.24			
PROPERTIES AND APPEARANCE:	Decomposes at 190–200°C. Colourless to white crystals; characteristic taste. Insoluble in ether			
SOLUBILITY % AT VARIOUS TEMPERATURE/pH COMBINATIONS:				
in water:	**@ 20°C**	Soluble in cold water	**@ 100°C**	Very soluble in hot water
in ethanol solution:	**@ 100%**	Slightly soluble in alcohol		
FUNCTION IN FOODS:	Flavour enhancer			
ALTERNATIVES:	Sodium inosinate; L-glutamic acid; monoammonium glutamate; monopotassium glutamate; monosodium glutamate; ammonium glutamate; potassium glutamate			
FOOD SAFETY ISSUES:	Moderately toxic by intraperitoneal, subcutaneous and intravenous routes; mildly toxic by ingestion. Mutagenic data. When heated to decomposition, will emit fumes of POx, NOx and Na_2O			

LEGISLATION:	**USA:** FDA – 21CFR 172.530: used as flavour enhancer; use at level not in excess of the amount reasonably required to accomplish the intended effect USDA – 9CFR 318.7, 381.147: sufficient for purpose. 21CFR 155.120, 155.200, 155.170, 155.130, 145.131	**UK and EUROPE:** Europe: listed UK: approved for use **AUSTRALIA/PACIFIC RIM:** Japan: approved for use
REFERENCE:	Lewis, R. J., Sr. (1989) *Food Additives Handbook*. Van Nostrand Reinhold, New York.	
ANY OTHER RELEVANT INFORMATION:	Manufacturer/Distributor: Ajinomoto USA., Inc. 9 West 57th Street, New York, NY. 10019 USA. Telex: 232220 (AJI UR); Tel: (212) 688-8360	

NAME:	Lactic acid
CATEGORY:	Flavour enhancer
FOOD USE:	Cheese spreads/ Egg (dry powder)/ Olives/ Poultry/ Salad dressing mix/ Wine/ Confectionery/ Cultured dairy products/ Frozen desserts/ Jams, jellies, preserves/ Seasoning, sauces/ Seafood/ Soft drinks
SYNONYMS:	CAS 50-21-5/ CAS 598-82-3 (=)/ CAS 79-33-4 (=)/ CAS 10326-41-7 (=)/ EINECS 200-018-0/ EINECS 209-954-4 (=)/ EINECS 201-296-2 (=)/ EINECS 233-713-2 (=)/ E270/ FEMA 2611/ Purac/ 2-Hydroxypropionic acid/ Acetonic acid/ Ethylidenelactic acid/ 1-Hydroxyethane 1-carboxylic acid/ 2-Hydroxypropanoic acid/ α-Hydroxypropionic acid/ Kyselina milecna (Czech)/ DL-lactic acid/ Milchsaure (German)/ Milk acid/ Ordinary lactic acid/ Racemic lactic acid
FORMULA:	CH3 CHOH COOH
MOLECULAR MASS:	90.09
ALTERNATIVE FORMS:	Lactic acid, monosodium salt/ Calcium lactate/ Sodium lactate/ Potassium lactate
PROPERTIES AND APPEARANCE:	Yellow to colourless crystals or syrupy 50% liquid. Odourless, acid taste. Volatile with superheated steam. Soluble in water, alcohol, furfurol; slightly soluble in ether; insoluble in chloroform, petroleum ether, carbon disulphide. Miscible in water, alcohol + ether, glycerol, furfural
BOILING POINT IN °C AT VARIOUS PRESSURES (INCLUDING 760 mm Hg):	122 @ 15 mm
MELTING RANGE IN °C:	16.8
FLASH POINT IN °C:	>110
DENSITY AT 20°C (AND OTHER TEMPERATURES) IN g/l:	1.209
FUNCTION IN FOODS:	Acidulant in beverages; antimicrobial agent; buffer; chelating agent; curing agent; flavour enhancer; flavouring adjuvant; flavouring agent; pH control agent; pickling agent; preservatives; raising agent;

	solvent; texture modifier; vehicle (for cheese, confectionery, cultured dairy products, olives, poultry, wine)
ALTERNATIVES:	Citric acid/ L-Tartaric acid/ Malic acid/ Sodium citrate/ Succinic acid/ Tannic acid
TECHNOLOGY OF USE IN FOODS:	Due to its mild acidic taste, it will not be effective in masking mild, aromatic flavours. When added to packing brine for green olives and onions, will ensure clarity of brine and improve flavour
SYNERGISTS:	Lactic acid in conjunction with sodium lactate will provide an excellent buffer system which is beneficial in manufacturing of confections as it provides acid taste without risk of sucrose inversion. Improves antimicrobial properties of other organic acids (i.e. benzoic, propionic, etc.)
ANTAGONISTS:	Incompatible with oxidising agents, iodides, HNO_3, albumin
FOOD SAFETY ISSUES:	Moderately toxic by ingestion and rectal routes. Mutagenic data available. Severe skin and eye irritant. An FDA over-the-counter drug. Mixture with nitric acid + hydrofluoric acid may react vigorously and is a storage hazard. When heated to decomposition, will emit acrid smoke and irritating fumes
LEGISLATION:	**USA:** FDA 21CFR 131.144, 131, 150.141, 150.161, 172.814; FDA – 21CFR 184.1061: approved for use in foods as antimicrobial, curing/pickling agent, flavouring agent/adjuvant, pH control agent, solvent/vehicle at GRAS quantities. GRAS: when used at a level not in excess of the amount reasonably required to accomplish the intended effect USDA – 9CFR 318.7, 381.147: sufficient for purpose BATF – 27CFR 240.1051 GRAS: when used in accordance with GMP Not for use in infant foods **UK and EUROPE:** Europe: listed UK: approved for use EEC Regulations (E270): approved for use as preservatives, antioxidant synergist, acidulant, flavouring in malting process. No ADI limits **CANADA** Approved for use as pH-adjusting or water-correcting agents **AUSTRALIA/PACIFIC RIM:** Japan: approved for use as acidity regulator, food acid, raising agent

REFERENCE:	Lewis, R. J., Sr. (1989) *Food Additives Handbook*. Van Nostrand Reinhold, New York.
ANY OTHER RELEVANT INFORMATION:	Manufacturer/Distributor: ADM Food Additives Division, 4666 Faries Parkway, Decatur, IL 62526. Tel: (217) 424-5387; Fax: (217) 424-2473

NAME:	Licorice root extract
CATEGORY:	Flavour enhancer
FOOD USE:	Bacon/ Baked goods/ Beverages (alcoholic)/ Beverages (non-alcoholic)/ Candy (hard)/ Candy (soft)/ Chewing gum/ Cocktail mixes/ Herbs/ Ice-cream/ Imitation whipped products/ Plant protein products/ Seasonings/ Soft drinks/ Syrups/ Vitamin or mineral dietary supplements
SYNONYMS:	CAS 8008-94-4/ Glycyrrhiza/ Glycyrrhizae (Latin)/ Glycyrrhiza extract/ Glycyrrhizina/ Kanzo (Japanese)/ Licorice root/ Licorice/ Licorice extract
ALTERNATIVE FORMS:	*Glycyrrhiza glabra* extract/ Licorice/ Licorice root/ Licorice extract
FUNCTION IN FOODS:	Flavour enhancer; flavouring agent; surface-active agent
ALTERNATIVES:	Licorice extract; thaumatin; ammoniated glycyrrhizin
FOOD SAFETY ISSUES:	Moderately toxic by intraperitoneal and subcutaneous routes; mildly toxic by ingestion. Mutagenic data available. When heated to decomposition, it emits acrid smoke and irritating fumes
LEGISLATION:	**USA:** FDA – 21CFR 184.1408: approved for use in foods as flavour enhancer, flavouring agent, surface-active agent at GRAS quantities GRAS: with limitation of (as glycyrrhizin) 0.05% in baked goods; 0.1% in alcoholic beverages; 0.15% in non-alcoholic beverages; 1.1% in chewing gum; 16.0% in hard candy; 0.15% in herbs and seasonings; 0.15% in plant protein products; 3.1% in soft candy; 0.5% in vitamin or mineral dietary supplements; 0.1% in all other foods except sugar substitutes when used in accordance with GMP Not permitted to be used as a non-nutritive sweetener in sugar substitutes
REFERENCE:	Lewis, R. J., Sr. (1989) *Food Additives Handbook*. Van Nostrand Reinhold, New York.

NAME:	**Magnesium sulphate**
CATEGORY:	Flavour enhancer
FOOD USE:	Various
SYNONYMS:	CAS 10034-99-8/ E518/ Magnesium sulfate/ Sulfuric acid magnesium salt (1 : 1)/ Epsom salts/ Bitter salts/ Mg 5 – Sulfat
FORMULA:	$MgSO_4$
MOLECULAR MASS:	120.37
PROPERTIES AND APPEARANCE:	Opaque needles or granular crystalline powder; odourless with cooling, bitter, salt taste. Slightly soluble in glycerin; slightly soluble in alcohol. Non-combustible; decomposes at 1124°C
DENSITY AT 20°C (AND OTHER TEMPERATURES) IN g/L:	2.65
SOLUBILITY % AT VARIOUS TEMPERATURE/pH COMBINATIONS:	
in water:	**@ 20°C** 71 g/100 ml **@ 40°C** 91 g/100 ml
FUNCTION IN FOODS:	Dietary supplement; fermentation aid; flavour enhancer; nutrient; processing aid; tofu coagulant (Japan); yeast nutrient
ALTERNATIVES:	Sodium lactate; ammonium chloride; calcium chloride; potassium chloride
FOOD SAFETY ISSUES:	Moderately toxic by ingestion, intraperitoneal, and subcutaneous routes; an experimental teratogen. Parenteral use or use in presence of renal insufficiency may lead to magnesium intoxication. Potentially explosive reaction when heated with ethoxyethynyl alcohols (e.g. 1-ethoxy-3-methyl-1-butyn-3-ol). When heated to decomposition, will emit toxic fumes of SOx

LEGISLATION:

USA:
FDA 21CFR 182.5443: approved for use in foods as dietary supplement at GRAS quantities.
FDA 21CFR 184.1443: approved for use in foods as flavour enhancer, nutrient supplement, processing aid at GRAS quantities.
GRAS: when used at a level not in excess of the amount reasonably required to accomplish the intended effect

CANADA:
Approved for use in the following: as pH-adjusting or water-correcting agent; as starch-modifying agent at 0.4%; to restore functional properties in egg albumen

UK and EUROPE:
Europe: approved for use
UK: approved for use
EEC Regulations (E518): approved for use as dietary supplement, firming agent, and for use in beer making

AUSTRALIA/PACIFIC RIM:
Japan: approved for use as coagulant for tofu, fermentation aid, yeast nutrient

REFERENCE:

Lewis, R. J., Sr. (1989) *Food Additives Handbook.* Van Nostrand Reinhold, New York.

NAME:	**Malic acid**
CATEGORY:	Flavour enhancer
FOOD USE:	Beverages (dry mix)/ Candy (hard)/ Candy (soft)/ Chewing gum/ Fats (chicken)/ Fillings/ Fruit juices/ Fruits (processed)/ Gelatins/ Jams/ Jellies/ Lard/ Non-alcoholic beverages/ Puddings/ Shortening/ Soft drinks/ Wine/ Fruit-flavour-based mints/ Pharmaceuticals
SYNONYMS:	CAS 6915-15-7/ CAS 97-67-6 (L)/ CAS 617-48-1 (DL)/ EINECS 202-601-5/ FEMA 2655/ E296/ Apple acid/ 1-Hydroxy-1,2-ethanedicarboxylic acid/ *N*-hydroxysuccinic acid/ Kyselina jablecna/ Hydroxybutanedioic acid/ Pomalus acid
FORMULA:	COOH CH2 CH(OH) COOH
MOLECULAR MASS:	134.10
ALTERNATIVE FORMS:	D(+) form/ L(–) form (apple acid)
PROPERTIES AND APPEARANCE:	White or colourless crystals; strong acid taste. Exhibits isomeric forms (DL, L and D). Very soluble in water and alcohol; slightly soluble in ether; practically insoluble in benzene (L form)
BOILING POINT IN °C AT VARIOUS PRESSURES (INCLUDING 760 mm Hg):	150 (DL)/ 140 (D or L) – decomposition
MELTING RANGE IN °C:	131–132 (DL); 101 (D); 100 (L)
DENSITY AT 20°C (AND OTHER TEMPERATURES) IN g/L:	1.595 (D or L); 1.601 (DL)
PURITY %:	Not less than 99.0% and not more than 100.5% of $C_4H_6O_5$
SOLUBILITY % AT VARIOUS TEMPERATURE/pH COMBINATIONS:	
in water:	**@ 20°C** 55.8 **@ 50°C** 0.1 – pH 2.78; 1.0 – pH 2.39; 3.0 – pH 2.04; 5.0 – pH 1.92; 10.0 – pH 1.78; 20.0 – pH 1.61; 50.0 – pH 0.96 **@ 100°C** 68.9
in ethanol solution:	**@ 25°C** 39.16

FUNCTION IN FOODS:	Flavour adjuvant; flavour enhancer; flavouring agent; food acidulant; pH control agent; synergist for antioxidant; creates a well-balanced sensation between mint and fruit flavour
ALTERNATIVES:	In beverages: citric acid, tartaric acid. In canned fruits and vegetables: lactic acid. Fumaric acid; tannic acid; succinic acid
TECHNOLOGY OF USE IN FOODS:	Increases product yield in cheese production. 1 gram is soluble in 1.4 ml of ethanol. Store at low humidity, low temperature, and air-tight container to avoid caking. Tart taste builds gradually and lasts longer than other food acids; the bitter "aftertaste" of some sugar substitutes is suppressed by malic acid. Calcium salt form of this acid has a better solubility than other food acids, which allows it to be used with hard water. In fruit drinks: enhances the flavour of fruits and extends the flavour sensation without masking the natural note; addition of malic acid in soft drinks will lower the pH creating an unfavourable environment for microorganisms thus may be considered a preservative; level of acid required varies inversely with the degree of carbonation of soft drink; used to correct the natural deficiencies of fruit when added directly to fruit/juice or during fermentation of wine. In bakery products: used to enhance fruit flavour in fruit fillings and aid in colour retention; when malic acid is used, the amount of acid and pectin added for gel formation in fillings is decreased; the slightly higher pH of malic acid helps prevent meltdown of icing or frostings when it is applied to warm baked products; malic acid in combination with lactic, acetic and propionic acid will give a sour dough flavour in sour dough bakery products. In gelatin desserts: allows for use of a lower level of gelatin to set the protein gel system. In puddings: the slightly higher pH prevents weeping of free water from starch gels formed when cooking. In hard candy: the lower melting point of this acid makes it easy to disperse in the syrup melt and thus requiring minimum folding and kneading. In canned fruits and vegetables: addition of this acid causes the pH to decrease thus reducing the processing time and temperature needed to eliminate the threat of microbial contamination, and greatly improves quality and nutritive value
SYNERGISTS:	Works synergistically with aspartame to reduce the amount of sugar substitute in diet products without affecting sweetness, while extending the sweetness potency. Works in combination with ascorbic acid to stabilise beverages and prevent discoloration. Can be used to increase the effectiveness of benzoates and sorbates as preservatives in fruit drinks

FOOD SAFETY ISSUES:

Moderately toxic by ingestion; a skin and severe eye irritant. When heated to decomposition, will emit acrid smoke and irritating fumes

LEGISLATION:

USA:
FDA 21CFR 146.113, 150.141, 150.161, 169.115, 169.140, 169.150, 131.144, 184.1069: approved for use in foods as flavour enhancer, flavouring agent and adjuvant, pH control agent at GRAS quantities.
GRAS: with limitation of 3.4% in non-alcoholic beverages – 3.0% in chewing gum – 0.8% in gelatins, puddings and fillings – 6.9% in hard candy – 2.6% jams and jellies – 3.5% in processed fruits and fruit juices – 3.0% in soft candy – 0.7% in all other foods when used in accordance with GMP USDA – 9CFR 318.7: limitation of 0.01% on basis of total weight in combination with antioxidants in lard and shortening
BATF – 27CFR 240.1051
Not for use in baby foods

UK and EUROPE:
Europe: listed
UK: approved for use
ECC Regulations (E296): malic acid in DL or L forms is approved for use as acid and flavouring

CANADA:
Approved for use in the following: as pH-adjusting or water-correcting agent for fruits, fruit jam/jelly, and cheeses

AUSTRALIA/PACIFIC RIM:
Japan: approved for use

REFERENCE:

Lewis, R. J., Sr. (1989) *Food Additives Handbook.* Van Nostrand Reinhold, New York.

ANY OTHER RELEVANT INFORMATION:	Manufacturers/Distributors: – Haarmann & Reimer Corp., Food Ingredients Division, 1884 Miles Avenue (P. O. Box 932) Elkhart, IN 46514 (46515); Tel: (800) 348-7414/(219) 264-8716; Fax: (219) 262-6747 – Browning Chemical Corporation, 707 Westchester Avenue, White Plains, NY 10604-3104; Tel: (914) 686-0300; Fax: (914) 686-0310; Telex: 23 5039

NAME:	**Ammonium glutamate**
CATEGORY:	Flavour enhancer
FOOD USE:	Meat/ Poultry
SYNONYMS:	CAS 7558-63-6/ Monoammonium glutamate/ Ammoniumglutaminat (German)/ MAG/ Monoammonium L-glutamate
FORMULA:	C5 H9 N O4 H3 N
MOLECULAR MASS:	164.19
ALTERNATIVE FORMS:	Monoammonium glutamate/ Monosodium glutamate/ Monopotassium glutamate
PROPERTIES AND APPEARANCE:	White crystalline powder; odourless. Soluble in water; insoluble in common organic solvents
FUNCTION IN FOODS:	Flavour enhancer; salt substitute; multipurpose food ingredient
ALTERNATIVES:	Monosodium glutamate; monoammonium glutamate; monopotassium glutamate; disodium guanylate; disodium inosinate; L-glutamic acid
FOOD SAFETY ISSUES:	Moderately toxic by intraperitoneal route. When heated to decomposition, will emit toxic fumes including NOx and NH_3
LEGISLATION:	**USA:** FDA 21CFR 182.1500: GRAS when used in accordance with GMP USDA 9CFR 318.7, 381.147: sufficient for purpose
REFERENCE:	Lewis, R. J., Sr. (1989) *Food Additives Handbook.* Van Nostrand Reinhold, New York.
ANY OTHER RELEVANT INFORMATION:	Manufacturer/Distributor: Ajinomoto USA., Inc., 500 Frank W. Burr Blvd., Teaneck, NJ 07666 USA. Tel: (201) 448-1212; Fax: (201) 488-6282 (NJ Head Office)

NAME:	Potassium glutamate		
CATEGORY:	Flavour enhancer		
FOOD USE:	Meat		
SYNONYMS:	CAS 19473-49-5 (monohydrate)/ EINECS 363-737-7 (monohydrate)/ E622/ Monopotassium salt/ Monopotassium L-glutamate (FCC)/ MPG/ Potassium glutamate/ Monopotassium glutaminate/ Potassium hydrogen L-glutamate/ L-glutamic acid		
FORMULA:	KOOC CH2 CH2 CH(NH2)COOH·H2O (monohydrate)		
MOLECULAR MASS:	185.24 (anhydrous); 203.24 (monohydrate)		
ALTERNATIVE FORMS:	Monosodium glutamate/ Monoammonium glutamate/ Monopotassium glutamate/ Ammonium glutamate		
PROPERTIES AND APPEARANCE:	White, free-flowing, hygroscopic crystalline powder; practically odourless. Freely soluble in water; slightly soluble in alcohol		
FUNCTION IN FOODS:	Flavour enhancer; salt substitute		
ALTERNATIVES:	Monosodium glutamate; monopotassium glutamate; monoammonium glutamate; ammonium glutamate; disodium inosinate; disodium guanylate; L-glutamic acid		
FOOD SAFETY ISSUES:	Mildly toxic by ingestion and possibly other routes. Human systemic effects by ingestion: headache. When heated to decomposition, will emit toxic fumes of K_2O and NO_x		
LEGISLATION:	**USA:**	**UK and EUROPE:**	**AUSTRALIA/ PACIFIC RIM:**
	FDA 21CFR 182.1516: GRAS when used at a level not in excess of the amount reasonably required to accomplish the intended effect	Europe: listed UK: approved for use EEC Regulations (E622): approved for use as flavour enhancer, salt substitute	Japan: approved for use as flavouring

REFERENCE:	Lewis, R. J., Sr. (1989) *Food Additives Handbook.* Van Nostrand Reinhold, New York.
ANY OTHER RELEVANT INFORMATION:	Manufacturer/Distributor: Ajinomoto USA., Inc., 500 Frank W. Burr Blvd., Teaneck, NJ 07666 USA. Tel: (201) 448-1212; Fax: (201) 488-6282 (NJ Head Office)

NAME:	Monosodium glutamate
CATEGORY:	Flavour enhancer
FOOD USE:	Meat/ Poultry/ Sauces/ Soups
SYNONYMS:	CAS 142-47-2/ EINECS 205-538-1/ E621/ Sodium hydrogen L-glutamate/ Accent/ Chinese seasoning/ Glutacyl/ Glutamic acid, monosodium salt/ Glutammato monosodico (Italian)/ Glutavene/ Monosodioglutammato (Italian)/ Natriomglutaminat (German)/ Monosodium-L-glutamate (FCC)/ α-Monosodium glutamate/ MSG/ RL-50/ Sodium glutamate/ Sodium L-glutamate/ L(+) sodium glutamate/ vetsin/ Zest/ Glutamic acid monosodium salt monohydrate/ L-glutamate monohydrate/ Sodium glutamate monohydrate/ Monosodium L-glutamate monohydrate/ Asahi Aji
FORMULA:	C5 H8 N Na O4 H2O or HOOC CH(NH2)-CH2 CH2 COONa H2O
MOLECULAR MASS:	180.13
ALTERNATIVE FORMS:	Monoammonium glutamate/ Potassium glutamate/ Monopotassium glutamate/ Ammonium glutamate
PROPERTIES AND APPEARANCE:	Monosodium salt of L-form of glutamic acid. White or almost-white crystals or powder; free-flowing cystals or crystalline powder; forms rhombic prisms when crystallised from water. Below –8°C will recrystallise as a pentahydrate which after filtration and exposure to air will lose water of crystallisation and become the monohydrate. Slight peptone-like odour; meat-like taste. Very soluble in water; slightly soluble in alcohol.
PURITY %:	Not less than 99.0% of $C_5H_8NNaO_4$ H_2O
SOLUBILITY % AT VARIOUS TEMPERATURE/pH COMBINATIONS:	
in water:	**@ 25°C** 74.2 g/100 ml **@ 60°C** 101.4 g/100 ml
FUNCTION IN FOODS:	Flavour enhancer; dietary supplement (Japan)
ALTERNATIVES:	Disodium inosinate; disodium guanylate; monoammonium glutamate; monopotassium glutamate; L-glutamic acid; ammonium glutamate; potassium glutamate

TECHNOLOGY OF USE IN FOODS:

pH of a 5% solution is pH 6.7–7.2. 5% maximum weight loss on drying; 12.3% in sodium content. Bulk density is 11.0 to 13.8 g per cubic inch; 2.88 calories per gram. Generally added in concentrations ranging from 0.1–0.8% based on the weight of the food.

Works in harmony with salty and sour tastes, but contributes little or nothing to many sweet foods. Once the correct amount is used, additional use contributes little (if any) to the food flavour; in fact, it can result in a decline in palatibility

FOOD SAFETY ISSUES:

Moderately toxic by intravenous route; mildly toxic by ingestion and other routes. Human systemic effects by ingestion and intravenous routes include somnolence, hallucinations and distorted perception, headache, dyspnoea, nausea or vomiting. An experimental teratogen and reprotoxicity effects. When heated to decomposition, will emit fumes of NO_x and Na_2O

LEGISLATION:

USA:
FDA 21CFR 145.131, 155.120, 155.130, 155.170, 155.200, 158.170, 161.190, 169.115, 169.140, 169.150, 172.320, 182.1: GRAS when used at levels not in excess of the amount reasonably required to accomplish the intended effect
USDA 9CFR 318.7: sufficient for purpose
USDA 9CFR 381.147: sufficient for purpose

UK and EUROPE:
Europe: listed
EEC Regulation (E621): approved for use as flavour enhancer, ingredient where reduction of sodium intake is desired. Limited ADI is 0–120 mg/kg body weight
UK: approved for use, with the exception of baby food

CANADA:
Approved for use to accentuate flavour in vegetable, meat, and fish products in levels of up to 0.8%

AUSTRALIA/PACIFIC RIM:
Japan: approved for use as dietary supplement (amino acid) and flavourings

OTHER COUNTRIES:
FAO and World Health Organization: ADI as 0–120 mg/kg body weight

REFERENCE:	Lewis, R. J., Sr. (1989) *Food Additives Handbook.* Van Nostrand Reinhold, New York.
ANY OTHER RELEVANT INFORMATION:	Manufacturer/Distributor: Ajinomoto USA., Inc., 9, West 57th Street, New York, NY 10019 USA. Telex: 232220 (AJI UR); Tel: (212) 688-8360

NAME:	Potassium chloride
CATEGORY:	Flavour enhancer
FOOD USE:	Jelly (artificially sweetened)/ Meat (raw cuts)/ Poultry (raw cuts)/ Preserves (artifically sweetened)
SYNONYMS:	CAS 7447-40-7/ EINECS 231-211-8/ E508/ Tripotassium trichloride/ Chlorid draselny (Czech)/ Chloropotassuril/ Dipotassium dichloride/ Emplets potassium chloride/ Enseal/ Kalitabs/ Kaochlor/ Kaon-cl/ Kay ciel/ K-lor – Con/ Klotrix/ k-prendedome/ Pfiklor/ Potassium monochloride/ Potavescent/ Rekawan/ Slow-k/ Capshure KCL-140-50/ Capshure KCL-165-70/ Durkote Potassium Chloride/hydrogen vegetable oil/ Frimulsion 6G/ Merecol LK/ Morton Flour Lite Salt Mixt/ Morton Lite Salt Mixt/ Morton Lite Salt TFCMixt/ Kaskay/ Kayback/ K-Contin/ K-Norm/ K-Tab/ Leo – K/ Micro – K/ Nu – K/ Peter – Kal/ Repone K/ Span – K/ Chlorvescent/ Camcopot/ Diffu-K/ Kaleorid/ Kalium – Duriles
FORMULA:	KCl
MOLECULAR MASS:	74.55
ALTERNATIVE FORMS:	Ammonium chloride
PROPERTIES AND APPEARANCE:	Colourless or white crystals or powder; odourless with salty taste. Saline taste at low concentrations. Soluble in water, glycerin; slightly soluble in alcohol; insoluble in absolute alcohol, ether, acetone. Hydrochloric acid, sodium or magnesium chlorides diminish its solubility in water
MELTING RANGE IN °C:	773 (sublimes @ 1500)
DENSITY AT 20°C (AND OTHER TEMPERATURES) IN g/l:	1.987
SOLUBILITY % AT VARIOUS TEMPERATURE/pH COMBINATIONS:	
in water:	**@ 20°C** 1 g dissolves in 2.8 ml H_2O **@ 100°C** 1 g dissolves in 1.8 ml of boiling water
FUNCTION IN FOODS:	Direct food additive; dietary supplement; flavour enhancer; flavouring agent; gelling agent; nutrient; pH control agent; salt substitute; tissue-softening agent; yeast food; stabiliser; thickener

ALTERNATIVES: Ammonium chloride/ Calcium chloride/ Sodium lactate/ Magnesium sulphate

FOOD SAFETY ISSUES: Human poison by ingestion. Poison experimentally by ingestion, intravenous, and intraperitoneal routes; moderately toxic by subcutaneous route. Human systemic effects by ingestion include nausea, blood clotting changes, cardiac arrythmias. Also an eye irritant. Mutagenic data available. Explosive reaction with BrF_3; sulfuric acid + potassium permanganate. When heated to decomposition, will emit toxic fumes of K_2O and Cl^-

LEGISLATION:

USA:
FDA 21CFR 150.141, 150.161, 166.110: approved for use in margarine.
21CFR 182.5622: approved for use in foods as dietary supplement at GRAS quantities.
Preparations containing >100 mg potassium/tablet or >20 mg potassium/ml regulated as drugs.
21CFR 184.1622: approved for use in foods as flavour enhancer, flavouring agent, nutrient supplement, pH control agent, stabiliser, thickener at GRAS quantities.
GRAS when used in accordance with GMP.
Preparations containing equal to or greater than 100 mg of potassium per tablet are drugs covered by 21CFR 201.306, USDA 9CFR 318.7
USDA 9CFR 381.147: limitation of not more than 3% of a 2.0 molar solution.

UK and EUROPE:
Europe: listed
UK: approved for use
EEC Regulations (E508): approved for use as gellant, salt substitute, dietary supplement in malting process for beer making

CANADA:
Approved for use as the following: pH-adjusting or water-correcting agent in brewing at GMP levels; yeast food for brewing and unstandardised bakery products at GMP levels; emulsifying, gelling, stabilising or thickening agent for unstandardised foods at GMP levels

AUSTRALIA/PACIFIC RIM:
Japan: approved for use

(*conts*)

	(*USA: contd*) A solution of the approved inorganic chlorides injected into or applied to raw meat cuts shall not result in a gain of more than 3% above the weight of the untreated product
REFERENCE:	Lewis, R. J., Sr. (1989) *Food Additives Handbook.* Van Nostrand Reinhold, New York.

NAME:	Sodium alginate
CATEGORY:	Flavour enhancer
FOOD USE:	Candy (hard)/ Condiments/ Confections/ Edible films/ Frostings/ Fruit juices/ Fruits (processed)/ Gelatins/ Puddings/ Relishes/ Sauces/ Toppings/ Ice-cream/ Boiler water additive
SYNONYMS:	CAS 9005-38-3/ E401/ Algin/ Alginate KMF/ Algin (polysaccharide)/ Algipon L-1168/ Amnucol/ Antimigrant C45/ Cecalgine TBV/ Cohasal-1H/ Dariloid Q/ Dariloid QH/ Duckalgin/ Halltex/ K′-algiline/ Kelco gel LV/ Kelco HV/ Kelcosol/ Kelgin F/ Kelgin HV/ Kelgin LV/ Kelgin MV/ Kelgin QL/ Kelgin XL/ Kelgum/ Kelset/ Kimitsu/ Kimitsu Algin 1–2/ Kimitsu 1–3/ Kelvis/ Kelsize/ Keltex/ Keltone/ Keltone LV/ Keltone HV/ Lamitex/ Manucol/ Manucol DM/ Manucol LB/ Manucol DMF/ Manucol DH/ Manugel GHB/ Manugel GMB/ Manugel DJX/ Manugel DMB/ W-300 FG/ Manutex/ Meypralgin R/LV/ Minus/ Mosanon/ Nouralgine/ OG1/ Pectalgine/ Proctin BUS/ Prime F-25/ Prime F-40/ Prime F-400/ Prime F-600/ Protacell 8/ Protanal 686/ Protanal HF120M/ Protanal HFC60/ Protanal KC 119/ Protanal KP/ Protanal KPM/ Protanal LF 5/60/ Protanal LF 20/ Protanal LF 20/40/ Protanal LF 60/ Protanal LF 120M/ Protanal LF 200/ Protanal LFS 40/ Protanal SF 40/ Protanal SF 60/ Protanal SF 120/ Protanal SF 120M/ Protanal SP 5H/ Protanal VK 687/ Protanal VK 749/ Protanal VK 805 IMP/ Protanal VK 990/ Protanal VK 998/ Protanal VPM/ Protanal VSM/ Protatek/ Snow algin H/ Solberg FD 100 Range/ Sodium Alginate HV NF/FCC/ Sodium Alginate LV/ Sodium Alginate LVC/ Sodium Alginate MV NF/FCC/ Sodium polymannuronate/ Stipine/ Tagat/ Tragaya/
FORMULA:	$(C_6H_7O_6Na)_n$
MOLECULAR MASS:	198.11
PROPERTIES AND APPEARANCE:	Sodium salt of alginic acid. Colourless to slightly yellow filamentous or granular solid or powder; odourless and tasteless. In water, it forms a viscous colloidal solution. Insoluble in ether, alcohol, chloroform
FUNCTION IN FOODS:	Boiler water additive; emulsifier; firming agent; flavour enhancer; formulation aid; processing aid; stabiliser; surface-active agent; texturiser; thickener

TECHNOLOGY OF USE IN FOODS: Long-term viscosity is poor when pH is above 10

ANTAGONISTS: Incompatible with divalent cations (except magnesium) or other heavy metal ions, cationic quaternary amines, or chemicals which cause alkaline degradation or acid precipitation. Can be avoided by use of sequestrants or careful pH control

FOOD SAFETY ISSUES: Poison by intravenous and intraperitoneal routes. When heated to decomposition, will emit toxic fumes of Na_2O

LEGISLATION:

USA:
FDA 21CFR 133.133, 133.134, 133.162, 133.178, 133.179, 150.141, 150.161, 173.310, 21CFR 184.1724: approved for use in foods as texturiser, formulation aid, stabiliser, thickener, firming agent, flavour enhancer, flavour adjuvant, emulsifier, processing aid, surface active agent at GRAS quantities.
GRAS:
– with limitation of 1.0% in condiments and relishes except pimento ribbon for stuffed olives;
– 6.0% in confections and frostings;
– 4.0% in gelatins, puddings;
– 10.0% in hard candy; – 2.0% in processed fruits and fruit juices;
– 1.0% in all other foods when used in accordance with GMP.

UK and EUROPE:
Europe: listed
UK: approved for use
EEC Regulations (E401): approved for use as stabiliser, suspending agent, thickener, calcium source, gellant, copper firming agent in brewing.
ADI limited to 0–50 mg/kg body weight

CANADA:
Permitted for use in the following:
– as emulsifying, gelling, stabilising, or thickening agent in brewing and salad dressings at GMP levels
– in infant formula at 0.03% as consumed
– in combination with carrageenan or guar gum, but total should not exceed 0.03%

AUSTRALIA/PACIFIC RIM:
Japan: approved for use as thickener, stabiliser, gelling agent

REFERENCE:	Lewis, R. J., Sr. (1989) *Food Additives Handbook.* Van Nostrand Reinhold, New York.
ANY OTHER RELEVANT INFORMATION:	Manufacture/Distributor: Kelco Division of Merck & Co. Inc., 8355 Aero Drive, San Diego, California 92123 USA. Tel: (800) 535-2656; Fax: (619) 467-6520; Telex: 695228

NAME:	Succinic acid
CATEGORY:	Flavour enhancer
FOOD USE:	Beverages/ Condiments/ Meat products/ Relishes/ Sausages (hot)
SYNONYMS:	CAS 110-15-6/ EINECS 203-740-4/ E363/ Asuccin/ Amber acid/ Bernsteinsaure (German)/ Butanedioic acid/ 1,2-Ethanedicarboxylic acid/ Ethylenesuccinic acid/ 1,4-Butanedioic acid
FORMULA:	HOOC CH2 CH2 COOH
MOLECULAR MASS:	118.10
ALTERNATIVE FORMS:	Potassium succinate/ potassium salt trihydrate
PROPERTIES AND APPEARANCE:	Colourles or white crystals, monoclinic prisms; odourless with acid taste. Soluble in water; very soluble in alcohol, ether, acetone, glycerin; practically insoluble in benzene, carbon disulphide, carbon tetrachloride, petroleum ether. Decomposes at 235°C. Combustible
BOILING POINT IN °C AT VARIOUS PRESSURES (INCLUDING 760 mg Hg):	235
MELTING RANGE IN °C:	187
DENSITY AT 20°C (AND OTHER TEMPERATURES) IN g/l:	1.564
FUNCTION IN FOODS:	Flavour enhancer; miscellaneous and general-purpose food chemical; neutralising agent; pH control agent
ALTERNATIVES:	Malic acid; tannic acid; L-tartaric acid; lactic acid
TECHNOLOGY OF USE IN FOODS:	1 gram of additive will dissolve in the following: 13 ml of cold water; 1 ml of boiling water; 6.3 ml of methanol; 36 ml of acetone; 113 ml of ether. The pH of a 0.1 M aqueous solution is 2.7

FOOD SAFETY ISSUES:

Moderately toxic by subcutaneous route; a severe eye irritant. When heated to decomposition, will emit acrid smoke and irritating fumes

LEGISLATION:

USA:
FDA 21CFR 131.144
FDA 21CFR 184.1091: approved for use in foods as flavour enhancer, pH control agent at GRAS quantities.
GRAS: with limitation of 0.084% in condiments and relishes; 0.0061% in meat products when used in accordance with GMP

UK and EUROPE:
Europe: listed
UK: approved for use
EEC Regulations (E363): approved for use as acidulant, buffer, neutraliser

AUSTRALIA/PACIFIC RIM:
Japan: approved for use as acidity regulator, food acid, flavouring

REFERENCE:

Lewis, R. J., Sr. (1989) *Food Additives Handbook.* Van Nostrand Reinhold, New York.

NAME:	Tannic acid
CATEGORY:	Flavour enhancer
FOOD USE:	Apple juice/ Baked goods/ Beer/ Candy (hard)/ Candy (soft)/ Cough drops/ Fats (rendered)/ Fillings/ Frozen dairy desserts and mixes/ Gelatins/ Meat products/ Non-alcoholic beverages/ Puddings/ Wine
SYNONYMS:	CAS 1401-55-4/ EINECS 276-638-0/ EINECS 215-753/ Tannin/ D'acid tannique (French)/ Gallotannic acid/ Gallotannin/ Glycerite
FORMULA:	$C_{76}H_{52}O_{46}$
MOLECULAR MASS:	1701.28
PROPERTIES AND APPEARANCE:	Occurs in the bark and fruit of many plants (e.g. oak species, sumac). Yellowish-white or light-brown amorphous, bulky powder or flakes or spongy masses; faint characteristic odour with astringent taste. Gradually darkens on exposure to air and light. Decomposes at 210–215°C. Very soluble in water, alcohol, acetone; almost insoluble in benzene, chloroform, ether, petroleum ether, carbon disulphide, carbon tetrachloride
MELTING RANGE IN °C:	200
FLASH POINT IN °C:	199 (open cup); autoignition temperature: 527
FUNCTION IN FOODS:	Boiler water additive; clarifying agent; fat rendering aid; flavouring agent; flavouring enhancer; pH control agent
ALTERNATIVES:	Malic acid; succinic acid; L-tartaric acid; lactic acid
TECHNOLOGY OF USE IN FOODS:	Keep in tightly closed container and protected from light. 1 gram of additive will dissolve in the following: 0.35 ml of water; 1 ml of warm glycerol
ANTAGONISTS:	Incompatible with salts of heavy metals, alkaloids, gelatin, albumin, starch, oxidising substances (e.g. permanganates, chlorates), spirit nitrous ether, lime water

FOOD SAFETY ISSUES:

Poison by ingestion, intramuscular, intravenous, and subcutaneous routes; moderately toxic by parenteral route. An experimental carcinogen and tumorigen; also experimental reprotoxicity effects. Mutagenic data available. Combustible when exposed to heat or flame; to fight fire, use water. Incompatible with salts of heavy metals, oxidising materials. When heated to decomposition, will emit acrid smoke and irritating fumes

LEGISLATION:

USA:

FDA 21CFR 173.310: approved for use as boiler water additive
FDA 21CFR 184.1097: approved for use in foods as flavouring agent, flavouring adjuvant, flavour enhancer, processing aid, pH control agent at GRAS quantities.
GRAS – with limitation of 0.01% in baked goods; – 0.015% in alcoholic beverages; – 0.005% in non-alcoholic beverages and for gelatins, puddings, and fillings; – 0.04% in frozen dairy desserts and mixes; – 0.04% in soft candy; – 0.013% in hard candy and cough drops; – 0.001% in meat products; when used in accordance with GMP
USDA 9CFR 318.7: sufficient for purpose
BATF 27CFR 240.1051: limitation of 3.0 g/l calculated as gallic acid equivalents (GAE) in apple juice or wine; limitation of 0.8 g/l in white wine and 0.3 g/l in rosé wine

CANADA:

Approved for use as the following: fining agent in cider, honey wine, wine (200 ppm); to reduce adhesion in chewing gum at GMP levels

REFERENCE:

Lewis, R. J., Sr. (1989) *Food Additives Handbook*, Van Nostrand Reinhold, New York.

NAME:	**L-Tartaric acid**
CATEGORY:	Flavour enhancer
FOOD USE:	Baking powder/ Effervescent beverages (grape and lime flavoured)/ Jellies (grape flavoured)/ Poultry/ Wine/ Baked goods/ Gelatin desserts/ Pharmaceuticals (i.e. effervescent salts)
SYNONYMS:	CAS 000526830/ EINECS 201-766-0/ FEMA 3044/ E334/ 2,3-Dihydrosuccinic acid/ 2,3-Dihydroxybutanedioic acid/ L-(+)-tartaric acid/ L-2,3-Dihydroxybutanedioic acid/ D-α,β-Dihydroxysuccinic acid/ Weinsteinsaure (German)/ Dextrotartaric acid
FORMULA:	COOH·CHOH·CHOH·COOH
MOLECULAR MASS:	150.10
PROPERTIES AND APPEARANCE:	Colourless to translucent crystals or white powder; odourless with a strong acid taste; stable to air and light. Soluble in water, alcohol, glycerin; insoluble in chloroform. Occurs in juice of grapes and few other fruits and plants (i.e. leaves of *Bauhinia reticulata*). Residues of wine industry, not as widely distributed as citric or malic acid
MELTING RANGE IN °C:	168–170
SOLUBILITY % AT VARIOUS TEMPERATURE/pH COMBINATIONS:	
in water:	**@ 20°C** 139 g/100 ml **@ 50°C** 195 g/100 ml **@ 100°C** 343 g/100 ml
in ethanol solution:	**100%** **@ 18°C** 20.4 g/100 g
FUNCTION IN FOODS:	Direct food additive; acidulant; firming agent; flavour enhancer; flavouring agent; humectant; pH control agent; sequestrant; dough conditioner; to correct acid deficiency in wine making during poor harvest years
ALTERNATIVES:	Malic acid (as acidulant in carbonated beverages); citric acid; phosphoric acid; fumaric acid; succinic acid; tannic acid; lactic acid

TECHNOLOGY OF USE IN FOODS: 1 gram of additive will dissolve in the following: 0.75 ml water at room temperature; 0.5 ml boiling water; 1.7 ml methanol; 3 ml ethanol; 10.5 ml propanol; 250 ml ether

FOOD SAFETY ISSUES: Strong organic acid. Moderately toxic by intravenous route; mildly toxic by ingestion. Strong solutions are mildly irritating to humans. When heated to decomposition, will emit acrid smoke and irritating fumes

LEGISLATION:

USA:
FDA 21CFR 150.161, 150.141, 131.144, 184.1099: approved for use in foods as firming agent, flavour enhancer, flavouring agent/adjuvant, processing aid, pH control agent at GRAS quantities.
GRAS when used in accordance with GMP.
USDA 9CFR 318.7
USDA 9CFR 381.147: sufficient for purpose
BATF 27CFR 240.1051: use as prescribed in 27CFR 240.364, 240.512
BATF 27CFR 240.364, 240.512

UK and EUROPE:
Europe: listed
UK: approved for use
EEC Regulations (E334): approved for use at ADI of 0–30 mg/kg body weight

CANADA:
Approved for use in the following: as Class IV preservative for fats and oils and unstandardised foods at GMP level; as pH-adjusting or water-correcting agent at GMP levels

AUSTRALIA/PACIFIC RIM:
Japan: approved for use

REFERENCE:

Lewis, R. J., Sr. (1989) *Food Additives Handbook*. Van Nostrand Reinhold, New York.

ANY OTHER RELEVANT INFORMATION: Manufacturer/Distributor: Browning Chemical Corporation, 707 Westchester Avenue, White Plains, NY 10604-3104. Tel: (914) 686-0300; Fax: (914) 686-0310; Telex: 23 5039

NAME:	Ethyl maltol
CATEGORY:	Flavour enhancer
FOOD USE:	Chocolate/ Desserts/ Wine
SYNONYMS:	CAS 4940-11-8/ FEMA 3487/ E637/ 3-Hydroxyl-2-ethyl-4-pyrone/ 2-Ethyl-3-hydroxy-4H-pyran-4-one/ 2-Ethyl pyromeconic acid/ Veltol®/ Veltol®-Plus
FORMULA:	$C_7H_8O_3$
MOLECULAR MASS:	140.14
PROPERTIES AND APPEARANCE:	White crystalline powder; caramel sweet odour; sweet fruity taste. Soluble in water, alcohol, propylene glycol, chloroform
MELTING RANGE IN °C:	88–90
FUNCTION IN FOODS:	Synthetic flavouring agent imparting sweet taste; flavour enhancer; processing aid; sweetness enhancement; increased creaminess; masked bitterness; reduced acid bite
ALTERNATIVES:	Aspartame/ glycine/ sucralose/ thaumatin/ ammoniated glycyrrhizin
TECHNOLOGY OF USE IN FOODS:	Veltol®-Plus. Soluble in alcohol – 1 gram of product will dissolve in: 55 ml of water; 10 ml alcohol; 17 ml propylene glycol; 5 ml chloroform
SYNERGISTS:	Flavour/fragrance modifier/enhancer for food and beverages will provide sweetness enhancement of aspartame, allowing reduced usage levels
FOOD SAFETY ISSUES:	Moderately toxic by ingestion, subcutaneous routes. Mutagenic data available. When heated to decomposition, will emit acrid smoke and fumes

LEGISLATION:	**USA:** FDA 21CFR 172.515: approved for use as flavouring agents in synthetic flavouring substances and adjuvants. Used at a level not in excess of the amount reasonably required to accomplish the intended effect BATF 27CFR 240.1051: limitation of 100 ml/l	**UK and EUROPE:** Europe: listed EEC regulation (E637): approved for use as a flavouring agent to impart sweet taste and as flavour enhancer. ADI 0–2 mg/kg of body weight UK: approved for use
REFERENCE:	Lewis, R. J., Sr. (1989) *Food Additives Handbook*, Van Nostrand Reinhold, New York.	
ANY OTHER RELEVANT INFORMATION:	Manufacturer/Distributor: Pfizer Canada Inc., Food Science Group, P.O. Box 800, Point-Claire/Dorval, Quebec H9R 4V2. Tel: (800) 363-7928/(514) 426-6888; Fax: (800) 265-9503/(514) 426-6895; Pfizer Canada, Inc., 17300 Trans Canada Highway, Kirkland, Quebec H9J 2M5. Tel: (800) 363-7928	

NAME:	**Glycine**
CATEGORY:	Flavour enhancer
FOOD USE:	Beverages/ Rendered fats/ Beverage base
SYNONYMS:	CAS 56-40-6/ EINECS 200-272-2/ FEMA 3287/ Hampshire glycine/ Aminoacetic acid/ Glycocoll/ Glycolixir/ Aminoethanoic acid/ Gyn-Hydralin/ Glycosthene
FORMULA:	H2N·CH2·COOH
MOLECULAR MASS:	75.08
ALTERNATIVE FORMS:	Sodium glycinate
PROPERTIES AND APPEARANCE:	Simplest amino acid, and the principal amino acid in sugar cane. White crystals; odourless, sweet taste. Soluble in water; insoluble in alcohol, ether. Decomposes at 233°C, and completely sintered at 290°C
MELTING RANGE IN °C:	232–236 (decomposition)
DENSITY AT 20°C (AND OTHER TEMPERATURES) IN g/l:	1.1607
SOLUBILITY % AT VARIOUS TEMPERATURE/pH COMBINATIONS:	
in water:	**@ 50°C** 39.1 g/100 ml **@ 100°C** 67.2 g/100 ml
FUNCTION IN FOODS:	Buffering agent; flavour enhancer; sweetener; stabiliser; nutrient; dietary supplement; reduces bitter taste of saccharin; retards rancidity in animal and vegetable fats; chicken feed additive
ALTERNATIVES:	Aspartame; ethyl maltol; sucralose; thaumatin; ammoniated glycyrrhizin
FOOD SAFETY ISSUES:	Moderately toxic by intravenous route; mildly toxic by ingestion. When heated to decomposition, will emit toxic fumes of NO_x

LEGISLATION:

USA:
FDA 21CFR 170.50: no longer GRAS for use in human food.
21CFR 582.5049: GRAS for animal feed.
21CFR 172.320: limitation of 3.5% by weight.
21CFR 172.812: approved for use as a masking agent for bitter aftertaste of saccharin in beverages and stabiliser in mono- and diglycerides. Limitations of 0.2% (of finished beverage) and 0.02% (of mono- and diglyceride)
USDA 9CFR 318.7: 0.01% in rendered animal

CANADA:
Approved for use as sequestering agent for mono- and diglycerides (0.02%)

AUSTRALIA/PACIFIC RIM:
Japan: approved for use as dietary supplement (amino acid) and flavourings

REFERENCE:

Lewis, R. J., Sr. (1989) *Food Additives Handbook*, Van Nostrand Reinhold, New York.

NAME:	L-Glutamic acid
CATEGORY:	Flavour enhancer
SYNONYMS:	CAS 56-86-0/ EINECS 200-293-7/ FEMA 3285/ E620/ α-Glutamic acid/ L-2-aminoglutaric acid/ 2-Aminopentanedioic acid/ α-Aminoglutaric acid/ L-Aminopropane-1,3-dicarboxylic acid/ Glusate/ Glutacid/ Glutamic acid/ d-Glutam-iensuur/ Glutaminic acid/ L-Glutaminic acid/ Glutaminol/ Glutaton
FORMULA:	$C_5H_9NO_4$
MOLECULAR MASS:	147.15
ALTERNATIVE FORMS:	Monosodium glutamate/ Monopotassium glutamate/ Monoammonium glutamate
PROPERTIES AND APPEARANCE:	An essential amino acid present in all complete proteins. White crystal or crystal powder; virtually odourless; decomposes at 247–249°C; sublimes at 200°C Slightly soluble in water; practically insoluble in methanol, ethanol, ether, acetone, cold glacial acetic acid
MELTING RANGE IN °C:	194 (DL); 224–225 (L)
DENSITY AT 20°C (AND OTHER TEMPERATURES) IN g/L:	1.4601 (DL); 1.538 (L)
SOLUBILITY % AT VARIOUS TEMPERATURE/pH COMBINATIONS:	
in water:	**@ 50°C** 21.86 g/l **@ 100°C** 140.00 g/l
FUNCTION IN FOODS:	Flavour enhancer; nutrient; dietary supplement; salt substitute
ALTERNATIVES:	Ammonium glutamate; disodium inosinate; disodium guanylate; monoammonium glutamate; monopotassium glutamate; monosodium glutamate; potassium glutamate
FOOD SAFETY ISSUES:	Human systemic effects by ingestion and intravenous route (headache, vomiting or nausea). When heated to decomposition, will emit toxic fumes of NOx

LEGISLATION:	**USA:** FDA 21CFR 172.320 – 12.4% maximum: approved for use as dietary and nutritional additive at level of 12.4% FDA 21CFR 182.1045 – GRAS when used in accordance with GMP. Approved for use as GRAS food substance and salt substitute	**UK and EUROPE:** Europe: listed UK: approved for use EEC regulation (E620): approved for use as dietary supplement, flavour enhancer, salt substitute at acceptable daily intake of 0–120 mg/kg body weight **AUSTRALIA/PACIFIC RIM:** Japan: approved for use as dietary supplement (amino acid) and flavourings
REFERENCE:	Lewis, R. J., Sr. (1989) *Food Additives Handbook*. Van Nostrand Reinhold, New York.	
ANY OTHER RELEVANT INFORMATION:	Manufacturer/Distributor: Ajinomoto USA., Inc., 9 West 57th Street, New York, NY 10019 USA. Telex: 232220 (AJI UR); Tel: (212) 688-8360	

NAME:	Licorice	
CATEGORY:	Flavour enhancer	
FOOD USE:	Baked goods/ Alcoholic beverages/ Non-alcoholic beverages/ Chewing gum/ Hard candy/ Herbs/ Seasonings/ Plant protein products/ Soft candy/ Vitamin and mineral supplements	
SYNONYMS:	Glycyrrhiza/ Liquorice/ Licorice root	
PROPERTIES AND APPEARANCE:	Dried rhizome and roots of *Glycyrrhiza glabra*	
FUNCTION IN FOODS:	Natural flavouring agent; flavour enhancer; surface-active agent	
ALTERNATIVES:	Ammoniated glycyrrhizin; licorice root extract; licorice extract; thaumatin	
LEGISLATION:	**USA:** FDA 21CFR 184.1408: approved for use in foods as flavour enhancer, flavouring agent, surface active agent at GRAS quantities. Limitation for use at levels of 0.05–16% Limitations of 0.05% (baked goods), 0.1% (alcoholic beverages), 0.15% (non-alcoholic) beverages), 1.1% (chewing gum), 16% (hard candy), 0.15% (vitamin/mineral supplements), 0.1% (other foods) Not permitted as non-nutritive sweetener in sugar substitutes	**AUSTRALIA/PACIFIC RIM:** Japan: approved for use as natural flavouring
REFERENCE:	Lewis, R. J., Sr. (1989) *Food Additives Handbook*. Van Nostrand Reinhold, New York.	

NAME:	**Licorice extract**	
CATEGORY:	Flavour enhancer	
FOOD USE:	Baked goods/ Alcoholic beverages/ Non-alcoholic beverages/ Chewing gum/ Hard candy/ Herbs/ Seasonings/ Plant protein products/ Soft candy/ Vitamin and mineral supplements	
SYNONYMS:	*Glycyrrhiza glabra* extract/ CAS 977004311/ CAS 068916916/ CAS 977070624	
PROPERTIES AND APPEARANCE:	Extract of *Glycyrrhiza glabra*	
FUNCTION IN FOODS:	Flavour enhancer; natural flavouring agent; surface-active agent	
ALTERNATIVES:	Licorice root extract; thaumatin; ammoniated glycyrrhizin	
LEGISLATION:	**USA:** FDA 21CFR 184.1408: substance affirmed as GRAS for use as flavour enhancer, flavouring agent, surface-active agent at 0.05–16%	**AUSTRALIA/PACIFIC RIM:** Japan: approved for use
REFERENCE:	Lewis, R. J., Sr. (1989) *Food Additives Handbook*. Van Nostrand Reinhold, New York.	

NAME:	**Magnesium sulphate heptahydrate**		
CATEGORY:	Flavour enhancer		
FOOD USE:	Beer-making		
SYNONYMS:	CAS 10034-99-8/ EINECS 231-298-8/ E518/ Bitter salts/ Espom salts		
FORMULA:	MgO4S·7H2O		
MOLECULAR MASS:	246.48		
ALTERNATIVE FORMS:	Magnesium sulphate		
PROPERTIES AND APPEARANCE:	Efflorescent crystal or powder; bitter, saline, cooling taste		
DENSITY AT 20°C (AND OTHER TEMPERATURES) IN g/L:	1.670		
SOLUBILITY % AT VARIOUS TEMPERATURE/pH COMBINATIONS:			
in water:	**@ 20°C** 71 g/100 ml		
FUNCTION IN FOODS:	Firming agent; flavour enhancer; nutrient/dietary supplement; processing aid		
TECHNOLOGY OF USE IN FOODS:	Keep in well-closed container		
LEGISLATION:	**USA:**	**UK and EUROPE:**	**AUSTRALIA/ PACIFIC RIM:**
	FDA 21CFR 184.1443: affirmed as GRAS for use as flavour enhancer, nutrient supplement and processing aid	Europe: listed EEC regulation (E518): approved for use as dietary supplement, firming agent and in beer-making UK: approved for use	Japan: approved for use as coagulant for tofu, fermentation aid, and yeast nutrient
REFERENCE:	Lewis, R. J., Sr. (1989) *Food Additives Handbook*. Van Nostrand Reinhold, New York.		

NAME:	Sucralose
CATEGORY:	Flavour enhancer
FOOD USE:	Beverages/ Processed foods
SYNONYMS:	CAS 56038-13-2/ 1,6-Dichloro-1,6-dideoxy-β-D-fructofuranosyl-4-chloro-4-deoxy-α-D-galactopyranoside/ 4,1′,6′-Trichlorogalactosucrose/ 4′,1′,6′-Trichloro-4,1′,6′-trideoxy-galacto-sucrose/ TGS
FORMULA:	$C_{12}H_{19}Cl_3O_8$
MOLECULAR MASS:	397.64
PROPERTIES AND APPEARANCE:	White crystalline powder; odourless; sweet taste. Soluble in water, methanol, alcohol; slightly soluble in ethyl acetate. 600 times sweeter than sucrose. Chlorinated sucrose derivative with enhanced sweetness
MELTING RANGE IN °C:	130
FUNCTION IN FOODS:	Non-nutritive sweetener; flavour enhancer
ALTERNATIVES:	Aspartame; ethyl maltol; glycine; thaumatin; ammoniated glycyrrhizin
LEGISLATION:	**CANADA:** Approved for use as sweetener and flavour enhancer in the following at the levels specified: – tabletop sweetener (GMP) – breakfast cereal, 0.1% – beverages (concentrates and mixes) and desserts, 0.025% – chewing gum, breath freshener, and table syrups, 0.15% – salad dressing, condiments, and puddings, 0.04% – fruit spreads, 0.045% – confectionery and alcoholic beverages, 0.07% – baking mixes and bakery products, 0.065% as consumed – processed fruit and vegetable products, 0.015%

REFERENCE:	Lewis, R. J., Sr. (1989) *Food Additives Handbook.* Van Nostrand Reinhold, New York.
ANY OTHER RELEVANT INFORMATION:	Manufacturer/Distributor: Redpath Specialty Products, 95 Queen's Quay East, Toronto, Ontario, Canada M5E 1A3. Tel: (800) 267-1517

NAME:	Thaumatin
CATEGORY:	Flavour enhancer
FOOD USE:	Chewing gum/ Dairy products/ Animal feeds/ Pet foods
SYNONYMS:	CAS 53850-34-3/ Katemfe/ Talin
MOLECULAR MASS:	21,000–22,000
PROPERTIES AND APPEARANCE:	Sweet-tasting basic protein extracted from the fruit of the tropical plant *Thaumatococcus danielli*. 750–1600 times sweeter than sucrose on weight basis; 30,000–100,000 times sweeter than sucrose on a molar basis. Licorice aftertaste; strongly cationic. Relatively stable in solution and on heating. Non-carcinogenic; favourable gestatory properties Talin is extremely soluble in water; soluble in ethanol, propanol-2, glycerol and propylene glycol; insoluble in acetone, ether and toluene
FUNCTION IN FOODS:	Flavour enhancement of sweetness and certain flavour compounds (i.e. peppermint, ginger, cinnamon, and coffee). Reduces fiery, peppery, or bitter elements associated with certain flavour compounds (i.e. peppermint, ginger, cinnamon, and coffee). Non-nutritive sweetener. Savoury flavour; taste modification; masking of bitter and unpleasant taste of metallic ions (i.e. sodium, iron and potassium)
ALTERNATIVES:	For the licorice aftertaste and flavour enhancement associated with thaumatin: licorice root; licorice root extract; ammoniated glycyrrhizin. For the sweetness potency and flavour enhancement associated with thaumatin: sucralose; aspartame; ethyl maltol; glycine
TECHNOLOGY OF USE IN FOODS:	Protein loses sweetness on heating, on splitting of sulphide bridges and also at pH <2.5; 0.9 g of thaumatin produces sweetness equivalent to 1.5 kg of sugar. Temperature affects the sweetness of thaumatin, but will depend on the concentration, presence of oxygen, salts or polyelectrolytes, and on pH. Irreversible thermal denaturation occurred at: 75°C and pH 5; 55°C and pH 3.0; 65°C and pH 7.2. Talin can be used: in chewing gum, 50–150 mg/kg; in soft drinks, 10–15 mg can substitute 20–30% of sucrose. Can also be used in gelatin-based confectionery and tobacco flavour, coffee and tea. Also successfully masks the unpleasant taste of drugs in medicinal applications. Talin is stable at pH 2–10;

thermal stability is affected mainly by pH, presence of oxygen, and other soluble matter, especially polysaccharides. Talin can be pasteurised or UHT-sterilised in non-alcoholic beverages at pH 2.8–3.5

Thaumatin is also combined with saccharin and L-glucose, with xylitol and others. Thaumatin admixed with glycyrrhizin and amino acids is marketed in Japan as San Sweet T-1 and T-100. It will remain stable indefinitely when freeze-dried or spray-dried, if stored under ambient conditions. It will also remain stable at 120°C in canning operations and under pasteurisation and UHT conditions. Thaumatin will impart licorice-like taste at higher levels; therefore, its use as the only source of sweetness is limited to applications where the required sweetness is less than the equivalent of 10% sucrose.

SYNERGISTS:

Synergistic with saccharin, acesulfame K. Talin, when used with either sucrose or Acesulfam-K or stevioside, showed some degree of synergism. Thaumatin also synergises with monosodium glutamate and 5′-nucleotides to increase the flavour enhancement

FOOD SAFETY ISSUES:

No adverse effects in short-term tests. Not allergenic, mutagenic or teratogenic

LEGISLATION:

USA:
Permitted in USA for use as flavour enhancer in chewing gum

UK and EUROPE:
UK: approved for use

CANADA:
Approved for use as follows: as sweetener and flavour enhancer in chewing gum and breath freshener products at 500 ppm; as bitterness masking agent in salt substitutes at 400 ppm; as flavour enhancer and unstandardised flavour preparations at 100 ppm

AUSTRALIA/PACIFIC RIM:
Japan: approved for use and considered as a natural food

OTHER COUNTRIES:
World Health Organization – ADI not specified

REFERENCES:

Lewis, R. J., Sr. (1989) *Food Additives Handbook.* Van Nostrand Reinhold, New York.

Krutosikova, A., and Uher, M. (1992) *Natural and Synthetic Sweet Substances.* Ellis Horwood Ltd. New York.

Anonymous (1996) Thaumatin – the sweetest substance known to man has a wide range of food applications. *Food Technol.*, **50**(1), 74–75.

NAME:	Potassium lactate
CATEGORY:	Flavour enhancer
FOOD USE:	Confectionery/ Jams/ Jellies/ Marmalades/ Margarine
SYNONYMS:	CAS 996-31-6/ EINECS 213-631-3/ E326/ Potassium-L-2-hydroxypropionate/ Propanoic acid/ 2-Hydroxy-monopotassium salt/ Arlac P/ Purasal®P/USP 60
FORMULA:	CH3 CHOH COOK
MOLECULAR MASS:	128.17
ALTERNATIVE FORMS:	Potassium-L-2-hydroxypropionate/ 2-Hydroxy-monopotassium salt/ Propanoic acid
PROPERTIES AND APPEARANCE:	White solid, odourless, hygroscopic. Potassium salt of lactic acid
FUNCTION IN FOODS:	Direct food additive; flavour enhancer; flavouring agent; flavouring adjuvant; humectant; pH control agent; antioxidant synergist
FOOD SAFETY ISSUES:	When heated to decomposition, will emit acrid smoke and irritating fumes
LEGISLATION:	**USA:** FDA 21CFR 184.1639: affirmed as GRAS for use as flavour enhancers, flavouring agent/adjuvant, humectant, pH control agent. Not authorised for infant foods and infant formulas **UK and EUROPE:** Europe: listed EEC regulation (E326): approved for use in synergism with antioxidants and as buffer with no specified limits for ADI. UK: approved for use
REFERENCE:	Lewis, R. J., Sr. (1989) *Food Additives Handbook.* Van Nostrand Reinhold, New York.

NAME:	Sodium lactate
CATEGORY:	Flavour enhancer
FOOD USE:	Biscuits/ Fruits/ Meat products/ Hog carcasses/ Trip/ Vegetable/ Nuts/ Sponge cake/ Swiss roll/ Water (canned)/ Water (bottled)
SYNONYMS:	Arlac S/ Patlac NAL/ CAS 72-17-3/ Per-glycerin/ E325/ EINECS 200-772-0/ Lacolin/ 2-hydroxypropanoic acid monosodium salt/ Lactic acid sodium salt/ Sodium-L-2-hydroxypropionate/ Purasal S/SP 60
FORMULA:	CH3 CHOH COONa
MOLECULAR MASS:	112.07
ALTERNATIVE FORMS:	Potassium lactate/ Calcium lactate
PROPERTIES AND APPEARANCE:	Sodium salt of lactic acid. Colourless or yellowish syrupy liquid; odourless; slight salt taste. Very hygroscopic; miscible in water and alcohol. Combustible
MELTING RANGE IN °C:	17
FUNCTION IN FOODS:	Cooked-out juice retention aid; corrosion preventative; denuding agent; emulsifier; flavour enhancer; flavouring agent/adjuvant; food additive; glycerol substitute; hog scald agent; humectant; lye peeling agent; pH control agent; washing agent
ALTERNATIVE:	Potassium lactate; calcium lactate
FOOD SAFETY ISSUES:	Moderately toxic by intraperitoneal route; eye irritant. When heated to decomposition, will emit toxic fumes of Na_2O

LEGISLATION:

USA:
FDA 21CFR 184.1768: affirmed as GRAS for use as emulsifier, flavour enhancer, flavouring agent or adjuvant, humectant, and pH control.
Not permitted for infant food and infant formulas
USDA 9CFR 318.7: in meat products, where allowed, limitation of 5% of phosphate in pickle at 10% pump level; 0.5% of phosphate in product (only clear solution may beinjected into product)

CANADA:
Approved for use as pH-adjusting or water-correcting agent for margarine and unstandardised foods in accordance with GMP

UK and EUROPE:
Europe: listed
ECC regulation (E325): approved for use as humectant, glycerol substitiute, bodying agent, and in synergism with antioxidants.
No specified ADI limits
UK: approved for use

AUSTRALIA/PACIFIC RIM:
Japan: approved for use as acidity regulator, food acid, flavouring

REFERENCE:

Lewis, R. J., Sr. (1989) *Food Additives Handbook.* Van Nostrand Reinhold, New York.

ANY OTHER RELEVANT INFORMATION:

Manufacturer/Distributor: ADM Food Additives Division, 4666 Faries Parkway, Decatur, IL 62526.
Tel: (217) 424-5387; Fax: (217) 424-2473

Part 7

Flour Additives

NAME:	**Alpha-amylase**
CATEGORY:	Flour additive/ Enzyme
FOOD USE:	Bread/ Products of the brewing industry
HEAVY METAL CONTENT MAXIMUM IN ppm:	40
ARSENIC CONTENT MAXIMUM IN ppm:	3
FUNCTION IN FOODS:	Fungal alpha-amylase acts as a bread improver: it increases loaf volume and oven spring, and improves crumb texture. It also enhances the gassing power of breadmaking flour, and is used in brewing in conjunction with other enzymes to hydrolyse barley starch
TECHNOLOGY OF USE IN FOODS:	The optimum level of addition of fungal alpha-amylase to breadmaking flour is about 80 Farrand units. It is active at temperatures up to 55–65°C. Above 65°C, it is deactivated rapidly
SYNERGISTS:	Hemicellulase
FOOD SAFETY ISSUES:	No safety concerns in food, but there have been reports of occasional allergic reactions in bakery workers
LEGISLATION:	**USA:** Permitted in flour **UK and EUROPE:** UK: permitted for general food use, including bread and flour. Europe: permitted in bread in most EU countries **CANADA:** Permitted in various foods, including bread, flour, whole wheat flour and "unstandardized bakery products" **AUSTRALIA/PACIFIC RIM:** Permitted in bread

REFERENCES:	Pritchard, P. E. (1992) Studies on the bread-improving mechanism of fungal alpha-amylase. *J. Biol. Educ.*, **26**(1), 12–18. Johnson, J. A., and Miller, B. S. (1949) Studies on the role of alpha-amylase and proteinase in breadmaking. *Cereal Chem.*, **26**, 371–383. Chamberlain, N., Collins, T. H., and McDermott, E. E. (1981) Alpha-amylase and bread properties. *J. Fd. Technol.*, **16**, 127–152. Miller, B. S., Johnson, J. A., and Palmer, D. L. (1953) A comparison of cereal, fungal and bacterial alpha-amylases as supplements for breadmaking. *Fd. Technol., Chicago*, **7**, 38–42.
ANY OTHER RELEVANT INFORMATION:	Fungal alpha-amylase for baking is produced by a selected strain of *Aspergillus oryzae*

NAME:	Chlorine dioxide
CATEGORY:	Flour additive
FOOD USE:	Flour for breadmaking
SYNONYMS:	Dyox/ 926 (no E prefix)
FORMULA:	ClO_2
MOLECULAR MASS:	67.46
PROPERTIES AND APPEARANCE:	Yellow/ reddish-yellow gas
BOILING POINT IN °C AT VARIOUS PRESSURES (INCLUDING 760 mm Hg):	11
MELTING RANGE IN °C:	–59
DENSITY AT 20°C (AND OTHER TEMPERATURES) IN g/l:	1.64 (liquid)
PURITY %:	4 (Dyox)
SOLUBILITY % AT VARIOUS TEMPERATURE/pH COMBINATIONS:	
in water:	**@ 20°C** 0.30%
FUNCTION IN FOODS:	Flour improver; flour bleaching agent
ALTERNATIVES:	Ascorbic acid; cysteine hydrochloride (flour improvers). Benzoyl peroxide (flour bleaching agent)
TECHNOLOGY OF USE IN FOODS:	Treatment of flour with chlorine dioxide is normally carried out at the mill. The chlorine dioxide, in the form of dyox gas (maximum 4% chlorine dioxide) must be generated on site, usually by the reaction of sodium chlorite solution and chlorine gas in a stream of air. Flour treated with chlorine dioxide is particularly suitable for bulk fermentation (traditional) breadmaking

FOOD SAFETY ISSUES: Chlorine dioxide is mutagenic; toxicological studies on its effects on flour are awaited

LEGISLATION:

USA:
Permitted flour treatment agent

CANADA:
Permitted flour treatment agent

UK and EUROPE:
UK and Ireland: permitted flour treatment agent. Not permitted in other EU countries

AUSTRALIA/PACIFIC RIM:
Permitted flour treatment agent

REFERENCES:

Daniels, D. G. H., and Whitehead, J. K. (1957) Laboratory preparation of chlorine dioxide. *Chem. Ind.* Sept 7, 1214.

Moore, T., Sharman, I. M., and Ward, R. J. (1957) The destruction of vitamin E in flour by chlorine dioxide. *J. Sci. Food. Agric.*, **8**, 97–104.

Parker, H. K., and Fortmann, K. L. (1949) Methods for the laboratory-scale production of chlorine dioxide and the treatment of flour. *Cereal Chem.*, **26**(6), 479–490.

NAME:	Chlorine
CATEGORY:	Flour additive
FOOD USE:	Cakemaking flour
SYNONYMS:	925 (no E prefix)
FORMULA:	Cl2
MOLECULAR MASS:	70.91
PROPERTIES AND APPEARANCE:	Clear yellow liquid when packed under pressure, which on vaporising gives a greenish-yellow gas
BOILING POINT IN °C AT VARIOUS PRESSURES (INCLUDING 760 mm Hg):	–34.05
MELTING RANGE IN °C:	–101
DENSITY AT 20°C (AND OTHER TEMPERATURES) IN g/l:	1.41 (liquid) at 6.86 atmospheres
VAPOUR PRESSURE AT VARIOUS TEMPERATURES IN mm Hg (E.G. AT 20°C IN mm Hg; AT 120°C IN mm Hg):	At 16°C, 3666
PURITY %:	99.5 minimum
WATER CONTENT MAXIMUM IN %:	0.01
HEAVY METAL CONTENT MAXIMUM IN ppm:	30
ARSENIC CONTENT MAXIMUM IN ppm:	3
ASH MAXIMUM IN %:	0.02

SOLUBILITY % AT VARIOUS TEMPERATURE/pH COMBINATIONS:

in water: **@ 25°C** 0.14%

FUNCTION IN FOODS:

Used to treat flours for high-ratio cakes and fruit cakes; it modifies the properties of the starch, thereby preventing "collapse" of the cake. Also acts as a flour bleaching agent

TECHNOLOGY OF USE IN FOODS:

Chlorine treatment of flour is normally carried out at the mill using liquid chlorine.

The optimum level of chlorination is usually 1500–2000 mg/kg

FOOD SAFETY ISSUES:

Acceptable for use as a flour treatment agent (JECFA 1985).

Currently under evaluation by the (UK) Committee on Toxicity of Chemicals in Food, Consumer Products and the Environment and by the (EU) Scientific Committee for Food

LEGISLATION:

USA:
Permitted flour treatment agent

CANADA:
Permitted flour treatment agent

UK and EUROPE:
UK and Ireland: permitted in cake flour, except wholemeal. Not permitted in most other EU countries

AUSTRALIA/PACIFIC RIM:
Permitted flour treatment agent

REFERENCES:

Telloke, G. W. (1985) Chlorination of cake flour and its effects on starch gelatinisation. *Starke*, **37**, 17–22.

Wei, C. I., Ghanberi, H. A., Wheeler, W. B., and Kirk, J. R. (1984) Fate of chlorine during flour chlorination. *J. Fd. Sci.*, **49**, 1136–1138, 1153.

Guy, R. C. E., and Pithawala, H. R. (1981) Rheological studies of high-ratio cake batters to investigate the mechanism of improvement of flours by chlorination or heat treatment. *J. Fd. Technol.*, **16**, 153–166.

Johnson, A. C., Hoseney, R. C., and Ghiasi, K. (1980) Chlorine treatment of cake flours, V: oxidation of starch. *Cereal Chem.*, **57**, 94–96.

NAME:	Glucono-delta-lactone
CATEGORY:	Flour additive/ Sequestrant
FOOD USE:	Baked goods and frozen doughs/ Dairy products: set yogurt; sour cream; cottage cheese/ Salad dressings/ Meat products/ Pickled foods
SYNONYMS:	Delta lactone of D-gluconic acid/ D-2,3,4,5,6-pentahydroxycaproic acid/ D-Glucono-1,5-lactone/ E575
FORMULA:	CH2(OH)CHCH(OH)CH(OH)CH(OH)CO [ring closed through —O— linking CH and CO]
MOLECULAR MASS:	178.14
PROPERTIES AND APPEARANCE:	White, fine or coarse crystalline powder
PURITY %:	>99.0
HEAVY METAL CONTENT MAXIMUM IN ppm:	10
ARSENIC CONTENT MAXIMUM IN ppm:	2
ASH MAXIMUM IN %:	0.05
SOLUBILITY % AT VARIOUS TEMPERATURE/pH COMBINATIONS:	
in water:	**@ 25°C** 59
in ethanol solution:	**@ 100%** 1
FUNCTION IN FOODS:	Produces a slow release of acidity in aqueous solution. Acts as an acid, an acidulant and as a raising agent. Lacks the unpleasant after-taste associated with some raising agents
ALTERNATIVES:	Potassium tartrate; calcium tetrahydrogen diorthophosphate; disodium dihydrogen diphosphate; sodium aluminium phosphate

TECHNOLOGY OF USE IN FOODS:

When used as an acidulant for bakery products, 1 part glucono-delta-lactone will neutralise 0.472 parts sodium bicarbonate. Release of carbon dioxide is slow at room temperature, the greater part of the sodium bicarbonate remains unreacted until the oven stage. It produces a lighter crumb colour and a finer texture in many baked products. It also reduces processing time in the production of cottage cheese, and can be used as an aid in development of cure colour in processed meat products.

It can be used with a reduced quantity of vinegar to produce pickled foods of milder flavour

FOOD SAFETY ISSUES:

No safety concerns

LEGISLATION:

USA:
Permitted generally in foods

CANADA:
Permitted in "unstandardized foods"

UK and EUROPE:
Permitted generally in foods

AUSTRALIA/PACIFIC RIM:
Permitted generally in foods

REFERENCES:

Timm, R. G. (1988) Baking powder – factors affecting its performance. In: Bush, P. B., Clarke, I. R., Kort, M. J., and Smith, M. F. (Eds.), *Functionality of Ingredients in the Baking Industry*. Natal Technikon Printers, Durban.

Bayoumi, S., and Madkor, S. (1988) The use of glucono-delta-lactone (GDL) in the manufacture of yogurt. *Egyptian J. Dairy Science*, **16**, 233–238.

Fox, P. F. (1988) Direct acidification of dairy products. *Dairy Science Abstracts*, **40**(12), 727–732.

LaBell, F. M. (1981) Pickled foods with less vinegar offer new flavour possibilities. *Food Development*, **October**, 22–24.

NAME:	Disodium dihydrogen diphosphate
CATEGORY:	Flour additive/ Sequestrant
FOOD USE:	Baked goods/ Meat products/ Processed cheese/ Potato products
SYNONYMS:	Sodium acid pyrophosphate/ Disodium pyrophosphate/ E450(a)
FORMULA:	$Na_2H_2P_2O_7$
MOLECULAR MASS:	221.94
PROPERTIES AND APPEARANCE:	White powder or grains
MELTING RANGE IN °C:	Decomposes at 220
DENSITY AT 20°C (AND OTHER TEMPERATURES) IN g/l:	1.85
PURITY %:	95.0% minimum
WATER CONTENT MAXIMUM IN %:	0.5
HEAVY METAL CONTENT MAXIMUM IN ppm:	20
ARSENIC CONTENT MAXIMUM IN ppm:	3
SOLUBILITY % AT VARIOUS TEMPERATURE/pH COMBINATIONS:	
in water:	**@ 20°C** 13% **@ 50°C** 35%
FUNCTION IN FOODS:	Acid raising agent of slow, but variable reactivity. Used in sausages and luncheon meats as an emulsion stabiliser and to improve flavour and texture. Used as an emulsifying salt in processed cheese. Acts as a buffer in on-dairy coffee whiteners. Protects against discoloration of potato products
ALTERNATIVES:	Monocalcium orthophosphate; Sodium aluminium phosphate; Glucono-delta-lactone

TECHNOLOGY OF USE IN FOODS:

Sodium acid pyrophosphate leavening acids are available in several grades with different rates of reaction with sodium bicarbonate. The slowest reacting are suitable for products requiring a long production cycle or long storage life, such as frozen doughs.

A grade appropriate for the intended purpose should be selected. The neutralising value is 72. When used as an emulsion stabiliser in frankfurters and luncheon meat, an addition of 0.2–0.3% to the dry mix should be made

FOOD SAFETY ISSUES:

The maximum tolerable daily intake is 70 mg/kg body weight for phosphates, expressed as P, from all sources, including those naturally present in food and those derived from additives

LEGISLATION:

USA:
Permitted generally in foods and in self-raising flour

CANADA:
Permitted to GMP in baking powder and "unstandardized foods"

UK and EUROPE:
Permitted at up to specified maximum levels in a wide range of foods, including bakery products, dairy products and meat products

AUSTRALIA/PACIFIC RIM:
Permitted at specified levels in a range of foods

REFERENCES:

Timm, R. G. (1988) Baking powder – factors affecting its performance. In: Bush, P. B., Clarke, I. R., Kort, M. J., and M. F. Smith (Eds.), *Functionality of Ingredients in the Baking Industry*. Natal Technikon Printers, Durban.

Conn, J. F. (1981) Chemical leavening systems in flour products. *Cereal Foods World*, **26**(3), 119–123.

LaBaw, G. D. (1982) Chemical leavening agents and their use in bakery products. *Bakers Digest*, **56**(1), 16–18, 20–21.

Stauffer, C. E. (1994) Chemical leavening. In: K. Kulp (Ed.), *Cookie Chemistry and Technology*. American Institute of Baking, Kansas.

NAME:	Monopotassium tartrate
CATEGORY:	Flour additive
FOOD USE:	Baked goods/ Baking powder
SYNONYMS:	Potassium acid tartrate/ E336/ Potassium hydrogen tartrate/ Cream of tartar
FORMULA:	KOOC-CH(OH)CH(OH)COOH
MOLECULAR MASS:	188.18
PROPERTIES AND APPEARANCE:	White crystalline or granulated powder
DENSITY AT 20°C (AND OTHER TEMPERATURES) IN g/l:	1.984
PURITY %:	99.5
WATER CONTENT MAXIMUM IN %:	0.5
HEAVY METAL CONTENT MAXIMUM IN ppm:	10
ARSENIC CONTENT MAXIMUM IN ppm:	3
SOLUBILITY % AT VARIOUS TEMPERATURE/pH COMBINATIONS:	
in water:	**@ 20°C** 0.61% **@ 100°C** 6.25%
in ethanol solution:	**@ 100%** 0.01%
FUNCTION IN FOODS:	A fast-acting raising agent, with 70% of carbon dioxide generated in the first two minutes of mixing. It is also used as an acidity regulator
ALTERNATIVES:	Calcium tetrahydrogen diorthophosphate; disodium dihydrogen diphosphate; glucono-delta-lactone; sodium aluminium phosphate

TECHNOLOGY OF USE IN FOODS: This acid leavening agent reacts too quickly for most purposes. Its neutralising value is 45

FOOD SAFETY ISSUES: No safety concerns

LEGISLATION:

USA:
Permitted generally in foods

CANADA:
Permitted in baking powder, honey wine and "unstandardized foods"

UK and EUROPE:
Permitted generally in foods and in organic products

AUSTRALIA/PACIFIC RIM:
Permitted generally in foods

REFERENCES:

Conn, J. F. (1981) Chemical leavening systems in flour products. *Cereal Foods World*, **26**(3), 119–123.

LaBaw, G. D. (1982) Chemical leavening agents and their use in bakery products. *Bakers Digest*, **56**(1), 16–18, 20–21.

NAME:	Hemicellulase
CATEGORY:	Flour additive/ Enzyme
FOOD USE:	Breadmaking
SYNONYMS:	Pentosanase
HEAVY METAL CONTENT MAXIMUM IN ppm:	40
ARSENIC CONTENT MAXIMUM IN ppm:	3
FUNCTION IN FOODS:	Fungal hemicellulase, when used in breadmaking, increases loaf volume, improves crumb structure, crust colour and crumb softness. It also increases the starch yield in starch production from wheat flour
TECHNOLOGY OF USE IN FOODS:	Hemicellulase is active at temperatures up to 55–65°C; above 65°C, it is inactivated rapidly. The recommended dosage level in breadmaking is 10–30 g/100 kg of flour of an enzyme preparation having an activity of approx. 400 Specific Hemicellulase Units per gram. The practical pH range is 4–6.5
SYNERGISTS:	Alpha-amylase
FOOD SAFETY ISSUES:	No safety concerns

LEGISLATION:

USA:
Permitted for general food use

UK and EUROPE:
UK: permitted for general food use and in breadmaking
EU: permitted for general food use in most countries; permitted specifically in breadmaking in certain countries including France and the Netherlands

CANADA:
Permitted in distillers' mash, liquid coffee concentrate and mash destined for vinegar manufacture

AUSTRALIA/PACIFIC RIM:
Permitted in bread

REFERENCES:

Hammond, J. (1994) Breadmaking with hemicellulase: overcoming the legal hurdles. *Food Technology International Europe*, 1994, 19–23.

Weegels, P. L., Marseille, J. P., and Hamer, R. J. (1992) Enzymes as a processing aid in the separation of wheat flour into starch and gluten. *Starch/Starke*, **44**(2), 44–48.

Ter Haseborg, E., and Himmelstein, A. (1988) Quality problems with high-fibre breads solved by use of hemicellulase enzymes. *Cereal Foods World*, **33**, 419, 421–422.

NAME:	**Calcium tetrahydrogen diorthophosphate**
CATEGORY:	Flour additive
FOOD USE:	Baked goods/ Baking powder/ Dough conditioner/ Dry powder beverages/ Fruits/ Puddings/ Yoghurts
SYNONYMS:	Acid calcium phosphate/ Monocalcium phosphate/ Monocalcium orthophosphate/ E341/ Calcium phosphate, monobasic
FORMULA:	$Ca(H_2PO_4)_2$
MOLECULAR MASS:	234.05
PROPERTIES AND APPEARANCE:	White crystals or granules, or granular powder. Also occurs as the monohydrate
BOILING POINT IN °C AT VARIOUS PRESSURES (INCLUDING 760 mm Hg):	203
MELTING RANGE IN °C:	Decomposes at 200
Density at 20°C (AND OTHER TEMPERATURES) IN g/l:	2.22
PURITY %:	16.8–18.3% of Ca (anhydrous); 15.9–17.7% of Ca (monohydrate)
WATER CONTENT MAXIMUM IN %:	1.0
HEAVY METAL CONTENT MAXIMUM IN ppm:	30
ARSENIC CONTENT MAXIMUM IN ppm:	3
SOLUBILITY % AT VARIOUS TEMPERATURE/pH COMBINATIONS:	
in water:	**@ 20°C** 1.8%
FUNCTION IN FOODS:	Fast-acting raising agent; ingredient of baking powders; acts as an acid and acidity regulator; yeast nutrient in yeast-raised goods; dietary supplement in fortified foods. Used for its action as a buffer and acidulant in beverage powders

ALTERNATIVES: Sodium acid pyrophosphate; monopotassium tartrate; glucono-delta-lactone; sodium aluminium phosphate

TECHNOLOGY OF USE IN FOODS: A very fast-acting raising agent with a neutralising value of 80 (monohydrate) and 83 (anhydrous). Used alone, it reacts with sodium bicarbonate very rapidly, releasing 60–70% of the available carbon dioxide in a two-minute mixing of dough. The reaction is completed during baking. It is used at low concentration in combination with a slow-acting acid such as sodium acid pyrophosphate in double-acting baking powder. Coated anhydrous monocalcium phosphate gives a more delayed reaction.

Used at a level of 0.03% in direct-set cottage cheese to aid acidification of the milk and decrease processing time.

Sometimes added to instant pudding and "no-bake" cheesecake mixes to strengthen the gel. May be used in canned fruit to increase firmness

FOOD SAFETY ISSUES: Maximum tolerable daily intake of 70 mg/kg bodyweight, expressed as P, for phosphates from all sources, including those naturally present in food and derived from additives

LEGISLATION:

USA:
Permitted generally in foods and in self-raising flour

UK and EUROPE:
Permitted at specified maximum levels in a wide range of foods, including specified dairyproducts, meat products and bakery products

CANADA:
Permitted to GMP in ale, baking powder, beer, light beer, malt liquor, porter, stout and in "unstandardized foods"

AUSTRALIA/PACIFIC RIM:
Permitted in a wide range of foods up to specified maximum levels

REFERENCES:

Timm, R. G. (1988) Baking powder – factors affecting its performance. In: Bush, P. B., Clarke, I. R., Kort, M. J., and M. F. Smith (Eds.), *Functionality of Ingredients in the Baking Industry*. Natal Technikon Printers, Durban.

Conn, J. F. (1981) Chemical leavening systems in flour products. *Cereal Foods World*, **26**(3), 119–123.

LaBaw, G. D. (1982) Chemical leavening agents and their use in bakery products. *Bakers Digest*, **56**(1), 16–18, 20–21.

Stauffer, C. E. (1994) Chemical leavening. In: Kulp, K. (Ed.), *Cookie Chemistry and Technology*. American Institute of Baking, Kansas.

NAME:	L-Cysteine hydrochloride
CATEGORY:	Flour additive
FOOD USE:	Bread/ Biscuits
SYNONYMS:	L-Cysteine monohydrochloride/ L-2-Amino-3-mercaptopropanoic acid hydrochloride
FORMULA:	C3 H7 N O2 S.HCl.H2O
MOLECULAR MASS:	175.63
PROPERTIES AND APPEARANCE:	White crystalline powder or colourless crystals
MELTING RANGE IN °C:	Decomposes 175
PURITY %:	98.5% minimum
WATER CONTENT MAXIMUM IN %:	12.0
HEAVY METAL CONTENT MAXIMUM IN ppm:	20
ARSENIC CONTENT MAXIMUM IN ppm:	3
ASH MAXIMUM IN %:	0.1
SOLUBILITY % AT VARIOUS TEMPERATURE/pH COMBINATIONS:	
in water:	**@ 20°C** 100%
in ethanol solution:	**@ 5%** Freely soluble
FUNCTION IN FOODS:	Flour improver suited to use in the activated dough development breadmaking process. Treatment of biscuit flours – serves to reduce elasticity and produce uniformity in the dough
ALTERNATIVES:	Sulphur dioxide; sodium metabisulphite

TECHNOLOGY OF USE IN FOODS:

L-Cysteine hydrochloride at a level of about 35 mg/kg and in combination with another flour treatment agent, such as ascorbic acid, is suitable for use in the activated dough development breadmaking process, a low-energy mixing process used by small bakers.

Up to 300 mg/kg of L-cysteine hydrochloride is used in flour for semi-sweet biscuits, such as Osborne, Marie or Rich Tea biscuits. It acts to improve the structure and appearance of the finished biscuits. Elastic recovery in the sheeted dough is reduced, leading to more uniform results from flours which are inherently variable

SYNERGISTS:

Ascorbic acid

FOOD SAFETY ISSUES:

No safety concerns

LEGISLATION:

USA:
Classed as a special dietary and nutritional additive

CANADA:
Up to 90 ppm in bread, flour and wholewheat flour; "unstandardized bakery foods" to GMP

UK and EUROPE:
UK: Up to 75 mg/kg permitted in all flour and bread, except wholemeal; up to 300 mg/kg permitted in biscuit flour
Belgium; up to 50 mg/kg permitted in flour
Denmark: up to 25 mg/kg permitted in bread
Germany: up to 30 mg/kg in bread, flour weight basis
Holland: up to 75 mg/kg in bread, flour weight basis

AUSTRALIA/PACIFIC RIM:
Up to 75 mg/kg in bread-making flour

REFERENCES:

Ranum, P. (1992) Flour treatment and additives. *AIB Research Department Technical Bulletin*, May 1992, **14**(5), 6pp.

Henika, R. G., and Zenner, S. F. (1960) *Bakers Digest*, **343** 36, 37, 40.

Geittner, J. (1978) L-Cysteine for rationalising biscuit and puff pastry production. *J. Getreide Mehl u. Brot*, **32**(5), 124–126.

Grandvoinnet, P., and Berger, M. (1979) Ascorbic acid and cysteine in bakery products. *Industries Alimentaires et Agricoles*, 941–947.

NAME:	Sodium hydrogen carbonate
CATEGORY:	Flour additive
FOOD USE:	Baked goods/ Baking powder
SYNONYMS:	Sodium bicarbonate/ Bicarbonate of soda/ Baking soda/ E500
FORMULA:	$NaHCO_3$
MOLECULAR MASS:	84.01
PROPERTIES AND APPEARANCE:	White crystalline powder
MELTING RANGE IN °C:	Decomposes at 50
DENSITY AT 20°C (AND OTHER TEMPERATURES) IN g/ml:	2.22
PURITY %:	>99
WATER CONTENT MAXIMUM IN %:	0.25
HEAVY METAL CONTENT MAXIMUM IN ppm:	5
ARSENIC CONTENT MAXIMUM IN ppm:	2
SOLUBILITY % AT VARIOUS TEMPERATURE/pH COMBINATIONS:	
in water:	**@ 20°C** 9.5%
FUNCTION IN FOODS:	Raising agent; alkali. On reaction with an acid, sodium bicarbonate releases carbon dioxide gas. This reaction provides the basis for chemical leavening systems. The carbon dioxide produced aerates the dough or batter during mixing and baking, thereby giving the product an enhanced volume and a lighter texture
ALTERNATIVES:	Ammonium hydrogen carbonate (for biscuits); potassium hydrogen carbonate (for biscuits)

TECHNOLOGY OF USE IN FOODS: A satisfactory chemical leavening system requires a balance between the sodium leavening bicarbonate and the acid leavening agent. The number of parts of some common acid agents equivalent to 1 part of sodium bicarbonate are as follows:

Monopotassium tartrate	2.24
Pure SAPP (sodium acid pyrophosphate)	1.32
Commercial SAPP	1.39
Pure ACP (acid calcium phosphate)	1.045
Commercial ACP	1.25
Glucono-delta-lactone	2.12

As commercially available acid leavening agents vary in their rate of reaction, they should be selected according to the particular application for which they are intended

SYNERGISTS: Acidulants, such as potassium tartrate, calcium tetrahydrogen diorthophosphate (ACP), disodium dihydrogen diphosphate (SAPP), glucono-delta-lactone

FOOD SAFETY ISSUES: No safety concerns

LEGISLATION:

USA:	**UK and EUROPE:**	**CANADA:**	**AUSTRALIA/PACIFIC RIM:**
Permitted for general food use and in self-raising flour	Permitted for general food use	Permitted for general food use	Permitted for general food use

REFERENCES:

Lajoie, M. S., and Thomas, M. C. (1991) Versatility of bicarbonate leavening bases. *Cereal Foods World*, **36**(5), 420–423.

Timm, R. G. (1988) Baking powder – factors affecting its performance. In: Bush, P. B., Clarke, I. R., Kort, M. J., and M. F. Smith (Eds.), *Functionality of Ingredients in the Baking Industry*. Natal Technikon Printers, Durban.

Conn, J. F. (1981) Chemical leavening systems in flour products. *Cereal Foods World*, **26**(3), 119–123.

Janovsky, C. (1993) Encapsulated Ingredients for the Baking Industry. *Cereal Foods World*, **38**(1), 85–87.

Lajoie, M. S., and Thomas, M. C. (1994) Sodium bicarbonate particle size and neutralisation in sponge-dough systems. *Cereal Foods World*, **39**(9), 684–687.

NAME:	**Sodium alumininium phosphate, acidic**
CATEGORY:	Flour additive
FOOD USE:	Baked goods/ Baking powder
SYNONYMS:	E541
FORMULA:	$Na_3Al_2H_{15}(PO_4)_8$ or $NaAl_3H_{14}(PO_4)_8.4H_2O$
MOLECULAR MASS:	897.82 or 949.88
PROPERTIES AND APPEARANCE:	White powder
PURITY %:	95.0% minimum
HEAVY METAL CONTENT MAXIMUM IN ppm:	40
ARSENIC CONTENT MAXIMUM IN ppm:	3
SOLUBILITY % AT VARIOUS TEMPERATURE/pH COMBINATIONS:	
in water:	**@ 20°C** Insoluble Soluble in dilute HCl
FUNCTION IN FOODS:	Slow-acting acid raising agent
ALTERNATIVES:	Monocalcium orthophosphate; sodium acid pyrophosphate; glucono-delta-lactone
TECHNOLOGY OF USE IN FOODS:	The major leavening action takes place only when the product is heated. Particularly suitable for cake mixes. Also used in refrigerated and frozen dough or batter products and in some baking powders. Blends of SALP with anhydrous monocalcium phosphate or with monocalcium phosphate monohydrate are suitable for a variety of purposes, including self-raising flour, fine-textured cakes and refrigerated pancake batters. SALP has a neutralising value of 100, and does not give rise to the bitter flavour associated, in particular, with pyrophosphates

FOOD SAFETY ISSUES:

The provisional tolerable weekly intake is 7 mg/kg bodyweight, expressed as aluminium, from all sources, including natural sources and additive usage.

LEGISLATION:

USA:
Classed as a multiple purpose GRAS food substance and permitted for use as a leavening agent

UK and EUROPE:
Permitted in scones and sponge wares at up to 1 g/kg, expressed as aluminium

CANADA:
Permitted to GMP in "unstandardized foods"

AUSTRALIA/PACIFIC RIM:
Permitted in bakery products, excluding bread, and in baking powder

REFERENCES:

Timm, R. G. (1988) Baking powder – factors affecting its performance. In: Bush, P. B., Clarke, I. R., Kort, M. J., and M. F. Smith (Eds.), *Functionality of Ingredients in the Baking Industry*. Natal Technikon Printers, Durban.

Conn, J. F. (1981) Chemical leavening systems in flour products. *Cereal Foods World*, **26**(3), 119–123.

LaBaw, G. D. (1982) Chemical leavening agents and their use in bakery products. *Bakers Digest*, **56**(1), 16–18, 20–21.

Stauffer, C. E. (1994) Chemical leavening. In: Kulp, K. (Ed.), *Cookie Chemistry and Technology*. American Institute of Baking, Kansas.

Part 8

Gases

NAME:	**Argon**							
CATEGORY:	Packaging gas							
FORMULA:	Ar							
MOLECULAR MASS:	39.948							
PROPERTIES AND APPEARANCE:	Colourless, odourless, tasteless, inert gas							
BOILING POINT IN °C AT VARIOUS PRESSURES (INCLUDING 760 mm Hg):	−185.7							
MELTING RANGE IN °C:	−189.2							
DENSITY AT 20°C (AND OTHER TEMPERATURES) IN g/l:	1.784							
SOLUBILITY % AT VARIOUS TEMPERATURE/pH COMBINATIONS:								
in water:	@ **0°C**	5.6	@ **20°C**	3.36	@ **50°C**	3.0	@ **100°C**	Soluble
in vegetable oil:	@ **20°C**	Soluble	@ **50°C**	Soluble	@ **100°C**	Soluble		
in sucrose solution:	@ **10%**	Soluble	@ **40%**	Soluble	@ **60%**	Soluble		
in sodium chloride solution:	@ **5%**	Soluble	@ **10%**	Soluble	@ **15%**	Soluble		
in ethanol solution:	@ **5%**	Soluble	@ **20%**	Soluble	@ **95%**	Soluble	@ **100%**	Soluble
in propylene glycol:	@ **20°C**	Soluble	@ **50°C**	Soluble	@ **100°C**	Soluble		
REFERENCE:	Merck Index, Twelfth Edition (1996) Merck & Co., Inc., Whitehouse Station, NJ, USA.							

NAME:	**Hydrogen**
CATEGORY:	Packaging gas
SYNONYMS:	Protium
FORMULA:	H2
MOLECULAR MASS:	2.01588
PROPERTIES AND APPEARANCE:	Colourless, odourless, tasteless gas
BOILING POINT IN °C AT VARIOUS PRESSURES (INCLUDING 760 mm Hg):	–252.77
MELTING RANGE IN °C:	–259.34
DENSITY AT 20°C (AND OTHER TEMPERATURES) IN g/l:	0.08987
SOLUBILITY % AT VARIOUS TEMPERATURE/pH COMBINATIONS:	
in water:	**@ 0°C** 2.14 **@ 20°C** 2.0 **@ 25°C** 1.91 **@ 50°C** 1.89
in ethanol solution:	**@ 100°C** 6.925
REFERENCE:	*Merck Index*, Twelfth Edition (1996) Merck & Co., Inc., Whitehouse Station, NJ, USA.

NAME:	Nitrous oxide					
CATEGORY:	Packaging gas					
SYNONYMS:	Nitrogen monoxide/ Nitrogen oxide/ Dinitrogen monoxide/ Laughing gas/ CAS 10024-97-2					
FORMULA:	N_2O					
MOLECULAR MASS:	44.01					
PROPERTIES AND APPEARANCE:	Colourless gas, slightly sweet odour					
BOILING POINT IN °C AT VARIOUS PRESSURES (INCLUDING 760 mm Hg):	–88					
MELTING RANGE IN °C:	–91					
DENSITY AT 20°C (AND OTHER TEMPERATURES) IN g/l:	1.97					
PURITY %:	99					
WATER CONTENT MAXIMUM IN %:	150 mg/m^3					
SOLUBILITY % AT VARIOUS TEMPERATURE/pH COMBINATIONS:						
in water:	**@ 20°C**	70	**@ 50°C**	Soluble	**@ 100°C**	Soluble
in vegetable oil:	**@ 20°C**	Soluble	**@ 50°C**	Soluble	**@ 100°C**	Soluble
in sucrose solution:	**@ 10%**	Soluble	**@ 40%**	Soluble	**@ 60%**	Soluble
in sodium chloride solution:	**@ 5%**	Soluble	**@ 10%**	Soluble	**@ 15%**	Soluble
in ethanol solution:	**@ 5%**	Freely soluble	**@ 20%**	Freely soluble	**@ 95%**	Freely soluble
	@ 100%	Freely soluble				
in propylene glycol:	**@ 20°C**	Freely soluble	**@ 50°C**	Freely soluble	**@ 100°C**	Freely soluble
FUNCTION IN FOODS:	Direct food additive; propellant; aerating agent; gas. Used in dairy products					

LEGISLATION:	**USA:** FDA 21CFR § 184.1545, GRAS
REFERENCES:	*Food Chemicals Codex*, Fourth Edition (1996) National Academy Press, Washington, DC, USA. *Merck Index*, Twelfth Edition (1996) Merck & Co., Inc., Whitehouse Station, NJ, USA. Ash, M., and Ash, I. (1995) *Food Additives*. Gower Publishing Co., Brookfield, VT, USA.
ANY OTHER RELEVANT INFORMATION:	Precaution: does not burn, but will support combustion; oxidiser. Toxicology: asphyxiant at high concentrations

NAME:	Ozone
CATEGORY:	Gas
FOOD USE:	Bottled water/ Air
SYNONYMS:	Triatomic oxygen/ CAS 10028-15-6/ Ozon
FORMULA:	O3
MOLECULAR MASS:	47.9982
PROPERTIES AND APPEARANCE:	Unstable bluish gas or dark blue liquid; pungent characteristic odour
BOILING POINT IN °C AT VARIOUS PRESSURES (INCLUDING 760 mm Hg):	–111.9
MELTING RANGE IN °C:	–192.7
DENSITY AT 20°C (AND OTHER TEMPERATURES) IN g/l:	2.144 (gas at –183°C), 1.614 (liquid at –195.4°C)

SOLUBILITY % AT VARIOUS TEMPERATURE/pH COMBINATIONS:

in water:	**@ 0°C**	49	**@ 50°C**	Soluble	**@ 100°C**	Soluble		
in vegetable oil:	**@ 20°C**	Soluble	**@ 50°C**	Soluble	**@ 100°C**	Soluble		
in sucrose solution:	**@ 10%**	Soluble	**@ 40%**	Soluble	**@ 60%**	Soluble		
in sodium chloride solution:	**@ 5%**	Soluble	**@ 10%**	Soluble	**@ 15%**	Soluble		
in ethanol solution:	**@ 5%**	Soluble	**@ 20%**	Soluble	**@ 95%**	Soluble	**@ 100%**	Soluble
in propylene glycol:	**@ 20°C**	Soluble	**@ 50°C**	Soluble	**@ 100°C**	Soluble		

FUNCTION IN FOODS:	Used to disinfect water for direct consumption and water for use in surface treatment of fish, fruit and vegetables and other perishable food. Also used to fumigate and deodorise storage environments and processing equipment surfaces. Processing aid (Japan)

TECHNOLOGY OF USE IN FOODS:

Produced on-site by an ozone generator for air treatment. Added to water for direct consumption. Added to water to provide it with antimicrobial effect. Processing aid (Japan). Limitation 0.4 mg/l residual of bottled water. Concentrations of 0.015 ppm of ozone in air produce barely detectable odour. 1 ppm produces disagreeable, sulfur-like odour. Highly reactive oxidising agent

ANTAGONISTS:

Incompatible with rubber and dinitrogen tetraoxide

FOOD SAFETY ISSUES:

Human poison by inhalation. May have in large dosages, neoplastigen, tumorigenic, teratogenic and reprotoxicity effects (experimental). Human systemic effects by inhalation: visual field changes, eye lacrimation, headache, decreased pulse rate with fall in blood pressure, blood pressure decrease, skin dermatitis, cough, dyspnoea, respiratory stimulation and other pulmonary changes.

Human mutagenic data. Skin, eye, upper respiratory system and mucous membrane irritant.

Severe explosion hazard in liquid form when shocked and exposed to heat or flame or in concentrated form by chemical reaction with powerful reducing agent

LEGISLATION:

USA:	CANADA:	AUSTRALIA/PACIFIC RIM:
FDA – 21CFR 184.1563 Limitation of 0.4 mg/l in bottled water, affirmed as GRAS for use as an antimicrobial	Approved for use, according to GMP, as chemosterilant in water represented as mineral or spring water. Also approved as maturing agent in cider and wine, according to GMP	Japan: approved for use as a natural processing aid

REFERENCES:

Lewis, R. J., Sr. (1989) *Food Additives Handbook*. Van Nostrand Reinhold, New York.

Smoley, G. K. (1993) *Everything added to Food in the United States*. U.S. Food and Drug Administration.

Budavari, S. (1996) *The Merck Index,* 12th Edition.

Anonymous (1995) *1995 Encyclopedia of Food Ingredients.* Food in Canada.

ANY OTHER RELEVANT INFORMATION:

At 1 ppm, the sulfur-like odour may cause headache and irritation of eyes and upper respiratory tract; symptoms disappear on cessation of exposure

NAME:	Carbon dioxide
CATEGORY:	Gas
FOOD USE:	Carbonated beverages/ Fruit/ Meat/ Poultry/ Wine
SYNONYMS:	Anhydride carbonique/ Carbonic acid gas/ Carbonic anhydride/ CAS 124-38-9/ Kohlendioxyd/ Kohlensaure/ EINECS 204-696-9/ E 290
FORMULA:	CO2
MOLECULAR MASS:	44.01
PROPERTIES AND APPEARANCE:	Colourless, odourless, non-combustible gas. Faint acid taste. Solid form as dry ice.
MELTING RANGE IN °C:	Sublimes @ –78.5
DENSITY AT 20°C (AND OTHER TEMPERATURES) IN g/l:	Gas @ 0°C 1.976 g/l at 760 mm Liquid @ 0°C 0.914 g/l at 34.3 atm Solid @ –56.6°C 1.512 g/l at 760 mm
SOLUBILITY % AT VARIOUS TEMPERATURE/pH COMBINATIONS:	
in water:	**@ 760 mm** 0°C = 171 ml CO_2/100 ml H_2O/ 20°C = 88 ml CO_2/100 ml H_2O/ 60°C = 36 ml CO_2/100 ml H_2O
FUNCTION IN FOODS:	Aerating agent/ Carbonation/ Cooling agent/ Leavening agent/ Modified atmosphere for microbial control/ pH control agent/ Processing aid/ Propellant/ Gas/ Preservative/ Freezant
TECHNOLOGY OF USE IN FOODS:	Less soluble in alcohol and other neutral organic solvents. Absorbed by alkaline solutions with formation of carbonates.
ANTAGONISTS:	Incompatible with acrylaldehyde, aziridine, metal acetylides, sodium peroxide
FOOD SAFETY ISSUES:	Non-combustible. Is an asphyxiant and has teratogenic and reprotoxicity effects (experimental). Contact of CO_2 snow with skin can cause burns. In CO_2 atmosphere, dusts of magnesium, zirconium, titanium

and some magnesium-aluminium alloy will ignite and explode. In CO_2 atmosphere, dusts of aluminium, chromium and magnesium when heated will ignite and explode. Reacts vigorously with $(Al + Na_2O_2)$; $(S_2O$; $Mg(C_2H_5)_2$; Li; $(Mg + Na_2O_2)$; K; KHC; Na; Na_2C_2; NaK; Ti. CO_2 fire extinguishers can produce highly incendiary sparks of 5–15 mJ at 10–20 kV by electrostatic discharge

LEGISLATION:

USA:

FDA – 21CFR 184.1240.
GRAS when used in accordance with GMP; additive affirmed as GRAS for use as leavening agent, processing aid, propellant, aerating agent, gas.
FDA – 21CFR 193.45.
Approved for modified atmosphere for pest control.
USDA – 9CFR 318.7, 381.147.
Sufficient for purpose.
BATF – 21CFR 240.1051.
CO_2 content of finished wine shall not be increased during the transfer operation.
FDA – 21CFR 169.115, 169.140, 169.150, 169.120.

UK and EUROPE:

E290.
Approved for use as preservative; coolant; freezant (liquid form); packaging gas; aerator.
Approved for use in UK

CANADA:

Approved for use according to GMP as carbonation in carbonated (naming the fruit) juice; wine; malt liquors; beer; ale; cider; stout; porter; light beer; water (represented as mineral or spring water)
In unstandardised foods – approved for use as carbonation and pressure dispensing agent according to GMP
Also approved for use according to GMP as solvent for decaffeination of green coffee beans; spice extracts; natural extractives; flavour (naming the flavour); hop extract; pre-isomerised hop extract

AUSTRALIA/PACIFIC RIM:

Japan: approved for use as acidity regulator and food acid

REFERENCES:

Lewis, R. J., Sr. (1989) *Food Additives Handbook.* Van Nostrand Reinhold, New York.

Smoley, C. K. (1993) *Everything added to Food in the United States.* U.S. Food and Drug Administration.

Budavari, S. (1996) *The Merck Index.* 12th Edition.

Anonymous (1995) *1995 Encyclopedia of Food Ingredients.* Food in Canada.

ANY OTHER RELEVANT INFORMATION: Department of Transport requires gas to be labelled as non-flammable gas

NAME:	Nitrogen
CATEGORY:	Gas
FOOD USE:	Fruit/ Poultry/ Various food in sealed containers/ Wine
SYNONYMS:	Compressed nitrogen/ Nitrogen gas/ Refrigerated liquid nitrogen/ CAS 7727-37-9/ EINECS 231-783-9
FORMULA:	N2
MOLECULAR MASS:	28.02
PROPERTIES AND APPEARANCE:	Colourless gas/ Colourless liquid or cubic crystals at low temperature. Condenses to liquid. Slightly soluble in water; soluble in liquid ammonia. Odourless; flavourless; slightly soluble in alcohol
BOILING POINT IN °C AT VARIOUS PRESSURES (INCLUDING 760 mm Hg):	−195.79
MELTING RANGE IN °C:	−210
DENSITY AT 20°C (AND OTHER TEMPERATURES) IN g/l:	@ 0°C = 1.2506 g/l @ −195.8°C = 0.808 g/l (liquid)
SOLUBILITY % AT VARIOUS TEMPERATURE/pH COMBINATIONS:	
in water:	**@ 0°C** 100 volumes of H_2O absorbs 2.4 volumes of gas; **@ 20°C** 100 volumes of H_2O absorbs 1.6 volumes of gas
in ethanol solution:	**@ 5%** **@ 20°C** 100 volumes of ethanol dissolves 0.1124 volumes of nitrogen
FUNCTION IN FOODS:	Aerating agent/ modified atmospheres for insect/microbial control/ oxygen exclusion/ propellant/ processing aid/ food-freezing processes
TECHNOLOGY OF USE IN FOODS:	An odourless, flavourless gas that is slightly soluble in alcohol and water. In liquid form, use as a freezant in cryogenic freezing of food product to avoid formation of large ice crystals during freezing of food. In gas form, largely used to replace oxygen in the headspace, especially in packaged items that are susceptible to lipid oxidation during storage

FOOD SAFETY ISSUES:

Low toxicity. In high concentration is a simple asphyxiant. Release of nitrogen from solution in blood with formation of small bubbles can lead to symptoms of compressed air illness (caisson disease). Narcotic at high concentration and high pressure. Non-flammable gas, can react violently with lithium; neodymium; titanium under specificied conditions. Combustible

LEGISLATION:

USA:
FDA 21CFR 184.1540.
Substance affirmed GRAS for use as propellant, aerating agent and gas
USDA 9CFR 318.7, 381.147
Sufficient for purpose
FDA 21CFR 169.115, 169.140, 169.150

CANADA:
Approved for use according to GMP to improve spreadability in cream cheese and cream cheese spread; cream cheese and cream cheese spreads with (name added ingredients).
In unstandardised foods – approved for use as pressure dispensing agent according to GMP

AUSTRALIA/PACIFIC RIM:
Japan: approved for use as natural processing aid

REFERENCES:

Lewis, R. J., Sr. (1989) *Food Additives Handbook*. Van Nostrand Reinhold, New York.

Smoley, C. K. (1993) *Everything added to Food in the United States.* U.S. Food and Drug Administration.

Budavari S. (1996) *The Merck Index,* 12th Edition.

Anonymous (1995) *1995 Encyclopedia of Food Ingredients*. Food in Canada.

ANY OTHER RELEVANT INFORMATION:

Department of Transport requires gas to be labelled as non-flammable gas

Part 9

Nutritive Additives

NAME:	DL-α-Tocopherol
CATEGORY:	Antioxidants, Nutritive additives
FOOD USE:	Edible oils and fats/ Meat products/ Poultry products/ Eggs and egg products
SYNONYMS:	Vitamin E/ All-rac-α-tocopherol
FORMULA:	$C_{29}H_{50}O_2$
MOLECULAR MASS:	430.71
ALTERNATIVE FORMS:	DL-α-tocopheryl-acetate/ Natural tocopherols mix
PROPERTIES AND APPEARANCE:	Yellow, viscous oil
BOILING POINT IN °C AT VARIOUS PRESSURES (INCLUDING 760 mm Hg):	200–220
MELTING RANGE IN °C:	2.5–3.5
DENSITY AT 20°C (AND OTHER TEMPERATURES) IN g/l:	0.947–0.958 (25/25°C)
SOLUBILITY % AT VARIOUS TEMPERATURE/pH COMBINATIONS:	
in vegetable oil:	**@ 20°C** Soluble
in ethanol solution:	**@ 5%** Soluble in alcohol
FUNCTION IN FOODS:	As antioxidant in oils, fats, fat-based products, sausages. Nutrient, preservative, inhibitor of nitrosamine formation in pump-cured bacon.
TECHNOLOGY OF USE IN FOODS:	Soluble in fat and oil

LEGISLATION:

USA:
Regulations: FDA 21CFR § 182.3890, 182.5890, 182.8890, 184.1890, GRAS

CANADA:
Food and Drug Regulations: D.03.002

UK and EUROPE:
UK approved

AUSTRALIA/PACIFIC RIM:
Japan: restricted for purpose of antioxidation

REFERENCES:

Ash, M., and Ash, I. (1995) *Food Additives: Electronic Handbook.* Gower Publishing Company, Brookfield, VT, 05036, USA.

Augustin, J., and Scarbrough, F. E. (1990) Nutritional Additives. In: Branen, A. L., Davidson, P. M., and S. Salminen (Eds.), *Food Additives*, pp. 33–81.

Budavari, S. (1996) *The Merck Index*, 12th edition.

Health and Welfare Canada (1996) Part D: Vitamins, Minerals and Amino Acids. In: *Food and Drug Regulations*. Department of Health, Ottawa, ON, Canada.

NAME:	**DL-α-Tocopheryl acetate**
CATEGORY:	Nutritive additives
FOOD USE:	Dairy products/ Sugars, sugar preserves and confectionery/ Edible oils and fats/ Cereals and cereal products
SYNONYMS:	Vitamin E acetate
FORMULA:	C31 H52 O3
MOLECULAR MASS:	472.75
ALTERNATIVE FORMS:	DL-α-tocopherol/ Natural tocopherols
PROPERTIES AND APPEARANCE:	Oily blends: slightly yellow viscous oil. Dry powders: slight yellowish free-flowing powder
BOILING POINT IN °C AT VARIOUS PRESSURES (INCLUDING 760 mm Hg):	$bp_{0.01}$ 184; $bp_{0.025}$ 194; $bp_{0.3}$ 224
MELTING RANGE IN °C:	–27.5
DENSITY AT 20°C (AND OTHER TEMPERATURES) IN g/l:	0.9533 (21.3/4°C)
SOLUBILITY % AT VARIOUS TEMPERATURE/pH COMBINATIONS:	
in ethanol solution:	**@ 100%** Soluble
FUNCTION IN FOODS:	Fortification of infant formulae, sugar and cocoa confectionery, oils and fats, fruit drinks, flour, liquid milk and milk powder, breakfast cereals
TECHNOLOGY OF USE IN FOODS:	Viscous oils are soluble in fat and oil; homogenising in a small amount of liquid is necessary prior to adding to milk, infant formulae and drinks. Dry powders are water-dispersible

LEGISLATION:	**USA:** Regulations: FDA 21CFR § 182.5892, 182.8892, GRAS	**CANADA:** Food and Drug Regulations: D.03.002
REFERENCES:	Ash, M., and Ash, I. (1995) *Food Additives: Electronic Handbook.* Gower Publishing Company, Brookfield, VT, 05036, USA. Augustin, J., and Scarbrough, F. E. (1990) Nutritional Additives. In: Branen, A. L., Davidson, P. M., and S. Salminen (Eds.), *Food Additives*, pp. 33–81. Budavari, S. (1996) *The Merck Index*, 12th edition. Health and Welfare Canada (1996) Part D: Vitamins, Minerals and Amino Acids. In: *Food and Drug Regulations*. Department of Health, Ottawa, ON, Canada.	
ANY OTHER RELEVANT INFORMATION:	Oily blends: 1000 IU g^{-1} potency Dry powders: 0.25–0.50 × 10^{6} IU g^{-1} potency	

NAME:	Pyridoxine
CATEGORY:	Nutritive additives
SYNONYMS:	Vitamin B_6
ALTERNATIVE FORMS:	Pyridoxine hydrochloride
FUNCTION IN FOODS:	Fortification
TECHNOLOGY OF USE IN FOODS:	N/A
LEGISLATION:	**CANADA:** Food and Drug Regulations: D.03.002
REFERENCE:	Health and Welfare Canada (1996) Part D: Vitamins, Minerals and Amino Acids. In: *Food and Drug Regulations*. Department of Health, Ottawa, ON, Canada.
ANY OTHER RELEVANT INFORMATION:	Potency 100%

NAME:	Nicotinic acid
CATEGORY:	Nutritive additives; Colours
FOOD USE:	Cereals and cereal products/ Dairy products/ Sugars, sugar preserves and confectionery/ Soft drinks
SYNONYMS:	Niacin (INCI)/ Vitamin B_3/ 3-Picolinic acid/ Pyridine-3-carboxylic acid
FORMULA:	C6 H5 N O2
MOLECULAR MASS:	123.12
ALTERNATIVE FORMS:	Niacinamide
PROPERTIES AND APPEARANCE:	White crystalline powder
MELTING RANGE IN °C:	236
DENSITY AT 20°C (AND OTHER TEMPERATURES) IN g/l:	1.473
SOLUBILITY % AT VARIOUS TEMPERATURE/pH COMBINATIONS:	
in water:	**@ 20°C** Soluble
in ethanol solution:	**@ 100%** Soluble
FUNCTION IN FOODS:	Fortification of flour, breakfast cereals, infant formulae, fruit drinks, sugar and cocoa confectionery, pasta, meal replacements. Colour stabilisation in meat products (if permitted)
TECHNOLOGY OF USE IN FOODS:	Direct food additive

LEGISLATION:

USA:
Regulations: FDA 21CFR § 135.115, 137, 139, 182.5530, 184.1530; GRAS

CANADA:
Food and Drug Regulations: D.03.002; B.13.010.1(1)

UK and EUROPE:
Europe: listed
UK: approved

AUSTRALIA/PACIFIC RIM:
Japan: restricted

REFERENCES:

Ash, M., and Ash, I. (1995) *Food Additives: Electronic Handbook.* Gower Publishing Company, Brookfield, VT, 05036, USA.

Augustin, J., and Scarbrough, F. E. (1990) Nutritional Additives. In: Branen, A. L., Davidson, P. M., and S. Salminen (Eds.), *Food Additives*, pp. 33–81.

Budavari, S. (1996) *The Merck Index*, 12th edition.

Health and Welfare Canada (1996) Part D: Vitamins, Minerals and Amino Acids. In: *Food and Drug Regulations*. Department of Health, Ottawa, ON, Canada.

ANY OTHER RELEVANT INFORMATION:

Niacin may cause vasodilation; niacinamide does not.

Niacinamide may cake.

Potency 100%

NAME:	Pyridoxine hydrochloride	
CATEGORY:	Nutritive additives	
FOOD USE:	Cereals and cereal products/ Soft drinks/ Dairy products/ Sugars, sugar preserves and confectionery/ Edible oils and fats	
SYNONYMS:	Vitamin B_6 hydrochloride/ Pyridoxol hydrochloride/ 5-Hydroxy-6-methyl-3,4-pyridinedimethanol hydrochloride/ Pyridoxinium chloride/ Adermine hydrochloride/ 3-Hydroxy-4,5-dihydroxymethyl-2-methylpyridine HCl	
FORMULA:	C8H12ClNO3	
MOLECULAR MASS:	205.64	
ALTERNATIVE FORMS:	Pyridoxine	
PROPERTIES AND APPEARANCE:	White or almost white crystalline powder	
MELTING RANGE IN °C:	204–206	
SOLUBILITY % AT VARIOUS TEMPERATURE/pH COMBINATIONS:		
in water:	**@ 20°C**	Soluble
in ethanol solution:	**@ 100%**	Soluble
in propylene glycol:	**@ 100°C**	Soluble
FUNCTION IN FOODS:	Fortification of breakfast cereals, sugar and cocoa confectionery, infant formulae, fruit drinks, flour, cereal products, oils and fats, milk drinks, meal replacements	
TECHNOLOGY OF USE IN FOODS:	One gram dissolves in about 4.5 ml water, or 90 ml alcohol. Also soluble in propylene glycol	

LEGISLATION:	**USA:** Regulations: FDA 21 CFR § 182.5676, 184.1676, GRAS	**CANADA:** Food and Drug Regulations: D.03.002	**AUSTRALIA/PACIFIC RIM:** Japan: approved

REFERENCES:

Ash, M., and Ash, I. (1995) *Food Additives: Electronic Handbook.* Gower Publishing Company, Brookfield, VT, 05036, USA.

Augustin, J., and Scarbrough, F. E. (1990) Nutritional Additives. In: Branen, A. L., Davidson, P. M., and S. Salminen (Eds.), *Food Additives*, pp. 33–81.

Budavari, S. (1996) *The Merck Index*, 12th edition.

Health and Welfare Canada (1996) Part D: Vitamins, Minerals and Amino Acids. In: *Food and Drug Regulations*. Department of Health, Ottawa, ON, Canada.

ANY OTHER RELEVANT INFORMATION: Potency 82.0%; Coated form 33.3%

NAME:	Riboflavin
CATEGORY:	Nutritive additives; Colours
FOOD USE:	Baked goods/ Edible oils and fats/ Sugars, sugar preserves and confectionery/ Dairy products/ Beverages/ Cereals and cereal products/ Soft drinks
SYNONYMS:	Vitamin B_2/ 7,8-Dimethyl-10-ribitylisoalloxazine/ Lactoflavine/ 7,8-Dimethyl-10-(D-ribo-2,3,4,5-tetrahydroxypentyl)isoalloxazine/ Vitamin G/ Flavaxin
FORMULA:	C17 H20 N4 O6
MOLECULAR MASS:	376.37
ALTERNATIVE FORMS:	Riboflavin-5′-phosphate sodium salt
PROPERTIES AND APPEARANCE:	Yellow to orange-yellow powder
MELTING RANGE IN °C:	282
SOLUBILITY % AT VARIOUS TEMPERATURE/pH COMBINATIONS:	
in water:	**@ 20°C** Slightly soluble in water
in ethanol solution:	**@ 5%** Slightly soluble in alcohol
FUNCTION IN FOODS:	Coloration of ice-creams, desserts, instant beverages, sauces, soups, confectionery products, pasta. Fortification of flour, breakfast cereals, sugar and cocoa confectionery, soups, infant formulae, fruit drinks, oils and fats, desserts, milk drinks, meal replacements. *Usage level*: ADI 0–0.5 mg/kg (EEC)
TECHNOLOGY OF USE IN FOODS:	Riboflavin-5′-phosphate is more water-soluble than pure riboflavin, and is particularly suitable for coloration or instant products. Protect solutions from light. *Usage level*: ADI 0–0.5 mg/kg (EEC)

LEGISLATION:

USA:
Regulations: FDA 21CFR § 73.450, 136.115, 137, 139, 182.5695, 184.1695, GRAS

UK and EUROPE:
Europe: listed
UK: approved

CANADA:
Food and Drug Regulations: D.03.002

AUSTRALIA/PACIFIC RIM:
Japan: approved

REFERENCES:

Ash, M., and Ash, I. (1995) *Food Additives: Electronic Handbook.* Gower Publishing Company, Brookfield, VT, 05036, USA.

Augustin, J., and Scarbrough, F. E. (1990) Nutritional Additives. In: Branen, A. L., Davidson, P. M., and S. Salminen (Eds.), *Food Additives*, pp. 33–81.

Budavari, S. (1996) *The Merck Index*, 12th edition.

Health and Welfare Canada (1996) Part D: Vitamins, Minerals and Amino Acids. In: *Food and Drug Regulations*. Department of Health, Ottawa, ON, Canada.

ANY OTHER RELEVANT INFORMATION:

Potency 100%

NAME:	Riboflavin-5′-phosphate sodium salt
CATEGORY:	Nutritive additives; Colours
FOOD USE:	Baked goods/ Edible oils and fats/ Sugars, sugar preserves and confectionery/ Dairy products/ Beverages/ Cereals and cereal products/ Soft drinks
SYNONYMS:	Vitamin B_2 phosphate sodium/ Riboflavin 5′-monophosphate sodium salt dihydrate/ Riboflavin 5′-phosphate ester monosodium salt
FORMULA:	C17 H20 N4 O9 P Na·2H2O
MOLECULAR MASS:	514.36
ALTERNATIVE FORMS:	Riboflavin
PROPERTIES AND APPEARANCE:	Yellow to orange powder
SOLUBILITY % AT VARIOUS TEMPERATURE/pH COMBINATIONS:	
in water:	**@ 20°C** Soluble
FUNCTION IN FOODS:	Fortification and coloration. Fortification of flour, breakfast cereals, sugar and cocoa confectionery, soups, infant formulae, fruit drinks, oils and fats, desserts, milk drinks, meal replacements. Coloration of ice-creams, desserts, instant beverages, sauces, soups, confectionery products, pasta
TECHNOLOGY OF USE IN FOODS:	Riboflavin-5′-phosphate is more water-soluble than pure riboflavin, and is particularly suitable for coloration or instant products. Protect solutions from light. Usage level: ADI 0–0.5 mg/kg (EEC)
LEGISLATION:	**USA:** Regulations: FDA 21 CFR § 182.5697, 184.1697, GRAS **UK and EUROPE:** Europe: listed UK: approved **CANADA:** Food and Drug Regulations: D.03.002 **AUSTRALIA/PACIFIC RIM:** Japan: approved

REFERENCES:	Ash, M., and Ash, I. (1995) *Food Additives: Electronic Handbook.* Gower Publishing Company, Brookfield, VT, 05036, USA. Augustin, J., and Scarbrough, F. E. (1990) Nutritional Additives. In: Branen, A. L., Davidson, P. M., and S. Salminen (Eds.), *Food Additives*, pp. 33–81. Budavari, S. (1996) *The Merck Index*, 12th edition. Health and Welfare Canada (1996) Part D: Vitamins, Minerals and Amino Acids. In: *Food and Drug Regulations*. Department of Health, Ottawa, ON, Canada.
ANY OTHER RELEVANT INFORMATION:	Potency 78.7%; Coated forms 25–33.3% potency. Decomposed by light when in solution

NAME:	Thiamin hydrochloride		
CATEGORY:	Nutritive additives		
FOOD USE:	Cereals and cereal products/ Dairy products/ Sugars, sugar preserves and confectionery/ Edible oils and fats		
SYNONYMS:	Vitamin B_1 hydrochloride/ Thiamin chloride/ Thiamine dichloride/ Aneurine hydrochloride		
FORMULA:	C12 H17 ClN4 O S·HCl		
MOLECULAR MASS:	337.30		
ALTERNATIVE FORMS:	Thiamin/ Thiamin mononitrate		
PROPERTIES AND APPEARANCE:	White or almost white powder		
MELTING RANGE IN °C:	248		
SOLUBILITY % AT VARIOUS TEMPERATURE/pH COMBINATIONS:			
in water:	**@ 20°C** Soluble		
in ethanol solution:	**@ 100%** Slightly soluble		
FUNCTION IN FOODS:	Fortification of flour, breakfast cereals, infant formulae, soups, sugar and cocoa confectionery, oils and fats, milk drinks, pasta, meal replacements		
TECHNOLOGY OF USE IN FOODS:	Soluble in water, slightly soluble in alcohol. Thiamin hydrochloride is more water-soluble than thiamin mononitrate Usage level: Limitation 0.005 lb/1000 gal (wine)		
LEGISLATION:	**USA:** Regulations: FDA 21CFR § 182.5875, 184.1875, GRAS; BATF 27CFR § 240.1051	**CANADA:** Food and Drug Regulations: D.03.002; B.11.150; B.13.010.1	**AUSTRALIA/PACIFIC RIM:** Japan: approved

REFERENCES:	Ash, M., and Ash, I. (1995) *Food Additives: Electronic Handbook.* Gower Publishing Company, Brookfield, VT, 05036, USA. Augustin, J., and Scarbrough, F. E. (1990) Nutritional Additives. In: Branen, A. L., Davidson, P. M., and S. Salminen (Eds.), *Food Additives*, pp. 33–81. Budavari, S. (1996) *The Merck Index*, 12th edition. Health and Welfare Canada (1996) Part D: Vitamins, Minerals and Amino Acids. In: *Food and Drug Regulations*. Department of Health, Ottawa, ON, Canada.
ANY OTHER RELEVANT INFORMATION:	Potency 89.3%

NAME:	Thiamin mononitrate		
CATEGORY:	Flavour enhancers and modifiers		
FOOD USE:	Soups and sauces		
SYNONYMS:	Vitamin B_1 nitrate/ Thiamine nitrate/ Aneurine mononitrate/ 3-[(4-amino-2-methyl-5-pyrimidinyl)methyl]-4-(2-hydroxy ethyl)-4-methylthiazolium nitrate (salt)		
FORMULA:	C12 H17 N4 O S·NO3		
MOLECULAR MASS:	327.36		
ALTERNATIVE FORMS:	Thiamin/ Thiamin hydrochloride		
PROPERTIES AND APPEARANCE:	White crystal or crystalline powder		
MELTING RANGE IN °C:	196–200		
WATER CONTENT MAXIMUM IN %:	2.7 g/100 ml water		
SOLUBILITY % AT VARIOUS TEMPERATURE/pH COMBINATIONS:			
in water:	**@ 20°C** Soluble		
in ethanol solution:	**@ 100%** Slightly soluble		
FUNCTION IN FOODS:	Used as reaction flavour component in flavouring for soups and sauces (meat flavour). Coated form used in dry products to mask taste		
TECHNOLOGY OF USE IN FOODS:	Thiamin hydrochloride is more water-soluble than thiamin mononitrate		
LEGISLATION:	**USA:** Regulations: FDA 21CFR § 182.5878, 184.1878, GRAS	**CANADA:** Food and Drug Regulations: D.03.002; B.11.150; B.13.010.1	**AUSTRALIA/PACIFIC RIM:** Japan: approved

REFERENCES:	Ash, M., and Ash, I. (1995) *Food Additives: Electronic Handbook.* Gower Publishing Company, Brookfield, VT, 05036, USA. Augustin, J., and Scarbrough, F. E. (1990) Nutritional Additives. In: Branen, A. L., Davidson, P. M., and S. Salminen (Eds.), *Food Additives*, pp. 33–81. Budavari, S. (1996) *The Merck Index*, 12th edition. Health and Welfare Canada (1996) Part D: Vitamins, Minerals and Amino Acids. In: *Food and Drug Regulations*. Department of Health, Ottawa, ON, Canada.
ANY OTHER RELEVANT INFORMATION:	Potency 92.0%; coated form, 33.3%

NAME:	Calcium-D-pantothenate		
CATEGORY:	Nutritive additives		
FOOD USE:	Cereals and cereal products/ Dairy products/ Soft drinks/ Sugars, sugar preserves and confectionery		
SYNONYMS:	Calcium *N*-(2,4-dihydroxy-3,3-dimethyl-l-oxobutyl)-β-alanine/ Vitamin B_5 (calcium salt)/ Calcium pantothenate		
FORMULA:	$C9H16NO5 \cdot \frac{1}{2}Ca$		
MOLECULAR MASS:	490.63		
ALTERNATIVE FORMS:	Pantothenic acid		
PROPERTIES AND APPEARANCE:	White powder		
MELTING RANGE IN °C:	170–172		
SOLUBILITY % AT VARIOUS TEMPERATURE/pH COMBINATIONS:			
in water:	**@ 20°C** Soluble		
in ethanol solution:	**@ 5%** Insoluble in alcohol	**@ 100%** Insoluble	
FUNCTION IN FOODS:	Fortification of infant formulae, breakfast cereals, fruit drinks, sugar and cocoa confectionery, milk drinks, meal replacements		
TECHNOLOGY OF USE IN FOODS:	Direct food additive. Only D isomer has vitamin activity; both D and DL isomers used in food		
LEGISLATION:	**USA:** Regulations: FDA 21CFR § 182.5212, 184.1212, GRAS	**CANADA:** Food and Drug Regulations: D.03.002	**AUSTRALIA/PACIFIC RIM:** Japan approved (1% max. as calcium)

REFERENCES:	Ash, M., and Ash, I. (1995) *Food Additives: Electronic Handbook.* Gower Publishing Company, Brookfield, VT, 05036, USA. Augustin, J., and Scarbrough, F. E. (1990) Nutritional Additives. In: Branen, A. L., Davidson, P. M., and S. Salminen (Eds.), *Food Additives*, pp. 33–81. Budavari, S. (1996) *The Merck Index*, 12th edition. Health and Welfare Canada (1996) Part D: Vitamins, Minerals and Amino Acids. In: *Food and Drug Regulations*. Department of Health, Ottawa, ON, Canada.
ANY OTHER RELEVANT INFORMATION:	Potency 92.0%

NAME:	**Natural tocopherols (mixture of d-α-; d-β; d-γ; d-δ-tocopherol forms)**
CATEGORY:	Antioxidants
FOOD USE:	Edible oils and fats/ Fish and seafoods and products/ Meat, poultry and egg products/ Baked goods/ Vinegar, pickles and sauces/ Fruit, vegetables and nut products
SYNONYMS:	Vitamin E
ALTERNATIVE FORMS:	DL-α-Tocopherol/ DL-α-Tocopheryl acetate
PROPERTIES AND APPEARANCE:	Oily blends: red to reddish-brown, slightly viscous liquid. Dry powders: off-white to pale yellow; not free-flowing
FUNCTION IN FOODS:	As antioxidant in animal fat, margarine, sausages, poultry products, shrimps (breaded), pasta, bakery products, snack foods, confectionery products, sauces, dehydrated vegetables
TECHNOLOGY OF USE IN FOODS:	Usage level: 0.02–0.1% fat and oil content
LEGISLATION:	**USA:** Regulations: GRAS **CANADA:** Food and Drug Regulations: D.03.002
REFERENCES:	Ash, M., and Ash, I. (1995) *Food Additives: Electronic Handbook.* Gower Publishing Company, Brookfield, VT, 05036, USA. Augustin, J., and Scarbrough, F. E. (1990) Nutritional Additives. In: Branen, A. L., Davidson, P. M., and S. Salminen (Eds.), *Food Additives*, pp. 33–81. Budavari, S. (1996) *The Merck Index*, 12th edition. Health and Welfare Canada (1996) Part D: Vitamins, Minerals and Amino Acids. In: *Food and Drug Regulations*. Department of Health, Ottawa, ON, Canada.

ANY OTHER RELEVANT INFORMATION:	In some cases, unsaturated vegetable oils do not benefit from adding a supplement of tocopherols because of the optimal level of tocopherols naturally present in oils after refining and processing (0.05–0.1%). Potency: 50–70% (oily blends); potency 30% (dry powders)

NAME:	Beta-carotene
CATEGORY:	Colours; Nutritive additives
FOOD USE:	Baked goods/ Dairy products/ Beverages/ Soft drinks/ Edible oils and fats/ Vinegar, pickles and sauces/ Sugars, sugar preserves and confectionery
SYNONYMS:	Provitamin A/ Food Orange 5/ Natural Yellow 26/ C.I. 40800/ C.I. 75130
FORMULA:	$C_{40}H_{56}$
MOLECULAR MASS:	536.88
PROPERTIES AND APPEARANCE:	Oily suspensions: brick-red viscous oil stabilised with tocopherol, ascorbyl palmitate. Emulsions: oil-in-water, stabilised with tocopherol or BHA/BHT. Dry powders: red-brown, fine granular powder. Fine orange powder
MELTING RANGE IN °C:	187.5
SOLUBILITY % AT VARIOUS TEMPERATURE/pH COMBINATIONS:	
in water:	**@ 20°C** Insoluble in water **@ 100°C** Insoluble in water
in ethanol solution:	**@ 100%** Slightly soluble
FUNCTION IN FOODS:	Oily suspensions: coloration and fortification of oils and fats, dressings, butter, ice-cream, confectionery, fruit drinks. Emulsions: coloration and fortification of fruit drinks, dairy products, pasta, bakery products, snacks, confectionery, dressings. Dry powders: coloration and fortification of dairy products, pasta, snacks, bakery products, confectionery, beverages, soups, sauces, dressings. Fine orange powder: coloration and fortification of instant products, confectionery, dairy products, pasta, snacks, bakery products

TECHNOLOGY OF USE IN FOODS:

Oily suspensions: soluble in warm fat or oil (50°C); colour range, yellow to orange.

Emulsions: water-dispersible; colour range, yellow to orange.

Dry powders: dispersible in hot water (60°C) to make stock solutions; colour range, orange.

Fine orange powder: cold water-dispersible; colour range, yellow.

Usage level: 20–50 ppm (as food colorant)

LEGISLATION:

USA:
Regulations: FDA 21CFR § 73.95, 73.1095, 73.2095, 166.110, 182.5245, 184.1245, GRAS

UK and EUROPE:
UK: approved
Europe: listed

CANADA:
Food and Drug Regulations: D.03.002

AUSTRALIA/PACIFIC RIM:
Japan: restricted

REFERENCES:

Ash, M., and Ash, I. (1995) *Food Additives: Electronic Handbook.* Gower Publishing Company, Brookfield, VT, 05036, USA.

Augustin, J., and Scarbrough, F. E. (1990) Nutritional Additives. In: Branen, A. L., Davidson, P. M., and S. Salminen (Eds.), *Food Additives*, pp. 33–81.

Budavari, S. (1996) *The Merck Index*, 12th edition.

Health and Welfare Canada (1996) Part D: Vitamins, Minerals and Amino Acids. In: *Food and Drug Regulations*. Department of Health, Ottawa, ON, Canada.

ANY OTHER RELEVANT INFORMATION:

Oily suspensions: 20–30% potency. Emulsions: 0.5–5% potency.

Dry powders: 2.4–10% potency.

Fine orange powder: 1% potency

NAME:	**Phytonadione**
CATEGORY:	Nutritive additives
FOOD USE:	Dairy products/ Edible oils and fats
SYNONYMS:	Vitamin K_i/ Phylloquinone/ Phytodione
FORMULA:	C31 H46 O2
MOLECULAR MASS:	450.68
PROPERTIES AND APPEARANCE:	Oily blends: clear yellow to amber, viscous oil stabilised with tocopherol. Dry powders: off-white to yellow, free-flowing powder
FUNCTION IN FOODS:	Fortification of infant formulae, liquid milk, oils and fats, dietetic products
TECHNOLOGY OF USE IN FOODS:	Soluble in fat and oil; homogenising in a small amount of liquid is necessary prior to adding to milk and infant formulae. Dry powders: water-dispersible
LEGISLATION:	**CANADA:** Food and Drug Regulations: D.03.002
REFERENCES:	Ash, M., and Ash, I. (1995) *Food Additives: Electronic Handbook.* Gower Publishing Company, Brookfield, VT, 05036, USA. Augustin, J., and Scarbrough, F. E. (1990) Nutritional Additives. In: Branen, A. L., Davidson, P. M., and Salminen, S. (Eds.), *Food Additives*, pp. 31–81. Budavari, S. (1996) *The Merck Index*, 12th edition. Health and Welfare Canada (1996) Part D: Vitamins, Minerals and Amino Acids. In: *Food and Drug Regulations*. Department of Health, Ottawa, ON, Canada.
ANY OTHER RELEVANT INFORMATION:	Oily blends: 100% potency. Dry powders: 10–50 $mg\,g^{-1}$ potency

NAME:	**Pantothenic acid**
CATEGORY:	Not used in its pure form
SYNONYMS:	(*R*)-*N*-(2,4-Dihydroxy-3,3-dimethyl-1-oxobutyl)-β-alanine/ Vitamin B_5/ D(+)-*N*-(2,4-Dihydroxy-3,3-dimethylbutyryl)-β-alanine/ Chick antidermatitis factor
FORMULA:	C9H17NO5
MOLECULAR MASS:	219.24
ALTERNATIVE FORMS:	Calcium-D-pantothenate
FUNCTION IN FOODS:	N/A
TECHNOLOGY OF USE IN FOODS:	N/A
LEGISLATION:	**CANADA:** Food and Drug Regulations: D.03.002; B.13.010.1(1)
REFERENCES:	Budavari, S. (1996) *The Merck Index*, 12th edition. Health and Welfare Canada (1996) Part D: Vitamins, Minerals and Amino Acids. In: *Food and Drug Regulations*. Department of Health, Ottawa, ON, Canada.
ANY OTHER RELEVANT INFORMATION:	Potency 100%

NAME:	**Niacinamide**
CATEGORY:	Nutritive additives
FOOD USE:	Cereals and cereal products/ Dairy products
SYNONYMS:	Niacin/ Nicotinamide/ 3-Pyridinecarboxamide/ Nicotinic acid amide
FORMULA:	C5 H4 NCONH2
MOLECULAR MASS:	122.13
ALTERNATIVE FORMS:	Nicotinic acid
PROPERTIES AND APPEARANCE:	White crystalline powder
MELTING RANGE IN °C:	129
DENSITY AT 20°C (AND OTHER TEMPERATURES) IN g/l:	1.40
SOLUBILITY % AT VARIOUS TEMPERATURE/pH COMBINATIONS:	
in water:	**@ 20°C** Soluble
in ethanol solution:	**@ 100%** Soluble
FUNCTION IN FOODS:	Fortification of flour, breakfast cereals, infant formulas, fruit drinks, sugar and cocoa confectionery, pasta, meal replacements. Colour stabilisation in meat products (if permitted). Addition level 50–100 ppm
TECHNOLOGY OF USE IN FOODS:	One gram dissolves in about 1 ml water, in about 1.5 ml alcohol, and in 10 ml glycerol

LEGISLATION:	**USA:** FDA 21CFR § 182.5535, 184.1535, GRAS	**CANADA:** Food and Drug Regulations: D.03.002; B.13.010.1(1)
REFERENCES:	Ash, M., and Ash, I. (1995) *Food Additives: Electronic Handbook.* Gower Publishing Company, Brookfield, VT, 05036, USA. Augustin, J., and Scarbrough, F. E. (1990) Nutritional Additives. In: Branen, A. L., Davidson, P. M., and Salminen, S. (Eds.), *Food Additives*, pp. 33–81. Budavari, S. (1996) *The Merck Index*, 12th edition. Health and Welfare Canada (1996) Part D: Vitamins, Minerals and Amino Acids. In: *Food and Drug Regulations*. Department of Health, Ottawa, ON, Canada.	
ANY OTHER RELEVANT INFORMATION:	Niacin may cause vasodilation; niacinamide does not. Niacinamide may cake. Coated form: 33.3% potency	

NAME:	Thiamin
CATEGORY:	Nutritive additives
SYNONYMS:	Vitamin B_1/ 3-[(4-Amino-2-methyl-5-pyrimidinyl)-methyl]-5-(2-hydroxyethyl)-4-methylthiazolium chloride
FORMULA:	C12 H17 Cl N4 OS
MOLECULAR MASS:	300.81
ALTERNATIVE FORMS:	Thiamin hydrochloride/ Thiamin mononitrate
FUNCTION IN FOODS:	Enriched flours
TECHNOLOGY OF USE IN FOODS:	N/A
LEGISLATION:	**CANADA:** Food and Drug Regulations: D.03.002; B.11.150; B.13.010.1
REFERENCES:	Ash, M., and Ash, I. (1995) *Food Additives: Electronic Handbook.* Gower Publishing Company, Brookfield, VT, 05036, USA. Augustin, J., and Scarbrough, F. E. (1990) Nutritional Additives. In: Branen, A. L., Davidson, P. M., and Salminen, S. (Eds.), *Food Additives*, pp. 33–81. Budavari, S. (1996) *The Merck Index*, 12th edition. Health and Welfare Canada (1996) Part D: Vitamins, Minerals and Amino Acids. In: *Food and Drug Regulations*. Department of Health, Ottawa, ON, Canada.
ANY OTHER RELEVANT INFORMATION:	Potency 100%

NAME:	L-Ascorbic acid
CATEGORY:	Nutritive additive
FOOD USE:	Bakery products/ Beer/ Canned fruits and vegetables/ Cereal products/ Cocoa confectionery and drink powders/ Dairy products/ Dietary supplements/ Flour and bread/ Fruit juices and drinks/ Infant formulae/ Meal replacements/ Milk modifiers/ Evaporated milk/ Potato products/ Sausages and other comminuted meats/ Skim milk/ Soft drinks/ Sugar confectionery/ Vegetable juices and wine
SYNONYMS:	Ascorbate/ L-ascorbic acid/ Cevitamic acid/ Vitamin C/ CAS 50-81-7/ EINECS 200-066-2/ FEMA 2109/ E300
FORMULA:	C6 H8 O6
MOLECULAR MASS:	176.13
ALTERNATIVE FORMS:	Ascorbyl palmitate/ Calcium ascorbate/ Dehydroascorbic acid/ Nicotinamide-ascorbic complex/ Sodium ascorbate
PROPERTIES:	Pleasant, sharp acidic taste. Various granulations available; fat-coated and ethyl cellulose-coated forms available. Stable to air when dry; not heat-stable at neutral pH
APPEARANCE:	White to slightly yellow crystalline powder
MELTING RANGE IN °C:	190–192
IONISATION CONSTANT AT 25°C:	$pK_1 = 4.17$ $pK_2 = 11.57$
DENSITY AT 20°C (AND OTHER TEMPERATURES) IN g/l:	1.65

SOLUBILITY % AT VARIOUS TEMPERATURE/pH COMBINATIONS:					
in water:		1 g in ~3 ml	**@ 45°C** 40%	**@ 100°C** 80%	
in vegetable oil:	Insoluble				
in ethanol solution:	**@ 95%**	1 g in ~30 ml	**@ 100%** 1 g in ~50 ml		
in propylene glycol:	**@ 20°C**	1 g in ~20 ml			
FUNCTION IN FOODS:	Dry form used for fortification of infant formulae, meal replacements, fruit juices and drinks, cereal products, dairy products, sugar and cocoa confectionery. Antioxidant in fruit juices, soft drinks, beer, wine, canned fruit and vegetables, potato products and dairy products. Flour and bread improver. Curing agent and nitrosamine inhibitor in sausage and other comminuted meats. Fat-coated and ethyl cellulose-coated forms used for fortification of milk modifiers, cocoa drink powders, bakery products; antioxidant in processed potatoes and sausages. Also used as antimicrobial agent, antioxidant, colour fixative, flavouring, oxidant, preservative, raising agent, reducing agent				
TECHNOLOGY OF USE IN FOODS:	Optimum pH is between 5 and 7; usage level as flour and bread improver is 50–200 ppm; usage level as a curing agent and nitrosamine inhibitor in sausage and other comminuted meats is 300–500 ppm				
ANTAGONISTS:	Oxygen; light; minerals; heat-labile in neutral environments, but less so in highly acidic environments				
FOOD SAFETY ISSUES:	Toxicology: LD_{50} (IV, mouse) = 518 mg/kg. Some adverse effects can occur with extremely high repeated doses in the 500 mg to 10 g range				

LEGISLATION:

USA:
GRAS for specified applications. For adults and children 4 years or older, Recommended Daily Intake is 60 mg of Vitamin C

UK and EUROPE:
Approved and listed, respectively. RDA = 60 mg of Vitamin C

CANADA:
For specific regulations, refer to Part D: Vitamins, Minerals and Amino Acids of the Food and Drug Regulations, Health and Welfare Canada.
For adults and children 2 years or older, Recommended Daily Intake is 60 mg of Vitamin C
For children less than 2 years old, Recommended Daily Intake is 20 mg of Vitamin C

AUSTRALIA/PACIFIC RIM:
Japan: approved

REFERENCES:

Ash, M., and Ash, I. (1995) *Food Additives: Electronic Handbook.* Gower Publishing Company, Brookfield, VT, 05036, USA.

Augustin, J., and Scarbrough, F. E. (1990) Nutritional Additives. In: Branen, A. L., Davidson, P. M., and S. Salminen (Eds.), *Food Additives*, pp. 33–81.

Budavari, S. (1996) *The Merck Index*, 12th edition.

Health and Welfare Canada (1996) Part D: Vitamins, Minerals and Amino Acids. In: *Food and Drug Regulations*. Department of Health, Ottawa, ON, Canada.

ANY OTHER RELEVANT INFORMATION:

Potency of powder form is 100%; potency of fat-coated and ethyl cellulose-coated is 96% and 97.5%, respectively. One unit (U.S.P. or international) is the Vitamin C activity of 0.05 mg of the U.S.P. ascorbic acid reference standard

NAME:	Ascorbyl palmitate
CATEGORY:	Nutritive additive
FOOD USE:	Beverages/ Baked goods/ Breads/ Dietary supplements/ Dietetic foods/ Evaporated milk/ Extruded cereals/ Fats and oils/ Fat-based products/ Formulated liquid diets/ Infant formulae/ Meal replacements/ Processed potatoes/ Skim milk/ Uncured frozen sausages
SYNONYMS:	L-Ascorbic acid 6-hexadecanoate/ Palmitoyl L-ascorbic acid/ Vitamin C/ CAS 137-66-6/ EINECS 205-305-4/ E304
FORMULA:	$C_{22}H_{38}O_7$
MOLECULAR MASS:	414.52
ALTERNATIVE FORMS:	Calcium ascorbate/ Sodium ascorbate
PROPERTIES:	Citrus odour. Not heat-stable
APPEARANCE:	White to yellowish crystalline powder
MELTING RANGE IN °C:	113–114
SOLUBILITY % AT VARIOUS TEMPERATURE/pH COMBINATIONS:	
in vegetable oil:	30 mg per 100 ml
FUNCTION IN FOODS:	Enrichment, fortification and restoration. Also used as antioxidant for fats and oils, colour preservative, emulsifier, sequestrant, stabiliser
TECHNOLOGY OF USE IN FOODS:	Optimum pH is between 5 and 7
SYNERGISTS:	α-Tocopherol
ANTAGONISTS:	Oxidation; light; minerals; heat-labile in neutral environments, but less so in highly acidic environments

FOOD SAFETY ISSUES:

Some adverse effects can occur with extremely high repeated doses in the 500 mg to 10 g ranges

LEGISLATION:

USA:
GRAS for specified applications; 0.02% max. in margarine. For adults and children 4 years or older, Recommended Daily Intake is 60 mg of Vitamin C

UK and EUROPE:
Listed and approved, respectively. Recommended Daily Allowance is 60 mg Vitamin C

CANADA:
For specific regulations, refer to Part D: Vitamins, Minerals and Amino Acids of the Food and Drug Regulations, Health and Welfare Canada.
For adults and children 2 years or older, Recommended Daily Intake is 60 mg of Vitamin C
For children less than 2 years old, Recommended Daily Intake is 20 mg of Vitamin C

AUSTRALIA/PACIFIC RIM:
Japan: approved

REFERENCES:

Ash, M., and Ash, I. (1995) *Food Additives: Electronic Handbook.* Gower Publishing Company, Brookfield, VT, 05036, USA.

Augustin, J., and Scarbrough, F. E. (1990) Nutritional Additives. In: Branen, A. L., Davidson, P. M., and S. Salminen (Eds.), *Food Additives*, pp. 33–81.

Health and Welfare Canada (1996) Part D: Vitamins, Minerals and Amino Acids. In: *Food and Drug Regulations*. Department of Health, Ottawa, ON, Canada.

ANY OTHER RELEVANT INFORMATION:

Potency is 43%

NAME:	**Biotin**
CATEGORY:	Nutritive additive
FOOD USE:	Dietary supplements/ Dietetic foods/ Infant formulae/ Meal replacements/ Substitute foods
SYNONYMS:	D-Biotin/ Bios II B/ Coenzyme R/ Egg white injury factor/ Vitamin H/ CAS 58-85-5/ EINECS 200-399-3
FORMULA:	$C_{10}H_{16}N_2O_3S$
MOLECULAR MASS:	244.31
ALTERNATIVE FORMS:	There are eight stereoisomers, with naturally occurring D-biotin exhibiting the highest biological activity, while DL-biotin only contains half that activity and L-biotin is biologically inactive
PROPERTIES:	Heat stable
APPEARANCE:	D-biotin is commercially available in white crystalline and powder forms
MELTING RANGE IN °C:	232–233
SOLUBILITY % AT VARIOUS TEMPERATURE/pH COMBINATIONS:	
in water:	**@ 25°C** ~22 mg/100 ml
in ethanol solution:	**@ 95%** ~80 mg/100 ml
FUNCTION IN FOODS:	Enrichment, fortification, or restoration
TECHNOLOGY OF USE IN FOODS:	Serial dilution required; dilution can be made using dicalcium phosphate; pH is not critical for use; biotin becomes inactive when combined with avidin in raw egg-white
ANTAGONISTS:	Subject to oxidative deterioration when exposed to oxidising agents such as potassium permanganate or hydrogen peroxide; UV light
FOOD SAFETY ISSUES:	No toxic symptoms as a result of heavy biotin dosage have been reported in humans

LEGISLATION:

USA:
GRAS for specified applications. For adults and children 4 years or older, Recommended Daily Allowance is 300 µg of biotin

UK and EUROPE:
Recommended Daily Allowance is 0.15 mg biotin

CANADA:
For specific regulations, refer to Part D: Vitamins, Minerals and Amino Acids of the Food and Drug Regulations, Health and Welfare Canada.
Recommended Daily Intake not available

REFERENCES:

Ash, M., and Ash, I. (1995) *Food Additives: Electronic Handbook.* Gower Publishing Company, Brookfield, VT, 05036, USA.

Augustin, J., and Scarbrough, F. E. (1990) Nutritional Additives. In: Branen, A. L., Davidson, P. M., and S. Salminen (Eds.), *Food Additives*, pp. 33–81.

Budavari, S. (1996) *The Merck Index*, 12th edition.

Health and Welfare Canada (1996) Part D: Vitamins, Minerals and Amino Acids. In: *Food and Drug Regulations*. Department of Health, Ottawa, ON, Canada.

ANY OTHER RELEVANT INFORMATION:

Potency of pure compound is 100%. The pure compound is stable to air and temperature. Moderately acid and neutral solutions are stable for several months; alkaline solutions are less stable, but appear reasonably stable up to a pH of about 9. Aqueous solutions are very susceptible to mould growth.
Acidic solutions can be heat-sterilised

NAME:	Calcium ascorbate
CATEGORY:	Nutritive additive
FOOD USE:	Breakfast cereals/ Dietary supplements/ Dietetic foods/ Formulated liquid diets/ Infant formulae/ Low-sodium dietetic products/ Meal replacements
SYNONYMS:	Ascorbic acid calcium salt/ Vitamin C
FORMULA:	$C_{12}H_{14}CaO_{12}$
MOLECULAR MASS:	390.31
ALTERNATIVE FORMS:	Ascorbic acid/ Ascorbyl palmitate/ Sodium ascorbate
PROPERTIES:	Odourless, bitter taste. Not heat-stable at neutral pH
APPEARANCE:	White to slightly yellow powder
SOLUBILITY % AT VARIOUS TEMPERATURE/pH COMBINATIONS:	
in water:	Freely soluble
FUNCTION IN FOODS:	Enrichment, fortification, or restoration
TECHNOLOGY OF USE IN FOODS:	Optimum pH is between 5 and 7. Other uses include antioxidant, meat colour preservative, preservative
ANTAGONISTS:	Oxidation, light, minerals, heat-labile in neutral environments, but less so in highly acidic environments
FOOD SAFETY ISSUES:	Some adverse effects can occur with extremely high repeated doses in the 500 mg to 10 g range

LEGISLATION:

USA:
GRAS for specified applications. For adults and children 4 years or older, Recommended Daily Intake is 60 mg Vitamin C

UK and EUROPE:
Recommended Daily Allowance is 60 mg Vitamin C

CANADA:
For specific regulations, refer to Part D: Vitamins, Minerals and Amino Acids of the Food and Drug Regulations, Health and Welfare Canada. For adults and children 2 years or older, Recommended Daily Intake is 60 mg Vitamin C
For children less than 2 years old, Recommended Daily Intake is 20 mg Vitamin C

REFERENCES:

Ash, M., and Ash, I. (1995) *Food Additives: Electronic Handbook.* Gower Publishing Company, Brookfield, VT, 05036, USA.

Augustin, J., and Scarbrough, F. E. (1990) Nutritional Additives. In: Branen, A. L., Davidson, P. M., and S. Salminen (Eds.), *Food Additives*, pp. 33–81.

Budavari, S. (1996) *The Merck Index*, 12th edition.

Health and Welfare Canada (1996) Part D: Vitamins, Minerals and Amino Acids. In: *Food and Drug Regulations*. Department of Health, Ottawa, ON, Canada.

ANY OTHER RELEVANT INFORMATION:

Potency is 83%. Aqueous solutions are neutral and oxidise quickly

NAME:	**Calcium glycerophosphate**
CATEGORY:	Nutritive additive
FOOD USE:	Breads/ Cereals/ Cornmeal/ Dietary supplements/ Dietetic foods/ Farina/ Flour/ Formulated liquid diets/ Infant formulae/ Meal replacements/ Pasta/ Rice
SYNONYMS:	β-Glycerophosphate (calcium salt)/ Calcium glycerinophosphate/ Calcium phosphoglycerate/ CAS 126-95-4/ EINECS 248-328-5
FORMULA:	C3 H7 Ca O6 P
MOLECULAR MASS:	210.14
PROPERTIES:	Odourless; almost tasteless; slightly hygroscopic; slightly alkaline. Bioavailability unknown
APPEARANCE:	Fine white powder
MELTING RANGE IN °C:	Decomposes above 170
SOLUBILITY % AT VARIOUS TEMPERATURE/pH COMBINATIONS:	
in water:	Moderate **@ 20°C** Soluble in about 50 parts water **@ 100°C** Almost insoluble
in ethanol solution:	Insoluble
FUNCTION IN FOODS:	Enrichment, fortification, or restoration. Also used as a stabiliser
ALTERNATIVES:	Calcium carbonate, calcium chloride, calcium citrate, calcium hydroxide, calcium lactate pentahydrate, calcium oxide, calcium phosphate monobasic, calcium phosphate dibasic, calcium phosphate tribasic, calcium pyrophosphate, calcium sulphate, ground limestone

LEGISLATION:

USA:
GRAS for specified applications. For adults and children 4 years or older, Recommended Daily Intake is 1000 mg calcium

UK and EUROPE:
Recommended Daily Allowance is 800 mg calcium

CANADA:
For specific regulations, refer to Part D: Vitamins, Minerals and Amino Acids of the Food and Drug Regulations, Health and Welfare Canada. For adults and children 2 years or older, Recommended Daily Intake is 1100 mg calcium
For children less than 2 years old, Recommended Daily Intake is 500 mg of calcium

AUSTRALIA/PACIFIC RIM:
Japan: restricted (1% max. as calcium)

REFERENCES:

Ash, M., and Ash, I. (1995) *Food Additives: Electronic Handbook.* Gower Publishing Company, Brookfield, VT, 05036, USA.

Budavari, S. (1996) *The Merck Index*, 12th edition.

Health and Welfare Canada (1996) Part D: Vitamins, Minerals and Amino Acids. In: *Food and Drug Regulations*. Department of Health, Ottawa, ON, Canada.

ANY OTHER RELEVANT INFORMATION:

Contains 19.1% calcium

NAME:	Calcium lactate pentahydrate
CATEGORY:	Nutritive additive
FOOD USE:	Beverages/ Breads/ Cereals/ Cornmeal/ Dietary supplements/ Dietetic foods/ Farina/ Flour/ Formulated liquid diets/ Infant formulae/ Juices/ Meal replacements/ Pasta/ Rice
SYNONYMS:	CAS 814-80-2/ EINECS 212-406-7
FORMULA:	+ $C_6H_{10}CaO_6$
MOLECULAR MASS:	218 (anhydrous)
PROPERTIES:	Odourless; tasteless; neutral; hygroscopic. Bioavailability of pure compound is good, but may be altered in presence of other food components
APPEARANCE:	White effervescent granules or powder
PURITY %:	98
SOLUBILITY % AT VARIOUS TEMPERATURE/pH COMBINATIONS:	
in water:	Very soluble in hot water
FUNCTION IN FOODS:	Enrichment, fortification, or restoration. Also used as a preservative
ALTERNATIVES:	Calcium carbonate, calcium chloride, calcium citrate, calcium glycerophosphate, calcium hydroxide, calcium oxide, calcium phosphate monobasic, calcium phosphate dibasic, calcium phosphate tribasic, calcium pyrophosphate, calcium sulphate, ground limestone

LEGISLATION:

USA:
GRAS. For adults and children 4 years or older, Recommended Daily Intake is 1000 mg calcium

UK and EUROPE:
Recommended Daily Allowance is 800 mg calcium

CANADA:
For specific regulations, refer to Part D: Vitamins, Minerals and Amino Acids of the Food and Drug Regulations, Health and Welfare Canada. For adults and children 2 years or older, Recommended Daily Intake is 1100 mg calcium
For children less than 2 years, Recommended Daily Intake is 500 mg calcium

REFERENCES:

Ash, M., and Ash, I. (1995) *Food Additives: Electronic Handbook.* Gower Publishing Company, Brookfield, VT, 05036, USA.

Augustin, J., and Scarbrough, F. E. (1990) Nutritional Additives. In: Branen, A. L., Davidson, P. M., and S. Salminen (Eds.), *Food Additives*, pp. 33–81.

Health and Welfare Canada (1996) Part D: Vitamins, Minerals and Amino Acids. In: *Food and Drug Regulations*. Department of Health, Ottawa, ON, Canada.

ANY OTHER RELEVANT INFORMATION:

Anhydrous compound is 18.4% calcium; hydrous compound is 13.0% calcium

NAME:	Calcium carbonate
CATEGORY:	Nutritive additive
FOOD USE:	Breads/ Cornmeal/ Dietary supplements/ Dietetic foods/ Farina/ Flour/ Formulated liquid diets/ Infant formulae/ Meal replacements/ Pasta/ Rice
SYNONYMS:	Carbonic acid calcium salt (1 : 1)/ Limestone/ Precipitated calcium carbonate/ Precipitated chalk/ CAS 471-34-1/ EINECS 207-439-9/ E170
FORMULA:	CaCO3
MOLECULAR MASS:	100.09
PROPERTIES:	Odourless, tasteless powder or crystals. Bioavailability of pure compound is fair, but may be altered in presence of other food components
APPEARANCE:	White powder
MELTING RANGE IN °C:	825 (decomposes)
DENSITY AT 20°C (AND OTHER TEMPERATURES) IN g/l:	2.7–2.95
SOLUBILITY % AT VARIOUS TEMPERATURE/pH COMBINATIONS:	
in water:	Not soluble
FUNCTION IN FOODS:	Enrichment, fortification, or restoration. Also used for neutralisation
ALTERNATIVES:	Calcium chloride, calcium citrate, calcium glycerophosphate, calcium hydroxide, calcium lactate pentahydrate, calcium oxide, calcium phosphate monobasic, calcium phosphate dibasic, calcium phosphate tribasic, calcium pyrophosphate, calcium sulphate, ground limestone
TECHNOLOGY OF USE IN FOODS:	Incompatible with acids

ANTAGONISTS: Acids

FOOD SAFETY ISSUES: Toxicology LD_{50} (oral, rat) = 6450 mg/kg

LEGISLATION: WHO limitation: 40 g/kg (cheese); 200 mg/kg (jams, jellies)

USA:

GRAS for specified applications. For adults and children 4 years or older, Recommended Daily Intake is 1000 mg calcium

UK and EUROPE:

Approved and listed, respectively. Recommended Daily Allowance is 800 mg calcium

CANADA:

For specific regulations, refer to Part D: Vitamins, Minerals and Amino Acids of the Food and Drug Regulations, Health and Welfare Canada. For adults and children 2 years or older, Recommended Daily Intake is 1100 mg calcium

For children less than 2 years old, Recommended Daily Intake is 500 mg calcium

AUSTRALIA/PACIFIC RIM:

Japan: approved (1–2%)

REFERENCES:

Ash, M., and Ash, I. (1995) *Food Additives: Electronic Handbook.* Gower Publishing Company, Brookfield, VT, 05036, USA.

Augustin, J., and Scarbrough, F. E. (1990) Nutritional Additives. In: Branen, A. L., Davidson, P. M., and S. Salminen (Eds.), *Food Additives*, pp. 33–81.

Budavari, S. (1996) *The Merck Index*, 12th edition.

Health and Welfare Canada (1996) Part D: Vitamins, Minerals and Amino Acids. In: *Food and Drug Regulations*. Department of Health, Ottawa, ON, Canada.

ANY OTHER RELEVANT INFORMATION: 40.0% calcium

NAME:	Calcium phosphate dibasic
CATEGORY:	Nutritive additive
FOOD USE:	Breads/ Cereals/ Cornmeal/ Dietary supplements/ Dietetic foods/ Farina/ Flour/ Formulated liquid diets/ Infant formulae/ Meal replacements/ Pasta/ Rice
SYNONYMS:	Calcium hydrogen orthophosphate/ Calcium hydrogen phosphate anhydrous/ Calcium monohydrogen phosphate/ DCP-0/ Dicalcium orthophophate/ Dicalcium phosphate/ Phosphoric acid calcium salt (1:1)/ Secondary calcium phosphate/ CAS 7757-93-9/ EINECS 231-826-1/ E341b
FORMULA:	CaHPO4
MOLECULAR MASS:	136.06
ALTERNATIVE FORMS:	Calcium phosphate tribasic/ Calcium phosphate monobasic
PROPERTIES:	Odourless; tasteless; alkaline
APPEARANCE:	White crystalline powder
DENSITY AT 20°C (AND OTHER TEMPERATURES) IN g/l:	2.306
SOLUBILITY % AT VARIOUS TEMPERATURE/pH COMBINATIONS:	
in water:	Not soluble
FUNCTION IN FOODS:	Enrichment, fortification or restoration. Also used as antioxidant synergist, dough conditioner, firming agent, stabiliser, yeast food for baked goods
ALTERNATIVES:	Calcium carbonate, calcium chloride, calcium citrate, calcium glycerophosphate, calcium hydroxide, calcium lactate pentahydrate, calcium oxide, calcium phosphate monobasic, calcium phosphate tribasic, calcium pyrophosphate, calcium sulphate, ground limestone

TECHNOLOGY OF USE IN FOODS: Affects Vitamin B_1 stability

LEGISLATION:

USA:
GRAS for specified applications. For adults and children 4 years or older, Recommended Daily Intake is 1000 mg calcium

UK and EUROPE:
Approved and listed respectively. Recommended Daily Allowance is 800 mg calcium

CANADA:
For specific regulations, refer to Part D: Vitamins, Minerals and Amino Acids of the Food and Drug Regulations, Health and Welfare Canada. For adults and children 2 years or older, Recommended Daily Intake is 1100 mg calcium
For children less than 2 years old, Recommended Daily Intake is 500 mg calcium

AUSTRALIA/PACIFIC RIM:
Japan: approved (1% max. as calcium)

REFERENCES:

Ash, M., and Ash, I. (1995) *Food Additives: Electronic Handbook.* Gower Publishing Company, Brookfield, VT, 05036, USA.

Augustin, J., and Scarbrough, F. E. (1990) Nutritional Additives. In: Branen, A. L., Davidson, P. M., and S. Salminen (Eds.), *Food Additives*, pp. 33–81.

Budavari, S. (1996) *The Merck Index*, 12th edition.

Health and Welfare Canada (1996) Part D: Vitamins, Minerals and Amino Acids. In: *Food and Drug Regulations*. Department of Health, Ottawa, ON, Canada.

ANY OTHER RELEVANT INFORMATION: Anhydrous compound is 30% calcium; also a source of phosphorus

NAME:	Calcium phosphate tribasic
CATEGORY:	Nutritive additive
FOOD USE:	Breads/ Cereals/ Cornmeal/ Dietary supplements/ Dietetic foods/ Farina/ Flour/ Formulated liquid diets/ Infant formulae/ Meal replacements/ Pasta/ Rice
SYNONYMS:	Calcium hydroxide phosphate/ Calcium orthophosphate/ Calcium phosphate tertiary/ Precipitated calcium phosphate/ TCP/ Tertiary calcium phosphate/ Tricalcium orthophosphate/ Tribasic calcium phosphate/ Tricalcium phosphate/ CAS 7758-87-4/ EINECS 231-840-8/ E341c
FORMULA:	Ca3 O8 P2
MOLECULAR MASS:	310.18
ALTERNATIVE FORMS:	Calcium phosphate dibasic/ Calcium phosphate monobasic
PROPERTIES:	Odourless; tasteless; bioavailability of pure compound is fair and may be altered in presence of other food components; alkaline
APPEARANCE:	White crystalline powder
MELTING POINT IN °C:	1670
DENSITY AT 20°C (AND OTHER TEMPERATURES) IN g/l:	3.14
PURITY %:	96
SOLUBILITY % AT VARIOUS TEMPERATURE/pH COMBINATIONS:	
in water:	Practically not soluble
FUNCTION IN FOODS:	Enrichment, fortification or restoration. Also used as anticaking agent, buffering agent, chewing gum base, clarifying agent, emulsifier (Japan), fat rendering aid, leavening agent, stabiliser, yeast nutrient

ALTERNATIVES: Calcium carbonate, calcium chloride, calcium citrate, calcium glycerophosphate, calcium hydroxide, calcium lactate pentahydrate, calcium oxide, calcium phosphate monobasic, calcium phosphate dibasic, calcium pyrophosphate, calcium sulphate, ground limestone

TECHNOLOGY OF USE IN FOODS: Affects Vitamin B_1 stability

LEGISLATION:

AUSTRALIA/PACIFIC RIM:
Japan: approved (1% max. as calcium), restricted

OTHER COUNTRIES:
FAO/WHO: ADI 0–70 mg/kg, total phosphorus intake

CANADA:
For specific regulations, refer to Part D: Vitamins, Minerals and Amino Acids of the Food and Drug Regulations, Health and Welfare Canada. For adults and children 2 years or older, Recommended Daily Intake is 1100 mg calcium For children less than 2 years old, Recommended Daily Intake is 500 mg calcium

USA:
GRAS for specified applications. For adults and children 4 years or older, Recommended Daily Intake is 1000 mg calcium

UK and EUROPE:
Approved and listed, respectively. Recommended Daily Allowance is 800 mg calcium

REFERENCES:

Ash, M., and Ash, I. (1995) *Food Additives: Electronic Handbook.* Gower Publishing Company, Brookfield, VT, 05036, USA.

Augustin, J., and Scarbrough, F. E. (1990) Nutritional Additives. In: Branen, A. L., Davidson, P. M., and S. Salminen (Eds.), *Food Additives*, pp. 33–81.

Budavari, S. (1996) *The Merck Index*, 12th edition.

Health and Welfare Canada (1996) Part D: Vitamins, Minerals and Amino Acids. In: *Food and Drug Regulations*. Department of Health, Ottawa, ON, Canada.

ANY OTHER RELEVANT INFORMATION: Anhydrous compound is 38.8% calcium

NAME:	Cholecalciferol
CATEGORY:	Nutritive additive
FOOD USE:	Breakfast cereals/ Cereal products/ Cornmeal/ Dietary supplements/ Dietetic foods/ Egg products/ Farina/ Fats and oils/ Formulated liquid diets/ Infant formulae/ Liquid milk/ Margarine/ Meal replacements/ Milk powder/ Pasta/ Rice
SYNONYMS:	Calciol/ 5,7-Cholestadien-3-β-ol/ Colecalciferol/ Activated 7-dehydrocholesterol/ Oleovitamin D_3/ Vitamin D_3/ CAS 67-97-0/ EINECS 200-673-2
FORMULA:	C27 H44 O
MOLECULAR MASS:	384.65
ALTERNATIVE FORMS:	Calciferol/ Provitamin 7-dehydrocholesterol
PROPERTIES:	Unstable in light and air; oily blends stabilised with tocopherol
APPEARANCE:	Oily blends: clear, colourless to slightly yellow. Dry blends: off-white to yellowish fine granular powder
MELTING RANGE IN °C:	84–85
SOLUBILITY % AT VARIOUS TEMPERATURE/pH COMBINATIONS:	
in water:	Practically insoluble
in vegetable oil:	Slightly soluble
in ethanol solution:	Soluble
FUNCTION IN FOODS:	Enrichment, fortification or restoration
ALTERNATIVES:	Vitamin D_2
TECHNOLOGY OF USE IN FOODS:	Frequently diluted for use by preparing in either an oily blend or a dry powder. Oily blends are soluble in fats and oils; homogenising in a small amount of liquid is necessary prior to adding to milk and infant formulae. Dry powders are water-dispersible

ANTAGONISTS: Air, humidity, oxygen

FOOD SAFETY ISSUES: Toxicology: LD_{50} (oral, rat) = 42 mg/kg. Repeated daily dosages exceeding 400 IU for children and 1000 IU for adults should be avoided

LEGISLATION:

USA:
GRAS in specified application. For adults and children over 4 years, Recommended Daily Intake is 400 IU Vitamin D

UK and EUROPE:
Recommended Daily Allowance is 5 µg Vitamin D

CANADA:
For specific regulations, refer to Part D: Vitamins, Minerals and Amino Acids of the Food and Drug Regulations, Health and Welfare Canada. For adults and children over 2 years, Recommended Daily Intake is 5 µg Vitamin D
For children under 2 years old, Recommended Daily Intake is 10 µg Vitamin D

AUSTRALIA/PACIFIC RIM:
Japan: approved for use

REFERENCES:

Ash, M., and Ash, I. (1995) *Food Additives: Electronic Handbook.* Gower Publishing Company, Brookfield, VT, 05036, USA.

Augustin, J., and Scarbrough, F. E. (1990) Nutritional Additives. In: Branen, A. L., Davidson, P. M., and S. Salminen (Eds.), *Food Additives*, pp. 33–81.

Budavari, S. (1996) *The Merck Index*, 12th edition.

Health and Welfare Canada (1996) Part D: Vitamins, Minerals and Amino Acids. In: *Food and Drug Regulations*. Department of Health, Ottawa, ON, Canada.

ANY OTHER RELEVANT INFORMATION: 1 µg cholecalciferol = 40 IU Vitamin D

Deterioration of pure crystal Vitamin D_3 is negligible after one year of storage in amber evacuated ampoules at refrigerated temperatures

NAME:	**Cupric gluconate**
CATEGORY:	Nutritive additive
FOOD USE:	Dietary supplements/ Dietetic foods/ Formulated liquid diets/ Infant formulae/ Meal replacements
SYNONYMS:	Bis(D-gluconato) copper/ Copper (II) gluconate/ Cupric gluconate/ CAS 527-09-3/ EINECS 208-408-2
FORMULA:	C12 H22 Cu O14
MOLECULAR MASS:	453.85
ALTERNATIVE FORMS:	Cupric gluconate hydrate
PROPERTIES:	Odourless crystals; astringent taste. Good bioavailability, but may be altered in presence of other food components
APPEARANCE:	Light blue to blue-green crystalline powder
SOLUBILITY % AT VARIOUS TEMPERATURE/pH COMBINATIONS:	
in water:	Very soluble **@ 25°C** 30 g/100 ml
in ethanol solution:	Slightly soluble
FUNCTION IN FOODS:	Enrichment, fortification or restoration
ALTERNATIVES:	Cupric sulphate
TECHNOLOGY OF USE IN FOODS:	Affects Vitamin C stability

LEGISLATION:

USA:
GRAS for specified applications. For adults and children 4 or older, Recommended Daily Intake is 2 mg copper

CANADA:
For specific regulations, refer to Part D: Vitamins, Minerals and Amino Acids of the Food and Drug Regulations, Health and Welfare Canada. Recommended Daily Intake is not available

UK and EUROPE:
Recommended Daily Allowance is not available

AUSTRALIA/PACIFIC RIM:
Japan: approved (0.6 mg/l as copper in milk)

REFERENCES:

Ash, M., and Ash, I. (1995) *Food Additives: Electronic Handbook.* Gower Publishing Company, Brookfield, VT, 05036, USA.

Augustin, J., and Scarbrough, F. E. (1990) Nutritional Additives. In: Branen, A. L., Davidson, P. M., and S. Salminen (Eds.), *Food Additives*, pp. 33–81.

Budavari, S. (1996) *The Merck Index*, 12th edition.

Health and Welfare Canada (1996) Part D: Vitamins, Minerals and Amino Acids. In: *Food and Drug Regulations*. Department of Health, Ottawa, ON, Canada.

ANY OTHER RELEVANT INFORMATION:

Anhydrous form is 14% copper; hydrous form is 13.5% copper

NAME:	Cyanocobalamin
CATEGORY:	Nutritive additive
FOOD USE:	Dietary supplement/ Dietetic foods/ Egg products/ Formulated liquid diets/ Heat-and-serve dinners/ Infant formulae/ Meal replacments
SYNONYMS:	Vitamin B_{12}/ Cyanocon (III) alamin/ α-(5,6-Dimethylbenzimidazolyl) cyanocobamide/ Extrinsic factor/ CAS 68-19-9/ EINECS 200-680-0
FORMULA:	C63 H88 Co N14 O14 P
MOLECULAR MASS:	1355.38
ALTERNATIVE FORMS:	Methylcobalamin/ Cobalamide (bioactive forms)
PROPERTIES:	Very hygroscopic; fairly heat stable; odourless and tasteless
APPEARANCE:	Dark red crystalline powder; can be diluted with sugar starch or dicalcium phosphate to give a fine pink powder
MELTING RANGE IN °C:	>300
WATER CONTENT MAXIMUM IN %:	12
SOLUBILITY % AT VARIOUS TEMPERATURE/pH COMBINATIONS:	
in water:	1 g dissolves in 80 ml
in ethanol solution:	Soluble
FUNCTION IN FOODS:	Enrichment, fortification or restoration
TECHNOLOGY OF USE IN FOODS:	Serial dilution of pure compound required. Commercially available in crystalline form as spray-dried powders in low and medium concentrations of 0.1 and 1.0%, respectively; as triturate products in mannitol or dicalcium phosphate, or encapsulated in gelatin coating. The spray-dried low concentrate appears to be the form most used in food applications. Optimum pH for use is 4–5

ANTAGONISTS: Light; reducing agents

FOOD SAFETY ISSUES: There appears to be no hazard to humans due to excessive intake in foods

LEGISLATION:

USA:
GRAS for specified applications. For adults and children 4 or older, Recommended Daily Allowance is 6 μg of Vitamin B_{12}

UK and EUROPE:
Recommended Daily Allowance is 1 μg of Vitamin B_{12}

CANADA:
For specific regulations, refer to Part D: Vitamins, Minerals and Amino Acids of the Food and Drug Regulations, Health and Welfare Canada. For adults and children 2 or older, Recommended Daily Allowance is 2 μg of Vitamin B_{12}
For children less than 2 years old, Recommended Daily Allowance is 0.3 μg of Vitamin B_{12}

AUSTRALIA/PACIFIC RIM:
Japan: approved

REFERENCES:

Ash, M., and Ash, I. (1995) *Food Additives: Electronic Handbook.* Gower Publishing Company, Brookfield, VT, 05036, USA.

Augustin, J., and Scarbrough, F. E. (1990) Nutritional Additives. In: Branen, A. L., Davidson, P. M., and S. Salminen (Eds.), *Food Additives*, pp. 33–81.

Budavari, S. (1996) *The Merck Index*, 12th edition.

Health and Welfare Canada (1996) Part D: Vitamins, Minerals and Amino Acids. In: *Food and Drug Regulations*. Department of Health, Ottawa, ON, Canada.

ANY OTHER RELEVANT INFORMATION: Potency of pure compound is 100%

NAME:	Ergocalciferol
CATEGORY:	Nutritive additive
FOOD USE:	Breakfast cereals/ Cereal products/ Cornmeal/ Dairy products/ Dietary supplements/ Dietetic foods/ Egg products/ Farina/ Fats and oils/ Formulated liquid diets/ Infant formulae/ Margarine/ Meal replacements/ Pasta/ Rice
SYNONYMS:	Calciferol/ Ercalciol/ Ergosterol, activated or irradiated/ Oleovitamin D_2/ Viosterol/ Vitamin D/ CAS 50-14-6/ EINECS 200-014-9
FORMULA:	C28 H44 O
MOLECULAR MASS:	396.66
ALTERNATIVE FORMS:	7-dehydrocholesterol/ Ergosterol/ 7-Procholesterol
PROPERTIES:	Odourless
APPEARANCE:	Crystalline white powder
MELTING RANGE IN °C:	115–118
SOLUBILITY % AT VARIOUS TEMPERATURE/pH COMBINATIONS:	
in water:	Insoluble
in vegetable oil:	Slightly soluble
in propylene glycol:	Soluble
FUNCTION IN FOODS:	Enrichment, fortification or restoration
ALTERNATIVES:	Vitamin D_2
ANTAGONISTS:	Air; heat; light especially excess UV irradiation; oxygen
FOOD SAFETY ISSUES:	Repeated daily dosages exceeding 400 IU for children and 1000 IU for adults should be avoided

LEGISLATION:

USA:
GRAS in specified applications. For adults and children over 4 years, Recommended Daily Intake is 400 IU Vitamin D

UK and EUROPE:
Recommended Daily Allowance is 5 µg Vitamin D

CANADA:
For specific regulations, refer to Part D: Vitamins, Minerals and Amino Acids of the Food and Drug Regulations, Health and Welfare Canada. For adults and children over 2 years, Recommended Daily Intake is 5 µg Vitamin D
For children under 2 years old, Recommended Daily Intake is 10 µg Vitamin D

AUSTRALIA/PACIFIC RIM:
Japan: approved for use

REFERENCES:

Ash, M., and Ash, I. (1995) *Food Additives: Electronic Handbook.* Gower Publishing Company, Brookfield, VT, 05036, USA.

Augustin, J., and Scarbrough, F. E. (1990) Nutritional Additives. In: Branen, A. L., Davidson, P. M., and S. Salminen (Eds.), *Food Additives*, pp. 33–81.

Budavari, S. (1996) *The Merck Index*, 12th edition.

Health and Welfare Canada (1996) Part D: Vitamins, Minerals and Amino Acids. In: *Food and Drug Regulations*. Department of Health, Ottawa, ON, Canada.

ANY OTHER RELEVANT INFORMATION:

May be kept without deterioration for 9 months in amber evacuated ampoules at refrigerated temperatures. 1 IU Vitamin D_2 = 0.0258 µg ergocalciferol

NAME:	**Ferrous fumarate**
CATEGORY:	Nutritive additive
FOOD USE:	Breads/ Cereals/ Cornmeal/ Dietary supplements/ Dietetic foods/ Egg products/ Farina/ Flour/ Formulated liquid diets/ Fruit-flavoured drinks and bases/ Heat-and-serve dinners/ Infant cereals/ Infant formulae/ Meal replacements/ Rice/ Pasta/ Peanut spreads/ Simulated meat and poultry products/ Waffles
SYNONYMS:	Iron (II) fumarate/ CAS 141-01-5/ EINECS 205-447-7
FORMULA:	$C_4H_2FeO_4$
MOLECULAR MASS:	169.90
PROPERTIES:	Odourless; almost tasteless. Bioavailability of pure compound is excellent, but may be altered in presence of other food components
APPEARANCE:	Reddish-orange to reddish-brown granular powder
MELTING POINT (°C):	Not melted at 280
DENSITY:	2.435 at 25°C
SOLUBILITY % AT VARIOUS TEMPERATURE/pH COMBINATIONS:	
in water:	**@ 25°C** 0.14 g/100 ml
in ethanol solution:	**@ 95%** <0.01 g/100 ml
FUNCTION IN FOODS:	Enrichment, fortification, or restoration
ALTERNATIVES:	Ferric choline citrate/ Ferric orthophosphate/ Ferric pyrophosphate/ Ferrous gluconate/ Ferrous lactate/ Ferrous sulphate/ Reduced elemental iron/ Sodium iron pyrophosphate
TECHNOLOGY OF USE IN FOODS:	Affects Vitamin C stability. Phytic acid, fibres, phosphates, polyphenolics, some proteins, and organic acids can adversely affect iron absorption, whereas other compounds such as ascorbic acid and some amino acids enhance the absorption of this mineral

FOOD SAFETY ISSUES: Toxicology: LD_{50} (oral, rat) = 3850 mg/kg

LEGISLATION:

USA:
GRAS for specified applications. For adults and children 4 years or older, Recommended Daily Intake is 18 mg iron

UK and EUROPE:
Recommended Daily Allowance is 14 mg iron

CANADA:
For specific regulations, refer to Part D: Vitamins, Minerals and Amino Acids of the Food and Drug Regulations, Health and Welfare Canada. For adults and children 2 years or older, Recommended Daily Intake is 14 mg iron. For children less than 2 years old, Recommended Daily Intake is 7 mg iron

REFERENCES:

Ash, M., and Ash, I. (1995) *Food Additives: Electronic Handbook*. Gower Publishing Company, Brookfield, VT, 05036, USA.

Augustin, J., and Scarbrough, F. E. (1990) Nutritional Additives. In: Branen, A. L., Davidson, P. M., and S. Salminen (Eds.), *Food Additives*, pp. 33–81.

Budavari, S. (1996) *The Merck Index*, 12th edition.

Health and Welfare Canada (1996) Part D: Vitamins, Minerals and Amino Acids. In: *Food and Drug Regulations*. Department of Health, Ottawa, ON, Canada.

ANY OTHER RELEVANT INFORMATION: Contains minimum 31.3% total iron

NAME:	**Ferric orthophosphate**
CATEGORY:	Nutritive additive
FOOD USE:	Breads/ Cereals/ Cornmeal/ Dietary supplements/ Dietetic foods/ Egg products/ Farina/ Flour/ Formulated liquid diets/ Fruit-flavoured drinks and bases/ Heat-and-serve dinners/ Infant cereals/ Infant formulae/ Meal replacements/ Rice/ Pasta/ Peanut spreads/ Simulated meat and poultry products
SYNONYMS:	Iron phoshate/ Iron (III) phosphate/ Ferric phosphate/ Ferric (III) phosphate/ Ferriphosphate/ CAS 10045-86-0 anhydrous, 14940-41-1 tetrahydrate/ EINECS 233-149-7 anhydrous, 239-018-0 tetrahydrate
FORMULA:	$FePO4 \cdot x\ H2O$, $x = 1–4$
MOLECULAR MASS:	150.82 (anhydrous)
PROPERTIES:	Tasteless; bioavailability of pure compound is poor
APPEARANCE:	Dihydrate is white, greyish-white, or light pink crystals or amorphous powder
MELTING POINT (°C):	Loses water above 140
DENSITY:	2.87 (dihydrate)
SOLUBILITY % AT VARIOUS TEMPERATURE/pH COMBINATIONS:	
in water:	Practically insoluble in water
FUNCTION IN FOODS:	Enrichment, fortification, or restoration
ALTERNATIVES:	Ferric choline citrate/ Ferric pyrophosphate/ Ferrous fumarate/ Ferrous gluconate/ Ferrous lactate/ Ferrous sulphate/ Reduced elemental iron/ Sodium iron pyrophosphate
TECHNOLOGY OF USE IN FOODS:	Used when whiteness and inertness are priorities. Phytic acid, fibres, phosphates, polyphenolics, some proteins, and organic acids can adversely affect iron absorption, whereas other compounds such as ascorbic acid and some amino acids enhance the absorption of this mineral
FOOD SAFETY ISSUES:	When heated to decomposition, toxic fumes of PO*x* are emitted

LEGISLATION:

USA:
GRAS for specified applications. For adults and children 4 years or older, Recommended Daily Intake is 18 mg iron

UK and EUROPE:
Recommended Daily Allowance is 14 mg iron

CANADA:
For specific regulations, refer to Part D: Vitamins, Minerals and Amino Acids of the Food and Drug Regulations, Health and Welfare Canada. For adults and children 2 years or older, Recommended Daily Intake is 14 mg iron. For children less than 2 years old, Recommended Daily Intake is 7 mg iron

REFERENCES:

Ash, M., and Ash, I. (1995) *Food Additives: Electronic Handbook.* Gower Publishing Company, Brookfield, VT, 05036, USA.

Augustin, J., and Scarbrough, F. E. (1990) Nutritional Additives. In: Branen, A. L., Davidson, P. M., and S. Salminen (Eds.), *Food Additives*, pp. 33–81.

Budavari, S. (1996) *The Merck Index*, 12th edition.

Health and Welfare Canada (1996) Part D: Vitamins, Minerals and Amino Acids. In: *Food and Drug Regulations*. Department of Health, Ottawa, ON, Canada.

ANY OTHER RELEVANT INFORMATION:

29.9% iron

NAME:	Ferrous sulphate, anhydrous
CATEGORY:	Nutritive additive
FOOD USE:	Breads/ Cereals/ Cornmeal/ Dietary supplements/ Dietetic foods/ Egg products/ Farina/ Flour/ Formulated liquid diets/ Fruit-flavoured drinks and bases/ Heat-and-serve dinners/ Infant cereals/ Infant formulae/ Meal replacements/ Rice/ Pasta/ Peanut spreads/ Simulated meat and poultry products
SYNONYMS:	Iron sulphate (ous)/ Iron (II) sulphate (1 : 1)/ Iron vitriol/ Copperas/ Green vitriol/ Sal chalybis/ CAS 7720-78-7/ EINECS 231-753-5
FORMULA:	Fe O4 S
MOLECULAR MASS:	151.91
ALTERNATIVE FORMS:	Monohydrate, heptahydrate
PROPERTIES:	Metallic taste. Bioavailability of pure compound is excellent, but may be altered in presence of other food components
APPEARANCE:	Greyish white to yellow crystalline powder
MELTING POINT (°C):	64
DENSITY:	1.89
SOLUBILITY % AT VARIOUS TEMPERATURE/pH COMBINATIONS:	
in water:	Slowly soluble
in ethanol solution:	Insoluble
FUNCTION IN FOODS:	Enrichment, fortification, or restoration
ALTERNATIVES:	Ferric choline citrate/ Ferric orthophosphate/ Ferric pyrophosphate/ Ferrous fumarate/ Ferrous gluconate/ Ferrous lactate/ Reduced elemental iron/ Sodium iron pyrophosphate

TECHNOLOGY OF USE IN FOODS:

Affects Vitamin C stability. Phytic acid, fibres, phosphates, polyphenolics, some proteins, and organic acids can adversely affect iron absorption, whereas other compounds such as ascorbic acid and some amino acids enhance the absorption of this mineral

FOOD SAFETY ISSUES:

Toxicology: LD_{50} (oral, rat) = 319 mg/kg

LEGISLATION:

USA:
GRAS for specified applications. For adults and children 4 years and older, Recommended Daily Intake is 18 mg iron

UK and EUROPE:
Recommended Daily Allowance is 14 mg iron

CANADA:
For specific regulations, refer to Part D: Vitamins, Minerals and Amino Acids of the Food and Drug. Regulations, Health and Welfare Canada. For adults and children 2 years and older, Recommended Daily Intake is 14 mg iron. For children less than 2 years old, Recommended Daily Intake is 7 mg iron

AUSTRALIA/PACIFIC RIM:
Japan: approved

REFERENCES:

Ash, M., and Ash, I. (1995) *Food Additives: Electronic Handbook.* Gower Publishing Company, Brookfield, VT, 05036, USA.

Augustin, J., and Scarbrough, F. E. (1990) Nutritional Additives. In: Branen, A. L., Davidson, P. M., and S. Salminen (Eds.), *Food Additives*, pp. 33–81.

Budavari, S. (1996) *The Merck Index*, 12th edition.

Health and Welfare Canada (1996) Part D: Vitamins, Minerals and Amino Acids. In: *Food and Drug Regulations*. Department of Health, Ottawa, ON, Canada.

ANY OTHER RELEVANT INFORMATION:

36.8% iron

NAME:	Folic acid
CATEGORY:	Nutritive additive
FOOD USE:	Breakfast cereals/ Fruit drinks/ Dietary supplement/ Dietetic foods/ Egg products/ Flour/ Formulated liquid diets/ Infant formulae/ Meal replacements/ Milk drinks/ Sugar and cocoa confectionery/ Peanut spreads
SYNONYMS:	Folacin/ Folate/ Pteroylglutamic acid/ Vitamin M/ CAS 59-30-3/ EINECS 200-419-0
FORMULA:	C19 H19 N7 O6
MOLECULAR MASS:	441.4
ALTERNATIVE FORMS:	Folinic acid or citrovorum factor; pteroyltriglutamic acid; pteroylheptaglutamic acid; 7,8-dihydro- or 5,6,7,8-tetrahydrofolic
PROPERTIES:	Odourless; not heat-stable
APPEARANCE:	Yellow to orange crystalline powder
MELTING POINT (°C):	No melting point; darkens and chars at 250
SOLUBILITY % AT VARIOUS TEMPERATURE/pH COMBINATIONS:	
in water:	@ **25°C** Very slightly (0.0016 mg/ml) @ **100°C** about 1%
in ethanol solution:	Insoluble
FUNCTION IN FOODS:	Enrichment, fortification or restoration
TECHNOLOGY OF USE IN FOODS:	Serial dilution may be required; optimum pH is 6–9
ANTAGONISTS:	Oxygen; light; reducing agents; labile to acid and light when in solution; heat lability depends on the type of compound, the pH of the heating medium, and the presence or absence of reducing agents
FOOD SAFETY ISSUES:	Toxicology: LD_{50} (IV, rat) = 500 mg/kg; folacin has a low toxicity for humans

LEGISLATION:

USA:
For adults and children 4 years or older, the Recommended Daily Intake is 400 μg of folacin.
For pregnant or lactating women, the Recommended Daily Intake is 800 μg of folacin. For children under 4 years old, the Recommended Daily Intake is 300 μg of folacin. For infants, the Recommended Daily Intake is 100 μg of folacin

UK and EUROPE:
Recommended Daily Allowance is 200 μg

CANADA:
For specific regulations, refer to Part D: Vitamins, Minerals and Amino Acids of the Food and Drug Regulations, Health and Welfare Canada. For adults and children 2 or older, the Recommended Daily Intake is 220 μg of folacin. For children under 2 years old, the Recommended Daily Intake is 65 μg of folacin

AUSTRALIA/PACIFIC RIM:
Japan: approved for use

REFERENCES:

Ash, M., and Ash, I. (1995) *Food Additives: Electronic Handbook.* Gower Publishing Company, Brookfield, VT, 05036, USA.

Augustin, J., and Scarbrough, F. E. (1990) Nutritional Additives. In: Branen, A. L., Davidson, P. M., and S. Salminen (Eds.), *Food Additives*, pp. 33–81.

Budavari, S. (1996) *The Merck Index*, 12th edition.

Health and Welfare Canada (1996) Part D: Vitamins, Minerals and Amino Acids. In: *Food and Drug Regulations*. Department of Health, Ottawa, ON, Canada.

ANY OTHER RELEVANT INFORMATION:

Potency of pure compound is 90%

NAME:	Magnesium carbonate hydroxide
CATEGORY:	Nutritive additive
FOOD USE:	Alimentary pastes/ Breakfast cereals/ Dietary supplements/ Dietetic foods/ Egg products/ Flour/ Formulated liquid diets/ Infant formulae
SYNONYMS:	Magnesite/ Magnesium carbonate basic/ Magnesium hydroxide carbonate/ CAS 12125-28-9, 39409-82-0/ EINECS 235-192-7
FORMULA:	Approx. (MgCO3)4·Mg(OH)2·5H2O
MOLECULAR MASS:	Approx. 485.7
ALTERNATIVE FORMS:	Aluminium magnesium hydroxide carbonate
PROPERTIES:	Odourless, tasteless to chalky. Excellent bioavailability, but may be altered in presence of other food components; slightly effervescent in water
APPEARANCE:	White powder
SOLUBILITY % AT VARIOUS TEMPERATURE/pH COMBINATIONS:	
in water:	Soluble in about 3300 parts CO_2-free water; more soluble in water containing CO_2
FUNCTION IN FOODS:	Enrichment, fortification or restoration. Also used as anticaking agent, flour conditioner, lubricant, pH control agent, processing aid, release agent
ALTERNATIVES:	Magnesium carbonate/ Magnesium chloride/ Magnesium hydroxide/ Magnesium oxide/ Magnesium phosphate dibasic/ Magnesium stearate/ Magnesium sulphate

LEGISLATION:

USA:
GRAS for specified applications. For adults and children 4 or older, Recommended Daily Intake is 400 mg magnesium

UK and EUROPE:
Recommended Daily Allowance is 300 mg of magnesium

CANADA:
For specific regulations, refer to Part D: Vitamins, Minerals and Amino Acids of the Food and Drug. Regulations, Health and Welfare Canada. For adults and children 2 or older, Recommended Daily Intake is 250 mg magnesium. For children under 2 years old, Recommended Daily Intake is 55 mg

REFERENCES:

Ash, M., and Ash, I. (1995) *Food Additives: Electronic Handbook.* Gower Publishing Company, Brookfield, VT, 05036, USA.

Augustin, J., and Scarbrough, F. E. (1990) Nutritional Additives. In: Branen, A. L., Davidson, P. M., and S. Salminen (Eds.), *Food Additives*, pp. 33–81.

Budavari, S. (1996) *The Merck Index*, 12th edition.

Health and Welfare Canada (1996) Part D: Vitamins, Minerals and Amino Acids. In: *Food and Drug Regulations*. Department of Health, Ottawa, ON, Canada.

ANY OTHER RELEVANT INFORMATION:

Magnesium content varies from 25–30%

NAME:	Magnesium oxide, heavy
CATEGORY:	Nutritive additive
FOOD USE:	Alimentary pastes/ Breakfast cereals/ Dietary supplements/ Dietetic foods/ Egg products/ Flour/ Formulated liquid diets/ Infant formulae
SYNONYMS:	Magnesia/ Calcined magnesia/ Magnesia usta/ Periclase/ CAS 1309-48-4/ EINECS 215-171-9/ E530
FORMULA:	MgO
MOLECULAR MASS:	40.30
PROPERTIES:	Chalky taste; odourless; hygroscopic; readily takes CO_2 and H_2O from air. Bioavailability is good, but may be altered in presence of other food components
APPEARANCE:	White, very fine powder
BOILING POINT IN °C AT VARIOUS PRESSURES (INCLUDING 760 mm Hg):	3600
MELTING POINT (°C):	2852
DENSITY:	3.58
PURITY %:	≥95
SOLUBILITY % AT VARIOUS TEMPERATURE/pH COMBINATIONS:	
in water:	Moderately soluble (≤0.4%)
FUNCTION IN FOODS:	Enrichment, fortification or restoration. Also used as anticaking agent, firming agent, lubricant, neutralising agent, pH control agent, processing aid (Japan), release agent
ALTERNATIVES:	Magnesium carbonate/ Magnesium carbonate hydroxide/ Magnesium chloride/ Magnesium hydroxide/ Magnesium phosphate dibasic/ Magnesium stearate/ Magnesium sulphate

TECHNOLOGY OF USE IN FOODS: Soluble in acids

LEGISLATION:

USA:
GRAS for specified applications. For adults and children 4 years or older, Recommended Daily Intake is 400 mg magnesium

UK and EUROPE:
Listed in Europe. Recommended Daily Allowance, is 300 mg of magnesium

CANADA:
For specific regulations, refer to Part D: Vitamins, Minerals and Amino Acids of the Food and Drug. Regulations, Health and Welfare Canada. For adults and children 2 years or older, Recommended Daily Intake is 250 mg magnesium. For children under 2 years old, Recommended Daily Intake is 55 mg

AUSTRALIA/PACIFIC RIM:
Japan: restricted use

REFERENCES:

Ash, M., and Ash, I. (1995) *Food Additives: Electronic Handbook.* Gower Publishing Company, Brookfield, VT, 05036, USA.

Augustin, J., and Scarbrough, F. E. (1990) Nutritional Additives. In: Branen, A. L., Davidson, P. M., and S. Salminen (Eds.), *Food Additives*, pp. 33–81.

Budavari, S. (1996) *The Merck Index*, 12th edition.

Health and Welfare Canada (1996) Part D: Vitamins, Minerals and Amino Acids. In: *Food and Drug Regulations*. Department of Health, Ottawa, ON, Canada.

ANY OTHER RELEVANT INFORMATION: Anhydrous compound contains 60.3% magnesium

NAME:	**Sodium ascorbate**
CATEGORY:	Nutritive additive
FOOD USE:	Cured meats/ Dietary supplements/ Dietetic foods/ Formulated liquid diets/ Infant formulae/ Meal replacements/ Milk products
SYNONYMS:	L(+)-Ascorbic acid sodium salt/ Monosodium ascorbate/ Vitamin C sodium salt/ CAS 134-03-2/ EINECS 205-126-1/ E301
FORMULA:	$C_6H_7NaO_6$
MOLECULAR MASS:	198.11
ALTERNATIVE FORMS:	Ascorbic acid/ Ascorbyl palmitate/ Calcium ascorbate/ Nicotinamide-ascorbic acid complex
PROPERTIES:	Not heat-stable at neutral pH
APPEARANCE:	White to slightly yellowish powder; various granulations available
MELTING POINT (°C):	Decomposes at 218
SOLUBILITY % AT VARIOUS TEMPERATURE/pH COMBINATIONS:	
in water:	**@ 25°C** Readily soluble 62% **@ 75°C** 78%
in ethanol solution:	Very slightly soluble
FUNCTION IN FOODS:	Enrichment, fortification or restoration. Also used as antioxidant, curing agent and nitrosamine inhibitor in cured meats, preservative
TECHNOLOGY OF USE IN FOODS:	Optimum pH is between 5 and 7; usage level is 300–500 ppm
ANTAGONISTS:	Oxidation; light; minerals; heat-labile in neutral environments, but less so in highly acidic environments
FOOD SAFETY ISSUES:	Some adverse effects can occur with extremely high repeated doses in the 500 mg to 10 g range

LEGISLATION:

USA:
GRAS for specified applications. For adults and children 4 years or older, Recommended Daily Intake is 60 mg of Vitamin C

UK and EUROPE:
Recommended Daily Allowance is 60 mg Vitamin C

CANADA:
For specific regulations, refer to Part D: Vitamins, Minerals and Amino Acids of the Food and Drug. Regulations, Health and Welfare Canada. For adults and children 2 years or older, Recommended Daily Intake is 60 mg of Vitamin C. For children less than 2 years old, Recommended Daily Intake is 20 mg of Vitamin C

REFERENCES:

Ash, M., and Ash, I. (1995) *Food Additives: Electronic Handbook.* Gower Publishing Company, Brookfield, VT, 05036, USA.

Augustin, J., and Scarbrough, F. E. (1990) Nutritional Additives. In: Branen, A. L., Davidson, P. M., and S. Salminen (Eds.), *Food Additives*, pp. 33–81.

Budavari, S. (1996) *The Merck Index*, 12th edition.

Health and Welfare Canada (1996) Part D: Vitamins, Minerals and Amino Acids. In: *Food and Drug Regulations*. Department of Health, Ottawa, ON, Canada.

ANY OTHER RELEVANT INFORMATION:

Potency is 88%, i.e. 1 mg of the sodium salt is equivalent to 0.8890 mg of ascorbic acid, or 1 mg of the acid is equivalent to 1.1248 mg of sodium ascorbate. The pH of aqueous solutions is 5.6–7.0 or higher. A 10% solution, made from commercial grade, may have a pH of 7.4–7.7. Aqueous solutions are unstable and subject to rapid oxidation by air at pH > 6.0

NAME:	Reduced elemental iron
CATEGORY:	Nutritive additive
FOOD USE:	Breads/ Cereals/ Cornmeal/ Dietary supplements/ Dietetic foods/ Egg products/ Farina/ Flour/ Formulated liquid diets/ Fruit-flavoured drinks and bases/ Heat-and-serve dinners/ Infant cereals/ Infant formulae/ Meal replacements/ Rice/ Pasta/ Peanut spreads/ Simulated meat and poultry products
SYNONYMS:	Carbonyl iron/ Electrolytic iron/ CAS 7439-89-6/ EINECS 231-096-4
FORMULA:	Fe
MOLECULAR MASS:	55.84
ALTERNATIVE FORMS:	Ferric orthophosphate/ Ferrous fumarate/ Ferrous sulphate
PROPERTIES:	Not water-soluble. Bioavailability of pure compound is good, but may be altered in the presence of other food components
APPEARANCE:	Grey-black powder
BOILING POINT IN °C AT VARIOUS PRESSURES (INCLUDING 760 mm Hg):	3000
MELTING POINT (°C):	1536
DENSITY:	7.87
FUNCTION IN FOODS:	Enrichment, fortification or restoration. Also used as processing aid in Japan
ALTERNATIVES:	Ferric choline citrate/ Ferric orthophosphate/ Ferric pyrophosphate/ Ferrous fumarate/ Ferrous gluconate/ Ferrous lactate/ Ferrous sulphate/ Sodium iron pyrophosphate
TECHNOLOGY OF USE IN FOODS:	Phytic acid, fibres, phosphates, polyphenolics, some proteins, and organic acids can adversely affect iron absorption, whereas other compounds such as ascorbic acid and some amino acids enhance the absorption of this mineral

FOOD SAFETY ISSUES: Poison by intraperitoneal route; potentially toxic by all forms and routes

LEGISLATION:

USA:
GRAS for specified applications. For adults and children 4 years or older, Recommended Daily Intake is 18 mg iron

UK and EUROPE:
Recommended Daily Allowance is 14 mg iron

CANADA:
For specific regulations, refer to Part D: Vitamins, Minerals and Amino Acids of the Food and Drug. Regulations, Health and Welfare Canada. For adults and children 2 years or older, Recommended Daily Intake is 14 mg iron. For children less than 2 years old, the Recommended Daily Intake is 7 mg iron

AUSTRALIA/PACIFIC RIM:
Japan: restricted use (can be used as a processing aid)

REFERENCES:

Ash, M., and Ash, I. (1995) *Food Additives: Electronic Handbook.* Gower Publishing Company, Brookfield, VT, 05036, USA.

Augustin, J., and Scarbrough, F. E. (1990) Nutritional Additives. In: Branen, A. L., Davidson, P. M., and S. Salminen (Eds.), *Food Additives*, pp. 33–81.

Budavari, S. (1996) *The Merck Index*, 12th edition.

Health and Welfare Canada (1996) Part D: Vitamins, Minerals and Amino Acids. In: *Food and Drug Regulations*. Department of Health, Ottawa, ON, Canada.

ANY OTHER RELEVANT INFORMATION: 100% iron, i.e. 100% by weight of the particles must pass through a 100-mesh sieve and at least 95% by weight of the particles must pass through a 325-mesh sieve; ultrafine powder is potentially explosive

NAME:	Retinyl acetate
CATEGORY:	Nutritive additive
FOOD USE:	Breakfast cereals/ Dietary supplements/ Dietetic foods/ Egg products/ Fats and oils/ Formulated liquid foods/ Infant formulae/ Liquid milk/ Margarine/ Meal replacements/ Milk powder
SYNONYMS:	Acetic acid retinyl ester/ Vitamin A acetate/ Vitamin A alcohol acetate/ CAS 127-47-9/ EINECS 204-844-2
FORMULA:	$C_{22}H_{32}O_2$
MOLECULAR MASS:	328.5
ALTERNATIVE FORMS:	Retinol; retinyl palmitate
APPEARANCE:	Oily blends–yellow oil. Dry blends–light yellow, fine granular powder
MELTING RANGE (°C):	57–58
FUNCTION IN FOODS:	Enrichment, fortification or restoration
TECHNOLOGY OF USE IN FOODS:	Oily blends are soluble in oil and fat; homogenising in a small amount of liquid is necessary prior to adding to milk and infant formulae. Dry blends are water-dispersible.
ANTAGONISTS:	Agitation; light; oxygen; extreme heat
FOOD SAFETY ISSUES:	Toxicology: LD_{50} (oral, mouse, 10 day) = 4100 mg/kg
LEGISLATION:	**USA:** GRAS for specified applications. For adults and children 4 years or older, Recommended Daily Intake is 5000 IU **CANADA:** For specific regulations, refer to Part D: Vitamins, Minerals and Amino Acids of the Food and Drug. Regulations, Health and Welfare Canada. For adults and children 2 years or older, Recommended Daily Intake is 1000 RE Vitamin A. For children

	UK and EUROPE: Recommended Daily Allowance is 800 µg Vitamin A
	less than 2 years old, Recommended Daily Intake is 400 RE Vitamin A

REFERENCES:

Ash, M., and Ash, I. (1995) *Food Additives: Electronic Handbook.* Gower Publishing Company, Brookfield, VT, 05036, USA.

Augustin, J., and Scarbrough, F. E. (1990) Nutritional Additives. In: Branen, A. L., Davidson, P. M., and S. Salminen (Eds.), *Food Additives*, pp. 33–81.

Budavari, S. (1996) *The Merck Index*, 12th edition.

Health and Welfare Canada (1996) Part D: Vitamins, Minerals and Amino Acids. In: *Food and Drug Regulations*. Department of Health, Ottawa, ON, Canada.

ANY OTHER RELEVANT INFORMATION:

Oily blends are stabilised with tocopherol or BHA/BHT. Dry blends are stabilised with tocopherol, ascorbic acid, ascorbyl palmitate or BHA/BHT. Nutritive activity is measured in retinol equivalents (RE): 1 RE = 1 µg retinol = 1.147 µg retinyl acetate = 3.33 IU; Oily blends are 2–4 times more potent than dry blends. Biopotency 2.904×10^6 IU/g

NAME:	**Retinol**
CATEGORY:	Nutritive additive
FOOD USE:	Breakfast cereals/ Dietary supplements/ Dietetic foods/ Egg products/ Fats and oils/ Formulated liquid foods/ Infant formulae/ Liquid milk/ Margarine/ Meal replacements/ Milk powder
SYNONYMS:	All *trans*-retinol/ Axerophthol/ Vitamin A/ Vitamin A alcohol/ CAS 68-26-8
FORMULA:	$C_{20}H_{30}O$
MOLECULAR MASS:	286.46
ALTERNATIVE FORMS:	Retinal/ Retinyl acetate/ Retinyl palmitate
APPEARANCE:	Oily blends: yellow oil stabilised with tocopherol or BHA/BHT. Dry powder: light yellow fine granular powder with antioxidants
MELTING RANGE (°C):	54–58 / 62–64 (depending on form)
SOLUBILITY % AT VARIOUS TEMPERATURE/pH COMBINATIONS:	
in water:	Practically insoluble
in vegetable oil:	Soluble
in ethanol solution:	Soluble
in propylene glycol:	Soluble
FUNCTION IN FOODS:	Enrichment, fortification or restoration
TECHNOLOGY OF USE IN FOODS:	Oily blends may crystallise upon storage; homogenise in small amount of liquid prior to adding to milk and infant formulae. Dry powders dispersible in water
SYNERGISTS:	pH > 6
ANTAGONISTS:	Agitation; extreme heat; light; oxygen; minerals, especially in the presence of high humidity and moisture

FOOD SAFETY ISSUES:

Toxicology: LD_{50} (oral, rat) = 2000 mg/kg

Toxic levels: For infants = 100,000 IU single dose; for children = 300,000 IU single dose; for adults = 500,000 IU single dose. Chronic toxicity can develop from prolonged daily intake of 25,000 IU

LEGISLATION:

USA:
GRAS in specified applications. For adults and children over 4 years, Recommended Daily Intake is 5,000 IU Vitamin A

UK and EUROPE:
Recommended Daily Allowance is 800 μg

CANADA:
For specific regulations, refer to Part D: Vitamins, Minerals and Amino Acids of the Food and Drug. Regulations, Health and Welfare Canada. For adults and children over 2 years, Recommended Daily Intake is 1,000 RE Vitamin A. For children under 2 years old, Recommended Daily Intake is 400 RE Vitamin A

REFERENCES:

Ash, M., and Ash, I. (1995) *Food Additives: Electronic Handbook.* Gower Publishing Company, Brookfield, VT, 05036, USA.

Augustin, J., and Scarbrough, F. E. (1990) Nutritional Additives. In: Branen, A. L., Davidson, P. M., and S. Salminen (Eds.), *Food Additives*, pp. 33–81.

Budavari, S. (1996) *The Merck Index*, 12th edition.

Health and Welfare Canada (1996) Part D: Vitamins, Minerals and Amino Acids. In: *Food and Drug Regulations*. Department of Health, Ottawa, ON, Canada.

ANY OTHER RELEVANT INFORMATION:

Nutritive activity measured in retinol equivalents (RE): 1 RE = 1 μg retinol = 1.147 μg retinyl acetate = 3.33 IU Vitamin A

NAME:	Retinyl palmitate
CATEGORY:	Nutritive additive
FOOD USE:	Breakfast cereals/ Dietary supplements/ Dietetic foods/ Egg products/ Fats and oils/ Formulated liquid foods/ Infant formulae/ Liquid milk/ Margarine/ Meal replacements/ Milk powder
SYNONYMS:	Retinol, hexadecanoate/ Vitamin A palmitate/ CAS 79-81-2/ EINECS 201-228-5
FORMULA:	C36 H60 O2
MOLECULAR MASS:	524.9
ALTERNATIVE FORMS:	Retinol/ Retinyl acetate
APPEARANCE:	Oily blends: yellow oil. Dry blends: light yellow, fine granular powder
MELTING RANGE (°C):	28–29
FUNCTION IN FOODS:	Enrichment, fortification or restoration
TECHNOLOGY OF USE IN FOODS:	Oily blends are soluble in oil and fat; homogenising in a small amount of liquid is necessary prior to adding to milk and infant formulae. Dry blends are water-dispersible.
ANTAGONISTS:	Agitation; light; oxygen; extreme heat
FOOD SAFETY ISSUES:	Toxicology: LD_{50} (oral, rat, 10 day) = 7910 mg/kg
LEGISLATION:	**USA:** For adults and children 4 years or older, Recommended Daily Intake is 5000 IU Vitamin A **UK and EUROPE:** Recommended Daily Intake is 800 µg Vitamin A **CANADA:** For specific regulations, refer to Part D: Vitamins, Minerals and Amino Acids of the Food and Drug. Regulations, Health and Welfare Canada. For adults and children 2 years or older, Recommended Daily Intake is 1000 RE Vitamin A. For children less than 2 years old, Recommended Daily Intake is 400 RE Vitamin A

REFERENCES:

Ash, M., and Ash, I. (1995) *Food Additives: Electronic Handbook.* Gower Publishing Company, Brookfield, VT, 05036, USA.

Augustin, J., and Scarbrough, F. E. (1990) Nutritional Additives. In: Branen, A. L., Davidson, P. M., and S. Salminen (Eds.), *Food Additives*, pp. 33–81.

Budavari, S. (1996) *The Merck Index*, 12th edition.

Health and Welfare Canada (1996) Part D: Vitamins, Minerals and Amino Acids. In: *Food and Drug Regulations*. Department of Health, Ottawa, ON, Canada.

ANY OTHER RELEVANT INFORMATION:

Oily blends are stabilised with tocopherol or BHA/BHT. Dry blends are stabilised with tocopherol, ascorbic acid, ascorbyl palmitate or BHA/BHT.

Nutritive activity is measured in retinol equivalents (RE): 1 RE = 1 μg retinol = 1.147 μg retinyl acetate = 3.33 IU; Biopotency 1.817×10^6 IU/g

NAME:	Zinc sulphate monohydrate
CATEGORY:	Nutritive additive
FOOD USE:	Beverages/ Breakfast cereals/ Dietary supplements/ Dietetic foods/ Egg products/ Formulated liquid diets/ Infant formulae/ Meal replacements/ Peanut spreads
SYNONYMS:	Dried zinc sulfate/ Sulphuric acid zinc salt (1 : 1)/ White copperas/ White vitriol/ Zinc vitriol/ CAS 7446-20-0/ EINECS 231-793-3
FORMULA:	$ZnSO4\cdot H2O$
MOLECULAR MASS:	161.44 anhydrous
ALTERNATIVE FORMS:	Zinc sulphate heptahydrate (cakes readily)
PROPERTIES:	Astringent taste. Good bioavailability, but may be altered in presence of other food components
APPEARANCE:	White, effervescent powder or granules
MELTING POINT (°C):	740, decomposes
DENSITY AT 15°C:	3.74
SOLUBILITY % AT VARIOUS TEMPERATURE/pH COMBINATIONS:	
in water:	Very soluble
in ethanol solution:	Practically insoluble
FUNCTION IN FOODS:	Enrichment, fortification or restoration
ALTERNATIVES:	Zinc oxide
ANTAGONISTS:	Phytates and dietary fibre present in plant foods have an adverse effect on zinc absorption
FOOD SAFETY ISSUES:	Toxicology: LD_{50} (oral, rat) = 2949 mg/kg

LEGISLATION:

USA:
GRAS for specified applications. For adults and children 4 years or older, Recommended Daily Intake is 15 mg zinc

UK and EUROPE:
Recommended Daily Allowance is 15 mg zinc

CANADA:
For specific regulations, refer to Part D: Vitamins, Minerals and Amino Acids of the Food and Drug. Regulations, Health and Welfare Canada. For adults and children 2 years or older, Recommended Daily Intake is 9 mg. For children less than 2 years old, Recommended Daily Intake is 4 mg

AUSTRALIA/PACIFIC RIM:
Japan: approved, with restrictions

REFERENCES:

Ash, M., and Ash, I. (1995) *Food Additives: Electronic Handbook.* Gower Publishing Company, Brookfield, VT, 05036, USA.

Augustin, J., and Scarbrough, F. E. (1990) Nutritional Additives. In: Branen, A. L., Davidson, P. M., and S. Salminen (Eds.), *Food Additives*, pp. 33–81.

Budavari, S. (1996) *The Merck Index*, 12th edition.

Health and Welfare Canada (1996) Part D: Vitamins, Minerals and Amino Acids. In: *Food and Drug Regulations*. Department of Health, Ottawa, ON, Canada.

ANY OTHER RELEVANT INFORMATION:

Anhydrous form is 40.5% zinc; hydrous form is 36.4% zinc

Part 10

Polysaccharides

NAME:	**Agar**
CATEGORY:	Polysaccharides
FOOD USE:	Meats/ Poultry/ Pet foods/ Desserts/ Baked goods/ Confections/ Beverages/ Food packaging/ Edible films/ Coatings/ All food products
SYNONYMS:	CAS 9002-18-0/ EINECS 232-658-1/ FEMA 2012/ E406/ Gelose/ Bengal gelatin/ Japan isinglass/ Agar-agar
FORMULA:	Variable, a mixture of agaran (agarose) and agaropectin with variable methoxy and pyruvate substituents
MOLECULAR MASS:	Variable
ALTERNATIVE FORMS:	Agar is usually found as a calcium salt, although variable levels of sodium and magnesium salts are present in commerical preparations
PROPERTIES AND APPEARANCE:	White to pale yellow colour, either odourless or a slight charcoal odour
SOLUBILITY % AT VARIOUS TEMPERATURE/pH COMBINATIONS:	Solubility varies with molecular weight. Generally, agar is soluble in hot water, sucrose solutions, sodium chloride solution and ethanol solutions, but solubility in vegetable oils and propylene glycol is limited Above 45°C viscosity is relatively constant at pH 4.5 to pH 9
in water:	**@ 20°C** Generally insoluble **@ 100°C** Soluble
FUNCTION IN FOODS:	Stabiliser; emulsifier; thickener; drying agent; flavouring agent; surface finisher; formulation aid; humectant; antistaling in baking, confections, meats, poultry; gellant in desserts, beverages; protective colloid in foods
ALTERNATIVES:	Gellan; alginate; curdlan; gelatin (dependent on application)
TECHNOLOGY OF USE IN FOODS:	Agar should be dispersed in cold water and heated for 10 minutes at >85°C to melt and hydrate the polysaccharide. Typically agar preparations are autoclaved (retorted) at 121°C for at least 10 minutes.

Agar is a mixture of agaran (agarose) and agaropectin. Agaran is a linear gelling polysaccharide which has a disaccharide (agarobiose) repeating unit composed of (1 → 3) beta-D-galactopyranosyl (1 → 4) linked to a 3,6-anhydro-alpha-D-galactopyranosyl unit. Agaropectin is a closely related branched molecule. The gelling properties of agar are due to interacting strands of agaran.

This polysaccharide dissolved above 85°C to form a low-viscosity solution (pumpable) that remains low viscosity as the temperature is reduced until about 30°C to 40°C whereupon it sets to form a clear, brittle gel. The gel remelts at >85°C

SYNERGISTS: None known

ANTAGONISTS: None known

FOOD SAFETY ISSUES: Non-toxic. Limitations: 0.8% (baked goods/mixes); 2% (confections, frostings); 1.2% (soft candy); 0.25% (other foods)

The powder is combustible

LEGISLATION:

USA:
FDA 21CFR § 150.141, 150.161, 184.1115, GRAS;
USDA 9CFR § 318.7; GRAS (FEMA)

UK and EUROPE:
UK: approved. Europe: listed

AUSTRALIA/PACIFIC RIM:
Japan: approved for use as a natural thickener and stabiliser

REFERENCES:

Selby, H. H., and Whistler, R. L. (1993) Agar. In: Whistler, R. L., and J. N. BeMiller (Eds.), *Industrial Gums: Polysaccharides and Derivatives*. 3rd edition, pp. 87–104.

Ash, M., and Ash, I. (1996) *Food Additives.* Gower Publishing Co., Brookfield, VT.

ANY OTHER RELEVANT INFORMATION: Agars are natural polysaccharides that can be extracted from some species of a class of seaweed called Rhodophyceae (red algae). The most commercially important species belong to the genus *Gelidium*, although other species (*Gracilaria*, *Pterocladia*, etc.) are also used. Polysaccharides with structures similar to agar are termed agaroids

NAME:	Algin
CATEGORY:	Polysaccharides
FOOD USE:	All food products, edible films/ Coatings
SYNONYMS:	CAS 9005-32-7/ E407/ Sodium alginate/ Sodium polymannuronate
FORMULA:	Variable, depending on ratio of mannuronic to guluronic acid, and the presence of other substituents
MOLECULAR MASS:	Variable
ALTERNATIVE FORMS:	Alginic acid; ammonium alginate; calcium alginate; potassium alginate
PROPERTIES AND APPEARANCE:	White to yellow powders or hard flakes
SOLUBILITY % AT VARIOUS TEMPERATURE/pH COMBINATIONS:	Solubility in aqueous solution varies with differences in structure, molecular weight, pH and cations. Alginic acid (low pH) is insoluble in water, monovalent salts (sodium, potassium and ammonium) are also soluble but divalent salts (including calcium) except magnesium are insoluble in water. Solubility declines with increasing molecular weight and increasing concentration. Viscosity is not affected in the pH 5 to 11 range but increases at lower pHs (due to alginic acid formation)
FUNCTION IN FOODS:	Gelling agent; thickening agent; stabiliser; flocculant and bulking agent. It can be used to stabilise emulsions and to retard ice-crystal formation
ALTERNATIVES:	Carrageenans; gelatin; gellan; agar; xanthan; furcellaran (application-dependent)
TECHNOLOGY OF USE IN FOODS:	Alginates should be dispersed in cold deionised water and heated for 10 minutes at 80°C to hydrate the polysaccharide. A high shear mixer will aid dispersion, although prolonged shear of high molecular-weight polymers will reduce the viscosity. Calcium gels can be prepared by addition of alginate solutions to a calcium solution (e.g. maraschino cherries), addition of calcium to a hot solution of alginate and then cooling (gel is thermo-irreversible), or by diffusion. Gels can also be formed by lowering the pH (alginic acid), but this can be technically difficult.

Alginates are composed of blocks of polymannuronic acid, polyguluronic acid and mixed regions. Ribbon-like structures are formed in solution and alginates with a high percentage of guluronic blocks can be cross-linked (gelled) by divalent ions (e.g. calcium)

SYNERGISTS: Divalent ion, pectins

ANTAGONISTS: None known

FOOD SAFETY ISSUES: LD_{50} (intravenous, rat) 1,000 mg/kg; ADI 0–50 mg/kg

The powder is combustible and emits toxic fumes when heated

LEGISLATION: Usage level: Limitation 1% (condiments); 6% (pimento for stuffed olives); 0.3% (confections); 4% (gelatins/ puddings); 10% (hard candy); 2% (processed fruits); 1% (other foods)

USA:	**UK and EUROPE:**	**AUSTRALIA/PACIFIC RIM:**
FDA 21CFR § 133.133, 133.134, 133.162, 133.178, 133.179, 150.141, 150.161, 173.310, 184.1724. GRAS	UK: approved; Europe: listed	Japan: approved

REFERENCES:

Ash, M., and Ash, I. (1996) *Food Additives.* Gower Publishing Co., Brookfield, VT.

Clare, K. (1993) Algin. In: Whistler R. L. and J. N. BeMiller (Eds.), *Industrial Gums: Polysaccharides and Derivatives.* 3rd Edition, pp. 105–143.

ANY OTHER RELEVANT INFORMATION: Alginates are natural polysaccharides which are extracted from brown seaweeds (Phaeophyceae). *Macrocystis pyrifera, Laminera hyperborea*, *Laminaria digitata*, *Laminaria japonica*, *Ascophyllum nodosum*, *Ecklonia maxima* and *Lessonia nigrescens* are the most important commercial sources of alginates. Structurally distinct bacterial alginates can be obtained from *Pseudomonas* species and *Azotobacter* species. Although algin, alginic acid, ammonium alginate, calcium alginate, potassium alginate and sodium alginate are essentially different states of the same type of compound, for legislative purposes they are listed separately

NAME:	Alginic acid
CATEGORY:	Polysaccharides
FOOD USE:	All food products, edible films/ Coatings
SYNONYMS:	CAS 9005-32-7/ EINECS 232-680-1/ E400/ Polymannuronic acid/ Norgine
FORMULA:	Variable, depending on ratio of mannuronic to guluronic acid, and the presence of other substituents
MOLECULAR MASS:	Variable
ALTERNATIVE FORMS:	Algin; ammonium alginate; calcium alginate; potassium alginate
PROPERTIES AND APPEARANCE:	White to yellow powders or hard flakes
SOLUBILITY % AT VARIOUS TEMPERATURE/pH COMBINATIONS:	Solubility in aqueous solution varies with differences in structure, molecular weight, pH and cations. Alginic acid (low pH) is insoluble in water, monovalent salts (sodium, potassium and ammonium) are also soluble, but divalent salts (including calcium) except magnesium are insoluble in water. Solubility declines with increasing molecular weight and increasing concentration. Viscosity is not affected in the pH 5 to 11 range, but increases at lower pHs (due to alginic acid formation)
FUNCTION IN FOODS:	Gelling agent; thickening agent; stabiliser; flocculant and bulking agent. It can be used to stabilise emulsions and to retard ice-crystal formation
ALTERNATIVES:	Carrageenans; gelatin; gellan; agar; xanthan; furcellaran (application-dependent)
TECHNOLOGY OF USE IN FOODS:	Alginates should be dispersed in cold deionised water and heated for 10 minutes at 80°C to hydrate the polysaccharide. A high shear mixer will aid dispersion, although prolonged shear of high molecular-weight polymers will reduced the viscosity. Calcium gels can be prepared by addition of alginate solutions to a calcium solution (e.g. maraschino cherries), addition of calcium to a hot solution of alginate and then cooling (gel is thermo-irreversible), or by diffusion. Gels can also be formed by lowering the pH (alginic acid), but this can be technically difficult. Alginates are composed of blocks of polymannuronic acid, polyguluronic acid and mixed regions. Ribbon-like structures are formed in solution and alginates with a high percentage of guluronic blocks can be cross-linked (gelled) by divalent ions (e.g. calcium)

SYNERGISTS:	Divalent ion, pectins
ANTAGONISTS:	None known
FOOD SAFETY ISSUES:	Moderately toxic, LD_{50} (intraperitoneal, rat) 1,600 mg/kg; ADI 0–50 mg/kg. The powder is combustible and emits irritating fumes when heated
LEGISLATION:	Usage level: Limitation 1% (condiments); 6% (pimento for stuffed olives); 0.3% (confections); 4% (gelatins/puddings); 10% (hard candy); 2% (processed fruits); 1% (other foods)

USA:	**UK and EUROPE:**	**AUSTRALIA/PACIFIC RIM:**
FDA 21CFR § 184.1011. GRAS	UK: approved; Europe: listed	Japan: approved

REFERENCES:

Ash, M., and Ash, I. (1996) *Food Additives.* Gower Publishing Co., Brookfield, VT.

Clare, K. (1993) Algin. In: Whistler R. L. and J. N. BeMiller (Eds.), *Industrial Gums: Polysaccharides and Derivatives.* 3rd Edition, pp. 105–143.

ANY OTHER RELEVANT INFORMATION:

Alginates are natural polysaccharides which are extracted from brown seaweeds (Phaeophyceae). *Macrocystis pyrifera*, *Laminera hyperborea*, *Laminaria digitata*, *Laminaria japonica*, *Ascophyllum nodosum*, *Ecklonia maxima* and *Lessonia nigrescens* are the most important commercial sources of alginates. Structurally distinct bacterial alginates can be obtained from *Pseudomonas* species and *Azotobacter* species. Although algin, alginic acid, ammonium alginate, calcium alginate, potassium alginate and sodium alginate are essentially different states of the same type of compound, for legislative purposes they are listed separately

NAME:	Ammonium alginate
CATEGORY:	Polysaccharides
FOOD USE:	All food products/ Edible films/ Coatings
SYNONYMS:	CAS 9005-34-9/ E403/ Ammonium alginate/ Alginic acid-ammonium salt/ Ammonium polymannuronate
FORMULA:	Variable, depending on ratio of mannuronic to guluronic acid, and the presence of other substituents
MOLECULAR MASS:	Variable
ALTERNATIVE FORMS:	Alginic acid; algin; calcium alginate; potassium alginate
PROPERTIES AND APPEARANCE:	White to yellow powders or hard flakes
SOLUBILITY % AT VARIOUS TEMPERATURE/pH COMBINATIONS:	Solubility in aqueous solution varies with differences in structure, molecular weight, pH and cations. Alginic acid (low pH) is insoluble in water, monovalent salts (sodium, potassium and ammonium) are also soluble, but divalent salts (including calcium) except magnesium are insoluble in water. Solubility declines with increasing molecular weight and increasing concentration. Viscosity is not affected in the pH 5 to 11 range but increases at lower pHs (due to alginic acid formation)
FUNCTION IN FOODS:	Gelling agent; thickening agent; stabiliser; flocculant and bulking agent. It can be used to stabilise emulsions and to retard ice-crystal formation
ALTERNATIVES:	Carrageenans; gelatin; gellan; agar; xanthan; furcellaran (application-dependent)
TECHNOLOGY OF USE IN FOODS:	Alginates should be dispersed in cold deionised water and heated for 10 minutes at 80°C to hydrate the polysaccharide. A high shear mixer will aid dispersion although prolonged shear of high molecular-weight polymers will reduce the viscosity. Calcium gels can be prepared by addition of alginate solutions to a calcium solution (e.g. maraschino cherries), addition of calcium to a hot solution of alginate and then cooling (gel is thermo-irreversible), or by diffusion. Gels can also be formed by lowering the pH (alginic acid), but this can be technically difficult. Ammonium alginate forms particularly viscous solutions.

Alginates are composed of blocks of polymannuronic acid, polyguluronic acid and mixed regions. Ribbon-like structures are formed in solution and alginates with a high percentage of guluronic blocks can be cross-linked (gelled) by divalent ions (e.g. calcium)

SYNERGISTS: Divalent ion, pectins

ANTAGONISTS: None known

FOOD SAFETY ISSUES: LD_{50} (intravenous, rat) 1,000 mg/kg; ADI 0–50 mg/kg

The powder is combustible and emits toxic fumes when heated

LEGISLATION: Usage level: Limitation 0.4% (confections); 0.5% (fats/oils); 0.5% (gelatins/puddings); 0.4% (gravies/sauces); 0.5% (sweet sauces); 0.1% (other foods)

USA:	**UK and EUROPE:**	**AUSTRALIA/PACIFIC RIM:**
FDA 21CFR § 173.310, 184.1133, GRAS, USDA 9CFR 318.7	UK: approved; Europe: listed	Japan: approved

REFERENCES:

Ash, M., and Ash, I. (1996) *Food Additives.* Gower Publishing Co., Brookfield, VT.

Clare, K. (1993) Algin. In: Whistler R. L. and J. N. BeMiller (Eds.), *Industrial Gums: Polysaccharides and Derivatives.* 3rd Edition, pp. 105–143.

ANY OTHER RELEVANT INFORMATION: Alginates are natural polysaccharides which are extracted from brown seaweeds (Phaeophyceae). *Macrocystis pyrifera, Laminera hyperborea, Laminaria digitata, Laminaria japonica, Ascophyllum nodosum, Ecklonia maxima* and *Lessonia nigrescens* are the most important commercial sources of alginates. Structurally distinct bacterial alginates can be obtained from *Pseudomonas* species and *Azotobacter* species. Although algin, alginic acid, ammonium alginate, calcium alginate, potassium alginate and sodium alginate are essentially different states of the same type of compound, for legislative purposes they are listed separately

NAME:	Gum arabic
CATEGORY:	Polysaccharides
FOOD USE:	Confectionery/ Beverages/ Baked goods/ Emulsified foods
SYNONYMS:	CAS 9000-01-5/ EINECS 232-519-5/ E414/ Acacia gum/ Sudan gum/ Gum arabica/ Gum hashab/ Kordofan gum/ Arabic gum/ Kami
MOLECULAR MASS:	Variable, although typically 384,000
ALTERNATIVE FORMS:	Gum arabic is slightly acidic, so different salt forms are possible
PROPERTIES AND APPEARANCE:	White/off-white powder or amber to dark brown crystals
SOLUBILITY % AT VARIOUS TEMPERATURE/pH COMBINATIONS:	Gum arabic is soluble in water, sucrose solutions, sodium chloride solution and ethanol solutions, but solubility in vegetable oils and propylene glycol is limited. Viscosity is optimal at pH 6, although the variation across the pH range 3 to 10 is almost negligible
FUNCTION IN FOODS:	Emulsifier; flavouring agent; adjuvant; formulation aid; humectant; surface-finishing agent; stabiliser; thickener; firming agent; processing aid; texturiser; crystallisation inhibitor; for confections, flavour syrups; dietary fibre; foam stabiliser
ALTERNATIVES:	Gum karaya; gum ghatti; gum tragacanth; maltodextrins; pyrodextrins; polydextroses; methyl celluloses; milk proteins (application-dependent)
TECHNOLOGY OF USE IN FOODS:	Solution of gum arabic can be prepared by dispersing in aqueous solution at room temperature. Emulsions should be prepared with a high shear mixer. Gum arabic is a complex branched arabinogalactan with a (1 → 3) beta-D-galactopyranosyl main chain with numerous branched and unbranched side chains composed of D-galactose, L-arabinose, D-glucuronic acid, L-rhamnose and some methyl substituents. It forms a globular structure in aqueous solution with few interactions with other food ingredients. It is thought that small amounts of protein may be important in gum arabic emulsifying properties.

Large quantities of gum arabic are used in confectionery where it inhibits sugar crystallisation, emulsifies lipid and helps to bind water. Gum arabic is also used as an encapsulating agent, to emulsify flavour/colour in beverages, and as a general emulsifier.

In comparison with many other polysaccharides it is highly soluble in aqueous solutions and very low in viscosity. Viscosity increases with increasing molecular weight and concentration. The viscosity is pH-independent, although like most polysaccharides it starts to degrade at >pH 8 and <pH 4, with concomitant reductions in viscosity

SYNERGISTS: None known

ANTAGONISTS: None known

FOOD SAFETY ISSUES: Non-toxic. The powder is combustible

LEGISLATION: Usage level: Limitation 2% (beverages); 5.6% (chewing gum); 12.4% (confections); 1.3% (dairy products); 1.5% (fats, oils); 2.5% (gelatins, puddings); 46.5% (hard candy); 8.3% (nuts); 6% (frozen confections); 4% (snack foods); 85% (soft candy); 1% (other foods)

USA:	UK and EUROPE:	AUSTRALIA/PACIFIC RIM:
FDA 21CFR § 169.179, 169.182, 184.1330, GRAS	UK: approved. Europe: listed	Japan: approved for use as a natural thickener and stabiliser

REFERENCES:

Anderson, D. M. W, Gill, M. L. C., Jeffrey, A. M., and McDougall, F. J. (1985) *Phytochemistry*, **24**, 71.

Whistler, R. L. (1993) In: Whistler, R. L., and J. N. BeMiller (Eds.), Exudate gums, *Industrial Gums: Polysaccharides and Derivatives.* 3rd edition, pp. 311–318.

Ash, M., and Ash, I. (1995) *Food Additives.* Gower Publishing Co., Brookfield, VT.

ANY OTHER RELEVANT INFORMATION: This is a natural polysaccharide that is produced as a protective sticky exudate by acacia trees in response to a wound. The primary species are *Acacia senegal*, *Acacia laetia*, *Acacia famesiana* and *Acacia seyal*. Gum arabic has traditionally been produced in Sudan and neighbouring African countries. Droughts and wars have dramatically affected the price and supply of this product in recent years. Product quality has also proved variable. Alternative products that can adequately reproduce all of the functionalities of gum arabic have proved elusive, although some novel microbial polysaccharides are producing promising results

NAME:	Calcium alginate
CATEGORY:	Polysaccharides
FOOD USE:	All food products/ Edible films/ Coatings
SYNONYMS:	CAS 9005-35-0/ E404/ Calcium alginate/ Alginic acid-calcium salt/ Calcium polymannuronate
FORMULA:	Variable, depending on ratio of mannuronic to guluronic acid, and the presence of other substituents
MOLECULAR MASS:	Variable
ALTERNATIVE FORMS:	Alginic acid; algin; calcium alginate; potassium alginate
PROPERTIES AND APPEARANCE:	White to yellow powders or hard flakes
SOLUBILITY % AT VARIOUS TEMPERATURE/pH COMBINATIONS:	Solubility in aqueous solution varies with differences in structure, molecular weight, pH and cations. Alginic acid (low pH) is insoluble in water, monovalent salts (sodium, potassium and ammonium) are also soluble, but divalent salts (including calcium) except magnesium are insoluble in water. Solubility declines with increasing molecular weight and increasing concentration. Viscosity is not affected in the pH 5 to 11 range, but increases at lower pHs (due to alginic acid formation).
FUNCTION IN FOODS:	Gelling agent; thickening agent; stabiliser; flocculant and bulking agent. It can be used to stabilise emulsions and to retard ice-crystal formation
ALTERNATIVES:	Carrageenans; gelatin; gellan; agar, xanthan; furcellaran (application-dependent)
TECHNOLOGY OF USE IN FOODS:	Alginates should be dispersed in cold deionised water and heated for 10 minutes at 80°C to hydrate the polysaccharide. A high shear mixer will aid dispersion, although prolonged shear of high molecular-weight polymers will reduce the viscosity. Calcium gels can be prepared by addition of alginate solutions to a calcium solution (e.g. maraschino cherries), addition of calcium to a hot solution of alginate and then cooling (gel is thermo-irreversible), or by diffusion. Gels can also be formed by lowering the pH (alginic acid), but this can be technically difficult. Ammonium alginate forms particularly viscous solutions.

Alginates are composed of blocks of polymannuronic acid, polyguluronic acid and mixed regions. Ribbon-like structures are formed in solution and alginates with a high percentage of guluronic blocks can be cross-linked (gelled) by divalent ions (e.g. calcium)

SYNERGISTS: Divalent ion, pectins

ANTAGONISTS: None known

FOOD SAFETY ISSUES: ADI 0–25 mg/kg. The powder is combustible and emits toxic fumes when heated

LEGISLATION: Usage level: Limitation 0.002% (baked goods); 0.4% (alcoholic beverages); 0.4% (frostings); 0.6% (egg products); 0.5% (fats/oils); 0.25% (gelatins/puddings); 0.4% (gravies); 0.5% (sweet sauces); 0.3% other foods)

USA:	UK and EUROPE:	AUSTRALIA/PACIFIC RIM:
FDA 21CFR § 184.1187. GRAS, USDA 9CFR 318.7	UK: approved. Europe: listed	Japan: approved

REFERENCES:

Ash, M., and Ash, I. (1996) *Food Additives*. Gower Publishing Co., Brookfield, VT.

Clare, K. (1993) Algin. In: Whistler, R. L., and J. N. BeMiller (Eds.), *Industrial Gums: Polysaccharides and Derivatives*. 3rd edition, pp. 105–143.

ANY OTHER RELEVANT INFORMATION: Alginates are natural polysaccharides which are extracted from brown seaweeds (Phaeophyceae). *Macrocystis pyrifera*, *Laminera hyperborea*, *Laminaria digitata*, *Laminaria japonica*, *Ascophyllum nodosum*, *Ecklonia maxima* and *Lessonia nigrescens* are the most important commercial sources of alginates. Structurally distinct bacterial alginates can be obtained from *Pseudomonas* species and *Azotobacter* species. Although algin, alginic acid, ammonium alginate, calcium alginate, potassium alginate and sodium alginate are essentially different states of the same type of compound, for legislative purposes they are listed separately

NAME:	Iota-carrageenan
CATEGORY:	Polysaccharides
FOOD USE:	All food products/ Edible films and coatings
SYNONYMS:	CAS 9004-07-1/ EINECS 232-524-2/ E407/ E407a/ Carrageenin/ Carrageen/ Carragheenan/ Chondrus/ Irish Moss/ Processed Euchema Carrageenan (PES)/ Philippine Natural Carrageenan (PNG)/ Alternatively Refined Carrageenan (ARC)
FORMULA:	$(C_6H_{7.5}S_1O_{7.5})_n$. A polymer formed from a disaccharide repeating unit composed of a 4-*O*-sulfato-beta-D-galactopyranosyl unit and a 3,6 anhydro-2-*O*-sulfato-alpha-D-galactopyranosyl unit
MOLECULAR MASS:	Variable, but typically 700,000 to 5 million
ALTERNATIVE FORMS:	Blends including iota-carrageenans; polymers in which most but not all the units are of the iota-carrageenan type; sodium salts; potassium salts; calcium salts; ammonium salts, etc.
PROPERTIES AND APPEARANCE:	White to yellow powders or flakes
WATER CONTENT MAXIMUM IN %:	0–22
ASH MAXIMUM IN %:	19–42
SOLUBILITY % AT VARIOUS TEMPERATURE/pH COMBINATIONS:	The sodium salt of iota-carrageenans is soluble in cold water, but potassium and calcium salts must be heated to >60°C. Iota-carrageenan is slightly soluble in hot concentrated sugar solutions, soluble in hot concentrated salt solutions, but insoluble in 35% ethanol, vegetable oils and propylene glycol. Optimum solubility occurs at pH 9 and declines as the pH is reduced (<pH 4). Carrageenans degrade spontaneously at low pHs. Elevated temperatures significantly improve the dispersion of carrageenans, particularly potassium and calcium salts of iota-carrageenan which require temperatures above 60°C for dissolution
FUNCTION IN FOODS:	Gelling agent; thickening agent; stabiliser; flocculant and bulking agent. It can be used to stabilise emulsions and to retard ice-crystal formation
ALTERNATIVES:	Gelatin; gellan; agar; alginate; xanthan; furcellaran (application-dependent)

TECHNOLOGY OF USE IN FOODS:

Carrageenans should be dispersed in cold water and heated for 10 minutes at 80°C to hydrate the polysaccharide. A high shear mixer will aid dispersion, although prolonged shear of high molecular-weight polymers will reduce the viscosity.

Commercially available grades of carrageenan are complex mixtures of a wide range of carrageenans (kappa, iota, lambda, etc.) with different molecular weights, a range of different cations (chiefly potassium, sodium and calcium) and other constituents (e.g. polysorbate 80, celluloses in semi-refined carrageenans). Consequently, the functional properties of carrageenan extracts are extremely diverse.

Extracts with a high concentration of kappa-carrageenan form strong thermoreversible gels with high melting points, especially in combination with potassium ions or casein.

Iota-carrageenan-enriched carrageenans can be used for compliant gels, and form stronger gels with calcium ions than potassium ions. Iota-carrageenans can interact with proteins below the protein isoelectric point, and this interaction has been used to improve the stability of protein products

SYNERGISTS:

Casein; chitosan; locust bean gum; potassium; calcium

ANTAGONISTS:

Any positive or negatively charged molecule is a potential antagonist

FOOD SAFETY ISSUES:

Suspected carcinogen. May induce colonic ulcers. The powder is combustible

LEGISLATION:

USA:	**UK and EUROPE:**	**AUSTRALIA/PACIFIC RIM:**
FDA 21CFR § 172.620, 172.623, 172.626, 182.7255. GRAS. USDA 9CFR § 318.7. ADI = 75 mg/kg	UK: approved Europe: listed. Carrageenans extracted using traditional aqueous extraction methods are labelled E407. Carrageenans (PES, PNC, ARC) extracted using an alternative commercial method are labelled E407a	Japan: approved for use as a natural thickener and stabiliser

REFERENCES:

Ash, M., and Ash, I. (1996) *Food Additives*. Gower Publishing Co., Brookfield, VT.

Craighie, J. S., and Leigh, C. (1978) Carrageenans and Agars. In: *Handbook of Physiological and Biochemical Methods*. Cambridge University Press, p. 114.

Shepherd, R. (1996) Extraction, purification and characterization of novel carrageenans with potential applications in dairy products. NZBA.

Therkelsen, G. H. (1993) Carrageenan. In: Whistler, R. L., and J. N. BeMiller (Eds.), *Industrial Gums: Polysaccharides and Derivatives*. 3rd edition, pp. 145–180.

ANY OTHER RELEVANT INFORMATION:

Carrageenans are natural sulphated polysaccharides which are extracted from red seaweeds (Rhodophyceae). *Euchema* spp., *Chondrus crispus, Iridaea* spp. and *Gigartina* spp. are the most important commercial sources of carrageenan. There is concern that low molecular-weight fractions of carrageenan could be toxic, and many manufacturers have voluntarily stopped using carrageenans in baby foods.

Semi-refined carrageenans are not water extracts from seaweeds and as such did not match the legal definition of a carrageenan. They are extracted from seaweeds using an alternative process and differ from other carrageenans in that they contain significant amounts of celluloses.

See also: carrageenan, kappa-carrageenan, lambda-carrageenan and furcellaran

NAME:	Kappa-carrageenan
CATEGORY:	Polysaccharides
FOOD USE:	All food products/ Edible films and coatings
SYNONYMS:	CAS 9004-07-1/ EINECS 232-524-2/ E407/ E407a/ Carrageenin/ Carrageen/ Carragheenan/ Chondrus/ Irish Moss/ Processed Euchema Carrageenan (PES)/ Philippine Natural Carrageenan (PNG)/ Alternatively Refined Carrageenan (ARC)
FORMULA:	$(C_6H_8S_{0.5}O_6)_n$. A polymer formed from a disaccharide repeating unit composed of a 4-*O*-sulfato-beta-D-galactopyranosyl unit and a 3,6 anhydro-alpha-D-galactopyranosyl unit
MOLECULAR MASS:	Variable, but typically 700,000 to 5 million
ALTERNATIVE FORMS:	Blends including kappa-carrageenans; polymers in which most but not all the units are of the kappa-carrageenan type; sodium salts; potassium salts; calcium salts; ammonium salts, etc.
PROPERTIES AND APPEARANCE:	White to yellow powders or flakes
WATER CONTENT MAXIMUM IN %:	0–22
ASH MAXIMUM IN %:	19–42
SOLUBILITY % AT VARIOUS TEMPERATURE/pH COMBINATIONS:	The sodium salt of kappa-carrageenans is soluble in cold water, but potassium and calcium salts must be heated to >60°C. Kappa-carrageenan will dissolve in hot concentrated sugar solutions but is insoluble in concentrated salt solutions and 35% ethanol. It is insoluble in vegetable oils and propylene glycol. Optimum solubility occurs at pH 9 and declines as the pH is reduced (<pH 4). Carrageenans degrade spontaneously at low pHs. Elevated temperatures significantly improve the dispersion of carrageenans, particularly potassium and calcium salts of kappa-carrageenans which require temperatures above 60°C for dissolution
FUNCTION IN FOODS:	Gelling agent; thickening agent; stabiliser; flocculant and bulking agent. It can be used to stabilise emulsions and to retard ice-crystal formation

ALTERNATIVES: Gelatin; gellan; agar; alginate; xanthan; furcellaran (application-dependent)

TECHNOLOGY OF USE IN FOODS: Carrageenans should be dispersed in cold water and heated for 10 minutes at 80°C to hydrate the polysaccharide. A high shear mixer will aid dispersion although prolonged shear of high molecular-weight polymers will reduce the viscosity.

Commercially available grades of carrageenan are complex mixtures of a wide range of carrageenans (kappa, iota, lambda, etc.) with different molecular weights, a range of different cations (chiefly potassium, sodium and calcium) and other constituents (e.g. polysorbate 80, celluloses in semi-refined carrageenans). Consequently, the functional properties of carrageenan extracts are extremely diverse.

Extracts with a high concentration of kappa-carrageenan form strong thermoreversible gels with high melting points, especially in combination with potassium ions or casein.

Kappa-carrageenans are particularly suitable for dairy applications because the strong synergistic interaction with casein allows gel formation at very low carrageenan concentrations (e.g. 0.04% w/w). The flocculation properties of carrageenans (Irish moss) have traditionally been used to clarify beers and wines. Kappa-carrageenans can interact with proteins below the protein isoelectric point, and this interaction has been used to improve the stability of protein products

SYNERGISTS: Casein; chitosan; locust bean gum; potassium; calcium

ANTAGONISTS: Any positive or negatively charged molecule is a potential antagonist.

FOOD SAFETY ISSUES: Suspected carcinogen. May induce colonic ulcers. The powder is combustible

LEGISLATION:

USA:	UK and EUROPE:	AUSTRALIA/PACIFIC RIM:
FDA 21CFR § 172.620, 172.623, 172.626, 182.7255. GRAS. USDA 9CFR § 318.7. ADI = 75 mg/kg	UK: approved. Europe: listed. Carrageenans extracted using traditional aqueous extraction methods are labelled E407. Carrageenans (PES, PNC, ARC) extracted using an alternative commercial method are labelled E407a	Japan: approved for use as a natural thickener and stabiliser

REFERENCES:

Ash, M., and Ash, I. (1996) *Food Additives*. Gower Publishing Co., Brookfield, VT.

Craighie, J. S., and Leigh, C. (1978) Carrageenans and Agars. In: *Handbook of Physiological and Biochemical Methods*. Cambridge University Press, p. 114.

Shepherd, R. (1996) Extraction, purification and characterization of novel carrageenans with potential applications in dairy products. NZBA.

Therkelsen, G. H. (1993) Carrageenan. In: Whistler, R. L., and J. N. BeMiller (Eds.), *Industrial Gums: Polysaccharides and Derivatives*. 3rd edition, pp. 145–180.

ANY OTHER RELEVANT INFORMATION:

Carrageenans are natural sulphated polysaccharides which are extracted from red seaweeds (Rhodophyceae). *Euchema* spp., *Chondrus crispus, Iridaea* spp. and *Gigartina* spp. are the most important commercial sources of carrageenan. There is concern that low molecular-weight fractions of carrageenan could be toxic, and many manufacturers have voluntarily stopped using carrageenans in baby foods.

Semi-refined carrageenans are not water extracts from seaweeds and as such did not match the legal definition of a carrageenan. They are extracted from seaweeds using an alternative process and differ from other carrageenans in that they contain significant amounts of celluloses.

See also: carrageenan, iota-carrageenan, lambda-carrageenan and furcellaran

NAME:	Lambda-carrageenan
CATEGORY:	Polysaccharides
FOOD USE:	All food products/ Edible films and coatings
SYNONYMS:	CAS 9004-07-1/ EINECS 232-524-2/ E407/ E407a/ Carrageenin/ Carrageen/ Carragheenan/ Chondrus/ Irish Moss/ Processed Euchema Carrageenan (PES)/ Philippine Natural Carrageenan (PNG)/ Alternatively Refined Carrageenan (ARC)
FORMULA:	(C6 H8.5 S1.5 O9.5)n. A polymer formed from a disaccharide repeating unit composed of a 2-*O*-sulfato-beta-D-galactopyranosyl unit and a 2,6 sulfato-alpha-D-galactopyranosyl unit
MOLECULAR MASS:	Variable, but typically 700,000 to 5 million
ALTERNATIVE FORMS:	Blends including lambda-carrageenans; polymers in which most but not all the units are of the lambda-carrageenan type; sodium salts; potassium salts; calcium salts; ammonium salts, etc
PROPERTIES AND APPEARANCE:	White to yellow powders or flakes
WATER CONTENT MAXIMUM IN %:	0–22
ASH MAXIMUM IN %:	19–42
SOLUBILITY % AT VARIOUS TEMPERATURE/pH COMBINATIONS:	Salts of lambda-carrageenan are soluble in cold water, concentrated sugar solutions, concentrated salt solutions and 35% ethanol. Lambda-carrageenan is insoluble in vegetable oils and propylene glycol. Optimum solubility occurs at pH 9 and declines as the pH is reduced (<pH 4). Carrageenans degrade spontaneously at low pHs. Elevated temperatures significantly improve the dispersion of carrageenans
FUNCTION IN FOODS:	Thickening agent; stabiliser; flocculant and bulking agent. It can be used to stabilise emulsions and to retard ice-crystal formation
ALTERNATIVES:	Xanthan; furcellaran (application-dependent)

TECHNOLOGY OF USE IN FOODS:

Carrageenans should be dispersed in cold water and heated for 10 minutes at 80°C to hydrate the polysaccharide. A high shear mixer will aid dispersion, although prolonged shear of high molecular-weight polymers will reduce the viscosity.

Commercially available grades of carrageenan are complex mixtures of a wide range of carrageenans (kappa, iota, lambda, etc.) with different molecular weights, a range of different cations (chiefly potassium, sodium and calcium) and other constituents (e.g. polysorbate 80, celluloses in semi-refined carrageenans). Consequently the functional properties of carrageenan extracts are extremely diverse.

Extracts with a high concentration of lambda-carrageenans do not gel at all but can be useful thickeners. Lambda-carrageenans can interact with proteins below the protein isoelectric point, and this interaction has been used to improve the stability of protein products

SYNERGISTS:

Proteins; chitosan

ANTAGONISTS:

Any positive or negatively charged molecule is a potential antagonist

FOOD SAFETY ISSUES:

Suspected carcinogen. May induce colonic ulcers. The powder is combustible

LEGISLATION:

USA:	**UK and EUROPE:**	**AUSTRALIA/PACIFIC RIM:**
FDA 21CFR § 172.620, 172.623, 172.626, 182.7255. GRAS. USDA 9CFR § 318.7. ADI = 75 mg/kg	UK: approved. Europe: listed. Carrageenans extracted using traditional aqueous extraction methods are labelled E407. Carrageenans (PES, PNC, ARC) extracted using an alternative commercial method are labelled E407a	Japan: approved for use as a natural thickener and stabiliser

REFERENCES:

Ash, M., and Ash, I. (1996) *Food Additives*. Gower Publishing Co., Brookfield, VT.

Craighie, J. S., and Leigh, C. (1978) Carrageenans and Agars. In: *Handbook of Physiological and Biochemical Methods*. Cambridge University Press, p. 114.

Shepherd, R. (1996) Extraction, purification and characterization of novel carrageenans with potential applications in dairy products. NZBA.

Therkelsen, G. H. (1993) Carrageenan. In: Whistler, R. L., and J. N. BeMiller (Eds.), *Industrial Gums: Polysaccharides and Derivatives*. 3rd edition, pp. 145–180.

ANY OTHER RELEVANT INFORMATION:	Carrageenans are natural sulphated polysaccharides which are extracted from red seaweeds (Rhodophyceae). *Eurchema* spp., *Chondrus crispus, Iridaea* spp. and *Gigartina* spp. are the most important commercial sources of carrageenan. There is concern that low molecular-weight fractions of carrageenan could be toxic, and many manufacturers have voluntarily stopped using carrageenans in baby foods. Semi-refined carrageenans are not water extracts from seaweeds and as such did not match the legal definition of a carrageenan. They are extracted from seaweeds using an alternative process and differ from other carrageenans in that they contain significant amounts of celluloses. See also: carrageenan, kappa-carrageenan, lambda-carrageenan and furcellaran

NAME:	**Carrageenan**
CATEGORY:	Polysaccharides
FOOD USE:	All food products/ Edible films and coatings
SYNONYMS:	CAS 9004-07-1/ EINECS 232-524-2/ E407/ E407a/ Carrageenin/ Carrageen/ Carragheenan/ Chondrus/ Irish Moss/ Processed Euchema Carrageenan (PES)/ Philippine Natural Carrageenan (PNG)/ Alternatively Refined Carrageenan (ARC)
FORMULA:	$(C_6H_{7\text{-}10}S_{0\text{-}6}O_{4.5\text{-}14})_n$. Sulfated galactans with a disaccharide repeating unit composed of beta-D-$(1 \rightarrow 4)$ galactose and alpha-D-$(1 \rightarrow 3)$ galactose units and close derivatives
MOLECULAR MASS:	Variable, but typically 100,000 to 5 million
ALTERNATIVE FORMS:	Kappa-carrageenan; iota-carrageenan; lambda-carrageenan; theta-carrageenan; mu-carrageenan; nu-carrageenan; xi-carrageenan; sodium salts; potassium salts; calcium salts; ammonium salts; Processed Euchema Carrageenan (PES); etc.
PROPERTIES AND APPEARANCE:	White to yellow powders or hard flakes
WATER CONTENT MAXIMUM IN %:	0–20
ASH MAXIMUM IN %:	19–42
SOLUBILITY % AT VARIOUS TEMPERATURE/pH COMBINATIONS:	Solubility in aqueous solution varies with differences in structure, molecular weight and cations. Lambda-carrageenans are generally more soluble than iota-carrageenans, which tend to be more soluble than kappa-carrageenans. Solubility declines with increasing molecular weight and increasing concentration and generally, sodium salts are more soluble than calcium and potassium salts. Kappa-, iota- and lambda-carrageenans are soluble in hot concentrated sugar solutions, iota- and lambda-carrageenan are soluble in hot saturated salt solutions, and only the sodium salt of lambda carrageenan is soluble in 35% ethanol solution. Carrageenans are not soluble in non-polar solvents like vegetable oils and propylene glycol. Optimum solubility occurs at pH 9 and declines as the pH is reduced (<pH 4). Carrageenans degrade spontaneously at low pHs. Elevated temperatures significantly improve the dispersion of carrageenans,

	particularly potassium and calcium salts of kappa- and iota-carrageenans which require temperatures above 60°C for dissolution
FUNCTION IN FOODS:	Gelling agent; thickening agent; stabiliser; flocculant and bulking agent. It can be used to stabilise emulsions and to retard ice-crystal formation
ALTERNATIVES:	Gelatin; gellan; agar; alginate; xanthan; furcellaran (application-dependent)
TECHNOLOGY OF USE IN FOODS:	Carrageenans should be dispersed in cold water and heated for 10 minutes at 80°C to hydrate the polysaccharide. A high shear mixer will aid dispersion, although prolonged shear of high molecular-weight polymers will reduce the viscosity. Commercially available grades of carrageenan are complex mixtures of a wide range of carrageenans (kappa, iota, lambda, etc.) with different molecular weights, a range of different cations (chiefly potassium, sodium and calcium) and other constituents (e.g. polysorbate 80, celluloses in PES). Consequently, the functional properties of carrageenan extracts are extremely diverse. Extracts with a high concentration of kappa-carrageenan form strong thermoreversible gels with high melting points, especially in combination with potassium ions or casein. Iota-carrageenan-enriched carrageenans can be used for compliant gels and form stronger gels with calcium ions than potassium ions. Lambda-carrageenans do not gel at all, but can be useful thickeners. Carrageenans are particularly suitable for dairy applications because the strong synergistic interaction with casein allows gel formation at very low carrageenan concentrations (e.g. 0.04% w/w). The flocculation properties of carrageenans have traditionally been used to clarify beers and wines. Carrageenans can interact with proteins below the protein isoelectric point, and this interaction has been used to improve the stability of protein products
SYNERGISTS:	Casein; chitosan; locust bean gum; potassium; calcium
ANTAGONISTS:	Any positive or negatively charged molecule is a potential antagonist
FOOD SAFETY ISSUES:	Suspected carcinogen. May induce colonic ulcers. The powder is combustible

LEGISLATION:

USA:
FDA 21 CFR § 172.620, 172.623, 172.626, 182.7255. GRAS. USDA 9CFR § 318.7. ADI = 75 mg/kg

AUSTRALIA/PACIFIC RIM:
Japan: approved for use as a natural thickener and stabiliser

UK and EUROPE:
UK: approved.
Europe: listed. Carrageenans extracted using traditional aqueous extraction methods are labelled E407.
Carrageenans (PES, PNC, ARC) extracted using an alternative commercial method are labelled E407a

REFERENCES:

Ash, M., and Ash, I. (1996) *Food Additives.* Gower Publishing Co., Brookfield, VT.

Craighie, J. S., and Leigh C. (1978) Carrageenans and Agars. In: *Handbook of Physiological and Biochemical Methods*. Cambridge University Press, p. 114.

Shepherd R. (1996) Extraction, purification and characterization of novel carrageenans with potential applications in dairy products. NZBA.

Therkelsen, G. H. (1993) Carrageenan. In: Whistler, R. L., and J. N. BeMiller (Eds.), *Industrial Gums: Polysaccharides and Derivatives*. 3rd edition, pp. 145–180.

ANY OTHER RELEVANT INFORMATION:

Carrageenans are natural sulphated polysaccharides which are extracted from red seaweeds (Rhodophyceae). *Euchema* spp., *Chondrus crispus*, *Iridaea* spp. and *Gigartina* spp. are the most important commercial sources of carrageenans.

There is concern that low molecular-weight fractions of carrageenan could be toxic, and many manufacturers have voluntarily stopped using carrageenans in baby foods.

Semi-refined carrageenans are not water extracts from seaweeds and as such did not match the legal definition of a carrageenan. They are extracted from seaweeds using an alternative process and differ from other carrageenans in that they contain significant amounts of celluloses.

See also: kappa-carrageenan, iota-carrageenan, lambda-carrageenan and furcellaran.

NAME:	Cellulose
CATEGORY:	Polysaccharides
FOOD USE:	Dry formulated products/ Confectionery processing aid/ Low-calorie foods/ All food products
SYNONYMS:	Beta-D-(1 → 4) glucan/ CAS 9004-34-6/ EINECS 232-674-9/ E460/ Alpha-cellulose/ Bleached wood pulp/ Cotton fibre
FORMULA:	$(C_6H_{10}O_5)_n$
MOLECULAR MASS:	Variable, generally 160,000–560,000
ALTERNATIVE FORMS:	None
PROPERTIES AND APPEARANCE:	A white powder or crystals
SOLUBILITY % AT VARIOUS TEMPERATURE/pH COMBINATIONS:	Cellulose is insoluble in water and organic solvents
FUNCTION IN FOODS:	Can be used as a processing aid, dietary fibre, a texturising agent, bulking agent and an anticaking agent. It is not digestible, so it can be used in low-calorie preparations
ALTERNATIVES:	Starch hydrolysis products (SHPs) and polydextrose (application-dependent)
TECHNOLOGY OF USE IN FOODS:	Cellulose powder can be mixed in with the dry ingredients or dispersed in an aqueous preparation
SYNERGISTS:	None known
ANTAGONISTS:	None known
FOOD SAFETY ISSUES:	Non-toxic. The powder is combustible

LEGISLATION:	**USA:**	**UK and EUROPE:**	**AUSTRALIA/PACIFIC RIM:**
	FDA GRAS	UK: not permitted in baby foods. Europe: listed	Japan: approved for use as a natural processing aid

REFERENCE: Ash, M., and Ash, I. (1996) *Food Additives*. Gower Publishing Co., Brookfield, VT.

ANY OTHER RELEVANT INFORMATION: Cellulose is natural polysaccharide particles extracted from various plant sources

See also Microcrystalline cellulose

NAME:	Chitosan
CATEGORY:	Polysaccharides
FOOD USE:	Food packaging/ Edible films and coatings/ All food products
SYNONYMS:	2-Acetamido-2-deoxy beta-1,4-D-glucan/ CAS 9012-76-4/ Chitan/ Deacetylated chitin/ Profloc/ Seacure
FORMULA:	Variable, depending on degree of acetylation
MOLECULAR MASS:	Variable
ALTERNATIVE FORMS:	Chitosans with different molecular weights and acetyl contents will have different properties
PROPERTIES AND APPEARANCE:	White/off-white/brown powder or soft flakes
SOLUBILITY % AT VARIOUS TEMPERATURE/pH COMBINATIONS:	Solubility varies with molecular weight. Generally, chitosan is insoluble in water, sucrose solutions, sodium chloride solution, ethanol solutions, vegetable oils and propylene glycol. It is soluble in aqueous solutions of acids (acetic, citric, etc.). Solubility declines with increasing molecular weight and increasing concentration. It is pH-dependent (<pH 4). Colloidal suspensions can be prepared at higher pHs. Elevated temperatures will improve the dispersion of high molecular-weight chitosan
FUNCTION IN FOODS:	Viscosifier; stabilising agent; suspending agent; coating in food contact surfaces; bulking agent
ALTERNATIVES:	Pectins; xanthan; locust bean gum; guar gum; starches (application-dependent)
TECHNOLOGY OF USE IN FOODS:	Chitosan should be dispersed in an aqueous solution of an organic acid ($pH < 4$) and heated for 10 minutes at 80°C to hydrate the polysaccharide. A high shear mixer will aid dispersion, although prolonged shear of high molecular-weight polymers will reduce the viscosity. Chitosan is a partially deacetylated polymer of acetyl glucosamine (2-acetamido-2-deoxy-β-1,4-D-glucan). It is essentially a natural, water-soluble derivative of cellulose with unique properties. In comparison with many other polysaccharides it is relatively insoluble, although it forms viscous solutions at low pHs and colloidal suspensions of chitosan at higher pHs. Viscosity increases with increasing molecular weight and concentration. Like most polysaccharides, it starts to degrade at >pH 8 and <pH 3, with concomitant reductions in viscosity

SYNERGISTS:	Xanthan
ANTAGONISTS:	None known
FOOD SAFETY ISSUES:	Non-toxic. The powder is combustible
LEGISLATION:	**USA:** GRAS status applied for **AUSTRALIA/PACIFIC RIM:** Japan: approved for use as a natural thickener and stabiliser
REFERENCES:	Shepherd, R., Reader, S., and Falshaw, A. (1997) Chitosan functional properties. *Glycoconjugate J.*, **14**, 535–542. Whistler, R. L. (1993) Chitin. In: Whistler, R. L., and J. N. BeMiller (Eds.), *Industrial Gums: Polysaccharides and Derivatives*. 3rd edition, pp. 399–417. Ash, M., and Ash, I. (1996) *Food Additives*. Gower Publishing Co., Brookfield, VT.
ANY OTHER RELEVANT INFORMATION:	Chitosan is usually prepared from chitin found in a wide range of natural sources, crustaceans, fungi, insects, annelids, molluscs, coelenterata. However, chitosan is only manufactured from crustaceans (crab, krill and crayfish), primarily because a large amount of the crustacean exoskeleton is available as a byproduct of food processing. Chitosan from mushroom and squid pens may soon be available commercially

NAME:	Carboxymethylcellulose
CATEGORY:	Polysaccharides
FOOD USE:	Soups/ Syrups/ Beverages/ Dairy products/ Dressings/ Dry formulated products/ Food packaging/ Edible films and coatings/ All food products
SYNONYMS:	Carboxymethylated beta-D-(1 → 4) glucan/ CAS 9004-32-4/ E466/ CMC/ Sodium CMC/ Cellulose gum/ Sodium carboxymethylcellulose
FORMULA:	Variable, depending on the degree of substitution
MOLECULAR MASS:	Variable from 21,000 to 500,000
ALTERNATIVE FORMS:	None
PROPERTIES AND APPEARANCE:	A white powder
SOLUBILITY % AT VARIOUS TEMPERATURE/pH COMBINATIONS:	Solubility varies with molecular weight and carboxymethylation. Generally, carboxymethylcellulose is soluble in water, sucrose solutions, sodium chloride solution and ethanol solutions, but solubility in vegetable oils and propylene glycol is limited. Solubility declines with increasing molecular weight, decreasing methylation and increasing concentration. Solubility is pH-dependent with the more insoluble acid form present below pH4
FUNCTION IN FOODS:	Thickening agent; stabiliser; suspending agent; bulking agent
ALTERNATIVES:	Methyl ethyl cellulose (MEC); hydroxy propyl methyl cellulose (HPMC); hydroxy ethyl methylcellulose (HEMC); methyl cellulose (MC); proteins; gum arabic; xanthan, carrageenans; dextrans; starches; locust bean gum; guar gum; gelatin; agar; alginate (application-dependent)
TECHNOLOGY OF USE IN FOODS:	Carboxymethylcellulose should be dispersed in hot water and cooled to hydrate the polysaccharide. A high shear mixer will aid dispersion, although prolonged shear of high molecular-weight polymers will reduce the viscosity.

The viscosity of carboxymethyl celluloses is relatively constant between pH 4 and 10, but increases below pH 4 due to formation of the acid form of the molecule. Medium to high molecular-weight molecules with 0.9 to 1.2 substitutions per sugar unit exhibit pseudoplastic behaviour (shear-dependent thinning), whereas at 0.4 to 0.7% substitution the solutions exhibit thixotropy (time-dependent thinning). Gels can be formed in the presence of trivalent ions like aluminium

SYNERGISTS: Trivalent cations, non-ionic polymers

ANTAGONISTS: None known

FOOD SAFETY ISSUES: Non-toxic; ADI 0–25 mg/kg. The powder is combustible

LEGISLATION:

USA:	**UK and EUROPE:**	**AUSTRALIA/PACIFIC RIM:**
FDA 21CFR § 133.134, 133.178, 133.179, 1150.141, 150.161, 173.310, 175.105, 175.300, 182.70, 182.1745, GRAS, USDA 9CFR § 318.7	UK: approved. Europe: listed	Japan: restricted (2% max)

REFERENCES:

Ash, M., and Ash, I. (1996) *Food Additives.* Gower Publishing Co., Brookfield, VT.

Feedersen, R. L., and Thorp, S. N. (1993) Sodium carboxymethylcellulose. In: Whistler, R. L., and J. N. BeMiller (Eds.), *Industrial Gums: Polysaccharides and Derivatives*. 3rd edition, pp. 537–577.

ANY OTHER RELEVANT INFORMATION: Chemically carboxymethylated natural cellulose.

See also methyl ethyl cellulose (MEC); hydroxy propyl methyl cellulose (HPMC); hydroxy propyl cellulose (HPC); and methyl cellulose (MC)

NAME:	Dextran
CATEGORY:	Polysaccharides
FOOD USE:	Food packaging/ Edible films and coatings/ All food products
SYNONYMS:	Alpha-D-(1 → 6) glucan/ CAS 9004-54-0/ EINECS 232-677-5/ Macrose/ Dextraven/ Gentran/ Hemodex/ Intradex/ Onkotin/ Plavolex/ Polyglucin/ Promit
FORMULA:	$(C_6H_{10}O_5)_n$
MOLECULAR MASS:	Variable; ranges from 2,000 to 500 million.
ALTERNATIVE FORMS:	None
PROPERTIES AND APPEARANCE:	White/off-white powder or soft flakes
SOLUBILITY % AT VARIOUS TEMPERATURE/pH COMBINATIONS:	Solubility varies with molecular weight. Generally, dextran is soluble in water, sucrose solutions, sodium chloride solution and ethanol solutions, but solubility in vegetable oils and propylene glycol is limited. Solubility declines with increasing molecular weight and increasing concentration. It is pH-independent (pH 4 to pH 8). Elevated temperatures will significantly improve the dispersion of high molecular-weight dextrans; however technically, increasing the temperature slightly reduces the solubility of dextrans
FUNCTION IN FOODS:	Coating in food contact surfaces; thickening agent; stabiliser; bulking agent
ALTERNATIVES:	Dextrins; polydextrose; gum arabic and xanthan (application-dependent)
TECHNOLOGY OF USE IN FOODS:	Dextran should be dispersed in cold water and heated for 10 minutes at 80°C to hydrate the polysaccharide. A high shear mixer will aid dispersion, although prolonged shear of high molecular-weight polymers will reduce the viscosity. Dextran is a linear polymer of glucose molecules that adopts a random coil conformation in aqueous solution with minimal interactions between dextran molecules and no known synergistic interactions.

In comparison with many other polysaccharides it is relatively soluble in aqueous solutions and relatively low in viscosity. Viscosity increases with increasing molecular weight and concentration. The viscosity is pH-independent, although like most polysaccharides it starts to degrade at >pH 8 and <pH 4 with concomitant reductions in viscosity

SYNERGISTS: None known

ANTAGONISTS: None known

FOOD SAFETY ISSUES: Non-toxic. The powder is combustible

LEGISLATION:

USA:	**AUSTRALIA/PACIFIC RIM:**
FDA 21 CFR § 186, GRAS as indirect additive. Until 1977, dextran was permitted as a direct additive in the USA, but approval has lapsed as legislation has been updated	Japan: approved for use as a natural thickener and stabiliser

REFERENCES:

Jeanes, A. (1978) *Dextran Bibliography*, Misc. Pub. 1355, U.S. Department of Agriculture.

De Belder, A. N. (1993) Dextran. In: Whistler, R. L., and J. N. BeMiller (Eds.), *Industrial Gums: Polysaccharides and Derivatives*. 3rd edition, pp. 399–417.

Ash, M., and Ash, I. (1996) *Food Additives.* Gower Publishing Co., Brookfield, VT.

ANY OTHER RELEVANT INFORMATION:

Dextran is a natural polysaccharide produced by a wide range of microorganisms. It is generally obtained from cultures of the food-grade microorganism *Leuconostoc mesenteroides*.

Commercially available grades of dextran often contain some alpha-D-(1→3) links and other linkages that could affect functional properties

NAME:	**Furcellaran**
CATEGORY:	Polysaccharides
FOOD USE:	All food products/ Edible films and coatings
SYNONYMS:	CAS 9000-21-9/ EINECS 232-531-0/ E408/ Furcellaran/ Danish gum
FORMULA:	A polymer formed from a disaccharide repeating unit composed of a 4-*O*-sulfato-beta-D-galactopyranosyl unit and a 3,6 anhydro-alpha-D-galactopyranosyl unit
MOLECULAR MASS:	Variable
ALTERNATIVE FORMS:	Very similar in structure to kappa-carrageenan, differing primarily in level of sulphate substitution. In kappa-carrageenan the sulphated hydroxyl unit is in theory 100% sulphated, whereas in furcellaran the hydroxyl is in theory 50% sulphated
PROPERTIES AND APPEARANCE:	White to yellow powder or flakes
SOLUBILITY % AT VARIOUS TEMPERATURE/pH COMBINATIONS:	The sodium salt of furcellaran is soluble in cold water, but potassium and calcium salts must be heated. Furcellaran will dissolve in hot concentrated sugar solutions, but is insoluble in concentrated salt solutions and 35% ethanol. It is insoluble in vegetable oils and propylene glycol. Solubility declines as the pH is reduced (<pH 4). Furcellaran degrades spontaneously at low pHs. Elevated temperatures significantly improve the dispersion of furcellaran, particularly potassium and calcium salts of furcellarans
FUNCTION IN FOODS:	Gelling agent; thickening agent; stabiliser; flocculant and bulking agent. It can be used to stabilise emulsions and to retard ice-crystal formation
ALTERNATIVES:	Kappa-carrageenan; gelatin; gellan; agar; alginate; xanthan (application-dependent)
TECHNOLOGY OF USE IN FOODS:	Furcellaran should be dispersed in cold water and heated for 10 minutes at 80°C to hydrate the polysaccharide. A high shear mixer will aid dispersion, although prolonged shear of high molecular-weight polymers will reduce the viscosity. Commercially available grades of furcellaran are complex mixtures of a wide range of polysaccharides. Consequently, the functional properties of furcellaran extracts may vary

SYNERGISTS:	Casein; chitosan; locust bean gum; potassium; calcium
ANTAGONISTS:	Any positive or negatively charged molecule is a potential antagonist
FOOD SAFETY ISSUES:	Moderately toxic; LD_{50} (oral, rat) 5000 mg/kg. The powder is combustible
LEGISLATION:	**USA:** FDA 21CFR § 172.655 **AUSTRALIA/PACIFIC RIM:** Japan: approved
REFERENCE:	Ash, M., and Ash, I. (1996) *Food Additives.* Gower Publishing Co., Brookfield, VT.
ANY OTHER RELEVANT INFORMATION:	Furcellaran is a natural sulphated polysaccharide which is extracted from a red seaweed *Furcellaria fastigiata* (Rhodophyceae). See also: kappa-carrageenan

NAME:	Gellan
CATEGORY:	Polysaccharides
FOOD USE:	Water gels/ Dessert gels/ Pet foods/ Jams/ Dairy products/ Food packaging/ Edible films and coatings/ All food products
SYNONYMS:	Polymer with a tetrasaccharide repeating unit composed of a beta-D-(1 → 3) glucopyranosyl unit, beta-D-(1 → 4) glucuronopyranosyl unit, beta-D-(1 → 4) glucopyranosyl unit and alpha-L-(1 → 4) rhamnopyranosyl unit with some acetyl and glycerol groups attached/ CAS 71010-52-1/ EINECS 275-117-5/ E418/ Gum gellan/ Gelrite/ Kelcogel
MOLECULAR MASS:	$1–2 \times 10^6$
ALTERNATIVE FORMS:	Salts of potassium; sodium; calcium; magnesium, etc.
PROPERTIES AND APPEARANCE:	White/off-white powder or soft flakes
SOLUBILITY % AT VARIOUS TEMPERATURE/pH COMBINATIONS:	Solubility varies with molecular weight and associated cations. Generally, not fully soluble until about 100°C in water, sucrose or sodium chloride. Solubility declines with increasing molecular weight and increasing concentration. It is not pH-independent
FUNCTION IN FOODS:	Gelling agent; coating in food contact surfaces; thickening agent; stabiliser; bulking agent
ALTERNATIVES:	Gelatin; agar; alginate; curdlan; carrageenan (application-dependent)
TECHNOLOGY OF USE IN FOODS:	Gellan should be dispersed in cold water and autoclaved at 121°C for 10 minutes to hydrate the polysaccharide. A high shear mixer will aid dispersion, although prolonged shear of high molecular-weight polymers will reduce the viscosity. Gellan is a linear polymer of glucose molecules that forms stacked interacting double helices when heated and cooled in aqueous solution. Cations (particularly divalent cations) are required for gel formation. Synergistic interactions with gelatin and gum arabic.

Forms strong, brittle gels which are optically clearer than agar and have similar gel strengths at half the concentration of the equivalent agar gel. In comparison with agar, calcium gellan gels have improved thermostability; they melt above 80°C

SYNERGISTS: Gelatin; gum arabic

ANTAGONISTS: None known

FOOD SAFETY ISSUES: Non-toxic. The powder is combustible

LEGISLATION:

USA:	**UK and EUROPE:**	**CANADA:**	**AUSTRALIA/PACIFIC RIM:**
FDA 21CFR § 172.665	Europe: E418	approved	Japan: approved for use as a natural thickener and stabiliser

REFERENCES:

Kang, K. S., and Pettitt, D. J. (1993) Xanthan, Gellan, Welan and Rhamsan. In: Whistler, R. L., and J. N. BeMiller (Eds.), *Industrial Gums: Polysaccharides and Derivatives.* 3rd edition, pp. 341–395.

Ash, M., and Ash, I. (1996) *Food Additives.* Gower Publishing Co., Brookfield, VT.

ANY OTHER RELEVANT INFORMATION: Gellan is a natural polysaccharide obtained from cultures of the microorganism *Pseudomonas elodea*

NAME:	Gum ghatti
CATEGORY:	Polysaccharides
FOOD USE:	Confectionery/ Beverages/ Baked goods/ Emulsified foods
SYNONYMS:	Indian gum
MOLECULAR MASS:	Variable
ALTERNATIVE FORMS:	Gum ghatti is slightly acidic, so different salt forms are possible
PROPERTIES AND APPEARANCE:	White/off-white powder or amber to dark brown lumps
SOLUBILITY % AT VARIOUS TEMPERATURE/pH COMBINATIONS:	Gum ghatti is a complex mixture of which 10% are insoluble in water. The remainder are soluble in water, sucrose solutions, sodium chloride solution and ethanol solutions, but solubility in vegetable oils and propylene glycol is limited. Maximum viscosity is observed at pH 8
FUNCTION IN FOODS:	Emulsifier; flavouring agent; adjuvant; formulation aid; humectant; surface-finishing agent; stabiliser; thickener; firming agent; processing aid; texturiser; crystallisation inhibitor; for confections, flavour syrups, dietary fibre; foam stabiliser
ALTERNATIVES:	Gum arabic; karaya gum; gum tragacanth; maltodextrins; pyrodextrins; polydextroses; methyl celluloses; milk proteins (application-dependent)
TECHNOLOGY OF USE IN FOODS:	Solution of gum ghatti can be prepared by dispersing in aqueous solution at room temperature. Emulsions should be prepared with a high shear mixer. Gum ghatti is a complex branched arabinogalactan with a $(1 \rightarrow 6)$ beta-D-galactopyranosyl main chain with $(1 \rightarrow 4)$ beta-D-glucuronic acid, $(1 \rightarrow 2)$ mannose and some L-arabinose attached. In comparison with gum arabic it is more viscous, darker in colour and contains more insoluble material
SYNERGISTS:	None known
ANTAGONISTS:	None known

FOOD SAFETY ISSUES: Non-toxic. The powder is combustible

LEGISLATION: Usage level: Limitation 0.2% (alcoholic beverages); 0.1% (other foods)

USA:	**UK and EUROPE:**	**AUSTRALIA/PACIFIC RIM:**
FDA 21CFR § 184.1333, GRAS	UK: approved. Europe: listed	Japan: approved

REFERENCES:

Anderson, D. M. W., Gill, M. L. C., Jeffrey, A. M., and McDougall, F. J. (1985) *Phytochemisty* **24**, 71.

Whistler, R. L. (1993) Exudate gums. In: Whistler, R. L., and BeMiller, J. N. (Eds.), *Industrial Gums: Polysaccharides and Derivatives*. 3rd edition, pp. 311–318.

Ash, M., and Ash, I. (1996) *Food Additives.* Gower Publishing Co., Brookfield, VT.

ANY OTHER RELEVANT INFORMATION: This is a natural polysaccharide that is produced as a protective sticky exudate by *Anogeissus latifolia* in response to a wound. See also gum arabic and tragacanth gum

NAME:	Guar gum
CATEGORY:	Polysaccharides
FOOD USE:	Dairy products/ Pet foods/ Baked goods/ Confectionery/ Dry formulated products/ Food packaging/ Edible films and coatings/ All food products
SYNONYMS:	Beta-D-(1 → 4) mannan with some attached D-galactose molecules linked via alpha (1 → 6) bonds/ CAS 9000-30-0/ EINECS 232-536-8/ E412/ FEMA 2537/ Galactomannan/ Guar flour/ Jaguar gum/ Gum cyamopsis
FORMULA:	Variable, depending on the ratio of galactose to mannose
MOLECULAR MASS:	Variable. Generally 300,000 to 360,000
ALTERNATIVE FORMS:	Guar gum
PROPERTIES AND APPEARANCE:	White/brown powder
SOLUBILITY % AT VARIOUS TEMPERATURE/pH COMBINATIONS:	Solubility varies with molecular weight. Generally, guar gum is soluble in water, sucrose solutions and sodium chloride solution, but solubility in ethanol, vegetable oils and propylene glycol is limited. Solubility declines with increasing molecular weight and increasing concentration. It is pH-independent (pH 4 to pH 8)
FUNCTION IN FOODS:	Thickening agent; stabiliser; bulking agent; suspending agent
ALTERNATIVES:	Locust bean gum; tara gum; carrageenans; dextrans; starches; xanthan; gelatin; agar; alginate (application-dependent)
TECHNOLOGY OF USE IN FOODS:	Guar gum should be dispersed in cold water and heated to 80°C for 10 minutes to hydrate the polysaccharide. A high shear mixer will aid dispersion, although prolonged shear of high molecular-weight polymers will reduce the viscosity. Guar gum molecules have a high hydrodynamic volume and can entangle with adjacent molecules, resulting in viscous properties. Synergistic interactions with xanthan gum result in gel formation. The

synergistic interaction with xanthan is less pronounced with guar gum than locust bean gum, primarily because regions of the molecule that are not galactose-substituted are responsible for the interaction and locust bean gum has a lower galactose to mannose ratio than guar

SYNERGISTS: Xanthan; agar; carrageenan; anionic polymer; anionic surfactants

ANTAGONISTS: None known

FOOD SAFETY ISSUES: Mildly toxic. The powder is combustible.

LEGISLATION: Usage level: Limitation 0.35% (baked goods); 1.2% (cereal); 0.8% (cheese); 1% (dairy products); 2% (fats, oils); 1.2% (gravies); 1% (jam); 0.6% (milk products, nuts); 2% (processed vegetables); 0.8% (soups); 1% (sweet sauces); 0.5% (other foods)

USA:	**UK and EUROPE:**	**AUSTRALIA/PACIFIC RIM:**
FDA 21CFR § 133.124, 133.133, 133.134, 133.162, 133.178, 133.179, 150.141, 150.161, 184.1339, GRAS	UK: approved. Europe: listed	Japan: approved

REFERENCES:

Maier, H., Anderson, M., Kurt, C., and Magnusen, K. (1993) Guar, Locust Bean, Tara and Fenugreek gums. In: Whistler, R. L., and BeMiller, J. N. (Eds.), *Industrial Gums: Polysaccharides and Derivatives*. 3rd edition, pp. 188–226.

Ash, M., and Ash, I. (1996) *Food Additives*. Gower Publishing Co., Brookfield, VT.

ANY OTHER RELEVANT INFORMATION: Guar gum is a natural polysaccharide extracted from the endosperm of the seeds of the *Cyamopsis tetragonolobus*. Guar gum is similar in structure to tara gum and locust bean gum, differing primarily in that guar gum has a higher galactose : mannose ratio (approx. 1 : 1.8)

NAME:	Hydroxy ethyl cellulose
CATEGORY:	Polysaccharides
FOOD USE:	Food packaging/ Edible films and coatings
SYNONYMS:	Hydroxylated and ethylated beta-D-(1 → 4) glucan/ HEC
FORMULA:	Variable, depending on the degree of hydroxylation and ethylation
MOLECULAR MASS:	Variable
ALTERNATIVE FORMS:	None
PROPERTIES AND APPEARANCE:	A white powder or fibrous flakes
SOLUBILITY % AT VARIOUS TEMPERATURE/pH COMBINATIONS:	Solubility varies with molecular weight and hydroxylation/ethylation. Generally, hydroxy ethyl celluloses with molar substitutions (MS) greater than 1.6 are soluble in water, sucrose solutions, sodium chloride solution, but solubility in ethanol solutions, vegetable oils and propylene glycol is limited. Solubility declines with increasing molecular weight, decreasing hydroxylation/ethylation and increasing concentration. It is essentially pH-independent (pH 4 to pH 8)
FUNCTION IN FOODS:	Films and packaging associated with food
ALTERNATIVES:	Methyl cellulose (MEC); hydroxy propyl methyl cellulose (HPMC); hydroxyethyl methyl cellulose (HEMC); carboxy methyl cellulose (CMC); xanthan; carrageenans; dextrans; starches; locust bean gum; guar gum; gelatin; agar; alginate (application-dependent)
TECHNOLOGY OF USE IN FOODS:	Hydroxy ethyl cellulose should be dispersed in cold water and heated to 80°C for 10 minutes to hydrate the polysaccharide. A high shear mixer will aid dispersion, although prolonged shear of high molecular weight polymers will reduce the viscosity. Alternatively, it can be dispersed in a water-miscible non-solvent or with other dry powders prior to dispersion at room temperature.

	Hydroxy ethyl cellulose has useful properties in that it is non-ionic, and forms viscous, pseudoplastic (viscosity decreases as the shear rate increases) solutions that become more viscous as temperature is lowered. However, it is not currently permitted in foods as a direct additive
SYNERGISTS:	None known
ANTAGONISTS:	None known
FOOD SAFETY ISSUES:	Non-toxic. The powder is combustible
LEGISLATION:	**USA:** FDA 21CFR § 176.170, 176.180, 175.300.
REFERENCES:	Ash, M., and Ash, I. (1996) *Food Additives*. Gower Publishing Co., Brookfield, VT. Desmarais, A. J., and Wint, R. F. (1993) Hydroxy alkyl and ethyl ethers of cellulose. In: Whistler, R. L., and BeMiller, J. N. (Eds.), *Industrial Gums: Polysaccharides and Derivatives*. 3rd edition, pp. 505–535.
ANY OTHER RELEVANT INFORMATION:	Chemically hydroxylated and ethylated natural cellulose. See methyl cellulose (MEC); hydroxy propyl methyl cellulose (HPMC); hydroxy propyl cellulose (HPC); carboxy methyl cellulose (CMC)

NAME:	Hydroxy propyl cellulose
CATEGORY:	Polysaccharides
FOOD USE:	Soups/ Syrups/ Beverages/ Dairy products/ Dressings/ Dry formulated products/ Food packaging/ Edible films and coatings/ All food products
SYNONYMS:	Hydroxylated and propylated beta-D-(1 → 4) glucan/ CAS 9004-64-2/ E463/ Cellulose 2-hydroxypropyl ether/ Oxypropylated cellulose/ Propylene glycol ether of cellulose/ HPC
FORMULA:	Variable, depending on the degree of hydroxylation and propylation
MOLECULAR MASS:	Variable
ALTERNATIVE FORMS:	None
PROPERTIES AND APPEARANCE:	A white powder or fibrous flakes
SOLUBILITY % AT VARIOUS TEMPERATURE/pH COMBINATIONS:	Solubility varies with molecular weight and hydroxylation/propylation. Generally, commercially available hydroxy propyl celluloses have molar substitutions (MS) greater than 2. In contrast to hydroxy ethyl celluloses they are soluble not only in aqueous solution but also in a wide range of polar solvents. Hydroxy propyl celluloses are soluble in water, sucrose solutions, sodium chloride solutions, ethanol solutions and propylene glycol, but solubility in vegetable oils is limited. Also in contrast to hydroxyethyl celluloses they are insoluble in hot aqueous solutions (as are methyl celluloses). Solubility declines with increasing molecular weight, decreasing hydroxylation/propylation and increasing concentration. It is essentially pH-independent (pH 4 to pH 8)
FUNCTION IN FOODS:	Emulsifier; thickening agent; stabiliser; suspending agent; bulking agent
ALTERNATIVES:	Methyl ethyl cellulose (MEC); hydroxy propyl methyl cellulose (HPMC); hydroxy ethyl methylcellulose (HEMC); carboxy methyl cellulose (CMC); proteins; gum arabic; xanthan; carrageenans; dextrans; starches; locust bean gum; guar gum; gelatin; agar; alginate (application-dependent)

TECHNOLOGY OF USE IN FOODS:

Hydroxy propyl cellulose should be dispersed in hot water and cooled to hydrate the polysaccharide. A high shear mixer will aid dispersion, although prolonged shear of high molecular-weight polymers will reduce the viscosity.

Hydroxy propyl celluloses have similar properties to methyl cellulose, but they are soluble in a wider range of solvents. They can form aqueous solutions with a wide range of viscosities (variable thickening, stabilising and bulking properties); they are surface-active (emulsifying properties); they can have a high yield point (suspending properties); and they form gel-like structures at high temperatures that melt on cooling (stabilising agents for foods with heat-sensitive components such as the toppings in pizzas). Properties (viscosity, melting points, etc.) vary considerably with the molecular weight and molar substitution

SYNERGISTS:

None known

ANTAGONISTS:

None known

FOOD SAFETY ISSUES:

LD_{50} (oral, rat) 10,200 mg/kg; ADI 0–25 mg/kg. Slightly toxic. The powder is combustible

LEGISLATION:

USA:	**UK and EUROPE:**
FDA 21CFR § 172.870, 177.1200	UK: approved. Europe: listed

REFERENCES:

Ash, M., and Ash, I. (1996) *Food Additives*. Gower Publishing Co., Brookfield, VT.

Desmarais, A. J., and Wint, R. F. (1993) Hydroxy alkyl and ethyl ethers of cellulose. In: Whistler, R. L., and BeMiller, J. N. (Eds.), *Industrial Gums: Polysaccharides and Derivatives*. 3rd edition, pp. 505–535.

ANY OTHER RELEVANT INFORMATION:

Chemically hydroxylated and ethylated natural cellulose. See methyl cellulose (MEC); hydroxy propyl methyl cellulose (HPMC); hydroxy ethyl cellulose (HEC); carboxy methyl cellulose (CMC)

NAME:	Hydroxy propyl methyl cellulose
CATEGORY:	Polysaccharides
FOOD USE:	Soups/ Syrups/ Beverages/ Dairy products/ Dressings/ Dry formulated products/ Food packaging/ Edible films and coatings/ All food products
SYNONYMS:	Hydroxylated, methylated and propylated beta-D-(1→4) glucan/ CAS 9004-65-3/ E464/ Cellulose 2-hydroxypropyl methyl ether/ Hypromellose/ Propylene glycol ether of methyl cellulose/ HPMC
FORMULA:	Variable, depending on the degree of hydroxylation, methylation and propylation
MOLECULAR MASS:	Variable
ALTERNATIVE FORMS:	None
PROPERTIES AND APPEARANCE:	A white powder or fibrous flakes
SOLUBILITY % AT VARIOUS TEMPERATURE/pH COMBINATIONS:	Solubility varies with molecular weight and hydroxylation/propylation/methylation. Generally hydroxypropyl celluloses have similar properties to hydroxypropyl methyl celluloses. In contrast to hydroxy ethyl celluloses, they are not only soluble in aqueous solution but also in a wide range of polar solvents. Hydroxy propyl celluloses are soluble in water, sucrose solutions, sodium chloride solutions, ethanol solutions and propylene glycol, but solubility in vegetable oils is limited. Also in contrast to hydroxyethyl celluloses they are insoluble in hot aqueous solutions (as are methyl celluloses). Solubility declines with increasing molecular weight, decreasing hydroxylation/propylation/methylation and increasing concentration. It is essentially pH-independent (pH 4 to pH 8)
FUNCTION IN FOODS:	Emulsifier; thickening agent; stabiliser; suspending agent; bulking agent
ALTERNATIVES:	Methyl ethyl cellulose (MEC); hydroxy propyl cellulose (HPMC); hydroxy ethyl methylcellulose (HEMC); carboxy methyl cellulose (CMC); proteins; gum arabic; xanthan; carrageenans; dextrans; starches; locust bean gum; guar gum; gelatin; agar; alginate (application-dependent)

TECHNOLOGY OF USE IN FOODS:	Hydroxy propyl methyl cellulose should be dispersed in hot water and cooled to hydrate the polysaccharide. A high shear mixer will aid dispersion, although prolonged shear of high molecular-weight polymers will reduce the viscosity. Hydroxy propyl methyl celluloses have similar properties to methyl cellulose but they are soluble in a wider range of solvents. They can form aqueous solutions with a wide range of viscosities (variable thickening, stabilising and bulking properties); they are surface-active (emulsifying properties); they can have a high yield point (suspending properties); and they form gel-like structures at high temperatures that melt on cooling (stabilising agents for foods with heat-sensitive components such as the toppings in pizzas). Properties (viscosity, melting points, etc.) vary considerably with the molecular weight and molar substitution
SYNERGISTS:	None known
ANTAGONISTS:	None known
FOOD SAFETY ISSUES:	LD_{50} (intraperitoneal, rat) 5,200 mg/kg. Mildly toxic. The powder is combustible
LEGISLATION:	**USA:** FDA 21CFR § 172.87, 175.105, 175.300 **UK and EUROPE:** Europe: listed
REFERENCES:	Ash, M., and Ash, I. (1996) *Food Additives*. Gower Publishing Co., Brookfield, VT. Desmarais, A. J., and Wint, R. F. (1993) Hydroxy alkyl and ethyl ethers of cellulose. In: Whistler, R. L., and BeMiller, J. N. (Eds.), *Industrial Gums: Polysaccharides and Derivatives*. 3rd edition, pp. 505–535.
ANY OTHER RELEVANT INFORMATION:	Chemically hydroxylated, methylated and ethylated natural cellulose. See methyl cellulose (MEC); hydroxy propyl cellulose (HPC); hydroxy ethyl cellulose (HEC); carboxy methyl cellulose (CMC)

NAME:	Karaya gum
CATEGORY:	Polysaccharides
FOOD USE:	Meringue/ Sherbet/ Ices/ Cheese spreads/ Ground meat products/ French dressing
SYNONYMS:	CAS 9000-36-6/ EINECS 232-539-4/ E416/ Sterculia gum/ Sterculia urens gum/ India tragacanth/ Indian tragacanth/ Kadaya gum/ Gum karaya
MOLECULAR MASS:	9,500,000
ALTERNATIVE FORMS:	Karaya gum is slightly acidic, so different salt forms are possible, chiefly calcium and magnesium
PROPERTIES AND APPEARANCE:	White/off-white powder
SOLUBILITY % AT VARIOUS TEMPERATURE/pH COMBINATIONS:	Karaya gum is poorly soluble in water and sparingly soluble in other liquids. Optimum solubility is at pH 6–8. Viscosity is dependent primarily on initial particle size
FUNCTION IN FOODS:	Stabiliser; thickener; texturiser; crystallisation inhibitor
ALTERNATIVES:	Gum ghatti; gum tragacanth (application-dependent)
TECHNOLOGY OF USE IN FOODS:	Karaya gum swells in water; larger particles will produce textured pastes whereas smaller particle sizes yield more homogeneous, viscous preparations. Tannins in the karaya gum result in a lightening of the solution colour in acidic pHs and a darkening in alkaline pHs. Karaya gum is a complex polysaccharide aned the structure has not been fully determined. It contains galactose, rhamnose, D-glucuronic acid. It is acetylated (8%) and contains a substantial quantity of uronic acid (37%)
SYNERGISTS:	None known
ANTAGONISTS:	None known
FOOD SAFETY ISSUES:	Mildly toxic. ADI 0–20 mg/kg. The powder is combustible

LEGISLATION:	Usage level: Limitation 0.9% (soft candy); 0.02% (milk products); 0.3% (frozen dairy desserts); 0.002% (other foods) **USA:** FDA 21CFR § 133.133, 133.134, 133.162, 133.178, 133.179, 150.141, 150.161, 184.1349, GRAS **UK and EUROPE:** UK: approved. Europe: listed **AUSTRALIA/PACIFIC RIM:** Japan: approved for use as a natural thickener and stabiliser
REFERENCES:	Anderson, D. M. W., Gill, M. L. C., Jeffrey, A. M., and McDougall, F. J. (1985) *Phytochemistry* **24**, 71. Whistler, R. L. (1993) Exudate gums. In: Whistler, R. L., and J. N. BeMiller (Eds.), *Industrial Gums: Polysaccharides and Derivatives.* 3rd edition, pp. 311–318. Ash, M., and Ash, I. (1996) *Food Additives.* Gower Publishing Co., Brookfield, VT.
ANY OTHER RELEVANT INFORMATION:	This is a natural polysaccharide extracted from the trunks of the tree, *Sterculia urens*. Originally introduced as an alternative to tragacanth gum

NAME:	Locust bean gum
CATEGORY:	Polysaccharides
FOOD USE:	Dairy products/ Pet foods/ Baked goods/ Confectionery/ Dry formulated products/ Food packaging/ Edible films and coatings/ All food products
SYNONYMS:	Beta-D-(1 → 4) mannan with some attached D-galactose molecules linked via alpha (1 → 6) bonds/ CAS 9000-40-2/ EINECS 232-541-5/ E410/ Galactomannan/ Carob flour/ Carob bean gum/ St John's bread/ Algoroba
FORMULA:	Variable, depending on the ratio of galactose to mannose
MOLECULAR MASS:	Variable. Generally 300,000 to 360,000
PROPERTIES AND APPEARANCE:	White/brown powder
SOLUBILITY % AT VARIOUS TEMPERATURE/pH COMBINATIONS:	Solubility varies with molecular weight. Generally, locust bean gum is soluble in water, sucrose solutions and sodium chloride solution, but solubility in ethanol, vegetable oils and propylene glycol is limited. Solubility declines with increasing molecular weight and increasing concentration. It is pH-independent (pH 4 to pH 8)
FUNCTION IN FOODS:	Thickening agent; stabiliser; bulking agent; suspending agent
ALTERNATIVES:	Guar gum; tara gum; carrageenans; dextrans; starches; xanthan; gelatin; agar; alginate (application-dependent)
TECHNOLOGY OF USE IN FOODS:	Locust bean gum should be dispersed in cold water and heated to 80°C for 10 minutes to hydrate the polysaccharide. A high shear mixer will aid dispersion, although prolonged shear of high molecular-weight polymers will reduce the viscosity. Locust bean gum molecules have a high hydrodynamic volume and can entangle with adjacent molecules, resulting in viscous properties. Synergistic interactions with xanthan gum result in gel

formation. The synergistic interaction with xanthan is more pronounced with locust bean gum than guar, primarily because regions of the molecule that are not galactose-substituted are responsible for the interaction and locust bean gum has a lower galactose to mannose ratio than guar

SYNERGISTS: Xanthan; agar; carrageenan; anionic polymer; anionic surfactants

ANTAGONISTS: None known

FOOD SAFETY ISSUES: Mildly toxic. ADI not specified. LD_{50} (rat, oral) 13 g/kg. The powder is combustible

LEGISLATION: Usage level: Limitation 0.15% (baked goods); 0.25% (beverages); 0.8% (cheese); 0.75% (gelatins, puddings); 0.75% (jams/jellies); 0.5% (other foods)

USA:	UK and EUROPE:	AUSTRALIA/PACIFIC RIM:
FDA 21CFR § 133.133, 133.134, 133.162, 133.178, 133.179, 150.141, 150.161, 182.20, 184.1343, 240.1051, GRAS	UK: approved. Europe: listed	Japan: approved

REFERENCES:

Maier, H., Anderson, M., Kurt, C., and Magnusen, K. (1993) Guar, Locust Bean, Tara and Fenugreek gums. In: Whistler, R. L., and J. N. BeMiller (Eds.), *Industrial Gums: Polysaccharides and Derivatives.* 3rd edition, pp. 188–226.

Ash, M., and Ash, I. (1996) *Food Additives.* Gower Publishing Co., Brookfield, VT.

ANY OTHER RELEVANT INFORMATION: Locust bean gum is a natural polysaccharide extracted from the endosperm of the seeds of the carob tree, *Ceretonia siliqua*. Locust bean gum is similar in structure to guar gum, differing primarily in that locust bean gum has a lower galactose: mannose ratio (approx. 1:3.9)

NAME:	Methyl cellulose
CATEGORY:	Polysaccharides
FOOD USE:	Soups/ Syrups/ Beverages/ Dairy products/ Dressings/ Dry formulated products/ Food packaging/ Edible films and coatings/ All food products
SYNONYMS:	Methylated beta-D-(1→ 4) glucan/ CAS 9004-67-5/ E461/ MC/ Cellulose methyl ether/ Cologel/ Methocel
FORMULA:	Variable, depending on the degree of methylation
MOLECULAR MASS:	Variable
ALTERNATIVE FORMS:	None
PROPERTIES AND APPEARANCE:	A white powder
SOLUBILITY % AT VARIOUS TEMPERATURE/pH COMBINATIONS:	Solubility varies with molecular weight and methylation. Generally, methylcellulose is soluble in water, sucrose solutions, sodium chloride solution and ethanol solutions, but solubility in vegetable oils and propylene glycol is limited. Solubility declines with increasing molecular weight, decreasing methylation and increasing concentration. It is essentially pH-independent (pH 4 to pH 8). Methyl celluloses are soluble in cold aqueous solution, but not in hot aqueous solutions; hence, elevated temperatures can be used to produce a homogeneous dispersion prior to dissolution
FUNCTION IN FOODS:	Emulsifier; thickening agent; stabiliser; suspending agent; bulking agent
ALTERNATIVES:	Methyl ethyl cellulose (MEC); hydroxy propyl methyl cellulose (HPMC); hydroxy ethyl methylcellulose (HEMC); carboxy methyl cellulose (CMC); proteins; gum arabic; xanthan; carrageenans; dextrans; starches; locust bean gum; guar gum; gelatin; agar; alginate (application-dependent)
TECHNOLOGY OF USE IN FOODS:	Methyl cellulose should be dispersed in hot water and cooled to hydrate the polysaccharide. A high shear mixer will aid dispersion, although prolonged shear of high molecular-weight polymers will reduce the viscosity.

Methyl celluloses have four important properties. They can form aqueous solutions with a wide range of viscosities (variable thickening, stabilising and bulking properties); they are surface-active (emulsifying properties); they can have a high yield point (suspending properties); and they form gels ar high temperatures that melt on cooling (stabilising agents for foods with heat-sensitive components like the toppings in pizzas). Properties (viscosity, melting points, etc.) vary considerably with the molecular weight and degree of methylation

SYNERGISTS: None known

ANTAGONISTS: None known

FOOD SAFETY ISSUES: Non-toxic. The powder is combustible

LEGISLATION:

USA:	UK and EUROPE:	AUSTRALIA/PACIFIC RIM:
FDA 21CFR § 150.141, 150.161, 175.105, 175.210, 175.300, 176.200, 182.1480, GRAS, USDA 9CFR § 318.7 (0.15% max in meat and vegetable products)	UK: approved. Europe: listed	Japan: restricted (2% max)

REFERENCES:

Ash, M., and Ash, I. (1996) *Food Additives.* Gower Publishing Co., Brookfield, VT.

Grover, J. S. (1993) Methyl cellulose and its derivatives. In: Whistler, R. L., and J. N. BeMiller (Eds.), *Industrial Gums: Polysaccharides and Derivatives*. 3rd edition, pp. 475–504.

ANY OTHER RELEVANT INFORMATION: Chemically methylated natural cellulose. See also methyl ethyl cellulose (MEC); hydroxy propyl methyl cellulose (HPMC); hydroxy propyl cellulose (HPC); carboxy methyl cellulose (CMC)

NAME:	Methyl ethyl cellulose
CATEGORY:	Polysaccharides
FOOD USE:	Whipped toppings/ Soups/ Syrups/ Beverages/ Dairy products/ Dressings/ Dry formulated products/ Food packaging/ Edible films and coatings/ All food products
SYNONYMS:	Methylated and ethylated beta-D-(1 → 4) glucan/ E465/ MEC/ Ethylmethylcellulose
FORMULA:	Variable, depending on the degree of methylation and ethylation
MOLECULAR MASS:	Variable
ALTERNATIVE FORMS:	None
PROPERTIES AND APPEARANCE:	A white powder or fibrous flakes
SOLUBILITY % AT VARIOUS TEMPERATURE/pH COMBINATIONS:	Solubility varies with molecular weight and methylation/ethylation. Generally, methylethylcellulose is soluble in water, sucrose solutions, sodium chloride solution and ethanol solutions, but solubility in vegetable oils and propylene glycol is limited. Solubility declines with increasing molecular weight, decreasing methylation/ethylation and increasing concentration. It is essentially pH-independent (pH 4 to pH 8)
FUNCTION IN FOODS:	Emulsifier; foaming stabiliser; thickening agent; general stabiliser; bulking agent; suspending agent
ALTERNATIVES:	Methyl cellulose (MEC); hydroxy propyl methyl cellulose (HPMC); hydroxyethyl methyl cellulose (HEMC); carboxy methyl cellulose (CMC); proteins; gum arabic; xanthan; carrageenans; dextrans; starches; locust bean gum; guar gum; gelatin; agar; alginate (application-dependent)
TECHNOLOGY OF USE IN FOODS:	Some similarities to methyl cellulose and ethyl cellulose (not food-grade)
SYNERGISTS:	None known
ANTAGONISTS:	None known

FOOD SAFETY ISSUES:	Non-toxic. The powder is combustible. ADI 0–25 mg/kg body weight	
LEGISLATION:	**USA:** FDA 21CFR § 172.872.	**UK and EUROPE:** UK: approved. Europe: listed.
REFERENCES:	Ash, M., and Ash, I. (1996) *Food Additives.* Gower Publishing Co., Brookfield, VT. Grover, J. S. (1993) Methyl cellulose and its derivatives. In: Whistler, R. L., and J. N. BeMiller (Eds.), *Industrial Gums: Polysaccharides and Derivatives*. 3rd edition, pp. 475–504.	
ANY OTHER RELEVANT INFORMATION:	Chemically methylated and ethylated natural cellulose. See also methyl cellulose (MEC); hydroxy propyl methyl cellulose (HPMC); hydroxy propyl cellulose (HPC); carboxy methyl cellulose (CMC)	

NAME:	**Microcrystalline cellulose**
CATEGORY:	Polysaccharides
FOOD USE:	Low-fat salad creams/ Low-fat meat spreads/ Dry formulated products/ Low-calorie foods/ Confectionery processing aid/ All food products
SYNONYMS:	Beta-D-(1→4) glucan/ CAS 9004-34-6/ EINECS 232-674-9/ E460/ cellulose gel/ Avicel
FORMULA:	$(C_6H_{10}O_5)_n$
MOLECULAR MASS:	Variable
ALTERNATIVE FORMS:	None
PROPERTIES AND APPEARANCE:	A white powder or gel
SOLUBILITY % AT VARIOUS TEMPERATURE/pH COMBINATIONS:	Microcrystalline cellulose is insoluble in water, but commercial preparations are generally colloidal mixtures of microcrystalline cellulose and carboxymethyl cellulose that can be dispersed in aqueous solutions
FUNCTION IN FOODS:	Can be used as a processing aid; fat mimetic; dietary fibre; thickening agent; stabiliser; bulking agent. It is not digestible and so can be used in low-calorie preparations
ALTERNATIVES:	Starch hydrolysis products (SHPs); polydextrose; xanthan (application-dependent)
TECHNOLOGY OF USE IN FOODS:	Microcrystalline cellulose can be dispersed in cold water. Mixing with a high shear mixer could change the physical properties of the preparation. Commercial preparations of microcrystalline cellulose consist of fine particles of crystalline cellulose mixed with carboxymethyl cellulose. The fine cellulose particles simulate the mouthfeel of fat crystals, whereas the carboxy methyl cellulose lubricates the particles and provides thickening properties
SYNERGISTS:	None known
ANTAGONISTS:	None known

FOOD SAFETY ISSUES:	Non-toxic. The powder is combustible $LD_{50} > 5$ g/kg. No significant hazard
LEGISLATION:	**USA:** FDA GRAS. **UK and EUROPE:** UK: approved. Europe: listed **AUSTRALIA/PACIFIC RIM:** Japan: approved for use as a natural processing aid
REFERENCE:	Ash, M., and Ash, I. (1996) *Food Additives.* Gower Publishing Co., Brookfield, VT.
ANY OTHER RELEVANT INFORMATION:	Microcrystalline cellulose is natural crystalline polysaccharide particles extracted from plant cellulose (which is composed of chains of crystalline and amorphous cellulose). See also cellulose

NAME:	Pectin
CATEGORY:	Polysaccharides
FOOD USE:	All food products/ Edible films and coatings
SYNONYMS:	CAS 9005-69-5/ EINECS 232-553-0/ E440a/ Citrus pectin/ Pectinate/ Pectinic acids/ Pectic acids/ Pectates/ Polygalacturonic acid/ E440b/ Amidated pectin
FORMULA:	Variable, depending on degree of methanol esterification and the presence of other substituents (acetyl units, neutral sugar side chains, rhamnose, ferulic acid, etc.)
MOLECULAR MASS:	Variable, generally 100,000
ALTERNATIVE FORMS:	Polygalacturonic acids (low-methoxyl pectins) are called pectinic acids; salts of pectic acids are pectates (e.g. sodium pectate). Polygalacturonic acids esterified with methanol (high-methoxyl pectins) are called pectinic acids and salts are called pectinates. Amidated pectins and salts
PROPERTIES AND APPEARANCE:	White powder or syrup
SOLUBILITY % AT VARIOUS TEMPERATURE/pH COMBINATIONS:	Solubility in aqueous solution varies with differences in structure, molecular weight, pH and cations. Generally, pectins are soluble in water, sucrose solutions, sodium chloride solution and low concentration ethanol solutions, but solubility in vegetable oils and propylene glycol is limited. Pectins are insoluble in the presence of divalent or trivalent salts. Above pH 5 pectins start to deesterify spontaneously; below pH 2.5, hydrolysis occurs so pectins are primarily suited for acidic foods
FUNCTION IN FOODS:	Gelling agent; thickening agent; stabiliser; flocculant; bulking agent. It can be used to stabilise emulsions and to retard ice-crystal formation
ALTERNATIVES:	Alginates; carrageenans; chitosan; xanthan; furcellaran (application-dependent)
TECHNOLOGY OF USE IN FOODS:	Pectins should be dispersed in cold water and heated to 80°C for 10 minutes to hydrate the polysaccharide. A high shear mixer will aid dispersion, although prolonged shear of high molecular-weight polymers will reduce the viscosity. High-methoxyl pectins (pectinates/pectinic acid) can, in the presence of large quantities of material (usually sugar), spontaneously form gels in the absence of polyvalent ions. Gel strength increases with

	increasing molecular weight and decreasing pH (until the polymer starts to hydrolyse). Gelation rate is related to the degree of methoxylation. High-methoxyl pectins with 70–75% methoxylation are called "rapid-set", 65–70% are "medium-set", and 55–65% are "slow-set". Low-methoxyl pectins (<50% methoxyl substitution) require polyvalent ions (usually calcium) for gelation. High-methoxyl pectins generally do not form gels above pH 3.5, whereas low-methoxyl pectins can form gels up to pH 6.5. Low-methoxyl pectins set rapidly at the gelation temperature, whereas high-methoxyl pectins set slowly. Amidated low-methoxyl pectins (typically 15–22% amidated) generally have higher gelation temperatures than the un-amidated pectins and are less affected by natural variations in calcium ion content. Hence, they have been found to be more consistent in functional properties and thus more controllable. Amidation increases gel strength. Ferulic acid-containing pectins (sugar beet, spinach) can be enzymically gelled with a peroxidase and hydrogen peroxide
SYNERGISTS:	Proteins, polyvalent ions
ANTAGONISTS:	Proteins, polyvalent ions
FOOD SAFETY ISSUES:	Non-toxic. The powder is combustible and emits irritating fumes when heated
LEGISLATION:	**USA:** FDA 21CFR § 135.40, 145.150, 172.385, 184.1588, GRAS **UK and EUROPE:** UK: approved. Europe: listed **AUSTRALIA/PACIFIC RIM:** Japan: approved
REFERENCES:	Ash, M., and Ash, I. (1996) *Food Additives.* Gower Publishing Co., Brookfield, VT. Rolin, C. (1993) Pectin. In: Whistler, R. L., and J. N. BeMiller (Eds.), *Industrial Gums: Polysaccharides and Derivatives*. 3rd edition, pp. 105–143.
ANY OTHER RELEVANT INFORMATION:	High-methoxy pectins are natural polysaccharides extracted from plant material, usually from orange peel (byproduct of juice manufacture) or apple pomace (byproduct of cider manufacture). Low-methoxy pectins are chemically synthesised (deesterified) from high-methoxy pectins. Pectin can also be synthesised from sugar beet and sunflower seeds, but the higher acetylation of these pectins significantly alters functional properties. Sucrose is added to most commercial pectin preparations to standardise functional properties

NAME:	**Propyleneglycol alginate**
CATEGORY:	Polysaccharides
FOOD USE:	All food products/ Edible films and coatings
SYNONYMS:	CAS 9005-32-2/ E405/ PGA/ Hydroxypropyl alginate/ Alginic acid ester with 1,2-propanediol/ Propane-1,2-diol alginate
FORMULA:	Variable, depending on propylene glycol substitution, ratio of mannuronic to guluronic acid, and the presence of other substituents
MOLECULAR MASS:	Variable
ALTERNATIVE FORMS:	None
PROPERTIES AND APPEARANCE:	White powder
SOLUBILITY % AT VARIOUS TEMPERATURE/pH COMBINATIONS:	Solubility in aqueous solution varies with differences in structure, molecular weight, pH and cations. The primary property of propylene glycol is that, unlike other alginates, it is soluble at low pHs
FUNCTION IN FOODS:	Gelling agent; thickening agent; stabiliser; flocculant; bulking agent. It can be used to stabilise emulsions and to retard ice-crystal formation
ALTERNATIVES:	Alginates; carrageenans; methylcellulose; gum ghatti; xanthan; furcellaran (application-dependent)
TECHNOLOGY OF USE IN FOODS:	Propyleneglycol alginate should be dispersed in cold water and heated to 80°C for 10 minutes to hydrate the polysaccharide. A high shear mixer will aid dispersion, although prolonged shear of high molecular-weight polymers will reduce the viscosity. In addition to viscosity, propylene glycol alginate has surface-active properties due to the hydrophobic nature of parts of the molecule. This makes it a useful emulsifier
SYNERGISTS:	Proteins
ANTAGONISTS:	None known

FOOD SAFETY ISSUES:

Mildly toxic. LD_{50} (oral, rat) 7200 mg/kg. ADI 0–25 mg/kg. The powder is combustible and emits irritating fumes when heated

LEGISLATION:

Usage level: Limitation 0.6% (condiments); 6% (frozen dairy desserts); 0.5% (baked goods); 0.9% (cheese); 0.5% (gravies); 1.7% (seasonings); 0.4% (jams/jellies); 1.1% (fats/oils); 0.6% (gelatins/puddings); 0.3% (other foods)

USA:	UK and EUROPE:	AUSTRALIA/PACIFIC RIM:
FDA 21CFR § 133.133, 133.134, 133.162, 133.178, 133.179, 172.210, 172.820, 172.858, 173.340, 176.170. GRAS	UK: approved. Europe: listed	Japan: approved (1% max)

REFERENCES:

Ash, M., and Ash, I. (1996) *Food Additives.* Gower Publishing Co., Brookfield, VT.

Clare, K. (1993) Algin. In: Whistler, R. L., and J. N. BeMiller (Eds.), *Industrial Gums: Polysaccharides and Derivatives*. 3rd edition, pp. 105–143.

ANY OTHER RELEVANT INFORMATION:

Chemically synthesised using propylene oxide and the natural polysaccharide alginate. Although algin, alginic acid, ammonium alginate, calcium alginate, potassium alginate and sodium alginate are essentially different states of the same type of compound, for legislative purposes they are listed separately

NAME:	Potassium alginate
CATEGORY:	Polysaccharides
FOOD USE:	All food products/ Edible films and coatings
SYNONYMS:	CAS 9005-36-1/ E402/ Potassium alginate/ Alginic acid-potassium salt/ Potassium polymannuronate
FORMULA:	Variable, depending on ratio of mannuronic to guluronic acid, and the presence of other substituents
MOLECULAR MASS:	Variable
ALTERNATIVE FORMS:	Alginic acid; algin; calcium alginate; potassium alginate
PROPERTIES AND APPEARANCE:	White to yellow powders or hard flakes
SOLUBILITY % AT VARIOUS TEMPERATURE/pH COMBINATIONS:	Solubility in aqueous solution varies with differences in structure, molecular weight, pH and cations. Alginic acid (low pH) is insoluble in water, monovalent salts (sodium, potassium and ammonium) are also soluble, but divalent salts (including calcium) except magnesium are insoluble in water. Solubility declines with increasing molecular weight and increasing concentration. Viscosity is not affected in the pH 5 to 11 range, but increases at lower pHs (due to alginic acid formation)
FUNCTION IN FOODS:	Gelling agent; thickening agent; stabiliser; flocculant; bulking agent. It can be used to stabilise emulsions and to retard ice-crystal formation
ALTERNATIVES:	Carrageenans; gelatin; gellan; agar; xanthan; furcellaran (application-dependent)
TECHNOLOGY OF USE IN FOODS:	Alginates should be dispersed in cold deionised water and heated for 10 minutes at 80°C to hydrate the polysaccharide. A high shear mixer will aid dispersion, although prolonged shear of high molecular-weight polymers will reduce the viscosity. Calcium gels can be prepared by addition of alginate solutions to a calcium solution (e.g. maraschino cherries), addition of calcium to a hot solution of alginate and then cooling (gel is thermo-irreversible), or by diffusion. Gels can also be formed by lowering the pH (alginic acid), but this can be technically difficult. Ammonium alginate forms particularly viscous solutions.

	Alginates are composed of blocks of polymannuronic acid, polyguluronic acid and mixed regions. Ribbon-like structures are formed in solution and alginates with a high percentage of guluronic blocks can be cross-linked (gelled) by divalent ions (e.g. calcium)
SYNERGISTS:	Divalent ion, pectins
ANTAGONISTS:	None known
FOOD SAFETY ISSUES:	ADI 0–25 mg/kg. The powder is combustible and emits irritating fumes when heated
LEGISLATION:	Usage level: Limitation 0.01% (confections, frostings); 0.7% (gelatins/puddings); 0.25% (processed fruits, fruit juices); 0.01% (other foods)

USA:	**UK and EUROPE:**	**AUSTRALIA/PACIFIC RIM:**
FDA 21CFR § 184.1610. GRAS	UK: approved. Europe: listed	Japan: approved

EFERENCES:

Ash, M., and Ash, I. (1996) *Food Additives.* Gower Publishing Co., Brookfield, VT.

Clare, K. (1993) Algin. In: Whistler, R. L., and J. N. BeMiller (Eds.), *Industrial Gums: Polysaccharides and Derivatives*. 3rd edition, pp. 105–143.

ANY OTHER RELEVANT INFORMATION:

Alginates are natural polysaccharides which are extracted from brown seaweeds (Phaeophyceae). *Macrocystis pyrifera*, *Laminera hyperborea*, *Laminaria digitata*, *Laminaria japonica*, *Ascophyllum nodosum*, *Ecklonia maxima* and *Lessonia nigrescens* are the most important commercial sources of alginates. Structurally distinct bacterial alginates can be obtained from *Pseudomonas* species and *Azotobacter* species. Although algin, alginic acid, ammonium alginate, calcium alginate, potassium alginate and sodium alginate are essentially different states of the same type of compound, for legislative purposes they are listed separately

NAME:	Tragacanth gum
CATEGORY:	Polysaccharides
FOOD USE:	Salad dressings/ Sauces/ Baked goods/ Confectionery/ All food products
SYNONYMS:	CAS 9000-65-1/ EINECS 232-552-5/ E413/ Gum tragacanth/ Gum dragon
MOLECULAR MASS:	Variable, generally approx. 840,000
ALTERNATIVE FORMS:	Tragacanth gum is slightly acidic, so different salt forms are possible, chiefly calcium, sodium and magnesium
PROPERTIES AND APPEARANCE:	White/off-white powder, granules or yellowish ribbons
SOLUBILITY % AT VARIOUS TEMPERATURE/pH COMBINATIONS:	Tragacanth is composed of water-soluble tragacanthin and water-swellable bassorin. Generally, tragacanth gum is soluble in sucrose solutions, sodium chloride solution and partially soluble in 70% ethanol (tragacanthin). Solubility in vegetable oils and propylene glycol is limited. Maximum viscosity is observed at pH 8
FUNCTION IN FOODS:	Stabiliser; thickener; crystallisation inhibitor
ALTERNATIVES:	Gum ghatti; Karaya gum (application-dependent)
TECHNOLOGY OF USE IN FOODS:	Tragacanth gum should be dispersed in cold water and heated to 80°C for 10 minutes to hydrate the polysaccharide. A high shear mixer will aid dispersion, although prolonged shear of high molecular-weight polymers will reduce the viscosity. Tragacanth gum is a complex polysaccharide mixture composed of neutral arabinogalactan and polymers composed primarily of alpha-D- (1→4) galacturonic acid with galactose, xylose and fucose substituents. There is also thought to be some methylation and associated protein
SYNERGISTS:	None known
ANTAGONISTS:	None known
FOOD SAFETY ISSUES:	Mildly toxic. LD_{50} (oral, rat) 16.4 g/kg. The powder is combustible

LEGISLATION:	Usage level: Limitation 0.2% (baked goods); 0.7% (condiments); 1.3% (fats and oils); 0.8% (gravies, sauces); 0.2% (meat products); 0.2% (processed fruits, fruit juices); 0.1% (other foods)

USA:	**UK and EUROPE:**	**AUSTRALIA/PACIFIC RIM:**
FDA 21CFR § 133.133, 133.134, 133.162, 133.178, 133.179, 150.141, 150.161, 184.1351, GRAS	Europe: listed. UK: approved	Japan: approved

REFERENCES:

Anderson, D. M. W., Gill, M. L. C., Jeffrey, A. M., and McDougall, F. J. (1985) *Phytochemistry* **24**, 71.

Whistler, R. L. (1993) Exudate gums. In: Whistler, R. L., and J. N. BeMiller (Eds.), *Industrial Gums: Polysaccharides and Derivatives.* 3rd edition, pp. 311–318.

Ash, M., and Ash, I. (1996) *Food Additives.* Gower Publishing Co., Brookfield, VT.

ANY OTHER RELEVANT INFORMATION: This is a natural polysaccharide exudate from trees of the genus *Astragalus*. Occasionally used as an emulsifier

NAME:	Xanthan
CATEGORY:	Polysaccharides
FOOD USE:	Soups/ Syrups/ Beverages/ Dairy products/ Dressings/ Dry formulated products/ Food packaging/ Edible films and coatings/ All food products
SYNONYMS:	Beta-D-(1 $\rightarrow$ 4) glucose polymer with a trisaccharide side chain composed of alternate glucopyranosyl units. The side chain is composed of a beta-D-pyranosyl unit between two mannopyranosyl units with pyruvate and acetyl substituents/ CAS 11138-66-2/ EINECS 234-394-2/ E415/ Gum xanthan/ Corn sugar gum/ Alginoid/ Dariloid/ Merecol/ Merezan/ Rhodigel/ Ticaxan
MOLECULAR MASS:	$3–50 \times 10^6$
PROPERTIES AND APPEARANCE:	White/off-white powder or soft flakes
SOLUBILITY % AT VARIOUS TEMPERATURE/pH COMBINATIONS:	Solubility varies with molecular weight. Generally, xanthan is soluble in water, sucrose solutions, sodium chloride solution and ethanol solutions, but solubility in vegetable oils and propylene glycol is limited. Solubility declines with increasing molecular weight and increasing concentration. It is essentially pH-independent (pH 4 to pH 8). Elevated temperatures will significantly improve the dispersion of high molecular-weight xanthans
FUNCTION IN FOODS:	Thickening agent; stabiliser; bulking agent; suspending agent
ALTERNATIVES:	Carrageenans; dextrans; starches; locust bean gum; guar gum; gelatin; agar; alginate (application-dependent)
TECHNOLOGY OF USE IN FOODS:	Xanthan should be dispersed in cold water and heated to 80°C for 10 minutes to hydrate the polysaccharide. A high shear mixer will aid dispersion, although prolonged shear of high molecular-weight polymers will reduce the viscosity. Xanthan forms aggregates of single or double helices in solution with a weak gel interaction. This weak gel formation manifests as a high yield point which helps suspend particles (herbs, fruit chunks, emulsion droplets) in viscous solutions. Synergistic interactions with gelatin and gum arabic. Forms gels with locust bean gum and guar gum

SYNERGISTS: Guar; locust bean gum

ANTAGONISTS: None known

FOOD SAFETY ISSUES: Non-toxic. The powder is combustible

LEGISLATION:

USA:
FDA 21CFR § 133.124, 133.133, 133.134, 133.162, 133.178, 133.179, 172.695, USDA9CFR § 318.7, 381.147

UK and EUROPE:
Europe: E418. ADI 10 mg/kg

CANADA:
Approved

AUSTRALIA/PACIFIC RIM:
Japan: approved for use as a natural thickener and stabiliser

REFERENCES:

Kang, K. S., and Pettitt, D. J. (1993) Xanthan, Gellan, Welan and Rhamsan. In: Whistler, R. L., and J. N. BeMiller (Eds.), *Industrial Gums: Polysaccharides and Derivatives*. 3rd edition, pp. 341–395.

Ash, M., and Ash, I. (1996) *Food Additives.* Gower Publishing Co., Brookfield, VT.

ANY OTHER RELEVANT INFORMATION: Xanthan is a natural polysaccharide obtained from cultures of the microorganism *Xanthomonas campestris*

Part 11

Preservatives

NAME:	Sorbic acid
CATEGORY:	Antimicrobial preservative
FOOD USE:	Baked goods/ Beverages/ Bread, cake batters, cake fillings, cake toppings/ Cheese, cottage cheese/ Fish (smoked, salted)/ Fruit juices (fresh), fruit (dried)/ Margarine/ Pickled products/ Pie crusts, pie fillings/ Salad dressings, salads (fresh)/ Sausage/ Seafood cocktails/ Syrups/ Wine
SYNONYMS:	Hexadienoic acid/ Sorbistat/ 2,4-Hexadienoic acid/ 2-Propenylacrylic acid/ E200/ CAS 110-44-1/ 22500-92-1/ EINECS 203-768-7/ Hexadienic acid/ (2-Butenylidene) acetic acid
FORMULA:	CH3-CH=CHCH=CH-COOH
MOLECULAR MASS:	112.13
ALTERNATIVE FORMS:	Sodium, potassium, calcium salts
PROPERTIES AND APPEARANCE:	White or off-white powder, characteristic odour
BOILING POINT IN °C AT VARIOUS PRESSURES (INCLUDING 760 mm Hg):	at 760 mm Hg decomposes at 228°C, at 50 mm Hg = 143°C, at 10 mm Hg = 119°C
MELTING RANGE IN °C:	134.5 (range 132–135)
FLASH POINT IN °C:	127
IONISATION CONSTANT AT 25°C:	1.73 E^{-5}
DENSITY AT 20°C (AND OTHER TEMPERATURES) IN g/L:	1.36
HEAT OF COMBUSTION AT 25°C IN J/kg:	27718 kJ/kg
VAPOUR PRESSURE AT VARIOUS TEMPERATURES IN mm Hg:	@ 20°C <0.01 @ 120°C 10 @ 140°C 43

PURITY %:	>99 (anhydrous basis)
WATER CONTENT MAXIMUM IN %:	0.5
HEAVY METAL CONTENT MAXIMUM IN ppm:	10 (as Pb)
ARSENIC CONTENT MAXIMUM IN ppm:	3
ASH MAXIMUM IN %:	0.2

SOLUBILITY % AT VARIOUS TEMPERATURE/pH COMBINATIONS:

in water:	@ **20°C**	pH 3.1 = 0.15% pH 4.4 = 0.22% pH 5.9 = 1.02%					@ **30°C**	0.25
	@ **50°C**	0.55	@ **100°C**	4.00				
in vegetable oil:	@ **20°C**	0.80	@ **50°C**	2.00				
in sucrose solution:	@ **10%**	0.15	@ **40%**	0.10	@ **60%**	0.08		
in sodium chloride solution:	@ **5%**	0.11	@ **10%**	0.07	@ **15%**	0.04		
in ethanol solution:	@ **5%**	0.16	@ **20%**	0.29	@ **95%**	12.6–14.5	@ **100%**	12.9–14.8
in propylene glycol:	@ **20°C**	5.5						
in glacial acetic acid:	11.5							

FUNCTION IN FOODS:	Antimicrobial preservative: Fungi – broad spectrum; bacteria – mostly strict aerobes. Not lactic acid bacteria. Flavouring. Acidulant
ALTERNATIVES:	Sodium sorbate; potassium sorbate; calcium sorbate; benzoic acid; sodium benzoate; potassium benzoate; calcium benzoate; propionic acid; sodium propionate; potassium propionate; calcium propionate
TECHNOLOGY OF USE IN FOODS:	The pK_a is 4.8, so sorbic acid and its salts would normally be used at pH less than 4.8. It can, however, be used at up to neutral pH, but the activity reduces as the pH increases. In wine, it should only be used in conjunction with sulphur dioxide, otherwise a characteristic off-odour may result

SYNERGISTS: Sulphur dioxide, carbon dioxide, sodium chloride, sucrose, propionic acid, dehydration, nisin with polyphosphates, benzoates

ANTAGONISTS: Non-ionic surfactants

FOOD SAFETY ISSUES: This is not an additive which has been subject to great criticism over safety

LEGISLATION:

USA:
GRAS if used in accordance with Good Manufacturing Practice. 0.1% individually or 0.2% if used in combination with its salts or benzoic acid or its salts. Not allowed in cooked sausage. Limited to 300 mg/1000 gallons of wine. FDA 21CFR § 133, 146.115, 146.152, 146.154, 150.141, 150.161, 166.110, 181.22, 181.23, 182.3089, GRAS; USDA 9CFR § 318.7; BATF 27CFR § 240.1051

UK and EUROPE:
UK: Salad cream 1000 ppm; salad dressing 1000 ppm; mayonnaise 1000 ppm; wine 200 ppm

CANADA:
Permitted in selected foods

AUSTRALIA/PACIFIC RIM:
Japan: approved with limitations

REFERENCES:

Sofos, J. N., and Busta, F. F. (1993) Sorbic acid and sorbates. In: Davidson, P. M., and A. L. Branen (Eds.), *Antimicrobials in Foods*, Second Edition. Marcel Dekker, New York.

Food Chemicals Codex, Fourth Edition (1996) National Academy Press, Washington, DC.

Merck Index, Twelfth Edition (1996) Merck & Co., Inc., Whitehouse Station, NJ.

Ash, M., and Ash, I. (1995) *Food Additives.* Gower Publishing Co., Brookfield, VT.

ANY OTHER RELEVANT INFORMATION: Occurs naturally in the berries of the mountain ash *Sorbus aucuparia* L., *Rosaceae* (sorbapple, rowan). pK (25°C) = 4.76

NAME:	Sodium sorbate
CATEGORY:	Antimicrobial preservative
FOOD USE:	Baked goods/ Beverages/ Bread, cake batters, cake fillings, cake toppings/ Cheese, cottage cheese/ Fish (smoked, salted)/ Fruit juices (fresh), fruit (dried)/ Margarine/ Pickled products/ Pie crusts, pie fillings/ Salad dressings, salads (fresh)/ Sausage/ Seafood cocktails/ Syrups/ Wine
SYNONYMS:	Sodium hexadienoate/ 2,4-Hexadienoic acid sodium salt/ Sorbic acid sodium salt/ 2-Propenylacrylic acid sodium salt/ E201/ CAS 7757-81-5/ Hexadienic acid/ (2-Butenylidene) acetic acid, sodium salt
FORMULA:	CH3CH=CHCH=CHCOOHNa
MOLECULAR MASS:	134.12
ALTERNATIVE FORMS:	Sorbic acid (potassium and calcium salts)
PROPERTIES AND APPEARANCE:	White or fine crystalline powder, characteristic odour
DENSITY AT 20°C (AND OTHER TEMPERATURES) IN g/L:	1.36
PURITY %:	>98
WATER CONTENT MAXIMUM IN %:	1.0
HEAVY METAL CONTENT MAXIMUM IN ppm:	10
ARSENIC CONTENT MAXIMUM IN ppm:	3

SOLUBILITY % AT VARIOUS TEMPERATURE/pH COMBINATIONS:	
in water:	**@ 20°C** 58.2
in ethanol solution:	**@ 100%** 6.5
FUNCTION IN FOODS:	Antimicrobial preservative: Fungi – broad spectrum; bacteria – mostly strict aerobes. Not lactic acid bacteria. Flavouring. Acidulant
ALTERNATIVES:	Potassium sorbate; sorbic acid; calcium sorbate; benzoic acid; sodium benzoate; potassium benzoate; calcium benzoate; propionic acid; sodium propionate; potassium propionate; calcium propionate
TECHNOLOGY OF USE IN FOODS:	The pK_a is 4.8, so sorbic acid and its salts would normally be used at pH less than 4.8. It can, however, be used at up to neutral pH, but the activity reduces as the pH increases. In wine, it should only be used in conjunction with sulphur dioxide, otherwise a characteristic off-odour may result
SYNERGISTS:	Sulphur dioxide, carbon dioxide, sodium chloride, sucrose, propionic acid, dehydration, nisin with polyphosphates, benzoates
ANTAGONISTS:	Non-ionic surfactants
FOOD SAFETY ISSUES:	This is not an additive which has been subject to great criticism over safety

LEGISLATION:

USA:
GRAS if used in accordance with Good Manufacturing Practice. 0.1% individually or 0.2% if used in combination with its salts or benzoic acid or its salts. Not allowed in cooked sausage. Limited to 300 mg/1000 gallons of wine. FDA 21CFR § 182.90, 182.3759, 182.3089, GRAS; USDA 9CFR § 318.7

UK and EUROPE:
UK: Salad cream 1000 ppm; salad dressing 1000 ppm; mayonnaise 1000 ppm; wine 200 ppm

CANADA:
Permitted in selected foods

AUSTRALIA/PACIFIC RIM:
Japan: approved with limitations

EFERENCES:

Sofos, J. N., and Busta, F. F. (1993) Sorbic acid and sorbates. In: Davidson, P. M., and A. L. Branen (Eds.), *Antimicrobials in Foods*, Second Edition. Marcel Dekker, New York.

Ash, M., and Ash, I. (1995) *Food Additives.* Gower Publishing Co., Brookfield, VT.

NAME:	Potassium sorbate
CATEGORY:	Antimicrobial preservative
FOOD USE:	Baked goods/ Beverages/ Bread, cake batters, cake fillings, cake toppings/ Cheese, cottage cheese/ Fish (smoked, salted)/ Fruit juices (fresh), fruit (dried)/ Margarine/ Pickled products/ Pie crusts, pie fillings/ Salad dressings, salads (fresh)/ Sausage/ Seafood cocktails/ Syrups/ Wine
SYNONYMS:	Potassium hexadienoate/ 2,4-Hexadienoic acid potassium salt/ Sorbic acid potassium salt/ 2-Propenylacrylic acid potassium salt/ E202/ CAS 590-00-1/ 24634-61-5/ 246-376-1/ EINECS 203-768-7/ Hexadienic acid, potassium salt/ (2-Butenylidene) acetic acid, potassium salt
FORMULA:	CH3CH=CHCH=CHCOOH
MOLECULAR MASS:	150.22
ALTERNATIVE FORMS:	Sorbic acid (sodium, and calcium salts)
PROPERTIES AND APPEARANCE:	White or off-white powder, characteristic odour
BOILING POINT IN °C AT VARIOUS PRESSURES (INCLUDING 760 mm Hg):	Decomposes above 270°C
MELTING RANGE IN °C:	Decomposes above 270°C
DENSITY AT 20°C (AND OTHER TEMPERATURES) IN g/L:	1.36

PURITY %:	>98 (anhydrous basis)							
WATER CONTENT MAXIMUM IN %:	1.0							
HEAVY METAL CONTENT MAXIMUM IN ppm:	10 (as Pb)							
ARSENIC CONTENT MAXIMUM IN ppm:	3							
SOLUBILITY % AT VARIOUS TEMPERATURE/pH COMBINATIONS:								
in water:	**@ 20°C**	58.2 (pH 3.1)	**@ 50°C**	61	**@ 100°C**	64		
in vegetable oil:	**@ 20°C**	0.01	**@ 50°C**	0.03				
in sucrose solution:	**@ 10%**	58	**@ 40%**	45	**@ 60%**	28		
in sodium chloride solution:	**@ 5%**	47	**@ 10%**	34	**@ 15%**	12–15		
in ethanol solution:	**@ 5%**	57.4	**@ 20%**	54.6	**@ 50%**	45.3	**@ 95%**	6.5
	@ 100%	2.0						
in propylene glycol:	**@ 20°C**	55	**@ 50°C**	48	**@ 100°C**	20		
FUNCTION IN FOODS:	Antimicrobial preservative: Fungi – broad spectrum; bacteria – mostly strict aerobes. Not lactic acid bacteria. Flavouring. Acidulant							
ALTERNATIVES:	Sodium sorbate; sorbic acid; calcium sorbate; benzoic acid; sodium benzoate; potassium benzoate; calcium benzoate; propionic acid; sodium propionate; potassium propionate; calcium propionate							
TECHNOLOGY OF USE IN FOODS:	The pK_a is 4.8, so sorbic acid and its salts would normally be used at pH less than 4.8. It can, however, be used at up to neutral pH, but the activity reduces as the pH increases. In wine, it should only be used in conjunction with sulphur dioxide, otherwise a characteristic off-odour may result							
SYNERGISTS:	Sulphur dioxide, carbon dioxide, sodium chloride, sucrose, propionic acid, dehydration, nisin with polyphosphates, benzoates							

ANTAGONISTS: Non-ionic surfactants

FOOD SAFETY ISSUES: This is not an additive which has been subject to great criticism over safety

LEGISLATION:

USA:
GRAS if used in accordance with Good Manufacturing Practice. 0.1% individually or 0.2% if used in combination with its salts or benzoic acid or its salts. Not allowed in cooked sausage. Limited to 300 mg/1000 gallons of wine. FDA 21CFR § 133, 150.141, 150.161, 166.110, 181.90, 183.3640, GRAS; USDA 9CFR § 318.7; BATF 27CFR § 240.1051, GRAS (FEMA)

UK and EUROPE:
UK: Salad cream 1000 ppm; salad dressing 1000 ppm; mayonnaise 1000 ppm; wine 200 ppm

CANADA:
Permitted in selected foods

AUSTRALIA/PACIFIC RIM:
Japan: approved with limitations

REFERENCES:

Sofos, J. N., and Busta, F. F. (1993) Sorbic acid and sorbates. In: Davidson, P. M., and A. L. Branen (Eds.), *Antimicrobials in Foods*, Second Edition. Marcel Dekker, New York.

Food Chemicals Codex, Fourth Edition (1996) National Academy Press, Washington, DC.

Merck Index, Twelfth Edition (1996) Merck & Co., Inc., Whitehouse Station, NJ.

Ash, M., and Ash, I. (1995) *Food Additives.* Gower Publishing Co., Brookfield, VT.

ANY OTHER RELEVANT INFORMATION: Occurs naturally in the berries of the mountain ash *Sorbus auuparia* L., *Rosaceae*

NAME:	Calcium sorbate
CATEGORY:	Antimicrobial preservative
FOOD USE:	Baked goods/ Beverages/ Bread, cake batters, cake fillings, cake toppings/ Cheese, cottage cheese/ Fish (smoked, salted)/ Fruit juices (fresh), fruit (dried)/ Margarine/ Pickled products/ Pie crusts, pie fillings/ Salad dressings, salads (fresh)/ Sausage/ Seafood cocktails/ Syrups/ Wine
SYNONYMS:	Calcium hexadienoate/ 2,4-Hexadienoic acid calcium salt/ Sorbic acid calcium salt/ 2-Propenylacrylic acid calcium salt/ E203/ Hexadienic acid, calcium salt/ (2-Butenylidene) acetic acid, calcium salt/ CAS 7492-55-9
FORMULA:	(CH3CH=CHCH=CHCOO)2Ca
MOLECULAR MASS:	262.32
ALTERNATIVE FORMS:	Sorbic acid (potassium and sodium salts)
PROPERTIES AND APPEARANCE:	White or fine crystalline powder, characteristic odour
BOILING POINT IN °C AT VARIOUS PRESSURES (INCLUDING 760 mm Hg):	Decomposes >400°C
MELTING RANGE IN °C:	Decomposes >400°C
PURITY %:	>98 (anhydrous basis)
WATER CONTENT MAXIMUM IN %:	1.0
HEAVY METAL CONTENT MAXIMUM IN ppm:	10 (as Pb)

SOLUBILITY % AT VARIOUS TEMPERATURE/pH COMBINATIONS:

in water:	**@ 20°C**	58.2%				
in vegetable oil:	**@ 20°C**	Practically insoluble	**@ 50°C**	Practically insoluble	**@ 100°C**	Practically insoluble
in sucrose solution:	**@ 10%**	Sparingly soluble	**@ 40%**	Sparingly soluble	**@ 60%**	Sparingly soluble
in sodium chloride solution:	**@ 5%**	Sparingly soluble	**@ 10%**	Sparingly soluble	**@ 15%**	Sparingly soluble
in ethanol solution:	**@ 5%**	Practically insoluble	**@ 20%**	Practically insoluble	**@ 95%**	Practically insoluble
	@ 100%	6.5%				
in propylene glycol:	**@ 20°C**	Practically insoluble				
in glacial acetic acid:	Sparingly soluble					

FUNCTION IN FOODS:	Antimicrobial preservative: Fungi – broad spectrum; bacteria – mostly strict aerobes. Not lactic acid bacteria. Flavouring. Acidulant
ALTERNATIVES:	Sodium sorbate; sorbic acid; potassium sorbate; benzoic acid; sodium benzoate; potassium benzoate; calcium benzoate; propionic acid; sodium propionate; potassium propionate; calcium propionate
TECHNOLOGY OF USE IN FOODS:	The pK_a is 4.8, so sorbic acid and its salts would normally be used at pH less than 4.8. It can, however, be used at up to neutral pH, but the activity reduces as the pH increases. In wine, it should only be used in conjunction with sulphur dioxide, otherwise a characteristic off-odour may result
SYNERGISTS:	Sulphur dioxide; carbon dioxide; sodium chloride; sucrose; propionic acid; dehydration; nisin with polyphosphates; benzoates
ANTAGONISTS:	Non-ionic surfactants
FOOD SAFETY ISSUES:	This is not an additive which has been subject to great criticism over safety

LEGISLATION:	**USA:** GRAS if used in accordance with Good Manufacturing Practice. 0.1% individually or 0.2% if used in combination with its salts or benzoic acid or its salts. Not allowed in cooked sausage. Limited to 300 mg/1000 gallons of wine. FDA21CFR § 133, 182.3225, GRAS; USDA 9CFR § 318.7	**UK and EUROPE:** UK: Salad cream 1000 ppm; salad dressing 1000 ppm; mayonnaise 1000 ppm; wine 200 ppm	**CANADA:** Permitted in selected foods **AUSTRALIA/PACIFIC RIM:** Japan: approved with limitations

REFERENCES:

Sofos, J. N., and Busta, F. F. (1993) Sorbic acid and sorbates. In: Davidson, P. M., and A. L. Branen (Eds.), *Antimicrobials in Foods*, Second Edition. Marcel Dekker, New York.

Food Chemicals Codex, Fourth Edition (1996) National Academy Press, Washington, DC.

NAME:	Benzoic acid
CATEGORY:	Antimicrobial preservative
FOOD USE:	Beverages/ Margarine/ Bakery products/ Fish products/ Fruit juice, fruit pulp/ Jam/ Liquid egg, whole egg, egg yolk/ Mayonnaise/ Mustard/ Pickles, sauces, ketchup/ Sausage
SYNONYMS:	Benzenecarboxylic acid/ Phenylformic acid/ Dracylic acid/ E210/ CAS 65-85-0/ EINECS 200-618-2/ FEMA 2131
FORMULA:	C7H6O2
MOLECULAR MASS:	122.12
ALTERNATIVE FORMS:	Sodium benzoate, potassium benzoate, calcium benzoate
PROPERTIES AND APPEARANCE:	White crystals, needles or scales, phenolic taste
BOILING POINT IN °C AT VARIOUS PRESSURES (INCLUDING 760 mm Hg):	@ 760 mm Hg: 249.2 @ 400 mm Hg: 227.0 @ 200 mm Hg: 205.8 @ 100 mm Hg: 186.2 @ 60 mm Hg: 172.8 @ 40 mm Hg: 162.6 @ 20 mm Hg: 146.7 @ 10 mm Hg: 132.1
MELTING RANGE IN °C:	122.4
FLASH POINT IN °C:	121–131
IONISATION CONSTANT AT 25°C:	6.4 E-5
DENSITY AT 20°C (AND OTHER TEMPERATURES) IN g/L:	1.2659

PURITY %:	>99.5 (anhydrous basis)
WATER CONTENT MAXIMUM IN %:	0.7
HEAVY METAL CONTENT MAXIMUM IN ppm:	10 (as Pb)
ASH MAXIMUM IN %:	0.05
SOLUBILITY % AT VARIOUS TEMPERATURE/pH COMBINATIONS:	
in water:	**@ 4°C:** 0.18 **@ 18°C:** 0.27 **@ 20°C:** 0.29 **@ 50°C:** 0.95 **@ 75°C:** 2.2 **@ 95°C:** 6.8
FUNCTION IN FOODS:	Antimycotic and antibacterial
ALTERNATIVES:	Other microbials where suitable
TECHNOLOGY OF USE IN FOODS:	Effective versus yeasts and moulds, food-poisoning bacteria, spore-forming bacteria. Not effective against many spoilage bacteria. pK_a is 4.2, so should be used in foods with pH less than 4.2. Levels for general use: Carbonated beverages 0.03–0.05%; non-carbonated beverages 0.1%; beverage syrups 0.1%; fruit drinks 0.1%; fruit juices 0.1%; purées and concentrates 0.1%; cider 0.05–0.1%; salted margarine 0.1%; pie and pastry fillings 0.1%; icings 0.1%; soy sauce 0.1%; mincemeat 0.1%; salads 0.1%; salad dressings 0.1%; fruit salads 0.1%; pickles 0.1%; relishes 0.1%; fruit cocktails 0.1%; olives 0.1%; sauerkraut 0.1%; preserves 0.1%; dried fruits 0.1%; jams 0.1%; jellies 0.1%; fish 0.15–0.35%; dipping solutions 0.15–0.34%; ice glaze 0.15–0.30%
SYNERGISTS:	Sulphur dioxide; carbon dioxide; sodium chloride; sucrose; sorbates; thermal processing; refrigeration; boric acid
ANTAGONISTS:	Lipids (partitioning); ferric salts; anionic surfactants

LEGISLATION:

USA:
FDA 21CFR 150.141, 150.161, 166.40, 166.110, 175.300, 184.1021, 184.1733, GRAS 0.1% max. in foods., EPA registered

UK and EUROPE:
0.5% max

CANADA:
Permitted in selected foods

AUSTRALIA/PACIFIC RIM:
Japan: 0.2% max

REFERENCES:

Davidson, P. M., and Branen, A. L. (1993) *Antimicrobials in Foods*, Second Edition. Marcel Dekker, New York, pp. 11–48.
Food Chemicals Codex, Fourth Edition (1996) National Academy Press, Washington, DC.
Merck Index, Twelfth Edition (1996) Merck & Co., Inc., Whitehouse Station, NJ.
Ash, M., and Ash, I. (1995) *Food Additives.* Gower Publishing Co., Brookfield, VT.

ANY OTHER RELEVANT INFORMATION:

Effect of pH on the dissociation of benzoic acid:

pH	*% Undissociated acid*
3	93.5
4	59.3
5	12.8
6	1.44
7	0.144
pK	4.19

Most berries contain appreciable amounts of benzoic acid (ca. 0.05%). It is excreted mainly as hippuric acid by most vertebrates, except fowl

NAME:	Sodium benzoate
CATEGORY:	Antimicrobial preservative
FOOD USE:	Beverages/ Margarine/ Bakery products/ Fish products/ Fruit juice, fruit pulp/ Jam/ Liquid egg, whole egg, egg yolk/ Mayonnaise, mustard, pickles, sauces, ketchup/ Sausage
SYNONYMS:	Benzoic acid, sodium salt/ CAS 532-32-1/ EINECS 208-534-8/ E211/ Benzoate of soda/ Benzoate sodium
FORMULA:	C6 H5 COONa
MOLECULAR MASS:	144.11
ALTERNATIVE FORMS:	Other salts/ Benzoic acid
PROPERTIES AND APPEARANCE:	White granular or crystalline powder, odourless, sweetish astringent taste
DENSITY AT 20°C (AND OTHER TEMPERATURES) IN g/L:	1.44
PURITY %:	>99 (anhydrous basis)
WATER CONTENT MAXIMUM IN %:	1.5
HEAVY METAL CONTENT MAXIMUM IN ppm:	10 (as Pb)
SOLUBILITY % AT VARIOUS TEMPERATURE/pH COMBINATIONS:	
in water:	**@ 0°C** 62.8 **@ 20°C** 66.0 **@ 100°C** 74.2
in ethanol solution:	**@ 90%** 2.0 **@ 100%** 1.33
FUNCTION IN FOODS:	Preservative, antimocrobial agent, flavouring agent, adjuvant; antimycotic migrating from food packaging

TECHNOLOGY OF USE IN FOODS:	Usage level: 0.5% max.; 0.1% max. in food; 0.1% (in distilling materials). Most effective in acid foods. Effective versus yeasts and moulds, food-poisoning bacteria, spore-forming bacteria. Not effective against many spoilage bacteria. pK_a is 4.2, so should be used in foods with pH less than 4.2. Levels for general use: Carbonated beverages 0.03–0.05%; non-carbonated beverages 0.1%; beverage syrups 0.1%; fruit drinks 0.1%; fruit juices 0.1%; purées and concentrates 0.1%; cider 0.05–0.1%; salted margarine 0.1%; pie and pastry fillings 0.1%; icings 0.1%; soy sauce 0.1%; mincemeat 0.1%; salads 0.1%; salad dressings 0.1%; fruit salads 0.1%; pickles 0.1%; relishes 0.1%; fruit cocktails 0.1%; olives 0.1%; sauerkraut 0.1%; preserves 0.1%; dried fruits 0.1%; jams 0.1%; jellies 0.1%; fish 0.15–0.35%; dipping solutions 0.15–0.35%; ice glaze 0.15–0.30%
SYNERGISTS:	Sulphur dioxide; carbon dioxide; sodium chloride; sucrose; sorbates; thermal processing; refrigeration; boric acid
ANTAGONISTS:	Lipids (partitioning); ferric salts; anionic surfactants
LEGISLATION:	**USA:** FDA 21CFR § 146.152, 146.154, 150.141, 150.161, 166.40, 166.110, 181.22, 181.23, 184.1733, 582.3733; GRAS; USDA 9CFR § 318.7; BATF 27CFR § 240.1051; EPA reg.; GRAS (FEMA) **UK and EUROPE:** Europe: listed 0.5% as acid **CANADA:** Permitted in selected foods **AUSTRALIA/PACIFIC RIM:** Japan: approved with limitations
REFERENCES:	Davidson, P. M., and Branen, A. L. (1993) *Antimicrobials in Foods*, Second Edition. Marcel Dekker, New York, pp. 11–48. *Food Chemicals Codex*, Fourth Edition (1996) National Academy Press, Washington, DC. *Merck Index*, Twelfth Edition (1996) Merck & Co., Inc., Whitehouse Station, NJ.

ANY OTHER RELEVANT INFORMATION: Effect of pH on the dissociation of benzoic acid:

pH	*% Undissociated acid*
3	93.5
4	59.3
5	12.8
6	1.44
7	0.144
pK	4.19

NAME:	**Potassium benzoate**
CATEGORY:	Antimicrobial preservative
FOOD USE:	Beverages/ Margarine/ Bakery products/ Fish products/ Fruit juice, fruit pulp/ Jam/ Liquid egg, whole egg, egg yolk/ Mayonnaise, mustard, pickles, sauces, ketchup/ Sausage
SYNONYMS:	Benzoic acid, potassium salt/ E212/ CAS 582-25-2 (anhydrous)/ EINECS 209-481-3
FORMULA:	C6H5COOH
MOLECULAR MASS:	160.22
ALTERNATIVE FORMS:	Other salts, benzoic acid
PROPERTIES AND APPEARANCE:	Orthorhombic crystals or powder
MELTING RANGE IN °C:	>300
DENSITY AT 20°C (AND OTHER TEMPERATURES) IN g/L:	1.44
PURITY %:	>99 (anhydrous basis)
WATER CONTENT MAXIMUM IN %:	1.5
HEAVY METAL CONTENT MAXIMUM IN ppm:	10 (as Pb)
SOLUBILITY % AT VARIOUS TEMPERATURE/pH COMBINATIONS:	
in water:	**@ 20°C** 50
in ethanol solution:	**@ 90%** 2.0 **@ 100%** 1.33
FUNCTION IN FOODS:	Antimicrobial preservative for margarine; oleomargarine; wine

TECHNOLOGY OF USE IN FOODS: Effective versus yeasts and moulds, food-poisoning bacteria, spore-forming bacteria. Not effective against many spoilage bacteria. pK_a is 4.2, so should be used in foods with pH less than 4.2. Levels for general use: carbonated beverages 0.03–0.05%; non-carbonated beverages 0.1%; beverage syrups 0.1%; fruit drinks 0.1%; fruit juices 0.1%; purées and concentrates 0.1%; cider 0.05–0.1%; salted margarine 0.1%; pie and pastry fillings 0.1%; icings 0.1%; soy sauce 0.1%; mincemeat 0.1%; salads 0.1%; salad dressings 0.1%; fruit salads 0.1%; pickles 0.1%; relishes 0.1%; fruit cocktails 0.1%; olives 0.1%; sauerkraut 0.1%; preserves 0.1%; dried fruits 0.1%; jams 0.1%; jellies 0.1%; fish 0.15–0.35; dipping solutions 0.15–0.35%; ice glaze 0.15–0.30%

SYNERGISTS: SO_2; CO_2; NaCl; sucrose; sorbates; thermal processing; refrigeration; boric acid

ANTAGONISTS: Lipids (partitioning); ferric salts; anionic surfactants

LEGISLATION:

USA:
USDA 9CFR § 318.7 (limitation 0.1%), BATF 27CFR § 240.1051 (limitation 0.1% in wine)

UK and EUROPE:
Europe: listed

CANADA:
Permitted in selected foods

REFERENCES:

Davidson, P. M., and Branen, A. L. (1993) *Antimicrobials in Foods*, Second Edition. Marcel Dekker, New York, pp. 11–48.

Food Chemicals Codex, Fourth Edition (1996) National Academy Press, Washington, DC.

ANY OTHER RELEVANT INFORMATION: Effect of pH on the dissociation of benzoic acid:

pH	*% Undissociated acid*
3	93.5
4	59.3
5	12.8
6	1.44
7	0.144
pK	4.19

NAME:	Calcium benzoate
CATEGORY:	Antimicrobial preservative
FOOD USE:	Beverages/ Margarine/ Bakery products/ Fish products/ Fruit juice, fruit pulp/ Jam/ Liquid egg, whole egg, egg yolk/ Mayonnaise, mustard, pickles, sauces, ketchup/ Sausage
SYNONYMS:	Benzoic acid, calcium salt/ E213/ CAS 2090-05-3/ EINECS 218-235-4
FORMULA:	C14H10O4·3H2O
MOLECULAR MASS:	374.26
ALTERNATIVE FORMS:	Other salts, benzoic acid
PROPERTIES AND APPEARANCE:	Orthorhombic crystals or powder
DENSITY AT 20°C (AND OTHER TEMPERATURES) IN g/l:	1.44
TECHNOLOGY OF USE IN FOODS:	Effective versus yeasts and moulds, food-poisoning bacteria, spore-forming bacteria. Not effective against many spoilage bacteria. pK_a is 4.2, so should be used in foods with pH less than 4.2. Levels for general use: carbonated beverages 0.03–0.05%; non-carbonated beverages 0.1%; beverage syrups 0.1%; fruit drinks 0.1%; fruit juices 0.1%; purées and concentrates 0.1%; cider 0.05–0.1%; salted margarine 0.1%; pie and pastry fillings 0.1%; icings 0.1%; soy sauce 0.1%; mincemeat 0.1%; salads 0.1%; salad dressings 0.1%; fruit salads 0.1%; pickles 0.1%; relishes 0.1%; fruit cocktails 0.1%; olives 0.1%; sauerkraut 0.1%; preserves 0.1%; dried fruits 0.1%; jams 0.1%; jellies 0.1%; fish 0.15–0.35%; dipping solutions 0.15–0.35%; ice glaze 0.15–0.30%
SYNERGISTS:	SO_2; CO_2, NaCl; sucrose; sorbates; thermal processing; refrigeration; boric acid
ANTAGONISTS:	Lipids (partitioning); ferric salts; anionic surfactants

LEGISLATION:

USA:
FDA 21CFR § 166.110, 178.2010;
USDA 9CFR § 318.7

UK and EUROPE:
Europe: listed. Limitation 0.1% (alone); 0.2% (in combination with sorbic acid or its salts)

CANADA:
Permitted in selected foods

REFERENCES:

Davidson, P. M., and Branen, A. L. (1993) *Antimicrobials in Foods*, Second Edition. Marcel Dekker, New York, pp. 11–48.

Ash, M., and Ash, I. (1995) *Food Additives*. Gower Publishing Co., Brookfield, VT.

ANY OTHER RELEVANT INFORMATION:

Effect of pH on the dissociation of benzoic acid:

pH	*% Undissociated acid*
3	93.5
4	59.3
5	12.8
6	1.44
7	0.144
pK	4.19

NAME:	Ethyl-4-hydroxybenzoate
CATEGORY:	Antimicrobial preservative
FOOD USE:	Bakery products (cakes, crusts, pastries, toppings, fillings)/ Soft drinks/ Fish (marinated, smoked, jellied)/ Flavour extracts/ Fruit products/ Gelatin, jams, jellies, preserves/ Malt extracts/ Olives/ Pickles/ Salad dressings/ Syrups/ Wines
SYNONYMS:	CAS 120-47-8/ EINECS 204-399-4/ E214/ Ethyl paraben/ 4-Hydroxybenzoic acid ethyl ester/ Ethyl *p*-hydroxybenzoate/ Carboxyphenol
FORMULA:	HO C6 H4 CO2 C2H5
MOLECULAR MASS:	166.18
ALTERNATIVE FORMS:	Methyl, propyl, butyl and heptyl esters
PROPERTIES AND APPEARANCE:	Ivory to white powder
BOILING POINT IN °C AT VARIOUS PRESSURES (INCLUDING 760 mm Hg):	297–298
MELTING RANGE IN °C:	114–118
SOLUBILITY % AT VARIOUS TEMPERATURE/pH COMBINATIONS:	
in water:	**@ 10°C** 0.07 **@ 25°C** 0.17 **@ 80°C** 0.86
in vegetable oil:	**@ 25°C** 3.0 (olive oil), 1.0 (peanut oil)
in ethanol solution:	**@ 100%** 70
in propylene glycol:	**@ 20°C** 25
FUNCTION IN FOODS:	Antimicrobial preservative – especially effective against bacteria and fungi at around neutral pH
ALTERNATIVES:	Other parabens

TECHNOLOGY OF USE IN FOODS:	pH optimum is 3–8. Parabens are often used as blends, e.g. methyl and propyl at a ratio of 2–3 to 1. Dissolve initially in water, ethanol, propylene glycol or the food itself. Hot water is recommended. Parabens can also be dry-blended with other dry ingredients
LEGISLATION:	**USA:** Not permitted; **UK and EUROPE:** Permitted in selected foods; **CANADA:** Permitted in selected foods
REFERENCES:	Davidson, P. M., and Branen, A. L. (1993) *Antimicrobials in Foods*, Second Edition. Marcel Dekker, New York. Ash, M., and Ash, I. (1995) *Food Additives.* Gower Publishing Co., Brookfield, VT.

NAME:	Ethyl-4-hydroxybenzoate, sodium salt
CATEGORY:	Antimicrobial preservative
FOOD USE:	Bakery products (cakes, crusts, pastries, toppings, fillings)/ Soft drinks/ Fish (marinated, smoked, jellied)/ Flavour extracts/ Fruit products/ Gelatin, jams, jellies, preserves/ Malt extracts/ Olives/ Pickles/ Salad dressings/ Syrups/ Wines
SYNONYMS:	E215/ Ethyl paraben sodium salt/ 4-Hydroxybenzoic acid ethyl ester sodium salt/ Ethyl *p*-hydroxybenzoate sodium salt/ Carboxyphenol sodium salt
FORMULA:	NaO C6H4 CO2 C2H5
ALTERNATIVE FORMS:	Other parabens
PROPERTIES AND APPEARANCE:	Ivory to white powder
FUNCTION IN FOODS:	Antimicrobial preservative – especially effective against bacteria and fungi at around neutral pH
TECHNOLOGY OF USE IN FOODS:	pH optimum is 3–8. Parabens are often used as blends, e.g. methyl and propyl at a ratio of 2–3 to 1. Dissolve initially in water, ethanol, propylene glycol or the food itself. Hot water is recommended. Parabens can also be dry-blended with other dry ingredients
LEGISLATION:	**USA:** Not permitted **UK and EUROPE:** Permitted in certain foods **CANADA:** Permitted in certain foods
REFERENCE:	Davidson, P. M., and Branen, A. L. (1993) *Antimicrobials in Foods*, Second Edition. Marcel Dekker, New York.

NAME:	Propyl-4-hydroxybenzoate
CATEGORY:	Antimicrobial preservative
FOOD USE:	Bakery products (cakes, crusts, pastries, toppings, fillings)/ Soft drinks/ Fish (marinated, smoked, jellied)/ Flavour extracts/ Fruit products/ Gelatin, jams, jellies/ Preserves/ Malt extracts/ Olives/ Pickles/ Salad dressings/ Syrups/ Wines
SYNONYMS:	Propyl-para-hydroxybenzoate/ Benzoic acid/ 4-Hydroxy propyl ester/ 4-Hydroxybenzoic acid propyl ester/ CAS 94-13-3/ EINECS 202-307-7/ E216
FORMULA:	$HOC_6H_4CO_2C_3H_7$
MOLECULAR MASS:	180.20
ALTERNATIVE FORMS:	Salt or other parabens
PROPERTIES AND APPEARANCE:	Colourless crystals or white powder
MELTING RANGE IN °C:	95–98
DENSITY AT 20°C (AND OTHER TEMPERATURES) IN g/l:	1.0630
PURITY %	>99
WATER CONTENT MAXIMUM IN %:	0.5
HEAVY METAL CONTENT MAXIMUM IN ppm:	10 (as Pb)
ASH MAXIMUM IN %:	0.05

SOLUBILITY % AT VARIOUS TEMPERATURE/pH COMBINATIONS:

in water:	@ **10°C** 0.025	@ **20°C** 0.05	@ **80°C** 0.30
in vegetable oil:	@ **25°C** 1.4 (peanut); 5.2 (olive)		
in ethanol solution:	@ **10%** 0.1	@ **50%** 18	@ **100%** 95
in propylene glycol:	@ **25°C** aqueous solution: 10% 0.06 50% 0.9 100% 26.0		

FUNCTION IN FOODS: Antimicrobial preservative – especially effective against bacteria and fungi at around neutral pH

TECHNOLOGY OF USE IN FOODS: pH optimum is 3–8. Parabens are often used as blends, e.g. methyl and propyl at a ratio of 2–3 to 1. Dissolve initially in water, ethanol, propylene glycol or the food itself. Hot water is recommended. Parabens can also be dry-blended with other dry ingredients

LEGISLATION:

USA:	**UK and EUROPE:**	**CANADA:**
FDA 21 CFR § 150.141, 150.161, 172.515, 181.22, 181.23, 184.1670	UK: approved; EU: listed	Permitted in certain foods

REFERENCE:

Davidson, P. M., and Branen, A. L. (1993) *Antimicrobials in Foods*, Second Edition. Marcel Dekker, New York.

NAME:	Propyl-4-hydroxybenzoate, sodium salt
CATEGORY:	Antimicrobial preservative
FOOD USE:	Bakery products (cakes, crusts, pastries, topping, fillings)/ Soft drinks/ Fish (marinated, smoked, jellied)/ Flavour extracts/ Fruit products/ Gelatin, jams, jellies/ Preserves/ Malt extracts/ Olives/ Pickles/ Salad dressings/ Syrups/ Wines
SYNONYMS:	E217/ Propyl-para-hydroxybenzoate sodium salt/ Benzoic acid sodium salt/ 4-Hydroxy propyl ester sodium salt/ 4-Hydroxybenzoic acid propyl ester sodium salt
FORMULA:	NaO C6H4 CO2 C3H7
FUNCTION IN FOODS:	Antimicrobial preservative – especially effective against bacteria and fungi at around neutral pH
TECHNOLOGY OF USE IN FOODS:	pH optimum is 3–8. Parabens are often used as blends, e.g. methyl and propyl at a ratio of 2–3 to 1. Dissolve initially in water, ethanol, propylene glycol or the food itself. Hot water is recommended. Parabens can also be dry-blended with other dry ingredients
LEGISLATION:	**USA:** Not permitted **UK and EUROPE:** Permitted in certain foods **CANADA:** Permitted in certain foods
REFERENCE:	Davidson, P. M., and Branen, A. L. (1993) *Antimicrobials in Foods*, Second Edition. Marcel Dekker, New York.

NAME:	Methyl para-hydroxybenzoate
CATEGORY:	Antimicrobial preservative
FOOD USE:	Bakery products (cakes, crusts, pastries, toppings, fillings)/ Soft drinks/ Fish (marinated, smoked, jellied)/ Flavour extracts/ Fruit products/ Gelatin, jams jellies/ Preserves/ Malt extracts/ Olives/ Pickles/ Salad dressings/ Syrups/ Wines
SYNONYMS:	Methyl 4-hydroxybenzoate/ 4-Hydroxybenzoic acid, methyl ester/ Methyl *p*-hydroxybenzoate/ E218/ FEMA 2710/ CAS 99-76-3/ EINEC 202-785-7
FORMULA:	C6H4(OH)CO2 CH3
MOLECULAR MASS:	152.14
ALTERNATIVE FORMS:	Sodium salt and ethyl, propyl and heptyl parabens and their sodium salts
PROPERTIES AND APPEARANCE:	Colourless crystals or white crystalline powder, odourless or faint charcoal odour, slight burning taste
BOILING POINT IN °C AT VARIOUS PRESSURES (INCLUDING 760 mm Hg):	270–280 (decomposes)
MELTING RANGE IN °C:	131
PURITY %:	>99
WATER CONTENT MAXIMUM IN %:	0.5
HEAVY METAL CONTENT MAXIMUM IN ppm:	10 (as Pb)
ASH MAXIMUM IN %:	0.05

SOLUBILITY % AT VARIOUS TEMPERATURE/pH COMBINATIONS:			
in water:	@ **10°C** 0.20	@ **25°C** 0.25	@ **80°C** 2.0
in vegetable oil:	@ **25°C** 0.5 (peanut), 2.9 (olive)		
in ethanol solution:	@ **10%** 0.5	@ **50%** 18.0	@ **100%** 52.0
in propylene glycol:	@ **25°C** aqueous solution: 10% 0.3	50% 2.7	100% 22.0

FUNCTION IN FOODS: Antimicrobial agent, preservative, flavouring agent; for baked goods, beverages, food colours, milk, wine; antimycotic migrating from food packaging. Usage level: 0.1–1.0%; use in foods restricted to 0.1%; especially effective against bacteria and fungi at around neutral pH

TECHNOLOGY OF USE IN FOODS: Effective against yeasts and moulds. Bacteria: mainly Gram-positive. Heptyl parabens more active than propyl, then ethyl, then methyl. Used in similar foods in which benzoic acid is used, but active to higher pH (pK_a is 8.5). Has a characteristic phenolic taste. pH optimum is 3–8. Parabens are often used as blends, e.g. methyl and propyl at a ratio of 2–3 to 1. Dissolve initially in water, ethanol, propylene glycol or the food itself. Hot water is recommended. Parabens can also be dry-blended with other dry ingredients

ANTAGONISTS: Proteins; emulsifiers; lipids; polysaccharides

LEGISLATION:

USA:	**UK and EUROPE:**	**CANADA:**
Regulatory: FDA 21CFR § 150.141, 150.161, 172.515, 181.22, 181.23, 184.1490, GRAS, limitation 0.1%, 556.390, zero limitation in milk; USA CIR approved, EPA reg	Permitted in certain foods	Permitted in certain foods

REFERENCES:

Davidson, P. M., and Branen, A. L. (1993) *Antimicrobials in Foods*, Second Edition. Marcel Dekker, New York.

Ash, M., and Ash, I. (1995) *Food Additives.* Gower Publishing Co., Brookfield, VT.

NAME:	Methyl-4-hydroxybenzoate, sodium salt
CATEGORY:	Antimicrobial preservative
FOOD USE:	Bakery products (cakes, crusts, pastries, toppings, fillings)/ Soft drinks/ Fish (marinated, smoked, jellied)/ Flavour extracts/ Fruit products/ Gelatin, jams, jellies/ Preserves/ Malt extracts/ Olives/ Pickles/ Salad dressings/ Syrups/ Wines
SYNONYMS:	E219/ Sodium methyl para-hydroxybenzoate
FORMULA:	NaO C6H4 CO2 CH3
MOLECULAR MASS:	174
FUNCTION IN FOODS:	Antimicrobial preservative – especially effective against bacteria and fungi at around neutral pH
TECHNOLOGY OF USE IN FOODS:	pH optimum is 3–8. Parabens are often used as blends, e.g. methyl and propyl at a ratio of 2–3 to 1. Dissolve initially in water, ethanol, propylene glycol or the food itself. Hot water is recommended. Parabens can also be dry-blended with other dry ingredients
LEGISLATION:	**USA:** Not permitted **UK and EUROPE:** Permitted in certain foods **CANADA:** Permitted in certain foods
REFERENCE:	Davidson, P. M., and Branen, A. L. (1993) *Antimicrobials in Foods*, Second Edition. Marcel Dekker, New York.

NAME:	Sulphur dioxide
CATEGORY:	Antimicrobial preservative
FOOD USE:	Bleaching agent/ Antimicrobial/ Preservative/ Dough modifier/ Vitamin C stabiliser. Not for use in meats, sources of vitamin B1, raw fruits and vegetables, or fresh potatoes
SYNONYMS:	Sulphurous oxide/ Sulphurous anhydride/ CAS 7446-09-5/ EINECS 231-195-2/ E220
FORMULA:	SO2
MOLECULAR MASS:	64.06
ALTERNATIVE FORMS:	Sodium sulphite, sodium bisulphite, sodium metabisulphite, potassium metabisulphite, calcium sulphite, calcium bisulphite
PROPERTIES AND APPEARANCE:	Colourless gas condenses @ −10°C to colourless liquid
BOILING POINT IN °C AT VARIOUS PRESSURES (INCLUDING 760 mm Hg):	−10
MELTING RANGE IN °C:	−72
DENSITY AT 20°C (AND OTHER TEMPERATURES) IN g/l:	1.5 (liquid), 2.26 (vapour)
PURITY %:	>99 (by weight)
WATER CONTENT MAXIMUM IN %:	0.05
HEAVY METAL CONTENT MAXIMUM IN ppm:	30 (as Pb)
ARSENIC CONTENT MAXIMUM IN ppm:	3
ASH MAXIMUM IN %:	0.05

SOLUBILITY % AT VARIOUS TEMPERATURE/pH COMBINATIONS:	
in water:	**@ 20°C** 11
in ethanol solution:	**@ 100%** 25
FUNCTION IN FOODS:	Bleaching agent, antimicrobial preservative, dough modifier
ALTERNATIVES:	Multifunctional – so need a blend of additives to provide each of the functions
TECHNOLOGY OF USE IN FOODS:	Multifunctional – an antimicrobial preservative, an antioxidant and a bleaching agent
ANTAGONISTS:	Sulphur dioxide is very reactive and binds to many substances in foods including oxygen, aldehydes, some ketones, sugars, carbonyls, thiamin, nucleotides, colours, anthocyanins
FOOD SAFETY ISSUES:	Normal humans are reasonably tolerant to ingested sulphites, but some individuals are hypersensitive and may be subject to asthma attacks and even life-threatening anaphylaxis
LEGISLATION:	**USA:** FDA 21CFR 182.3862, GRAS: BATF 27CFR 240.1051 **UK and EUROPE:** Permitted in specified foods **CANADA:** Permitted in specified foods **AUSTRALIA/PACIFIC RIM:** Japan: approved 0.03–5 g/kg
REFERENCES:	Davidson, P. M., and Branen, A. L. (1993) *Antimicrobials in Foods*, Second Edition. Marcel Dekker, New York., pp. 137–190. *Food Chemicals Codex*, Fourth Edition (1996) National Academy Press, Washington, DC. Ash, M., and Ash, I. (1995) *Food Additives*. Gower Publishing Co., Brookfield, VT.

NAME:	**Sodium sulphite**
CATEGORY:	Antimicrobial preservative
FOOD USE:	Food preservative and antioxidant/ Boiler water additive. Not for use on meats, sources of vitamin B_1, raw fruits and vegetables, or fresh potatoes. Bleaching agent
SYNONYMS:	CAS 7757-83-7/ EINECS 231-821-4/ E221/ Sulfurous acid sodium salt (1:2)/ Sulfurous acid disodium salt/ Sodium sulfite (2:1)/ Sodium sulfite anhydrous
FORMULA:	Na2 S O3
MOLECULAR MASS:	126.04
ALTERNATIVE FORMS:	Sulphur dioxide, sodium bisulphite, sodium metabisulphite, potassium metabisulphite, calcium sulphite, calcium bisulphite
PROPERTIES AND APPEARANCE:	White powder or small hexagonal crystals, odourless, salty sulphurous taste
BOILING POINT IN °C AT VARIOUS PRESSURES (INCLUDING 760 mm Hg):	Decomposes
MELTING RANGE IN °C:	Decomposes
DENSITY AT 20°C (AND OTHER TEMPERATURES) IN g/l:	2.633 (15.4°C)
PURITY %:	95 (as Na_2SO_3) (or 48% SO_2)
HEAVY METAL CONTENT MAXIMUM IN ppm:	10 (as Pb)
ARSENIC CONTENT MAXIMUM IN ppm:	3

SOLUBILITY % AT VARIOUS TEMPERATURE/pH COMBINATIONS:

in water:	**@ 0°C** 12.54 **@ 40°C** 28 **@ 80°C** 28.3
FUNCTION IN FOODS:	Antimicrobial preservative, antioxidant and bleach
ALTERNATIVES:	Combination of antimicrobial preservative, antioxidant and bleach
TECHNOLOGY OF USE IN FOODS:	Multifunctional – an antimicrobial preservative, an antioxidant and a bleaching agent
ANTAGONISTS:	Sulphur dioxide is very reactive and binds to many substances in foods including oxygen, aldehydes, some ketones, sugars, carbonyls, thiamin, nucleotides, colours, anthocyanins
FOOD SAFETY ISSUES:	Normal humans are reasonably tolerant to ingested sulphites, but some individuals are hypersensitive and may be subject to asthma attacks and even life-threatening anaphylaxis
LEGISLATION:	**USA:** FDA 21CFR § 172.615, 173.310, 177.1200, 182.3798, GRAS **AUSTRALIA/PACIFIC RIM:** Japan: Approved (0.03–5 g/kg max. residual as sulphur dioxide)
REFERENCES:	Davidson, P. M., and Branen, A. L. (1993) *Antimicrobials in Foods*, Second Edition. Marcel Dekker, New York. *Food Chemicals Codex*, Fourth Edition (1996) National Academy Press, Washington, DC. Ash, M., and Ash, I. (1995) *Food Additives*. Gower Publishing Co., Brookfield, VT.

NAME:	**Sodium bisulphite**
CATEGORY:	Antimicrobial preservative
FOOD USE:	Preservative. Not for use on meats, sources of vitamin B_1, raw fruits and vegetables, or fresh potatoes
SYNONYMS:	CAS 7631-90-5/ EINECS 231-548-0/ E222 sodium acid sulfite/ Acid sodium sulfite/ Sodium bisulfite (1:1)/ Sulfurous acid monosodium salt/ Sodium hydrogen sulfite/ Sodium sulfhydrate
FORMULA:	Na H S O3 (commercially sodium bisulphite consists mainly of sodium metabisulfite Na2S2O5 with sodium bisulphite)
MOLECULAR MASS:	104.06
ALTERNATIVE FORMS:	Sodium sulphite, sulphur dioxide, sodium metabisulphite, potassium metabisulphite, calcium sulphite, calcium bisulphite
PROPERTIES AND APPEARANCE:	White crystalline powder, SO_2 odour, disagreeable taste
MELTING RANGE IN °C:	315
DENSITY AT 20°C (AND OTHER TEMPERATURES) IN g/l:	1.48
PURITY %:	61.6 (as sulphur dioxide)
HEAVY METAL CONTENT MAXIMUM IN ppm:	10 (as Pb)
SOLUBILITY % AT VARIOUS TEMPERATURE/pH COMBINATIONS:	
in water:	**@ 20°C** 300% **@ 50°C** Very soluble **@ 100°C** Very soluble
FUNCTION IN FOODS:	Antimicrobial preservative, antioxidant and bleach
ALTERNATIVES:	Combination of antimicrobial preservative, antioxidant and bleach

TECHNOLOGY OF USE IN FOODS: Multifunctional – an antimicrobial preservative, an antioxidant and a bleaching agent

SYNERGISTS: Combination of antimicrobial preservative, antioxidant and bleach

ANTAGONISTS: Sulphur dioxide is very reactive and binds to many substances in foods including oxygen, aldehydes, some ketones, sugars, carbonyls, thiamin, nucleotides, colours, anthocyanins

FOOD SAFETY ISSUES: Normal humans are reasonably tolerant to ingested sulphites but some individuals are hypersensitive and may be subject to asthma attacks and even life-threatening anaphylaxis

LEGISLATION:

USA:	UK and EUROPE:	CANADA:
FDA 21CFR § 161.173, 173.310, 182.3739, GRAS	Permitted in certain foods	Permitted in certain foods

REFERENCES:

Davidson, P. M., and Branen, A. L. (1993) *Antimicrobials in Foods*, Second Edition. Marcel Dekker, New York.

Ash, M., and Ash, I. (1995) *Food Additives*. Gower Publishing Co., Brookfield, VT.

NAME:	Sodium metabisulphite
CATEGORY:	Antimicrobial preservative
FOOD USE:	Preservative, antioxidant. Flavouring in cherries and shrimp/ Boiler water additive for food contact/ Bleaching agent. Not for use on meats, sources of vitamin B_1, raw fruits and vegetables, or fresh potatoes
SYNONYMS:	CAS 7681-57-4/ EINECS 231-673-0/ E223/ Disulfurous acid disodium salt/ Sodium pyrosulfite/ Disodium pyrosulfite/ Sodium bisulfite
FORMULA:	Na2 S2 O5
MOLECULAR MASS:	190.10
ALTERNATIVE FORMS:	Sodium sulphite, sodium bisulphite, sulphur dioxide, potassium metabisulphite, calcium sulphite, calcium bisulphite
PROPERTIES AND APPEARANCE:	Colourless crystalline or white to yellowish powder, SO_2 odour
MELTING RANGE IN °C:	Decomposes >150
DENSITY AT 20°C (AND OTHER TEMPERATURES) IN g/l:	1.4
PURITY %:	67.4 as SO_2
HEAVY METAL CONTENT MAXIMUM IN ppm:	10 (as Pb)
SOLUBILITY % AT VARIOUS TEMPERATURE/pH COMBINATIONS:	
in water:	**@ 20°C** 54 **@ 100°C** 81.7
FUNCTION IN FOODS:	Antimicrobial preservative, antioxidant and bleach
ALTERNATIVES:	Combination of antimicrobial preservative, antioxidant and bleach

TECHNOLOGY OF USE IN FOODS: Multifunctional – an antimicrobial preservative, an antioxidant and a bleaching agent

ANTAGONISTS: Sulphur dioxide is very reactive and binds to many substances in foods including oxygen, aldehydes, some ketones, sugars, carbonyls, thiamin, nucleotides, colours, anthocyanins

FOOD SAFETY ISSUES: Normal humans are reasonably tolerant to ingested sulphites, but some individuals are hypersensitive and may be subject to asthma attacks and even life-threatening anaphylaxis

LEGISLATION:

USA:	**UK and EUROPE:**	**CANADA:**
FDA 21CFR § 173.310, 177.1200, 182.3766, GRAS	Permitted in certain foods	Permitted in certain foods

REFERENCES:

Davidson, P. M., and Branen, A. L. (1993) *Antimicrobials in Foods*, Second Edition. Marcel Dekker, New York.

Ash, M., and Ash, I. (1995) *Food Additives*. Gower Publishing Co., Brookfield, VT.

NAME:	Potassium metabisulphite					
CATEGORY:	Antimicrobial preservative					
FOOD USE:	Food preservative/ Antioxidant/ Steriliser/ Brewing, wine making. Not for use in meats, sources of vitamin B_1, raw fruits and vegetables, or fresh potatoes					
SYNONYMS:	CAS 16731-55-8/ EINECS 240-795-3/ E224/ Disulfurous acid dipotassium salt/ Potassium pyrosulfite/ Potassium disulfite/ Dipotassium disulfite					
FORMULA:	K2 S2 O5					
MOLECULAR MASS:	222.32					
ALTERNATIVE FORMS:	Sodium sulphite, sodium bisulphite, sodium metabisulphite, sulphur dioxide, calcium sulphite, calcium bisulphite					
PROPERTIES AND APPEARANCE:	White granules or powder, pungent sharp odour					
MELTING RANGE IN °C:	>300, decomposes at 150–190					
DENSITY AT 20°C (AND OTHER TEMPERATURES) IN g/l:	2.34					
PURITY %:	57.6 as SO_2					
HEAVY METAL CONTENT MAXIMUM IN ppm:	10 (as Pb)					
ARSENIC CONTENT MAXIMUM IN ppm:	3					
SOLUBILITY % AT VARIOUS TEMPERATURE/pH COMBINATIONS:						
in water:	**@ 20°C**	25	**@ 50°C**	Slightly soluble	**@ 100°C**	Slightly soluble
in ethanol solution:	**@ 5%**	Slightly soluble	**@ 20%**	Slightly soluble	**@ 95%**	Slightly soluble
	@ 100%	Slightly soluble				

FUNCTION IN FOODS: Antimicrobial preservative, antioxidant and bleach

ALTERNATIVES: Combination of antimicrobial preservative, antioxidant and bleach

TECHNOLOGY OF USE IN FOODS: Multifunctional – an antimicrobial preservative, an antioxidant and a bleaching agent

ANTAGONISTS: Sulphur dioxide is very reactive and binds to many substances in foods including oxygen, aldehydes, some ketones, sugars, carbonyls, thiamin, nucleotides, colours, anthocyanins

FOOD SAFETY ISSUES: Normal humans are reasonably tolerant to ingested sulphites, but some individuals are hypersensitive and may be subject to asthma attacks and even life-threatening anaphylaxis

LEGISLATION:

USA:	UK and EUROPE:	CANADA:
FDA 21CFR § 182.3637, GRAS; BATF 27CFR § 240.1051	Permitted in certain foods	Permitted in certain foods

REFERENCES:

Davidson, P. M., and Branen, A. L. (1993) *Antimicrobials in Foods*, Second Edition. Marcel Dekker, New York.

Ash, M., and Ash, I. (1995) *Food Additives*. Gower Publishing Co., Brookfield, VT.

NAME:	Calcium sulphite
CATEGORY:	Antimicrobial preservative
FOOD USE:	Bleaching agent/ Antimicrobial/ Preservative/ Dough modifier/ Vitamin C stabiliser. Not for use in meats, sources of vitamin B_1, raw fruits and vegetables, or fresh potatoes
SYNONYMS:	E226
FORMULA:	Ca O3 S(1/2 H2O)
MOLECULAR MASS:	129.15
ALTERNATIVE FORMS:	Sodium sulphite, sodium bisulphite, sodium metabisulphite, potassium metabisulphite, sulphur dioxide, calcium bisulphite
PROPERTIES AND APPEARANCE:	Dihydrate, crystals or powder
MELTING RANGE IN °C:	−1/2 H20 >250
SOLUBILITY % AT VARIOUS TEMPERATURE/pH COMBINATIONS:	
in water:	**@ 20°C** 0.0043 **@ 100°C** 0.0011
FUNCTION IN FOODS:	Preservative in cider and fruit juices; disinfectant in brewing vats; in sugar manufacture
ALTERNATIVES:	Combination of antimicrobial preservative, antioxidant and bleach
TECHNOLOGY OF USE IN FOODS:	Multifunctional – an antimicrobial preservative, an antioxidant and a bleaching agent
ANTAGONISTS:	Sulphur dioxide is very reactive and binds to many substances in foods including oxygen, aldehydes, some ketones, sugars, carbonyls, thiamin, nucleotides, colours, anthocyanins
FOOD SAFETY ISSUES:	Normal humans are reasonably tolerant to ingested sulphites, but some individuals are hypersensitive and may be subject to asthma attacks and even life-threatening anaphylaxis

LEGISLATION:	**USA:** Not permitted	**UK and EUROPE:** Permitted in certain foods	**CANADA:** Permitted in certain foods

REFERENCE:

Davidson, P. M., and Branen, A. L. (1993) *Antimicrobials in Foods*, Second Edition. Marcel Dekker, New York.

NAME:	Calcium bisulphite					
CATEGORY:	Antimicrobial preservative					
FOOD USE:	Bleaching agent/ Antimicrobial/ Preservative/ Dough modifier/ Vitamin C stabiliser. Not for use in meats, sources of vitamin B_1, raw fruits and vegetables, or fresh potatoes					
SYNONYMS:	Dihydrogen sulfite/ Calcium hydrogen sulfite/ E227					
FORMULA:	Ca(HSO3)2					
MOLECULAR MASS:	202.21					
ALTERNATIVE FORMS:	Sodium sulphite, sodium bisulphite, sodium metabisulphite, potassium metabisulphite, calcium sulphite, sulphur dioxide					
PROPERTIES AND APPEARANCE:	Yellowish liquid, strong SO_2 odour					
SOLUBILITY % AT VARIOUS TEMPERATURE/pH COMBINATIONS:						
in water:	**@ 20°C**	Soluble	**@ 50°C**	Soluble	**@ 100°C**	Soluble
in sucrose solution:	**@ 10%**	Soluble	**@ 40%**	Soluble	**@ 60%**	Soluble
in sodium chloride solution:	**@ 5%**	Soluble	**@ 10%**	Soluble	**@ 15%**	Soluble
in ethanol solution:	**@ 5%**	Soluble	**@ 20%**	Soluble	**@ 95%**	Soluble
	@ 100%	Soluble				
FUNCTION IN FOODS:	Antimicrobial preservative, antioxidant and bleach					
ALTERNATIVES:	Combination of antimicrobial preservative, antioxidant and bleach					
TECHNOLOGY OF USE IN FOODS:	Multifunctional – an antimicrobial preservative, an antioxidant and a bleaching agent					
ANTAGONISTS:	Sulphur dioxide is very reactive and binds to many substances in foods including oxygen, aldehydes, some ketones, sugars, carbonyls, thiamin, nucleotides, colours, anthocyanins					

FOOD SAFETY ISSUES: Normal humans are reasonably tolerant to ingested sulphites, but some individuals are hypersensitive and may be subject to asthma attacks and even life-threatening anaphylaxis

LEGISLATION:

USA:	**UK and EUROPE:**	**CANADA:**
Not permitted	Permitted in certain foods	Permitted in certain foods

REFERENCE:

Davidson, P. M., and Branen, A. L. (1993) *Antimicrobials in Foods*, Second Edition. Marcel Dekker, New York.

NAME:	Biphenyl
CATEGORY:	Antimicrobial preservative
FOOD USE:	Citrus fruit wrappers
SYNONYMS:	CAS 92-52-4/ EINECS 202-163-5/ FEMA 3129/ E230/ PHPH/ Bibenzene/ Phenylbenzene/ Diphenyl/ Xenene/ 1,1′-Biphenyl
FORMULA:	$C_6H_5C_6H_5$
MOLECULAR MASS:	154.22
PROPERTIES AND APPEARANCE:	White scales; pleasant odour
BOILING POINT IN °C AT VARIOUS PRESSURES (INCLUDING 760 mm Hg):	760 mmHg: 256. 22 mm Hg: 145
MELTING RANGE IN °C:	69–71
FLASH POINT IN °C:	112.7
DENSITY AT 20°C (AND OTHER TEMPERATURES) IN g/l:	0.8660
SOLUBILITY % AT VARIOUS TEMPERATURE/pH COMBINATIONS:	
in water:	Insoluble
in vegetable oil:	Soluble
in sucrose solution:	Insoluble
in sodium chloride solution:	Soluble
in ethanol solution:	Soluble
in propylene glycol:	Soluble
FUNCTION IN FOODS:	Broad-spectrum fungicide

TECHNOLOGY OF USE IN FOODS:	Impregnated citrus wrap. Citrus surface application (70 ppm) 1–5 g/m in impregnated wrappers
SYNERGISTS:	Other fungicides
LEGISLATION:	**UK and EUROPE:** Europe: listed; UK: approved **AUSTRALIA/PACIFIC RIM:** Japan: approved
REFERENCE:	Ash, M., and Ash, I. (1995) *Food Additives*. Gower Publishing Co., Brookfield, VT.

NAME:	2-Hydroxybiphenyl
CATEGORY:	Antimicrobial preservative
FOOD USE:	Citrus fruit
SYNONYMS:	(1,1′-Biphenyl)-2-ol/ *o*-Phenylphenol/ 2-Phenylphenol/ *o*-Xenol/ E231/ *o*-Hydroxybiphenyl/ CAS 90-43-7/ EINECS 201-993-5
FORMULA:	$C_{12}H_{10}O$
MOLECULAR MASS:	170.22
PROPERTIES AND APPEARANCE:	Nearly white or light buff crystals, mild charcoal odour
BOILING POINT IN °C AT VARIOUS PRESSURES (INCLUDING 760 mm Hg):	760 mmHg: 280–284. 14 mm Hg: 145
MELTING RANGE IN °C:	56–58
DENSITY AT 20°C (AND OTHER TEMPERATURES) IN g/l:	1.217
FUNCTION IN FOODS:	Fungicide applied to surface of citrus fruit
ALTERNATIVES:	Other surface-applied fungicides, e.g. biphenyl, 2(thiazol-4-yl)-benzimidazole
TECHNOLOGY OF USE IN FOODS:	Fungicide applied to surface of citrus fruit
SYNERGISTS:	Other fungicides

LEGISLATION:	**USA:** EPA reg.; FDA 21CFR § 175.105, 176.210, 177.2600	**UK and EUROPE:** Approved
REFERENCE:	Ash, M., and Ash, I. (1995) *Food Additives.* Gower Publishing Co., Brookfield, VT.	

NAME:	Sodium *o*-phenylphenate		
CATEGORY:	Antimicrobial preservative		
FOOD USE:	Fruit		
SYNONYMS:	CAS 132-27-4/ EINECS 205-055-6/ E232/ *o*-Phenylphenol sodium salt/ Sodium *o*-phenylphenol/ Sodium biphenyl-2-yl oxide/ (1,1′-Biphenyl)-2-ol, sodium salt/ Sodium *o*-phenylphenolate		
FORMULA:	C6H4(C6H5)ONa·4HOH		
MOLECULAR MASS:	192.20		
PROPERTIES AND APPEARANCE:	White flakes		
SOLUBILITY % AT VARIOUS TEMPERATURE/pH COMBINATIONS:			
in water:	**@ 20°C** 122		
in propylene glycol:	**@ 20°C** 28		
FUNCTION IN FOODS:	Preservative, antimicrobial, mould inhibitor for apples, etc.		
TECHNOLOGY OF USE IN FOODS:	Surface application on fruit to prevent mould growth		
LEGISLATION:	**USA:** FDA 21CFR § 175.105, 175.300, 176.170, 176.210, 177.1210, 178.3120	**UK and EUROPE:** Europe: listed	**AUSTRALIA/PACIFIC RIM:** Japan: approved (0.01 g/kg residual)
REFERENCE:	Ash, M., and Ash, I. (1995) *Food Additives.* Gower Publishing Co., Brookfield, VT.		

NAME:	**Thiabendazole**
CATEGORY:	Antimicrobial preservative
FOOD USE:	Citrus, banana fruit
SYNONYMS:	CAS 148-79-8/ E233/ Thiaben/ 2-(Thiazol-4-yl) benzimidazole/ 2-(4-Thiazolyl)-1H-benzimidazole/ 4-(2-benzimidazolyl)thiazole/ MK-360/ Omnizole/ Thibenzole/ Bovizole/ Eprofil/ Equizole/ Mintezol/ Top Form Wormer/ Mertect/ Lombristop/ Minzolum/ Nemapan/ Polival/ TBZ/ Tecto
FORMULA:	$C_{10}H_7N_3S$
MOLECULAR MASS:	201.25
PROPERTIES AND APPEARANCE:	Colourless crystals
MELTING RANGE IN °C:	300 (sublimes)
FUNCTION IN FOODS:	Anthelmintic, fungicide
TECHNOLOGY OF USE IN FOODS:	Citrus surface application (3 ppm); banana surface application (1 ppm)
LEGISLATION:	**USA:** Not permitted **UK and EUROPE:** Approved
REFERENCE:	Ash, M., and Ash, I. (1995) *Food Additives.* Gower Publishing Co., Brookfield, VT.

NAME:	**Nisin**
CATEGORY:	Antimicrobial preservative
FOOD USE:	Pasteurised cheese and cheese spreads/ Canned fruits and vegetables
SYNONYMS:	E234/ CAS 1414-45-5/ EINECS 215-807-5/ *Streptococcus lactis* (*Lactococcus lactis*) bacteriocin
FORMULA:	$C_{143}H_{230}N_{42}O_{37}S_7$ (contains L-amino acids and the unusual sulphur-containing amino acids lanthionine and beta-methyllanthionine). There are five variants of nisin – A, B, C, D and E. The most active antimicrobially is variant A. The commercial product contains mainly variant A
MOLECULAR MASS:	3500 (usually occurs as the dimer – 7000)
ALTERNATIVE FORMS:	Nisaplin® is the tradename for nisin produced by Aplin & Barrett Ltd., Dorset, England
PROPERTIES AND APPEARANCE:	Crystals derived from pure culture fermentation of *Streptococcus lactis* (*Lactococcus lactis*)
PURITY %:	Not less than 900 international units per mg
WATER CONTENT MAXIMUM IN %:	3.0
HEAVY METAL CONTENT MAXIMUM IN ppm:	10
ASH MAXIMUM IN %:	Not less than 50% NaCl
SOLUBILITY % AT VARIOUS TEMPERATURE/pH COMBINATIONS:	
in water:	**@ pH 2.5** 12.0 **@ pH 5.0** 4.0 **@ >pH 7.0** Insoluble
FUNCTION IN FOODS:	As a preservative in cheese and canned fruits and vegetables. Effective against Gram-positive bacteria, lactic acid bacteria, *Streptococcus, Bacillus, Clostridium*
ALTERNATIVES:	Other anti-botulinal antimicrobial preservative treatments

TECHNOLOGY OF USE IN FOODS: Cheese (6–12 ppb); dairy products (6–12 ppb); canned foods (stable at low pH and high temperature) (6–12 ppb). Activity in foods is lost gradually upon storage

ANTAGONISTS: High pH, pancreatin, alpha-chymotrypsin, nisinase, penicillinase, high microbial load

LEGISLATION:

USA:	UK and EUROPE:	CANADA:
FDA 21CFR 184.1538, GRAS, limited to 250 ppm nisin in finished product	Permitted in certain foods	Permitted in certain foods

REFERENCES:

Hurst, A., and Hoover D. G. (1993) Nisin. In: Davidson, P. M., and A. L. Branen (Eds.), *Antimicrobials in Foods*, 2nd edition. Marcel Dekker, Inc., New York.

Ash, M., and Ash, I. (1995) *Food Additives*. Gower Publishing Co., Brookfield, VT.

NAME:	**Hexamethylenetetramine**
CATEGORY:	Antimicrobial preservative
FOOD USE:	Provolone cheese/ Marinated fish
SYNONYMS:	CAS 100-97-0/ EINECS 202-905-8/ E239/ HMTA/ Methenamine/ Aminoform/ Urotropine/ Hexamine
FORMULA:	(CH2)6N4
MOLECULAR MASS:	140.22
PROPERTIES AND APPEARANCE:	White crystalline powder or colourless lustrous crystals
DENSITY AT 20°C (AND OTHER TEMPERATURES) IN g/l:	1.27
FUNCTION IN FOODS:	Antimicrobial preservative; releases formaldehyde in solution. Setting agent for proteins including casein used in food packaging paper/paperboard
TECHNOLOGY OF USE IN FOODS:	Added to provolone cheese and marinated fish; releases formaldehyde which exerts its broad-spectrum antimicrobial effect
ANTAGONISTS:	High pH (formaldehyde is released in acid solution)
LEGISLATION:	**USA:** FDA 21CFR § 181.30 **UK and EUROPE:** UK: approved; Europe: listed
REFERENCE:	Ash, M., and Ash, I. (1995) *Food Additives.* Gower Publishing Co., Brookfield, VT.

NAME:	Potassium nitrite
CATEGORY:	Antimicrobial preservative
FOOD USE:	Cured meats/ Meat products/ Fish products
SYNONYMS:	Nitrous acid potassium salt/ E249/ CAS 7758-09-0/ EINECS 231-832-4
FORMULA:	KNO_2
MOLECULAR MASS:	85.10
ALTERNATIVE FORMS:	Sodium nitrite
PROPERTIES AND APPEARANCE:	White or slightly yellow deliquescent granules or rods
BOILING POINT IN °C AT VARIOUS PRESSURES (INCLUDING 760 mm Hg):	decomposes @ 320
MELTING RANGE IN °C:	441 (decomposition starts at 350)
DENSITY AT 20°C (AND OTHER TEMPERATURES) IN g/l:	1.915
PURITY %:	90
HEAVY METAL CONTENT MAXIMUM IN ppm:	0.002%, not more than 10 ppm Pb
SOLUBILITY % AT VARIOUS TEMPERATURE/pH COMBINATIONS:	
in water:	**@ 0°C** 281 **@ 100°C** 413
in ethanol solution:	**@ 5%** Soluble **@ 20%** Soluble **@ 95%** Soluble **@ 100%** Soluble
FUNCTION IN FOODS:	Antimicrobial preservative; colour fixative; flavour enhancer
ALTERNATIVES:	Multifunctional, so requires a blend of additives

TECHNOLOGY OF USE IN FOODS: 2 lb/100 gal (pickle), 1 oz/100 lb meat, 0.25 oz/100 lb (chopped meat), 200 ppm max nitrite as sodium nitrate

LEGISLATION:

USA:	UK and EUROPE:	AUSTRALIA/PACIFIC RIM:
FDA 21CFR, USDA 9CFR 318.7, 381.147	Permitted in certain foods UK: approved. Europe: listed	Japan: approved 0.005–0.07 g/kg

REFERENCE:

Ash, M., and Ash, I. (1995) *Food Additives*. Gower Publishing Co., Brookfield, VT.

NAME:	**Sodium nitrite**			
CATEGORY:	Antimicrobial preservative			
FOOD USE:	Cured meats/ Meat products/ Fish products			
SYNONYMS:	Nitrous acid sodium salt/ E250/ CAS 7632-00-0/ EINECS 231-555-9			
FORMULA:	$NaNO_2$			
MOLECULAR MASS:	69.00			
ALTERNATIVE FORMS:	Potassium nitrite			
PROPERTIES AND APPEARANCE:	White or slightly yellow hygroscopic granules rods or powder			
BOILING POINT IN °C AT VARIOUS PRESSURES (INCLUDING 760 mm Hg):	decomposes @ 320			
MELTING RANGE IN °C:	271			
DENSITY AT 20°C (AND OTHER TEMPERATURES) IN g/l:	2.168			
PURITY %	97			
WATER CONTENT MAXIMUM IN %:	0.25			
HEAVY METAL CONTENT MAXIMUM IN ppm:	0.002%, not more than 10 ppm Pb			
SOLUBILITY % AT VARIOUS TEMPERATURE/pH COMBINATIONS:				
in water:	**@ 15°C**	81.5	**@ 100°C**	163
in ethanol solution:	**@ 20%**	0.3	**@ 100%**	3.0
FUNCTION IN FOODS:	Antimicrobial preservative; colour fixative; flavour enhancer			

ALTERNATIVES:	Multifunctional, so requires a blend of additives
TECHNOLOGY OF USE IN FOODS:	Used in cured meats and fish as an antibotulinum agent as well as an antimicrobial preservative, colour fixative, flavour enhancer
LEGISLATION:	**USA:** FDA 21CFR 172.175, 172.177, 181.34, USDA 9CFR 318.79 381.147 **UK and EUROPE:** Permitted in certain foods **CANADA:** Permitted in certain foods **AUSTRALIA/PACIFIC RIM:** Japan: approved 0.005–0.07 g/kg
REFERENCE:	Ash, M., and Ash, I. (1995) *Food Additives.* Gower Publishing Co., Brookfield, VT.

NAME:	**Sodium nitrate**
CATEGORY:	Antimicrobial preservative
SYNONYMS:	CAS 7631-99-4/ EINECS 231-554-3/ E251/ Soda niter/ Cubic niter/ Chile saltpeter
FORMULA:	$NaNO_3$
MOLECULAR MASS:	84.99
PROPERTIES AND APPEARANCE:	Colourless transparent crystals, odourless
BOILING POINT IN °C AT VARIOUS PRESSURES (INCLUDING 760 mm Hg):	decomposes at 380
MELTING RANGE IN °C:	308
DENSITY AT 20°C (AND OTHER TEMPERATURES) IN g/l:	2.267
PURITY %:	99
HEAVY METAL CONTENT MAXIMUM IN ppm:	10 as Pb
SOLUBILITY % AT VARIOUS TEMPERATURE/pH COMBINATIONS:	
in water:	**@ 25°C** 92.1 **@ 100°C** 180
in ethanol solution:	**@ 5%** Soluble **@ 20%** Soluble **@ 95%** Soluble **@ 100%** Soluble
FUNCTION IN FOODS:	Antimicrobial agent, preservative. Source of nitrite, colour fixative in cured meats, fish, poultry; boiler water additive; curing salt
TECHNOLOGY OF USE IN FOODS:	Used in cured meats and fish as an antibotulinum agent as well as an antimicrobial preservative, colour fixative, flavour enhancer

LEGISLATION:	**USA:** FDA 21CFR § 171.170, 172.170, 172.177, 173.310, 181.33; USDA 9CFR § 318.7, 381.147 Limitation 500 ppm (smoked fish, meat curing), 200 ppm (smoked chub), 7 lb/100 gal (pickle), 3.5 oz/100 lb (meat), 2.75 oz/100 lb (chopped meat)	**UK and EUROPE:** UK: approved; Europe: listed	**CANADA:** Approved in certain foods
REFERENCE:	Ash, M., and Ash, I. (1995) *Food Additives.* Gower Publishing Co., Brookfield, VT.		

NAME:	Potassium nitrate
CATEGORY:	Antimicrobial preservative
SYNONYMS:	E252/ Nitre/ Niter/ Saltpetre /CAS 7757-79-1/ EINECS 231-818-8
FORMULA:	KNO3
MOLECULAR MASS:	101.11
PROPERTIES AND APPEARANCE:	Transparent colourless or white crystals or crystal powder, odourless, cooling pungent salty taste
BOILING POINT IN °C AT VARIOUS PRESSURES (INCLUDING 760 mm Hg):	Decomposes at 400
MELTING RANGE IN °C:	334
DENSITY AT 20°C (AND OTHER TEMPERATURES) IN g/l:	2.109
PURITY %:	99
WATER CONTENT MAXIMUM IN %:	1
HEAVY METAL CONTENT MAXIMUM IN ppm:	0.002
SOLUBILITY % AT VARIOUS TEMPERATURE/pH COMBINATIONS:	
in water:	**@ 20°C** 33 **@ 100°C** 200
in ethanol solution:	**@ 100%** 0.13
FUNCTION IN FOODS:	Formerly used as a curing agent – now superseded by nitrite
ALTERNATIVES:	Nitrite

TECHNOLOGY OF USE IN FOODS:	200 ppm cod roe; 7 lb/100 gallon pickles; 3.5 oz/100 lb meat; 2.75 oz/100 lb chopped meat
LEGISLATION:	**USA:** FDA 21CFR § 172.160, 181.33, USDA 9CFR § 318.7, 381.147
REFERENCE:	Ash, M., and Ash, I. (1995) *Food Additives.* Gower Publishing Co., Brookfield, VT.

NAME:	Propionic acid
CATEGORY:	Antimicrobial preservative
FOOD USE:	Baked products/ Cheese products
SYNONYMS:	Methylacetic acid/ Propanoic acid/ Ethylformic acid/ E280/ CAS 79-09-4/ EINECS 201-176-3
FORMULA:	C2 H5 COOH
MOLECULAR MASS:	74.09
PROPERTIES AND APPEARANCE:	Oily liquid, pungent, rancid odour
BOILING POINT IN °C AT VARIOUS PRESSURES (INCLUDING 760 mm Hg):	760 mmHg: 141.1. 400 mmHg: 122.0. 100 mmHg: 85.8. 10 mmHg: 41.6. 1 mmHg: 4.6
MELTING RANGE IN °C:	–21.5
FLASH POINT IN °C:	58
DENSITY AT 20°C (AND OTHER TEMPERATURES) IN g/l:	0.993
PURITY %:	99.5
WATER CONTENT MAXIMUM IN %:	0.15
HEAVY METAL CONTENT MAXIMUM IN ppm:	10
ASH MAXIMUM IN %:	0.01
FUNCTION IN FOODS:	Antimicrobial preservative; antimycotic; flavouring agent; preservative additive; mould inhibitor
ALTERNATIVES:	Other antimicrobial preservatives depending on the application

TECHNOLOGY OF USE IN FOODS:	1% (foods); limitation 0.32% (flour in white bread/rolls), 0.38% (whole wheat), 0.3% (cheese products)
ANTAGONISTS:	Calcium chloride
LEGISLATION:	**USA:** FDA 21CFR, 172.515, 184.1081, GRAS, GRAS(FEMA), EPA reg **UK and EUROPE:** UK: approved; Europe: listed **CANADA:** Approved in certain foods **AUSTRALIA/PACIFIC RIM:** Japan: restricted to flavouring use, limitation with sorbic acid 3 g/kg total
REFERENCES:	Ash, M., and Ash, I. (1995) *Food Additives.* Gower Publishing Co., Brookfield, VT. Davidson, P. M., and Branen, A. L. (Eds.) (1993) *Antimicrobials in Foods*, Second Edition. Marcel Dekker, New York.
ANY OTHER RELEVANT INFORMATION:	Refractive index: 1.3862 Surface tension 27.21 dynes/cm (15°C) Viscosity: 1.020 cP (15°C) pK = 4.87 Swiss cheese contains up to 1% propionic acid as a result of growth and metabolism of propionibacteria, and this limits mould growth

NAME:	Sodium propionate
CATEGORY:	Antimicrobial preservative
FOOD USE:	Baked products/ Cheese products
SYNONYMS:	Methylacetic acid/ Propanoic acid sodium salt/ Ethylformic acid sodium salt/ E281/ CAS 137-40-6/ EINECS 205-290-4
FORMULA:	C_2H_5COONa
MOLECULAR MASS:	96.07
ALTERNATIVE FORMS:	Other salts
PROPERTIES AND APPEARANCE:	Transparent crystals, granules. Deliquescent in moist air
MELTING RANGE IN °C:	287–289
PURITY %:	99
WATER CONTENT MAXIMUM IN %:	1
HEAVY METAL CONTENT MAXIMUM IN ppm:	10
FUNCTION IN FOODS:	Antimicrobial preservative; antimycotic; flavouring agent; preservative additive; mould inhibitor
ALTERNATIVES:	Other antimicrobial preservatives, depending on the application
TECHNOLOGY OF USE IN FOODS:	1% (foods); limitation 0.32% (flour in white bread/rolls), 0.38% (whole wheat), 0.3% (cheese products)
ANTAGONISTS:	Calcium chloride

LEGISLATION:

USA:
FDA 21CFR § 133.123, 133.124, 133.169, 133.173, 133.179, 150.141,150.161, 179.45, 181.22, 81.23, 184.1784, GRAS; USDA 9CFR § 318.7, 381.147

UK and EUROPE:
UK: approved; Europe: listed

CANADA:
Permitted in certain foods

AUSTRALIA/PACIFIC RIM:
Japan: restricted to flavouring use, limitation with sorbic acid 3 g/kg total

REFERENCES:

Ash, M., and Ash, I. (1995) *Food Additives.* Gower Publishing Co., Brookfield, VT.

Davidson, P. M., and Branen, A. L. (Eds.) (1993) *Antimicrobials in Foods*, Second Edition. Marcel Dekker, New York.

NAME:	Calcium propionate
CATEGORY:	Antimicrobial preservative
FOOD USE:	Baked products/ Cheese products
SYNONYMS:	Methylacetic acid/ Propanoic acid, calcium salt/ Ethylformic acid, calcium salt/ E282/ CAS 4075-81-4/ EINECS 223-795-8
FORMULA:	$(C_2H_5COO)_2Ca$
MOLECULAR MASS:	186.22
PROPERTIES AND APPEARANCE:	Transparent crystals, granules. Powder or monoclinic crystals, hygroscopic
MELTING RANGE IN °C:	>300
PURITY %:	98
WATER CONTENT MAXIMUM IN %:	5
HEAVY METAL CONTENT MAXIMUM IN ppm:	10
FUNCTION IN FOODS:	Antimicrobial preservative; antimycotic; flavouring agent; preservative additive; mould inhibitor
ALTERNATIVES:	Other antimicrobial preservatives depending on the application
TECHNOLOGY OF USE IN FOODS:	1% (foods); limitation 0.32% (flour in white bread/rolls), 0.38% (whole wheat), 0.3% (cheese products)
ANTAGONISTS:	Calcium chloride

LEGISLATION:	**USA:** FDA 21CFR § 133.123, 133.124, 133.173, 133.179, 136.110, 136.115, 136.130, 136.160, 136.180, 150.141, 150.161, 179.45, 181.22, 181.23, 184.1221, GRAS; USDA 9CFR § 318.7, 0.32% max. on weight of flour, 381.147, 0.3% max. on weight of flour in fresh pie dough	**UK and EUROPE:** UK: approved; Europe: listed **CANADA:** Approved in certain foods **AUSTRALIA/PACIFIC RIM:** Japan: restricted to flavouring use, limitation with sorbic acid 3 g/kg total
REFERENCES:	Ash, M., and Ash, I. (1995) *Food Additives.* Gower Publishing Co., Brookfield, VT. Davidson, P. M., and Branen, A. L. (Eds.) (1993) *Antimicrobials in Foods*, Second Edition. Marcel Dekker, New York.	
ANY OTHER RELEVANT INFORMATION:	pH of a 1 in 10 solution = 8–10	

NAME:	**Potassium propionate**
CATEGORY:	Antimicrobial preservative
FOOD USE:	Baked products/ Cheese products
SYNONYMS:	E283
FORMULA:	K C2H5 COO
ALTERNATIVE FORMS:	Other salts
FUNCTION IN FOODS:	Antimicrobial preservative; antimycotic; flavouring agent; preservative additive; mould inhibitor
ALTERNATIVES:	Other antimicrobial preservatives, depending on the application
TECHNOLOGY OF USE IN FOODS:	1% (foods); limitation 0.32% (flour in white bread/rolls), 0.38% (whole wheat), 0.3% (cheese products)
ANTAGONISTS:	Calcium chloride
LEGISLATION:	**USA:** FDA 21CFR, 172.515, 184.1081, GRAS, GRAS(FEMA), EPA reg **UK and EUROPE:** UK: approved; Europe: listed **CANADA:** Approved in certain foods **AUSTRALIA/PACIFIC RIM:** Japan: restricted to flavouring use, limitation with sorbic acid 3 g/kg total
REFERENCES:	Ash, M., and Ash, I. (1995) *Food Additives.* Gower Publishing Co., Brookfield, VT. Davidson, P. M., and Branen, A. L. (Eds.) (1993) *Antimicrobials in Foods*, Second Edition. Marcel Dekker, New York.

NAME:	Heptyl paraben	
CATEGORY:	Antimicrobial preservative	
FOOD USE:	Antioxidant/ Preservative/ Antimicrobial agent for beer, fermented malt beverages/ Non-carbonated soft drinks and fruit drinks	
SYNONYMS:	*n*-Heptyl *p*-hydroxybenzoate/ CAS 1085-12-7	
FORMULA:	C14 H20 O3	
MOLECULAR MASS:	236.31	
PROPERTIES AND APPEARANCE:	Colourless small crystals or white crystalline powder, odourless or faint charcoal odour, slight burning taste	
MELTING RANGE IN °C:	48–51	
PURITY %:	>99	
WATER CONTENT MAXIMUM IN %:	0.5	
HEAVY METAL CONTENT MAXIMUM IN ppm:	10 (as Pb)	
ASH MAXIMUM IN %:	0.05	
SOLUBILITY % AT VARIOUS TEMPERATURE/pH COMBINATIONS:		
in water:	**@ 20°C**	1.5
in ethanol solution:	**@ 100%**	Readily soluble
FUNCTION IN FOODS:	Antioxidant, preservative; antimicrobial agent for beer, fermented malt beverages, non-carbonated soft drinks and fruit drinks. Especially effective against bacteria and fungi at around neutral pH	

TECHNOLOGY OF USE IN FOODS: Antioxidant, preservative, antimicrobial agent for beer, fermented malt beverages, non-carbonated soft drinks and fruit drinks. Especially effective against bacteria and fungi at around neutral pH

LEGISLATION:

USA:	**UK and EUROPE:**	**CANADA:**
FDA 21CFR § 172.145, 12 ppm max. in fermented malt beverages, 20 ppm max. in soft/fruit drinks; BATF 27CFR § 240.1051, 12 ppm max. in wine	Not permitted	Not permitted

REFERENCES:

Davidson, P. M., and Branen, A. L. (1993) *Antimicrobials in Foods*, Second Edition. Marcel Dekker, New York.

Ash, M., and Ash, I. (1995) *Food Additives.* Gower Publishing Co., Brookfield, VT.

Part 12

Sequestrants

NAME:	Calcium acetate
CATEGORY:	Sequestrant
FOOD USE:	Edible caseinates/ Instant puddings/ Sweet sauces/ Baked goods/ Gelatins/ Syrups/ Cake mixes/ Fillings/ Toppings
SYNONYMS:	Brown acetate of lime/ Lime acetate/ Sorbocalcion/ Vinegar salts/ Grey acetate of lime/ INS 263/ E263
FORMULA:	C4 H6 Ca O4 or C4 H6 Ca O4 · H2O
MOLECULAR MASS:	158.17 anhydrous or 176.18 monohydrate
ALTERNATIVES FORMS:	Calcium diacetate
PROPERTIES AND APPEARANCE:	White hygroscopic bulky crystalline solid with slight bitter taste. Monohydrates – needles, granules or powder
BOILING POINT IN °C AT VARIOUS PRESSURES (INCLUDING 760 mm Hg):	160, decomposes to acetone and $CaCO_3$
MELTING RANGE IN °C:	160
DENSITY AT 20°C (AND OTHER TEMPERATURES) IN g/l:	1.5
PURITY %:	98–99
WATER CONTENT MAXIMUM IN %:	7–11
HEAVY METAL CONTENT MAXIMUM IN ppm:	25–30
ARSENIC CONTENT MAXIMUM IN ppm:	3

SOLUBILITY % AT VARIOUS TEMPERATURE/pH COMBINATIONS:

in water:	**@ 20°C**	Slightly soluble in water
in ethanol solution:	**@ 5%**	Insoluble in ethanol

FUNCTION IN FOODS: Corrosion inhibitor in metal containers. Medically, as a source of calcium. Used in the manufacture of acetic acid and acetone; also used as an emulsifier and firming agent. Functions as a mould-control agent

ALTERNATIVES: Sodium acetate; potassium acetate; sodium diacetate

TECHNOLOGY OF USE IN FOODS: pH 6.9 in a 1 in 10 solution. Store in well-closed containers. Avoid excessive heat, sparks or open flame >160°C

FOOD SAFETY ISSUES: Low oral toxicity

LEGISLATION:

USA:
FDA generally recognised as safe (GRAS) if used in accordance with good manufacturing practice (GMP). 0.2% in baked goods; 0.2% in gelatins/ puddings and fillings; 0.15% in sweet sauces, toppings and syrups; 1 ppm in beverages

UK and EUROPE:
Authorised without limitation

CANADA:
GMP in alcoholic beverages

AUSTRALIA/PACIFIC RIM:
Japan: GMP

NAME:	**Calcium chloride**
CATEGORY:	Sequestrant
FOOD USE:	Sliced apples/ Canned fruit/ Coffee and tea/ Apple pie mix/ Canned milks/ Processed fruit/ Jams and jellies/ Milk powders/ Fruit juice/ Canned tomatoes and vegetables/ Baked goods/ Cheeses/ Beverages
SYNONYMS:	Superflake anhydrous/ Peladow/ Dowflake/ INS 509/ Snomelt/ Calpus/ Caltac/ E509
FORMULA:	Anhydrous – CaCl2
MOLECULAR MASS:	110.99 (dihydrate 147.02; hexahydrate 219.08)
ALTERNATIVE FORMS:	Dihydrate/ hexahydrate
PROPERTIES AND APPEARANCE:	Anhydrous – white deliquescent lumps; dihydrate – white hard deliquescent fragments. Hexahydrate – colourless, very deliquescent crystals
BOILING POINT IN °C AT VARIOUS PRESSURES (INCLUDING 760 mm Hg):	1600; hexahydrate decomposes at 200
MELTING RANGE IN °C:	Anhydrous 772; hexahydrate 30
DENSITY AT 20°C (AND OTHER TEMPERATURES) IN g/l:	Anhydrous 2.152; dihydrate 0.84; hexahydrate 1.68
VAPOUR PRESSURE AT VARIOUS TEMPERATURES IN mm Hg:	0.01
PURITY %:	Anhydrous >93; dihydrate 99; hexahydrate 98
HEAVY METAL CONTENT MAXIMUM IN ppm:	20–40
ARSENIC CONTENT MAXIMUM IN ppm:	3

SOLUBILITY % AT VARIOUS TEMPERATURE/pH COMBINATIONS:

in water: **@ 20°C** 59 g/100 ml at 0°C
Freely dissolves in water

FUNCTION IN FOODS: General-purpose food additive. Used as a firming agent in canned fruit and vegetables. In evaporated milk – adjust salt balance to prevent clotting. Flavour protectant in pickles. Gelling enhancer

TECHNOLOGY OF USE IN FOODS: Store in airtight containers to avoid moisture

SYNERGISTS: Calcium gluconate/ calcium lactate/ disodium EDTA

ANTAGONISTS: Methyl vinyl ether/ strong acids

LEGISLATION:

USA:
FDA GRAS if used in accordance with GMP. Miscellaneous general-purpose food additive: 0.3% in baked goods; in evaporated milk up to 0.1% by weight of finished product; in canned vegetables up to 260–360 ppm; 0.22% in non-alcoholic beverages; 0.2% in cheese and processed fruits/fruit juice; 0.32% in coffee/tea; 0.2% in gravies

UK and EUROPE:
GMP. 2000 ppm in canned milk products/creams. Denmark: 5000 ppm in canned milk products

CANADA:
260 ppm in deep-frozen apples

AUSTRALIA/PACIFIC RIM:
Japan: 2% as calcium as dietary supplement

NAME:	Calcium citrate
CATEGORY:	Sequestrant
FOOD USE:	Confections/ Flour/ Milk powders/ Jams and jellies/ Canned vegetables/ Processed cheese/ Saccharin/ Evaporated milk/ Cream/ Condensed milk/ Edible ices/ Frozen apples
SYNONYMS:	Tricalcium citrate/ Tricalcium salt of 2-hydroxy-1,2,3 propanetricarboxylic acid/ INS 333/ E333
FORMULA:	$Ca3(C6H5O7)2 \cdot 4H2O$
MOLECULAR MASS:	498.44–570.5
PROPERTIES AND APPEARANCE:	Fine white powder; odourless
MELTING RANGE IN °C:	120
PURITY %:	97.5
WATER CONTENT MAXIMUM IN %:	10–14
HEAVY METAL CONTENT MAXIMUM IN ppm:	20
ARSENIC CONTENT MAXIMUM IN ppm:	3
SOLUBILITY % AT VARIOUS TEMPERATURE/pH COMBINATIONS:	
in water:	**@ 20°C** 85 g/100 ml
in ethanol solution:	**@ 5%** Insoluble in ethanol
FUNCTION IN FOODS:	Used in the production of citric acid and other citrates. Used to improve baking properties of flours. Soluble in fats – good stabiliser
TECHNOLOGY OF USE IN FOODS:	Store in well-closed containers. When heated to decomposition, emits acrid smoke and irritating fumes

FOOD SAFETY ISSUES:	Citrates may interfere with laboratory tests, including pancreatic function, liver function and blood alkalinity-acidity. Citrate esters showed no adverse effects on rats in a 2-year feeding study. Poses no hazard to consumers
LEGISLATION:	**USA:** FDA GRAS with no limitations if used in accordance with GMP **CANADA:** 260 ppm in deep-frozen apples (sliced or not sliced) **UK and EUROPE:** Authorised without limitation Finland: 2000 ppm in frozen fruit **AUSTRALIA/PACIFIC RIM:** Japan: GMP

NAME:	**Calcium gluconate**
CATEGORY:	Sequestrant
FOOD USE:	Pickled cucumbers/ Cured meats/ Sugar substitute/ Coffee powders/ Gelatin and puddings/ Dairy product analogues/ Canned fruit and vegetables/ Baked goods/ Jams and jellies
SYNONYMS:	D-Gluconic acid calcium salt/ Calciofon/ Caglucon/ Calcium di-gluconate/ Calcium di-D-gluconate monohydrate/ INS 578/ E578
FORMULA:	Ca(C6 H11 O7)2 · H2O
MOLECULAR MASS:	430.38–448.39
PROPERTIES AND APPEARANCE:	White crystalline granules
MELTING RANGE IN °C:	120, loses H_2O
PURITY %:	98
WATER CONTENT MAXIMUM IN %:	3
HEAVY METAL CONTENT MAXIMUM IN ppm:	10–20
ARSENIC CONTENT MAXIMUM IN ppm:	3
SOLUBILITY % AT VARIOUS TEMPERATURE/pH COMBINATIONS:	
in water:	**@ 20°C** 3.3
FUNCTION IN FOODS:	Used in fruits and vegetables as a firming agent
TECHNOLOGY OF USE IN FOODS:	Store in well-closed containers. pH 6–7 aqueous solution

FOOD SAFETY ISSUES:	May cause gastrointestinal and cardiac disturbances	
LEGISLATION:	**USA:** FDA GRAS in accordance with GMP Used at 1.75% in baked goods; 0.4% in dairy analogues; 4.5% in gelatins and puddings; 0.01% in sugar substitutes	**UK and EUROPE:** GMP in canned fruit and vegetables **AUSTRALIA/PACIFIC RIM:** Japan: 1% as calcium in cured meat products

NAME:	Calcium sulphate
CATEGORY:	Sequestrant
FOOD USE:	Yeast food/ Brewing, sherry and wine/ Artificially sweetened fruit/ Cottage cheese/ Bread (rolls and buns)/ Jelly/ Canned tomatoes/ Cheese/ Cereal flours/ Canned potatoes/ Canned sweet peppers/ Soft ice-cream/ Frozen apples/ Confections/ Puddings
SYNONYMS:	INS 516/ E516
FORMULA:	Ca O4 S
MOLECULAR MASS:	136.14
ALTERNATIVE FORMS:	Calcium sulphate dihydrate
PROPERTIES AND APPEARANCE:	Crystals are orthorhombic, yellow to white powder
BOILING POINT AT VARIOUS PRESSURES (INDLUDING 760 mm Hg):	N/A
MELTING RANGE IN °C:	>120
FLASH POINT IN °C:	Non-flammable
DENSITY AT 20°C (AND OTHER TEMPERATURES) IN g/l:	2.964
VAPOUR PRESSURE AT VARIOUS TEMPERATURES IN mm Hg	Unknown
PURITY %:	98
WATER CONTENT MAXIMUM IN %:	<1.5: dihydrate 19–23
HEAVY METAL CONTENT MAXIMUM IN ppm:	10–20

ARSENIC CONTENT MAXIMUM IN ppm: 3

FUNCTION IN FOODS: Used as a carrier in bleaching, bleaching agent in flour. Used in cottage cheese as an alkali.

Used as a firming agent in canned vegetables. Used as a dough conditioner/ strengthener in breadmaking, and as a calcium source for reaction with alginates to form dessert gels. Also used as a nutritional supplement

TECHNOLOGY OF USE IN FOODS: Store in well-closed containers. Reacts violently with aluminium when heated.

Mixtures with phosphorus ignite at high temperatures. When heated to decomposition, it emits toxic fumes of SOx

SYNERGISTS: Alginates, ammonium chloride/ benzoyl peroxide/ dicalcium phosphate

LEGISLATION:

USA:
FDA GRAS if used in accordance with GMP
Used in cottage cheese (5 g/kg); canned tomatoes (800 mg/kg); in specific vegetables GMP up to 700 ppm. Also used in bakery products (0.25% w/w flour), in confections and frostings (3%), 1.3% in baked goods, in preserved seafood (up to 700 ppm) and 0.35% in processed vegetables. Used in canned milks and chocolate drinks (up to 700 ppm) and in beverages (700 ppm)

UK and EUROPE:
Authorised without limitation

CANADA:
GMP in alcoholic beverages. Used in deep-frozen apples (260 ppm) and dairy desserts (5000 ppm)

AUSTRALIA/ PACIFIC RIM:
Japan: GMP

NAME:	Citric acid
CATEGORY:	Sequestrant
FOOD USE:	Fruit juice drinks and beverages/ Frozen dairy products/ Preserves/ Wines and cider/ Evaporated milks/ Curing meats/ Jams and jellies/ Cheese and cheese spreads/ Pie fillings/ Jelly candies/ Sherbet and ice-cream/ Canned apples/ Canned fruit/ Confections/ Canned sardines/ Carbonated beverages/ Canned figs/ Canned crab and shrimp/ Frozen fruit and dried fruit/ Mayonnaise/ Cocoa powder/ Canned vegetables/ Salad dressing/ Table olives/ Fats and oils/ Canned chilli/ Instant potatoes/ Pickled cucumbers/ Bouillons and consommes/ Canned baby food/ Infant formula/ Cereal-based foods for infants
SYNONYMS:	Citric acid anhydrous/ 2-Hydroxy-1,2,3 propane tricarboxylic acid/ Citric acid monohydrate/ Beta-hydroxytricarballytic acid/ Citro/ INS 330/ E330
FORMULA:	C6H8O7
MOLECULAR MASS:	192.12
ALTERNATIVE FORMS:	Ammonium citrate/ Isopropyl citrate/ Stearyl citrate/ Calcium citrate/ Potassium citrate/ Sodium citrate/ Triethyl citrate
PROPERTIES AND APPEARANCE:	White or colourless crystalline solid; strongly acid taste
MELTING RANGE IN °C:	153
FLASH POINT IN °C:	100
IONISATION CONSTANT AT 25°C:	K_1 8.2×10^{-4}, K_2 1.8×10^{-5}, K_3 3.9×10^{-6}
DENSITY AT 20°C (AND OTHER TEMPERATURES) IN g/l	1.665
HEAT OF COMBUSTION AT 25°C IN J/kg:	474.5 kcal/mole
PURITY %:	99.5

WATER CONTENT MAXIMUM IN %: 0.5

HEAVY METAL CONTENT MAXIMUM IN ppm: 10

ARSENIC CONTENT MAXIMUM IN ppm: 3

ASH MAXIMUM IN %: 0.05

SOLUBILITY % AT VARIOUS TEMPERATURE/pH COMBINATIONS:

in water: @ **20°C** 60% w/w; very soluble @ **50°C** 70% @ **100°C** 84%

in ethanol solution: @ **5%** Freely soluble @ **95%** >100 mg/ml @22°C

FUNCTION IN FOODS: Collects and deactivates metal contaminants, increases the effect of preservatives. Used to maintain freshness and prevent rancidity. Soluble in fats – good stabilisers. Also used as a sequestrant to adjust acid-alkali balance. Mixed with erythorbic acid to prevent browning. 1% solution used in the canning of crabmeat. Used to decrease turbidity in wines, ciders, vinegars. Mixed with ascorbic acid in seafood dip to prevent discoloration. Prevents colour and flavour changes in canned fruit and vegetables and fish

ALTERNATIVES: L-Tartaric acid; Phosphoric acid; EDTA; Acetic acid

TECHNOLOGY OF USE IN FOODS: Dissolve to makes solution first, e.g. 1% solutions to prevent black spot formation in seafood.
The dilute aqueous solution may ferment on standing.
Combustible liquid; potentially explosive reaction to metal nitrates

SYNERGISTS:
With amino acids – serves to increase water retention of products.
With erythorbic acid – retards browning of bananas.
With gelatin, salt, vitamin C, glucose, carageenates.
With antioxidants – inhibits rancidity of foods containing fats and oils

ANTAGONISTS: Potassium tartrate/ alkali and alkaline earth carbonates and bicarbonates/ acetates and sulphides/ strong bases

FOOD SAFETY ISSUES:

Very low levels of toxicity. Digested and metabolised normally by the body. Citrates are normal constituents of many foods; they are present in far higher amounts in the body than amounts added to processed foods. May cause severe eye and moderate skin irritation

LEGISLATION:

USA:
FDA GRAS with no limitations if used in accordance with GMP
Citric acid – single or mixed with antioxidants (100 ppm)
Meat Inspection Division – 0.01% in shortening/lard. 0.01% in fats and oils; 0.003% in dry sausage/dried meat

UK and EUROPE:
Authorised without limitation – GMP
Finland: 1000 ppm quick-frozen fruit. 3 g/l in fruit juices

CANADA:
Citric acid – GMP

AUSTRALIA/PACIFIC RIM:
Japan: GMP

REFERENCE:

Bouchard, E. F., and Merrit, E. G. (1979) *Kirk-Othmer Encyclopedia of Chemical Toxicology*, Volume 6, pp. 150–179.

NAME:	**Disodium ethylenediaminetetraacetate (disodium EDTA)**
CATEGORY:	Sequestrant
FOOD USE:	Lard/ Canned legumes/ Creamed turkey/ Fats and oils/ Potatoes and french fries/ Dried banana/ Fruit juices/ Canned seafood/ Cereal/ Ham/ Bacon/ Frankfurters/ Milks/ Strawberry pie filling/ Canned corn/ Egg custards/ Mayonnaise/ Canned apples/ Cheese/ Frozen ground beef/ Processed vegetables and fruit
SYNONYMS:	Edate disodium/ *N,N'*-1,2-ethanediyl-*bis*[*N*(carboxymethyl)glycine]disodium salt/ EDTA disodium salt/ Disodium hydrogen EDTA/ Ethylenedinitrolo-tetraacetic acid disodium salt/ INS 386
FORMULA:	C10 H14 N2 Na2 O8
MOLECULAR MASS:	336.21–372.24
ALTERNATIVE FORMS:	Sodium EDTA
PROPERTIES AND APPEARANCE:	White dihydrate crystals
BOILING POINT IN °C AT VARIOUS PRESSURES (INCLUDING 760 mm Hg):	Decomposes at 250
VAPOUR PRESSURE AT VARIOUS TEMPERATURES IN mm Hg:	Negligible
PURITY %:	99
HEAVY METAL CONTENT MAXIMUM IN ppm:	20
ARSENIC CONTENT MAXIMUM IN ppm:	3
SOLUBILITY % AT VARIOUS TEMPERATURE/pH COMBINATIONS:	
in water:	**@ 20°C** 10 g/100 ml H_2O

FUNCTION IN FOODS:	Stabilises lard and also vitamins in food systems. Prevents undesirable colour changes. Decreases the poor flavour in milk; decreases the grey discoloration on meat surfaces. Acts as a preservative; promotes colour and flavour retention in foods. Decreases crystal formation	
ALTERNATIVES:	Citrates; BHT; phosphates and pyrophosphates; calcium disodium EDTA	
TECHNOLOGY OF USE IN FOODS:	Does not contribute any flavour. pH 4.5 in a 1% solution. Store in well-closed containers	
SYNERGISTS:	Ascorbates	
LEGISLATION:	**USA:** FDA – 145 ppm maximun in canned black eye peas; 165 ppm cooked chick peas/kidney beans; 315 ppm in dried banana; 500 ppm in strawberry pie filling; 100 ppm in frozen potatoes; 36 ppm in coated sausage; 25 ppm in alcoholic beverages; 75 ppm in standardised dressing/mayonnaise	**CANADA:** 25 ppm in alcoholic beverages

NAME:	Disodium pyrophosphate
CATEGORY:	Sequestrant
FOOD USE:	Evaporated milk/ Cheese/ Biscuits/ Pasta products/ Cured meats/ Bologna/ Flour/ Cake mixes/ Doughnuts/ Poultry products/ Processed potato/ Canned fish products
SYNONYMS:	Disodium dihydrogen diphosphate/ Disodium dihydrogen pyrophosphate/ Acid sodium pyrophosphate/ Sodium pyrophosphate/ INS 450 (i)/ E450a/ Disodium salt diphosphoric acid
FORMULA:	Na2 H2 P2 O7
MOLECULAR MASS:	221.94
PROPERTIES AND APPEARANCE:	White crystalline powder or granules
MELTING RANGE IN °C:	220, decomposition
DENSITY AT 20°C (AND OTHER TEMPERATURES) IN g/l:	1.862
WATER CONTENT MAXIMUM IN %:	0.5
HEAVY METAL CONTENT MAXIMUM IN ppm:	20
ARSENIC CONTENT MAXIMUM IN ppm:	3
SOLUBILITY % AT VARIOUS TEMPERATURE/pH COMBINATIONS:	
in water:	**@ 20°C** 15 g/100 ml H_2O
FUNCTION IN FOODS:	Increases water-holding capacity in meat. Emulsifier in cheese. Decreases cooked-out juices. Buffering agent
TECHNOLOGY OF USE IN FOODS:	1 in 100 solution — pH 8

LEGISLATION:	**USA:** FDA – GRAS sequestrant. 5% phosphate in pickle; 0.1% in evaporated milk; emulsifier in cheese 3% by weight; macaroni/noodles 0.5–1%; 0.5% final meat product and total poultry product	**UK and EUROPE:** Authorised without limitations. 2000 ppm in canned milk **AUSTRALIA/PACIFIC RIM:** Japan: GMP

NAME:	**Glucono-delta-lactone**
CATEGORY:	Sequestrant
FOOD USE:	Jelly powder/ Dessert mixes/ Fish/ Sausages/ Luncheon meat/ Frankfurters/ Cooked chopped meat
SYNONYMS:	δ-D-Gluconolactone/ Gluconolactone/ D-Gluconic acid, δ-lactone/ D-Glucono-1,5-lactone/ GDL/ INS 575/ E575
FORMULA:	C6H10O6
MOLECULAR MASS:	178.14
PROPERTIES AND APPEARANCE:	Fine white, odourless powder
BOILING POINT:	4351 (decomposes at 153°C)
DENSITY AT 20°C (AND OTHER TEMPERATURES) IN g/l:	61.7
VAPOUR PRESSURE AT VARIOUS TEMPERATURES IN mm Hg:	Negligible
PURITY %:	99
WATER CONTENT MAXIMUM IN %:	1
HEAVY METAL CONTENT MAXIMUM IN ppm:	20
ARSENIC CONTENT MAXIMUM IN ppm:	3
FUNCTION IN FOODS:	Leavening agent in jelly powder and soft drink powder. Used in the dairy industry to prevent milk stone. Used in breweries to prevent milk stone. Speeds up colour fixing process in smoked meats. Acidulant in dessert mixes

TECHNOLOGY OF USE IN FOODS: 1% solution pH of 3.6 and decreases to 2.5 within 2 hours.
Store in well-closed containers

ANTAGONISTS: Strong oxidisers

LEGISLATION:

USA:
FDA – GRAS in accordance with GMP.
Cleared by the USDA Meat Inspection Division; 3 g/kg in cooked meat and luncheon meat; limit of 8 oz/100 lb meat

CANADA:
GMP in cooked sausage
5000 ppm

UK and EUROPE:
GMP
Finland: 9000 ppm in preserved sausage; 3000 ppm in boiled sausage
Denmark: 5000 ppm in smoked herring
Germany: 1% in semi-preserved anchovy and fish

AUSTRALIA/PACIFIC RIM:
Japan: authorised without limitation in meat products and fish products

NAME:	Isopropyl citrate
CATEGORY:	Sequestrant
FOOD USE:	Oleomargarine/ Fats and oils/ Premier jus
SYNONYMS:	Citric acid mixture of 2-propanol/ Isopropyl citrate mixture
ALTERNATIVE FORMS:	Monoisopropyl citrate 27 parts by weight/ Diisopropyl citrate 9 parts by weight/ Triisopropyl citrate 2 parts by weight
HEAVY METAL CONTENT MAXIMUM IN ppm:	30
ARSENIC CONTENT MAXIMUM IN ppm:	10
ASH MAXIMUM IN %:	3
FUNCTION IN FOODS:	Added to margarine to protect flavour. Chelates ions in fats/oils to prevent rancidity
ALTERNATIVES:	Citric acid; Phosphoric acid
TECHNOLOGY OF USE IN FOODS:	When heated to decomposition, emits acrid smoke and irritating fumes
SYNERGISTS:	Phosphoric acid/ monoglyceride citrate
LEGISLATION:	**USA:** FDA – GRAS with no limitations if used in accordance with GMP Meat Inspection Division – up to 0.02% in oleomargarine; 100 ppm in fats and oils; 100 ppm in margarine

NAME:	Potassium phosphate dibasic
CATEGORY:	Sequestrant
FOOD USE:	Bouillons and consommes/ Cured ham and chopped meat/ Milk powder and cream powder/ Luncheon meat/ Evaporated milks/ Processed cheese/ Edible ices and ice mixes/ Low-sodium products/ Caramels/ Sparkling wines/ Yeast food/ Cream/ Cured pork shoulder/ Condensed milk/
SYNONYMS:	Dipotassium hydrogen phosphate/ Dipotassium orthophosphate/ Dipotassium monophosphate/ Dibasic potassium phosphate/ Dipotassium phosphate/ DKP/ Dipotassium phosphate/ Dipotassium acid phosphate
FORMULA:	K2HPO4
MOLECULAR MASS:	174.18
ALTERNATIVE FORMS:	Monobasic/ Pyrophosphate/ Tribasic forms
PROPERTIES AND APPEARANCE:	Colourless or white granular powder, crystals or masses. Deliquescent
PURITY %:	>98
WATER CONTENT MAXIMUM IN %:	2–5
HEAVY METAL CONTENT MAXIMUM IN ppm:	20–30
ARSENIC CONTENT MAXIMUM IN ppm:	3
SOLUBILITY % AT VARIOUS TEMPERATURE/pH COMBINATIONS:	
in water:	**@ 20°C** 6 g/100 ml H_2O
FUNCTION IN FOODS:	Used as a yeast food in the brewing industry. Excellent acidifying agent. Able to form soluble complexes with alkali and alkali-earth metal ions; these ions interfere with food processing reactions, e.g. beverage industry. Decreases amount of cooked-out juices. Increases hydration in freeze–thaw products. Stabilises meat emulsions

ALTERNATIVES: Sodium phosphates

TECHNOLOGY OF USE IN FOODS: Strongly alkaline with a pH of 12 (pH 8.7–9.3 for 1 in 100 solution). Less astringent than their sodium counterparts. Must adjust the formulation to account for water of hydration in the phosphate. Store in air-tight containers; avoid moisture, heat and strong oxidisers, sparks and flames

FOOD SAFETY ISSUES: No known toxicity

LEGISLATION:

USA:
FDA – GRAS with no limitations except GMP. Consommes/bouillons – 1000 ppm; luncheon meat 3 g/kg; 5% phosphate in pickles; 0.5% in products, 0.5% in total poultry product

UK and EUROPE:
2000 ppm in canned milks

CANADA:
GMP

AUSTRALIA/PACIFIC RIM:
Japan: GMP

NAME:	**Potassium phosphate monobasic**
CATEGORY:	Sequestrant
FOOD USE:	Milk/ Whole eggs/ Caramels/ Meat products/ Yeast food/ Poultry food products/ Low-sodium products/ Sparkling wine
SYNONYMS:	Potassium dihydrogen orthophosphate/ Monopotassium monophosphate/ Potassium acid phosphate/ Potassium biphosphate
FORMULA:	K H2 P O4
MOLECULAR MASS:	136.09
ALTERNATIVE FORMS:	Monobasic/ Tribasic/ Metaphosphates
PROPERTIES AND APPEARANCE:	Colourless white granular salt
MELTING RANGE IN °C:	253; @ 400 loses water
FLASH POINT IN °C:	None
DENSITY AT 20°C (AND OTHER TEMPERATURES) IN g/l:	2.34
PURITY %:	>98
WATER CONTENT MAXIMUM IN %:	1
HEAVY METAL CONTENT MAXIMUM IN ppm:	20
ARSENIC CONTENT MAXIMUM IN ppm:	3
SOLUBILITY % AT VARIOUS TEMPERATURE/pH COMBINATIONS:	
in water:	**@ 20°C** 20 g/100 ml H_2O **@ 25°C** 33

FUNCTION IN FOODS:	Preserves colour in whole eggs. Low-sodium products typical usage range 0.1–0.5%. Used as a yeast food in brewing industry. Excellent acidifying agent. Complexes metal ions and precipitates them out in beverage industry. Decreases amount of cooked-out juices in meat products; increases hydration in freeze–thaw products; stabilises meat emulsions
ALTERNATIVES:	Sodium phosphate
TECHNOLOGY OF USE IN FOODS:	pH mild acid 4–5. Usually less astringent than their sodium counterparts. Before use, must adjust formulation to account for amount of water of hydration in the phosphate. Deliquescent when exposed to moist air. Store in air-tight containers; avoid excessive heat; avoid aluminium and steel (may corrode these); avoid strong bases as may react violently
FOOD SAFETY ISSUES:	No known toxicity
LEGISLATION:	**USA:** FDA – GRAS food additive with no limitations except GMP. 0.5% phosphate in product; 0.5% in total poultry product; 5000 ppm in frozen whole eggs **UK and EUROPE:** 2000 ppm in canned milks **AUSTRALIA/PACIFIC RIM:** Japan: GMP

NAME:	**Potassium sodium tartrate**
CATEGORY:	Sequestrant
FOOD USE:	Cheese products/ Citrus marmalade/ Margarines/ Minced meat/ Jams and jellies/ Edible ices and ice mixes/ Sausage casings/ Bouillons and consommes
SYNONYMS:	Rochelle salt/ Seignette salt/ Potassium sodium dextro-tartrate/ Potassium sodium D-tartrate/ Potassium sodium dextro-tartrate/ INS 337/ E337
FORMULA:	C4 H4 K Na O6 · 4H2O
MOLECULAR MASS:	210.16–282.23
ALTERNATIVE FORMS:	Cream of tartar (acid monopotassium salt)/ Tartaric acid/ Diacetyl glyceryl tartrate
PROPERTIES AND APPEARANCE:	Translucent crystals or white crystal powder
BOILING POINT IN °C AT VARIOUS PRESSURES (INCLUDING 760 mm Hg):	760 mm Hg 220°C, decomposes 225°C
MELTING RANGE IN °C	70–80 (100°C, loses 3 H_2O; at 130–140°C, becomes anhydrous)
DENSITY AT 20°C (AND OTHER TEMPERATURES) IN g/l:	1.79
PURITY %:	99
WATER CONTENT MAXIMUM IN %:	21–26
HEAVY METAL CONTENT MAXIMUM IN ppm:	20
ARSENIC CONTENT MAXIMUM IN ppm:	3
SOLUBILITY % AT VARIOUS TEMPERATURE/pH COMBINATIONS:	
in water:	**@ 20°C** 100 g/100 ml H_2O

FUNCTION IN FOODS:	Sequestrant, pH control agent and a buffering agent. Prevents rancidity with antioxidants. Stabiliser in minced meat
TECHNOLOGY OF USE IN FOODS:	pH 6.5–7.5 for a 1 in 10 solution. Avoid excessive heat – 1% solution has a pH of 10
SYNERGISTS:	Fumaric acid and their salts/ With antioxidants
ANTAGONISTS:	Acid lead salts/ Calcium, magnesium sulphate/ Silver nitrate
LEGISLATION:	**USA:** FDA – GRAS if used in accordance with GMP. Jams/jellies up to 0.19% **UK and EUROPE:** Authorised without limitation. GMP in fruits and vegetables **CANADA:** GMP **AUSTRALIA/PACIFIC RIM:** Japan: GMP

NAME:	Manganese citrate
CATEGORY:	Sequestrant
FOOD USE:	Baked goods/ Fish products/ Milk products/ Beverages/ Infant formula/ Poultry products/ Dairy product analogues/ Meat products
FORMULA:	$Mn_3(C_6H_5O_7)_2$
MOLECULAR MASS:	543.02
PROPERTIES AND APPEARANCE:	Pale orange or pinkish white powder
MELTING RANGE IN °C:	160
FUNCTION IN FOODS:	Serves as a nutritional supplement and sequestrant in foods
TECHNOLOGY OF USE IN FOODS:	When heated to decomposition it emits acrid smoke and irritating fumes. pH of 2% solution = 6.3. Avoid heat, flames and moisture; store in an air-tight container
LEGISLATION:	**USA:** FDA – GRAS if used in accordance with GMP

NAME:	**Sodium monohydrogen phosphate 2:1:1**	
CATEGORY:	Sequestrant	
FOOD USE:	Coffee whitener/ Meat food products/ Puddings/ Cream sauce/ Evaporated milks/ Whipped products/ Broths/ Poultry/ Macaroni/ Hog carcasses	
SYNONYMS:	Dibasic sodium phosphate/ Disodium hydrogen phosphate/ Disodium monohydrogen phosphate/ Sodium phosphate dibasic anhydrous/ DSP/ INS 399(ii)/ Disodium orthophosphate/ Disodium phosphate/ Disodium phosphoric acid/ Phosphate of soda/ E399/ Sodium hydrophosphate	
FORMULA:	Na2HPO4	
MOLECULAR MASS:	141.96	
ALTERNATIVE FORMS:	Dihydrate – Na2HPO4.2H2O/ Heptahydrate	
PROPERTIES AND APPEARANCE:	Hygroscopic white powder or crystalline granules	
MELTING RANGE IN °C:	240	
PURITY %:	>98	
WATER CONTENT MAXIMUM IN %:	<5% (dihydrate 18–22%)	
HEAVY METAL CONTENT MAXIMUM IN ppm:	10–30	
ARSENIC CONTENT MAXIMUM IN ppm:	3	
SOLUBILITY % AT VARIOUS TEMPERATURE/pH COMBINATIONS:		
in water:	**@ 25°C**	12 g/100 ml
in ethanol solution:	**@ 5%**	Insoluble in ethanol

FUNCTION IN FOODS:	Used as a buffer, emulsifier and sequestrant in food. Decreases cooked-out juices in meat products. Used as a dietary supplement. Used to increase the pH of food products and stabilise the optimum pH in a food system. Used in evaporated milks as a buffer, and also prevents gelation. Decreases setting time in instant puddings. Used as a dispersant in producing a swelling of protein
TECHNOLOGY OF USE IN FOODS:	pH of 1% aqueous solution at 25°C = 9.1. Store in air-tight containers, as extremely hygroscopic; store in cool, dry vented areas. Avoid acids. When heated to decomposition, emits highly toxic fumes of oxides of phosphorus and sodium
FOOD SAFETY ISSUES:	Mildly toxic by ingestion. Skin and eye irritant, or on mucous membranes
LEGISLATION:	**USA:** FDA – GRAS if used in accordance with GMP. 5% phosphate in pickle; 0.5% phosphate in product; 0.5% in poultry products

NAME:	Sodium polyphosphate
CATEGORY:	Sequestrant
FOOD USE:	Processed cheeses/ Cream powders and cream/ Bouillons and consommes/ Sweetened condensed milk/ Luncheon meat and cured ham/ Cured pork/ Milk powders/ Frozen seafood/ Edible ices and ice mixes/ Dried eggs/ Evaporated milk/ Canned fruit and vegetables/ Carbonated beverages/ Alcoholic beverages
SYNONYMS:	Sodium hexametaphosphate/ Sodium tetraphosphate/ Graham salt/ Sodium metaphosphate/ INS 452/ Insoluble sodium metaphosphate/ Sodium polyphosphate, glassy/ Sodium polyphosphate/ Sodium phymetaphosphate/ E450c
FORMULA:	$(NaPO_3)_x$ x > 2 terminated by Na_2PO_4
MOLECULAR MASS:	101.96–611.2
PROPERTIES AND APPEARANCE:	Colourless, glassy transparent platelets or powders
MELTING RANGE IN °C:	628
FLASH POINT IN °C:	Non-combustible
PURITY %:	60–71
HEAVY METAL CONTENT MAXIMUM IN ppm:	20
ARSENIC CONTENT MAXIMUM IN ppm:	3
SOLUBILITY % AT VARIOUS TEMPERATURE/pH COMBINATIONS:	
in water:	**@ 20°C** Very soluble
FUNCTION IN FOODS:	Decreases amount of cooked-out juice. Prevents staining on exterior of canned goods. Inactivates metallic ions; sequestering in canning of fruits and vegetables. Retards oxidation of unsaturated fats in aqueous food systems. Decreases growth of microbes; softens water for preparation of alcoholic

beverages; decreases loss of carbonation in beverages; precipitates casein beta-lactoglobulin in milk (stabiliser effect); decreases turbidity in wines/ciders/vinegars

ALTERNATIVES: EDTA (ethylenediaminetetraacetate); citrates; phosphates

TECHNOLOGY OF USE IN FOODS: Store in air-tight containers; extremely hygroscopic, protect from moisture. Avoid strong oxidising agents; avoid high heat. 1% solution = pH 10

FOOD SAFETY ISSUES: Dust may be irritating to skin and eyes and mucous membranes

LEGISLATION:

USA:
FDA – GRAS sequestrant – miscellaneous food additive.
5% of phosphate in pickles; in food starches, 0.4% of calculated phosphates; 0.5% in total poultry product

UK and EUROPE:
2000 ppm in canned milk

CANADA:
2000 ppm in raw meat products; 5000 ppm in frozen fish

AUSTRALIA/PACIFIC RIM:
Japan: GMP

REFERENCE:

VanWaser, J. R. (1958) *Phosphorus and its Components*, Volume 1. Interscience, New York, pp. 601–800.

NAME:	**Sodium tartrate**
CATEGORY:	Sequestrant
FOOD USE:	Meat products/ Sausage casings/ Cheeses
SYNONYMS:	Disodium tartrate/ Disodium D-tartrate/ Sodium dextrotartrate/ Disodium (+)-2,3-dihydroxybutanedioic acid/ Disodium L-tartrate/ INS 335/ E335
FORMULA:	C4 H4 Na2 O6. 2H2O
MOLECULAR MASS:	194.05–230.08
PROPERTIES AND APPEARANCE:	Translucent, colourless, odourless crystals
DENSITY AT 20°C (AND OTHER TEMPERATURES) IN g/l:	1.82
PURITY %:	99
WATER CONTENT MAXIMUM IN %:	14–17
HEAVY METAL CONTENT MAXIMUM IN ppm:	10–20
ARSENIC CONTENT MAXIMUM IN ppm:	3
SOLUBILITY % AT VARIOUS TEMPERATURE/pH COMBINATIONS:	
in water:	**@ 20°C** 33 g/100 ml H_2O
FUNCTION IN FOODS:	Used in laxatives. Stabiliser in cheese. Used in artificially sweetened jelly. Used as a chemical reactant
TECHNOLOGY OF USE IN FOODS:	Store in air-tight containers. pH 7–7.5; 1% solution pH 10

LEGISLATION:	USA:	UK and EUROPE:
	USA: FDA – GRAS sequestrant. Emulsifier up to 3% by weight in specific cheeses; jams/jellies up to 0.19% **CANADA:** GMP	**UK and EUROPE:** Authorised without limitations. GMP in fruits and vegetables **AUSTRALIA/PACIFIC RIM:** Japan: GMP

NAME:	Oxystearin
CATEGORY:	Sequestrant
FOOD USE:	Canola oil/ Sunflowerseed oil/ Sesame seed oil/ Soyabean oils/ Maize oil/ Safflower oil/ Cottonseed oil/ Rapeseed oil/ Mustard seed oil/ Beet sugar/ Yeast
SYNONYMS:	INS 387
PROPERTIES AND APPEARANCE:	Tan to light brown, fatty wax-like substance. Mixture of glycerides of partially oxidised stearic acid and other fatty acids
HEAVY METAL CONTENT MAXIMUM IN ppm:	10
ARSENIC CONTENT MAXIMUM IN ppm:	3
SOLUBILITY % AT VARIOUS TEMPERATURE/pH COMBINATIONS:	
in water:	**@ 20°C** Insoluble in water
in ethanol solution:	**@ 5%** Soluble in ethanol
FUNCTION IN FOODS:	Crystallisation inhibitor in vegetable oils and salad oils. Defoaming agent in beet sugar and yeast processing
TECHNOLOGY OF USE IN FOODS:	Oxystearin is a mixture of partially oxidised stearic acid and other fatty acids. Store in well-closed containers. Refractive index between 1.465–1.467 at 48°C. Soluble in ether, hexane and chloroform
FOOD SAFETY ISSUES:	Limited information available, but may represent a hazard to the public when used at current levels. Uncertainty exists requiring additional studies
LEGISLATION:	**USA:** FDA – GRAS in edible fats and oils at 1250 ppm

NAME:	Phosphoric acid
CATEGORY:	Sequestrant
FOOD USE:	Colas and root beer/ Jams and jellies/ Frozen dairy products/ Carbonated beverages/ Caramel candies/ Candies/ Canned seafood/ Cocoa powder/ Chocolate/ Processed cheeses/ Rendered fat, fats and oils, vegetable oils/ Brewing industry
SYNONYMS:	Orthophosphoric acid/ INS 338/ *o*-Phosphoric acid/ E338
FORMULA:	H3 O4 P
MOLECULAR MASS:	98
ALTERNATIVE FORMS:	Monocalcium phosphate/ Dicalcium phosphate/ Sodium aluminium phosphate/ Sodium tripolyphosphate/ Sodium acid pyrophosphate/ Ammonium phosphate/ Sodium phosphate mono, di, tribasic
PROPERTIES AND APPEARANCE:	Clear, syrupy liquid
BOILING POINT IN °C AT VARIOUS PRESSURES (INCLUDING 760 mm Hg):	158
MELTING RANGE IN °C:	21
FLASH POINT IN °C:	42.4
IONISATION CONSTANT AT 25°C:	K_1 = 7.107 E-3
DENSITY AT 20°C (AND OTHER TEMPERATURES) IN g/l:	1.874 (100%); 1.68 (85%); 1.33 (50%); 1.05 (10%)
VAPOUR PRESSURE AT VARIOUS TEMPERATURES IN mm Hg:	0.0285 mm @ 20°C 0.03–2.16
PURITY %:	>75

HEAVY METAL CONTENT MAXIMUM IN ppm:	10
ARSENIC CONTENT MAXIMUM IN ppm:	2–3
FUNCTION IN FOODS:	Used as an acidulant in processed cheese. Used to clarify/acidity collagen in the production of gelatin/ fats and oils. pH of 0.1 N aqueous solution = 1.5. Antioxidant and sequestrant in vegetable and animal fats; synergist with antioxidants
ALTERNATIVES:	Disodium phosphate; citric acid; L-tartaric acid
TECHNOLOGY OF USE IN FOODS:	Food-grade phosphoric acid is supplied at 75, 80 and 85% aqueous solutions. An acid containing about 88% H_3PO_4 will frequently crystallise on prolonged cooling. Store in suitable stainless steel containers. Do not mix with nitromethane as the mixture becomes explosive. Do not store in stainless steel containers in contact with chlorides, as forms H_2 gas. Do not mix with sodium tetrahydroborate. Avoid excessive heat
SYNERGISTS:	Antioxidants/ Monoglyceride citrate, isopropyl citrate in fats and oils
ANTAGONISTS:	Chlorides/ strong alkalis/ metals, sulphides and sulphites
FOOD SAFETY ISSUES:	Concentrated solutions are irritating to skin and mucous membranes. LD_{50} = 3500–4400 mg/kg
LEGISLATION:	**USA:** FDA – GRAS in accordance with GMP. Miscellaneous all-purpose additive. Used at 0.01% to increase effectiveness of antioxidants in lard/shortening. 10.9 ppm total phosphorus in cheese; 2000–5000 ppm in jams and jellies **UK and EUROPE:** GMP **CANADA:** GMP

NAME:	Potassium dihydrogen citrate
CATEGORY:	Sequestrant
FOOD USE:	Bouillons and consommes/ Minarine/ Processed cheese/ Evaporated milk/ Margarine/ Edible ices and ice mix/ Condensed milk/ Jams and jellies/ Edible caseinates/ Milk powders/ citrus marmalade
SYNONYMS:	Monopotassium citrate/ Potassium citrate monobasic/ Monopotassium salt of 2-hydroxypropan-1,2,3-tricarboxylic acid/ INS 332(i)/ E332
FORMULA:	C6H7KO7
MOLECULAR MASS:	230.21
PROPERTIES AND APPEARANCE:	Transparent crystals/white powder, slight acid taste
PURITY %:	99
WATER CONTENT MAXIMUM IN %:	0.5
HEAVY METAL CONTENT MAXIMUM IN ppm:	10
ARSENIC CONTENT MAXIMUM IN ppm:	3
FUNCTION IN FOODS:	Chelates and deactivates metal contaminants. Increases the effect of preservatives. Soluble in fats – good stabilisers
TECHNOLOGY OF USE IN FOODS:	pH of 0.05 molal solution – pH scale standard at 25°C = 3.776. Store in air-tight containers

LEGISLATION:

USA:	UK and EUROPE:	CANADA:	AUSTRALIA/PACIFIC RIM:
FDA – GRAS miscellaneous or general-purpose ingredient. Used in accordance with GMP	Authorised without limitation	GMP	Japan: GMP

Part 13

Solvents

NAME:	Acetic acid
CATEGORY:	Solvent
FOOD USE:	Baked goods/ Chewing gum/ Resins/ Volatile oils/ Fats and oils/ Condiments: catsup, pickles, relish, mayonnaise/ Salad dressings/ Sauces, gravies/ Dairy products: cheese/ Meat products
SYNONYMS:	Ethanoic acid/ E260/ CAS 64-19-7
FORMULA:	CH3-COOH
MOLECULAR MASS:	60.05
PROPERTIES AND APPEARANCE:	Clear, colourless liquid, pungent vinegar odour
BOILING POINT IN °C AT VARIOUS PRESSURES (INCLUDING 760 mm Hg):	118
MELTING RANGE IN °C:	16.7
FLASH POINT IN °C:	39
DENSITY AT 20°C (AND OTHER TEMPERATURES) IN g/l:	1.053
PURITY %:	99% as glacial acetic acid
HEAVY METAL CONTENT MAXIMUM IN ppm:	10
ARSENIC CONTENT MAXIMUM IN ppm:	3

SOLUBILITY % AT VARIOUS TEMPERATURE/pH COMBINATIONS:		
in water:	**@ 20°C**	Completely soluble in water
in vegetable oil:	**@ 20°C**	Immiscible
in sucrose solution:	**@ 10%**	Miscible
in sodium chloride solution:	**@ 5%**	Miscible
in ethanol solution:	**@ 5%**	Miscible
in propylene glycol:	**@ 20°C**	Miscible
FUNCTION IN FOODS:	Normally used to acidify foods and as a flavouring agent. Used as a solvent for gums, resins, baked goods, fats and oils to carry flavourings and colouring agents	
TECHNOLOGY OF USE IN FOODS:	Used as a pH-adjusting agent in amounts consistent with GMP. $pK_a = 4.74$, therefore would normally be used at pH less than 4.7. pH of aqueous solutions 1.0 M = 2.4; 0.1 M = 2.9; 0.01 M = 3.4	
FOOD SAFETY ISSUES:	Acetic acid is not an additive that has been subject to great criticism over safety. FAO/WHO recognises that acetic acid is a normal constituent of foods and has set no limit on the daily acceptable intake for humans. As a concentrated solution it is caustic and can cause burns and is a severe eye and skin irritant	

LEGISLATION:

USA:

GRAS when used at levels in accordance with Good Manufacturing Practice. (21CFR 182.1005)

Current GMP results in the following:

0.8% for cheese and dairy products	0.5% in fats and oils
0.5% in chewing gum	0.25% for baked goods
9.0% in condiments and relishes	0.6% in meat products
3% in gravies and sauces	0.15% for all other foods

CANADA:

Listed as a pH-adjusting agent for use in cream cheese spread; processed cheese, processed cheese food, processed cheese spread, cold-pack cheese, cold-pack cheese food, whey cheese; canned asparagus, gelatin and unstandardised foods according to GMP.

Listed as a class I preservative for use in preserved fish, preserved meat; preserved meat by-product, preserved poultry meat, preserved poultry meat by-product, pumping pickle, cover pickle and dry cure employed in the curing of preserved meat or preserved meat by-product according to GMP

REFERENCES:

Doores, S. (1990) pH Control agents and acidulants. In: Branen, A. P., Davidson, P. M. and S. Salminen (Eds.), *Food Additives*. Marcel Dekker, Inc., New York.

Lewis, R. J. (1989) *Food Additives Handbook*. Van Nostrand Reinhold, New York.

NAME:	**Acetylated monoglycerides**		
CATEGORY:	Solvent		
FOOD USE:	Soft drinks/ Baked goods/ Chewing gums, caramels/ Frozen desserts, ice-cream/ Meat products/ Margarine, oleomargarine, shortening, cake shortening/ Nuts, peanut butter/ Puddings/ Whipped toppings		
FORMULA:	Varies		
MOLECULAR MASS:	Varies, esters of glycerin with acetic acid and edible fat-forming fatty acids		
PROPERTIES AND APPEARANCE:	May be white to pale yellow liquids or solids with a bland taste		
SOLUBILITY % AT VARIOUS TEMPERATURE/pH COMBINATIONS:			
in water:	**@ 20°C**	Insoluble in water	
in vegetable oil:	**@ 20°C**	Soluble	
in sucrose solution:	**@ 10%**	Insoluble	
in sodium chloride solution:	**@ 5%**	Insoluble	
in ethanol solution:	**@ 5%**	Soluble	
FUNCTION IN FOODS:	Solvent for antioxidants and spices. Also used as coating agent, emulsifier, lubricant or texture-modifying agent		
TECHNOLOGY OF USE IN FOODS:	Emits an acrid smoke and irritating fumes when heated to decomposition		
FOOD SAFETY ISSUES:	Not a food additive that has been subject to concern		
LEGISLATION:	**USA:** Used at a level not in excess of the amount reasonably required to produce its intended effect in food. (21 CFR 172.828) USDA regulation permits 0.5% in oleomargarine or margarine (9CFR 318.7)		**UK and EUROPE:** UK: GMP in soft drinks

REFERENCES:	Lewis, R. J. (1989) *Food Additives Handbook.* Van Nostrand Reinhold, New York. FAO/WHO (1992) *Compendium of Food Additive Specifications Vol. I, II.* Joint Expert Committee on Food Additives. FAO Food and Nutrition Paper 52/1.

NAME:	**Amyl acetate**
CATEGORY:	Solvent
FOOD USE:	Information available was very limited. Is listed as a solvent for flavourings, but food uses not clarified.
SYNONYMS:	Isoamyl acetate/ Acetic acid esters of amyl alcohol/ 3-Methylbutyl acetate/ Isopentyl alcohol acetate/ CAS 123-92-2/ Amylacetic ester/ Banana oil/ Pear oil
FORMULA:	CH3 COO CH2 CH2 CH(CH3)2
MOLECULAR MASS:	130.19
PROPERTIES AND APPEARANCE:	Colourless, clear liquid with fruit-like odour
BOILING POINT IN °C AT VARIOUS PRESSURES (INCLUDING 760 mm Hg):	120–142
FLASH POINT IN °C:	33
DENSITY AT 20°C (AND OTHER TEMPERATURES) IN g/l:	0.868–0.878
PURITY %:	98
HEAVY METAL CONTENT MAXIMUM IN ppm:	10
ARSENIC CONTENT MAXIMUM IN ppm:	3

SOLUBILITY % AT VARIOUS TEMPERATURE/pH COMBINATIONS:

in water:	**@ 20°C**	0.25%	**@ 50°C**	Slightly soluble	**@ 100°C**	Slightly soluble
in vegetable oil:	**@ 20°C**	Miscible	**@ 50°C**	Miscible	**@ 100°C**	Miscible
in sucrose solution:	**@ 10%**	Slightly soluble	**@ 40%**	Slightly soluble	**@ 60%**	Slightly soluble
in sodium chloride solution:	**@ 5%**	Slightly soluble	**@ 10%**	Slightly soluble	**@ 15%**	Slightly soluble
in ethanol solution:	**@ 5%**	Miscible	**@ 20%**	Miscible	**@ 95%**	Miscible
in propylene glycol:	**@ 20°C**	Insoluble	**@ 50°C**	Insoluble	**@ 100°C**	Insoluble

FUNCTION IN FOODS: Flavouring agent and carrier solvent in foods

TECHNOLOGY OF USE IN FOODS: Miscible with ether, ethyl acetate, amyl alcohol, and most oils. Insoluble in 1,2-propanediol and glycerin

FOOD SAFETY ISSUES: Exposure to concentrations greater than 1000 ppm for 1 hour can cause headaches, fatigue, pulmonary irritation and toxic effects. Mildly toxic by ingestion

LEGISLATION:

USA:	**CANADA:**
Used in minimum quantity required to produce intended effect (21CFR 172.515)	Not listed as a solvent for food use. Flavouring agent only

REFERENCES:

Lewis, R. J. (1989) *Food Additives Handbook.* Van Nostrand Reinhold, New York.

FAO/WHO (1992) *Compendium of Food Additive Specifications Vol. I, II.* Joint Expert Committee on Food Additives. FAO Food and Nutrition Paper 52/1.

NAME:	Benzyl alcohol
CATEGORY:	Solvent
FOOD USE:	Various
SYNONYMS:	Benzenemethanol/ Phenylcarbinol/ Phenylmethyl alcohol/ Alpha-hydroxy toluene/ CAS 100-51-6
FORMULA:	C6 H5 CH2 OH
MOLECULAR MASS:	108.14
PROPERTIES AND APPEARANCE:	Colourless clear liquid with faint aromatic odour
BOILING POINT IN °C AT VARIOUS PRESSURES (INCLUDING 760 mm Hg):	at 760 mm Hg: 204.7 at 400 mm Hg: 183.0 at 200 mm Hg: 160.0 at 100 mm Hg: 141.7 at 60 mm Hg: 129.3 at 40 mm Hg: 119.8 at 20 mm Hg: 105.8 at 10 mm Hg: 92.6 at 5 mm Hg: 58.0
MELTING RANGE IN °C:	–15.19
FLASH POINT IN °C:	100.5–104.4
DENSITY AT 20°C (AND OTHER TEMPERATURES) IN g/l:	@ 20°C: 1.04535; @ 25°C: 1.0456
PURITY %:	98
HEAVY METAL CONTENT MAXIMUM IN ppm:	10

ARSENIC CONTENT MAXIMUM IN ppm: 3

SOLUBILITY % AT VARIOUS TEMPERATURE/pH COMBINATIONS:

in water:	**@ 20°C**	4	**@ 50°C**	Soluble	**@ 100°C**	Soluble
in ethanol solution:	**@ 50%**	67				

FUNCTION IN FOODS: A carrying solvent for flavouring agents

TECHNOLOGY OF USE IN FOODS: Miscible with chloroform and ether. Water-white liquid with faint, aromatic odour and sharp burning taste

FOOD SAFETY ISSUES: Poisonous if ingested. Moderately toxic by inhalation, skin contact and subcutaneous routes

LEGISLATION:

USA:	**CANADA:**
Use at a level not in excess of the amount reasonably required to accomplish the intended effect. (21CFR 172.515)	GMP for flavours and flavouring preparations

REFERENCES:

Lewis, R. J. (1989) *Food Additives Handbook*. Van Nostrand Reinhold, New York.

FAO/WHO (1992) *Compendium of Food Additive Specifications Vol. I, II*. Joint Expert Committee on Food Additives. FAO Food and Nutrition Paper 52/1.

NAME:	1,3-Butanediol
CATEGORY:	Solvent
FOOD USE:	Bread/ flour/ Mixes/ Ices
SYNONYMS:	1,3-Butylene glycol/ 1,3-Dihydroxybutane/ β-Butyleneglycol/ Methyltrimethylene glycol/ Butane-1,3-diol/ CAS 107-88-0
FORMULA:	CH3 CHOH CH2 CH2 OH
MOLECULAR MASS:	90.12
PROPERTIES AND APPEARANCE:	Colourless, viscous liquid
BOILING POINT IN °C AT VARIOUS PRESSURES (INCLUDING 760 mm Hg):	207.5
MELTING RANGE IN °C:	Melts below 50
FLASH POINT IN °C:	121
DENSITY AT 20°C (AND OTHER TEMPERATURES) IN g/l:	1.006
VAPOUR PRESSURE AT VARIOUS TEMPERATURES IN mm Hg	0.06 mm Hg @ 20°C
PURITY %:	99
WATER CONTENT MAXIMUM IN %:	0.5
HEAVY METAL CONTENT MAXIMUM IN ppm:	10
ARSENIC CONTENT MAXIMUM IN ppm:	3

SOLUBILITY % AT VARIOUS TEMPERATURE/pH COMBINATIONS:								
in water:	**@ 20°C**	Miscible	**@ 50°C**	Miscible	**@ 100°C**	Miscible		
in vegetable oil:	**@ 20°C**	Soluble	**@ 50°C**	Soluble	**@ 100°C**	Soluble		
in sucrose solution:	**@ 10%**	Miscible	**@ 40%**	Miscible	**@ 60%**	Miscible		
in sodium chloride solution:	**@ 5%**	Miscible	**@ 10%**	Miscible	**@ 15%**	Miscible		
in ethanol solution:	**@ 5%**	Soluble	**@ 20%**	Soluble	**@ 95%**	Soluble	**@ 100%**	Soluble

FUNCTION IN FOODS: Used as a solvent for colours, flavours and flavour enhancers

ALTERNATIVES: Glycerol/ Propylene glycol/ Sorbitol

TECHNOLOGY OF USE IN FOODS: Very hygroscopic; keep sealed. Miscible with ether and acetone. Soluble in ether

FOOD SAFETY ISSUES: Not a food additive that has been subject to criticism.

LEGISLATION:

USA:	**CANADA:**
May by used as a solvent for natural and synthetic flavourings if the substance is used in the minimum amount required to perform its intended effect	GMP in flavours

NAME:	Castor oil
CATEGORY:	Solvent, release agent
FOOD USE:	Butter/ Margarine/ Hard candy/ Vitamin and mineral tablets
SYNONYMS:	Ricinus oil/ Oil of Palma Christi/ Tangantangan oil/ Neoloid/ INS 1503/ CAS 8001-79-4/ Phorbyol/ Aromatic castor oil
FORMULA:	N/A
PROPERTIES AND APPEARANCE:	Pale yellow or almost colourless clear liquid
BOILING POINT IN °C AT VARIOUS PRESSURES (INCLUDING 760 mm Hg):	313F (156°C)
MELTING RANGE IN °C:	–12
FLASH POINT IN °C:	230
DENSITY AT 20°C (AND OTHER TEMPERATURES) IN g/l:	0.952–0.966
HEAVY METAL CONTENT MAXIMUM IN ppm:	10
ARSENIC CONTENT MAXIMUM IN ppm:	3
SOLUBILITY % AT VARIOUS TEMPERATURE/pH COMBINATIONS:	
in vegetable oil:	**@ 20°C** Soluble in fixed oils
in ethanol solution:	**@ 95%** 100
in propylene glycol:	Slightly soluble in light petroleum, miscible in glacial acetic acid, chloroform and ether
FUNCTION IN FOODS:	Carrier solvent for colouring agents: oil-soluble annatto butter colour, annatto butter colour, annatto margarine colour. Also used as a release agent and anti-sticking agent in the manufacture of confections

TECHNOLOGY OF USE IN FOODS:	Excellent stability; does not turn rancid unless exposed to excessive heat	
FOOD SAFETY ISSUES:	Not subject to concerns. Moderately toxic by ingestion. Also an allergen	
LEGISLATION:	**USA:**	**CANADA:**
	Max 500 ppm in hard candy	GMP as solvent for annatto colourings
ANY OTHER RELEVANT INFORMATION:	Composed of 87% ricinoleic, 7% oleic, 3% linoleic, 2% palmitic, 1% stearic fatty acids	

NAME:	Diethyl tartrate	
CATEGORY:	Solvent	
FOOD USE:	Unspecified	
SYNONYMS:	2,3-Dihydroxy butanedioic acid diethyl ester/ CAS 87-91-2/ Ethyl tartrate	
FORMULA:	C2H5-COO-CHOH-CHOH-COO-C2H5	
MOLECULAR MASS:	206.18	
PROPERTIES AND APPEARANCE:	Colourless, thick oily liquid with wine-like odour	
BOILING POINT IN °C AT VARIOUS PRESSURES (INCLUDING 760 mm Hg):	280 (at 11 mm Hg: 150°C)	
MELTING RANGE IN °C:	17	
DENSITY AT 20°C (AND OTHER TEMPERATURES) IN g/l:	1.204–1.207	
PURITY %:	99	
HEAVY METAL CONTENT MAXIMUM IN ppm:	10	
ARSENIC CONTENT MAXIMUM IN ppm:	1	
SOLUBILITY % AT VARIOUS TEMPERATURE/pH COMBINATIONS:		
in water:	@ 20°C	Slightly soluble in water
in ethanol solution:	@ 5%	Miscible with alcohol
in propylene glycol:	@ 100°C	Soluble in fixed oils, ether
FUNCTION IN FOODS:	Carrier solvent for flavouring agents	

TECHNOLOGY OF USE IN FOODS:	Slightly soluble in water. Miscible with alcohol or ether
FOOD SAFETY ISSUES:	Not subject to concern
LEGISLATION:	**CANADA:** Not listed as solvent

NAME:	Ethanol
CATEGORY:	Solvent
FOOD USE:	Baked goods/ Beverages/ Confections/ Ice-cream/ Liquors/ Sauces/ Sprayable vegetable oils/ Pizza crust
SYNONYMS:	Ethyl alcohol/ Ethyl hydroxide/ Absolute ethanol/ Alcohol/ CAS 64-17-5/ Grain alcohol/ Ethyl hydrate/ Spirits of wine, spirit
FORMULA:	CH3 CH2 OH
MOLECULAR MASS:	46.07
PROPERTIES AND APPEARANCE:	Clear colourless liquid
BOILING POINT IN °C AT VARIOUS PRESSURES (INCLUDING 760 mm Hg):	78.5
FLASH POINT IN °C:	13
DENSITY AT 20°C (AND OTHER TEMPERATURES) IN g/l:	0.789
VAPOUR PRESSURE AT VARIOUS TEMPERATURES IN mm Hg:	40 mm Hg at 20°C
PURITY %:	94.9% for 95% ethanol
HEAVY METAL CONTENT MAXIMUM IN ppm:	1

SOLUBILITY % AT VARIOUS TEMPERATURE/pH COMBINATIONS:		
in water:	**@ 20°C**	Miscible
in sucrose solution:	**@ 10%**	Miscible
in sodium chloride solution:	**@ 5%**	Miscible
in propylene glycol:	**@ 20°C**	Miscible

FUNCTION IN FOODS: Carrier solvent for additives. Also used as extraction solvent. Antimicrobial agent

ALTERNATIVES: Isopropanol; methanol; *n*-octyl alcohol

TECHNOLOGY OF USE IN FOODS: Available in different concentrations: 99.9%, 95%, etc. Absorbs water rapidly from air. Keep tightly closed in cool dark place. Flammable when exposed to flame, can react vigorously with oxidisers

FOOD SAFETY ISSUES: Not subject to safety concerns

LEGISLATION:

USA:	UK and EUROPE:	CANADA:
GRAS – Limit of 2% in pizza crusts. Used in accordance with GMP	Finland and UK: GMP in soft drinks.Belgium and Sweden: GMP for biscuits	GMP in spice extracts, natural extracts, unstandardised flavouring preparations, colour mixtures and preparations, meat- and egg-marking inks, food additive preparations and hop extracts

NAME:	**Ethyl acetate**
CATEGORY:	Solvent
FOOD USE:	Breads/ Confections/ Flour/ Ices/ Mixes/ Powders/ Decaffeination of coffee, tea/ Fruits/ Vegetables
SYNONYMS:	Acetic acid ethyl ester/ Acetic ether/ Vinegar naphtha/ CAS 141-78-6/ Ethyl ethanoate/ Acetoxyethane
FORMULA:	CH3-COOC2H5
MOLECULAR MASS:	88.11
PROPERTIES AND APPEARANCE:	Clear volatile liquid with fruity odour
BOILING POINT IN °C AT VARIOUS PRESSURES (INCLUDING 760 mm Hg):	77
MELTING RANGE IN °C:	–83
FLASH POINT IN °C:	7.2
DENSITY AT 20°C (AND OTHER TEMPERATURES) IN g/l:	0.897–0.907 at 20°C; 0.894–0.901 at 25°C
VAPOUR PRESSURE AT VARIOUS TEMPERATURES IN mm Hg:	100 mm Hg at 27°C
PURITY %:	97
SOLUBILITY % AT VARIOUS TEMPERATURE/pH COMBINATIONS:	
in water:	**@ 20°C** 10% (less soluble at higher temperatures) **@ 50°C** Miscible in water at 54°C
in ethanol solution:	**@ 5%** Miscible in alcohol
FUNCTION IN FOODS:	Carrier solvent for spice extracts and flavouring agents. Used for decaffeination of coffee beans and tea

TECHNOLOGY OF USE IN FOODS:	Keep tightly closed in a cool place. Absorbs water. Highly flammable liquids, do not expose to heat and flame		
FOOD SAFETY ISSUES:	Not an additive that has been subject to criticism. Mildly toxic if ingested		
LEGISLATION:	**USA:**	**UK and EUROPE:**	**CANADA:**
	GMP in coffee as a solvent – FDA in decaffeination of coffee/tea	UK: GMP in soft drinks. Sweden: GMP in biscuits	GMP in flavour extracts. Roasted decaffeinated coffee beans: 10 ppm; tea: 50 ppm

NAME:	**Glycerin**
CATEGORY:	Solvent
FOOD USE:	Alcoholic beverages/ Baked goods/ Cured meats/ Egg products/ Ices/ Soft drinks/ Candy/ Marshmallows
SYNONYMS:	Glycerol/ 1,2,3-Propanetriol/ Glycerine/ Trihydroxypropane/ CAS 56-81-5/ E422/ INS 422/ Glycylalcohol
FORMULA:	CH2-OH-CHOH-CH2OH
MOLECULAR MASS:	92.09
PROPERTIES AND APPEARANCE:	Clear, or yellowish hygroscopic viscous liquid with slight odour and slightly sweet taste
BOILING POINT IN °C AT VARIOUS PRESSURES (INCLUDING 760 mm Hg):	290.0 at 760 mm Hg 263 at 400 mm Hg 240 at 200 mm Hg 208 at 60 mm Hg 182.2 at 20 mm Hg 167.2 at 10 mm Hg 153.8 at 5 mm Hg 125.5 at 1 mm Hg
MELTING RANGE IN °C:	17.8
FLASH POINT IN °C:	176
DENSITY AT 20°C (AND OTHER TEMPERATURES) IN g/l:	of 95% aqueous solution: / of 90% aqueous solution: / of 80%: / 50%: at 15°C 1.2570 / 1.23950 / 1.213 / 1.129 at 20°C 1.26362 / 1.23755 at 25°C 1.26201 / 1.23585
WATER CONTENT MAXIMUM IN %:	5

HEAVY METAL CONTENT MAXIMUM IN ppm:	5			
ARSENIC CONTENT MAXIMUM IN ppm:	3			
SOLUBILITY % AT VARIOUS TEMPERATURE/pH COMBINATIONS:				
in water:	**@ 20°C** Miscible in water			
in ethanol solution:	**@ 5%** Miscible in ethanol			
FUNCTION IN FOODS:	Carrier solvent for flavour extracts, colour mixes. Also used as a humectant in meats (sausage casings), as a glaze for preserved meats and as a sweetener			
ALTERNATIVES:	Butylene glycol; sorbitol; propylene glycol			
TECHNOLOGY OF USE IN FOODS:	Miscible with water and alcohol. 0.6 times sweeter than sucrose			
FOOD SAFETY ISSUES:	This is not an additive that has been subject to concerns regarding safety			
LEGISLATION:	**USA:**	**UK and EUROPE:**	**CANADA:**	**AUSTRALIA/ PACIFIC RIM:**
	GMP in alcoholic beverages, cured meats, and soft drinks. Allowed without limit in flour and baked goods but prohibited in bread. GRAS when used in accordance with GMP	Italy: 7000 ppm in alcoholic beverages: prohibited in egg products, ices, soft drinks, flour and bread UK: Allowed without limit in flour and baked goods but is prohibited in breads. GMP in cured meats, ices, soft drinks. Denmark: GMP in flour; prohibited in bread, cured meats, egg products, and alcoholic beverages; 5% in baked goods; 2% in ices;	GMP in baked goods, cured meats; prohibited in alcoholic beverages, ices, soft drinks, flour and bread	Japan: GMP in alcoholic beverages, egg products, ices, bread and baked goods. Allowed without limit in soft drinks, tea, coffee and chocolate drinks. Australia: Prohibited in flour and bread; GMP in egg products, ices and soft drinks

(*conts*)

(*UK and EUROPE: contd*)
5 g/l in soft drinks.
Finland: 5% in flour and baked goods; prohibited in breads, alcoholic beverages; GMP in cured meats.
France: 5% in baked goods; prohibited in alcoholic beverages, cured meats, ices, soft drinks, bread and flour.
Ireland: GMP in bread, baked goods, cake mixes; prohibited in flour. GMP in cured meats.
Germany: allowed without limit in flour, bread and baked goods.
Prohibited in alcoholic beverages, cured meats, ices.
GMP in soft drinks.
Spain: allowed without limit with special permission in baked goods.
Prohibited in ices, soft drinks.
Belgium, Luxembourg, the Netherlands: 1.5% in ices.
Sweden: 100 ppm in ices; GMP in soft drinks, cured meat casings. Prohibited in egg products, flour and bread

NAME:	Isopropanol
CATEGORY:	Solvent
FOOD USE:	Baked goods/ Beverages/ Spices/ Beet sugar/ Confectionery/ Food supplements in tablet form/ Gum/ Hops extract/ Lemon oil/ Yeast
SYNONYMS:	Isopropyl alcohol/ 2-Propanol/ Dimethylcarbinol
FORMULA:	CH3 CHOH CH3
MOLECULAR MASS:	60.10
PROPERTIES AND APPEARANCE:	Clear, colourless liquid
BOILING POINT IN °C AT VARIOUS PRESSURES (INCLUDING 760 mm Hg):	at 760 mm Hg: 82.5 at 60 mm Hg: 30.5 at 400 mm Hg: 67.8 at 40 mm Hg: 23.8 at 200 mm Hg: 53.0 at 20 mm Hg: 12.7 at 100 mm Hg: 39.5 at 10 mm Hg: 2.4
MELTING RANGE IN °C:	−88.5 to −89.5
FLASH POINT IN °C:	11.7
DENSITY AT 20°C (AND OTHER TEMPERATURES) IN g/l:	0.78505 at 20°C 0.78084 at 25°C 0.728 at 83°C
PURITY %:	99.5
WATER CONTENT MAXIMUM IN %:	<0.2
HEAVY METAL CONTENT MAXIMUM IN ppm:	1
ARSENIC CONTENT MAXIMUM IN ppm:	3

SOLUBILITY % AT VARIOUS TEMPERATURE/pH COMBINATIONS:				
in water:	**@ 20°C** Miscible			
in sodium chloride solution:	**@ 5%** Insoluble			
in ethanol solution:	**@ 5%** Miscible			
FUNCTION IN FOODS:	Carrier solvent for additives and inks			
ALTERNATIVES:	Ethanol/ methanol/ *n*-octyl alcohol			
TECHNOLOGY OF USE IN FOODS:	Miscible with water, alcohol, ether. Insoluble in salt solutions. Very flammable liquid, reacts with air to form dangerous peroxides			
FOOD SAFETY ISSUES:	Not subject to concern			
LEGISLATION:	**USA:**	**UK and EUROPE:**	**CANADA:**	**AUSTRALIA/ PACIFIC RIM:**
	50 ppm residue in spice oleoresins Maximum of 2.2% residue in hop extracts. Prohibited in bread. 6 ppm in lemon oil	UK: GMP in soft drinks. Norway: GMP in tea, coffee and chocolate drinks. Ireland: GMP in bread	50 ppm in spice extracts and natural extractives. GMP in flavour and meat and egg marking inks. 0.15% residue in fish protein. GMP in cured meats	Japan: GMP in bread

NAME:	Lactic acid (D, DL or L)
CATEGORY:	Solvent
FOOD USE:	Cheese and cheese products/ Ice-cream mix/ Baking powder/ Olives/ Bread/ Egg white, egg/powder/ Sherbert/ Salad dressing/ Poultry/ Margarine/ Mayonnaise/ Canned fruit/ Pickles/ Wine/ Alcoholic beverages
SYNONYMS:	2-Hydroxypropanoic acid/ α-Hydroxypropanoic acid (DL mixture)/ Paralactic acid (L form only)/ Sarcolactic acid (L form only)/ E270/ INS 270/ CAS No. 50-21-5/ L: 79-33-4; D: 10326-41-7; DL: 598-82-3/ Acetonic acid
FORMULA:	CH3-CHOH-COOH
MOLECULAR MASS:	90.08
PROPERTIES AND APPEARANCE:	Viscous, non-volatile, colourless hygroscopic liquid
BOILING POINT IN °C AT VARIOUS PRESSURES (INCLUDING 760 mm Hg):	122 at 14–15 mm Hg; 82–85 at 0.5–1 mm Hg
MELTING RANGE IN °C:	16.8
IONISATION CONSTANT AT 25°C:	0.000137
DENSITY AT 20°C (AND OTHER TEMPERATURES) IN g/l:	1.249 at 15°C
HEAT OF COMBUSTION AT 25°C IN J/kg:	3615 cal/kg
HEAVY METAL CONTENT MAXIMUM IN ppm:	10
ARSENIC CONTENT MAXIMUM IN ppm:	3

SOLUBILITY % AT VARIOUS TEMPERATURE/pH COMBINATIONS:

in water:	**@ 20°C**	Miscible in water
in ethanol solution	**@ 5%**	Miscible

FUNCTION IN FOODS: Primarily used as an acidifying agent. Also used as curing agent, flavour enhancer, flavouring agent, and pickling agent. Can be used as a solvent and carrier

TECHNOLOGY OF USE IN FOODS: Sold as 80% lactic acid in water

FOOD SAFETY ISSUES: Concern expressed regarding lethality for infants if D or DL forms used. L form is recommended for infant food use.

LEGISLATION: **USA:**
GRAS/ GMP/ Cannot be used in infant foods or formulas

ANY OTHER RELEVANT INFORMATION: FAO/WHO acceptable daily intake for D(−) isomer is 100 mg/kg body weight

NAME:	Methanol
CATEGORY:	Solvent
FOOD USE:	Spices oleoresins/ Hop extracts
SYNONYMS:	Methyl alcohol/ Carbinol/ Wood alcohol/ CAS 67-56-1/ Methyl hydroxide
FORMULA:	CH3 OH
MOLECULAR MASS:	32.04
PROPERTIES AND APPEARANCE:	Clear, colourless liquid
BOILING POINT IN °C AT VARIOUS PRESSURES (INCLUDING 760 mm Hg):	at 760 mm Hg: 64.7 at 400 mm Hg: 49.9 at 200 mm Hg: 34.8 at 100 mm Hg: 21.2 at 60 mm Hg: 12.1 at 50 mm Hg: 5.0 at 20 mm Hg: –6.0 at 10 mm Hg: –16.2 at 5 mm Hg: –25.3 at 1 mm Hg: –44
MELTING RANGE IN °C:	64.8
FLASH POINT IN °C:	12
DENSITY AT 20°C (AND OTHER TEMPERATURES) IN g/l:	at 25°C: 0.7866 at 20°C: 0.7915 at 15°C: 0.7960 at 0°C: 0.8100
VAPOUR PRESSURE AT VARIOUS	100 mm Hg at 20°C
TEMPERATURES IN mm Hg	

SOLUBILITY % AT VARIOUS TEMPERATURE/pH COMBINATIONS:			
in water:	**@ 20°C**	Miscible	
in sucrose solution:	**@ 10%**	Miscible	
in sodium chloride solution:	**@ 5%**	Miscible	
in ethanol solution:	**@ 5%**	Miscible	
in propylene glycol:	**@ 100°C**	Miscible in most organic solvents	
FUNCTION IN FOODS:	Carrier solvent for additives, labelling inks		
ALTERNATIVES:	Ethanol; propanol; *n*-octyl alcohol		
TECHNOLOGY OF USE IN FOODS:	Miscible with water, ethanol, ether. Unlike ethanol, it will dissolve many organic salts. Flammable liquid, can react vigorously with oxidising materials		
FOOD SAFETY ISSUES:	Poisoning can occur from ingestion of excess		
LEGISLATION:	**USA:**		**CANADA:**
	50 ppm maximum residue in spice oleoresins. 2.2% residue in hop extracts		50 ppm in spice extracts. 2.2% in hop extracts for use in malt liquors. GMP in meat and egg marking inks

NAME:	Mono- and diglycerides (glycerol monostearate is used as an example)	
CATEGORY:	Carrier solvent	
FOOD USE:	Annatto butter colour/ Annatto margarine colour/ Beverages/ Baked goods/ Fats/ Poultry/ Ice-cream/ Frozen desserts/ Cottage cheese/ Sausage casing/ Sherbert/ Fudge and fudge sauces/ Whipped topping/ Sour cream/ Cakes/ Caramel/ Chewing gum/ Coffee whiteners/ Peanut butter	
SYNONYMS:	INS 471/ E471	
FORMULA:	CH2OOCR-CHOH-CH2OH (CH2OOCR-CHOOCR-CH2OH)	
MOLECULAR MASS:	Monostearate 357.6; distearate 625.0	
ALTERNATIVE FORMS:	Glyceryl monopalmitate/ Glyceryl monostearin/ Glyceryl dioleate/ Glyceryl monooleate/ Glyceryl monopalmitin/ Glyceryl distearate/ Glyceryl monoolein/ Glyceryl dipalmitate/ etc.	
PROPERTIES AND APPEARANCE:	White or cream-coloured fat of waxy appearance	
MELTING RANGE IN °C:	56–58	
WATER CONTENT MAXIMUM IN %:	2	
HEAVY METAL CONTENT MAXIMUM IN ppm:	10	
ARSENIC CONTENT MAXIMUM IN ppm:	3	
SOLUBILITY % AT VARIOUS TEMPERATURE/pH COMBINATIONS:		
in water:	@ 20°C	Insoluble in water
in ethanol solution	@ 5%	Soluble in ethanol
in propylene glycol:	@ 100°C	Soluble in chloroform and benzene
FUNCTION IN FOODS:	Primarily used as emulsifier, texture modifying agent, humectant, release agent. Can be used as a solvent for antioxidants and spices in beverages. Consists of a mixture of glyceryl mono and diesters prepared from fats or oils of edible origin	

TECHNOLOGY OF USE IN FOODS:	Soluble in hot organic solvents such as alcohol		
FOOD SAFETY ISSUES:	These compounds have not been subject to concern regarding safety		
LEGISLATION:	**USA:**	**UK and EUROPE:**	**CANADA:**
	GMP – for lard and shortening. GRAS as a solvent	GMP in soft drinks	GMP in flavours, annatto butter colour

NAME:	**Octyl alcohol**
CATEGORY:	Solvent
FOOD USE:	Processing aid for extraction of citric acid/ Beverages/ Candy/ Gelatin desserts/ Ice-cream/ Pudding mixes
SYNONYMS:	Caprylic alcohol/ Alcohol-C8/ Octanol/ Heptyl carbinol/ *n*-Octanol/ l-Hydroxyoctane
FORMULA:	$C_8H_{18}O$
MOLECULAR MASS:	130.26
PROPERTIES AND APPEARANCE:	Colourless, viscous liquid
BOILING POINT IN °C AT VARIOUS PRESSURES (INCLUDING 760 mm Hg):	239.7
MELTING RANGE IN °C:	16.7
FLASH POINT IN °C:	81
DENSITY AT 20°C (AND OTHER TEMPERATURES) IN g/l:	0.910
SOLUBILITY % AT VARIOUS TEMPERATURE/pH COMBINATIONS:	
in water:	**@ 20°C** 0.068%
in ethanol solution:	**@ 5%** Miscible
FUNCTION IN FOODS:	Extraction solvent for citric acid produced during fermentation with *Aspergillus niger*. Combustible liquid if exposed to heat; flame can react with oxidising materials

TECHNOLOGY OF USE IN FOODS:	Used as a processing aid in the production of citric acid
LEGISLATION:	**USA:** Limited to 16 ppm residual in citric acid. Only used for encapsulation of essential oils

NAME:	1,2-Propanediol
CATEGORY:	Solvent
FOOD USE:	Margarine/ Baked goods/ Poultry/ Seasonings/ Beverages (alcoholic)/ Flavourings/ Wine/ Frostings/ Frozen dairy products/ Hog carcasses/ Confections/ Nut products
SYNONYMS:	Propylene glycol/ Methyl glycol/ Methylethylene glycol/ 1,2-Dihydroxypropane/ INS No. 1520/ CAS No. 57-55-6/ Trimethylglycol/ Propabe-1,2-diol
FORMULA:	$CH_3CHOHCH_2OH$
MOLECULAR MASS:	76.10
ALTERNATIVE FORMS:	L or D form available
PROPERTIES AND APPEARANCE:	Viscous, clear, colourless liquid; slight bitter taste
BOILING POINT IN °C AT VARIOUS PRESSURES (INCLUDING 760 mm Hg):	45.5 at 1.0 mm Hg
MELTING RANGE IN °C:	59
FLASH POINT IN °C:	99
DENSITY AT 20°C (AND OTHER TEMPERATURES) IN g/l:	1.0362 at 25°C
WATER CONTENT MAXIMUM IN %:	0.2
HEAVY METAL CONTENT MAXIMUM IN ppm:	10
ARSENIC CONTENT MAXIMUM IN ppm:	3

SOLUBILITY % AT VARIOUS TEMPERATURE/pH COMBINATIONS:

in water:	**@ 20°C**	Miscible
in ethanol solution:	**@ 5%**	Miscible

FUNCTION IN FOODS: Solvent for flavour extracts. Also used as an anti-caking agent in salt. Can be used as a humectant

ALTERNATIVES: Glycerol; butylene glycol; sorbitol

TECHNOLOGY OF USE IN FOODS: Miscible with water. Will dissolve many essential oils. Under normal conditions, is stable but when exposed to high temperatures, will oxidise. Combustible liquid when exposed to heat or flame can react with oxidising materials

FOOD SAFETY ISSUES: *Codex Alimentarius* acceptable daily intake is 25 mg/kg

LEGISLATION:

USA:	**UK and EUROPE:**	**CANADA:**	**AUSTRALIA/ PACIFIC RIM:**
GMP that results in maximum of 5% for alcoholic beverages; 24% for confections and frostings; 2.5% for frozen dairy products; 97% for seasonings and flavourings; 5% for nuts and nut products and 2% for all other food categories. GRAS	UK: allowed without limit in baked goods; GMP in soft drinks. Finland: 3000 ppm in baked goods; GMP in soft drinks. Sweden: GMP in baked goods. Norway: GMP in tea, coffee and chocolate drinks	GMP in flavour extracts, essence, oil-soluble annatto butter colour, annatto-coloured margarine; colour mixtures and preparations; and food additive preparations	Japan: GMP in baked goods; 600 ppm tea, coffee and chocolate drinks

NAME:	Triacetyl glycerin
CATEGORY:	Solvent
FOOD USE:	Baked goods and baking mixes/ Alcoholic beverages/ Fresh fruit and vegetables/ Chewing gum/ Confections and frostings/ Frozen dairy desserts/ Raisins/ Gelatins/ Puddings and fillings/ Non-alcoholic beverages/ Frostings
SYNONYMS:	Triacetin/ 1,2,3-Propanetriol triacetate/ Glycerintriacetate/ Enzactin/ Fungacetin/ INS 1518/ CAS 102-76-1
FORMULA:	C3H5(OCOCH3)3
MOLECULAR MASS:	218.21
PROPERTIES AND APPEARANCE:	Colourless, oily liquid; bitter taste
BOILING POINT IN °C AT VARIOUS PRESSURES (INCLUDING 760 mm Hg):	258–260
MELTING RANGE IN °C:	78
FLASH POINT IN °C:	138
DENSITY AT 20°C (AND OTHER TEMPERATURES) IN g/l:	at 25°C: 1.1562; at 20°C: 1.1596
PURITY %:	98.5
HEAVY METAL CONTENT MAXIMUM IN ppm:	10
ARSENIC CONTENT MAXIMUM IN ppm:	3

SOLUBILITY % AT VARIOUS TEMPERATURE/pH COMBINATIONS:

in water:	**@ 20°C**	7.14% (soluble)
in ethanol solution:	**@ 5%**	Miscible

FUNCTION IN FOODS: Solvent for additives, such as flavouring and colourings

TECHNOLOGY OF USE IN FOODS: Combustible when exposed to heat, flame and powerful oxidisers

FOOD SAFETY ISSUES: Not an additive that has been subject to concern

LEGISLATION:

USA:	**UK and EUROPE:**	**CANADA:**
GMP in baked goods, baking mixes, alcoholic beverages, non-alcoholic beverages and beverage bases, chewing gum, confections and frostings, frozen dairy desserts and mixes, gelatins, puddings and fillings, hard candy and soft candy. GRAS	Belgium: 1000ppm in egg powder. Ireland: GMP. Norway: GMP in tea, coffee and chocolate drinks. Switzerland: GMP in cake mixes. France: <3000ppm in raisins. UK: GMP in soft drinks	GMP in flavours and unstandardised flavouring preparations – cake mixes

Part 14

Sweeteners

NAME:	Acesulphame
CATEGORY:	Non-nutritive sweetener/ Flavour modifier
FOOD USE:	Baked goods (dry bases for mixes)/ Beverages (dairy beverages, instant tea, instant coffee, dry bases for mixes, fruit-based beverages)/ Soft drinks (colas, citrus-flavoured drinks, fruit-based soft drinks)/ Sugars, sugar preserves and confectionery (calorie-free dustings, frostings, icings, toppings, fillings, syrups)/ Alcoholic drinks (beer)/ Vinegar, pickles and sauces (sandwich spreads, salad dressings, pickles, sauces, toppings)/ Dairy products (yoghurt and yoghurt-type products, dry bases for puddings, desserts and dairy analogues, sugar-free ice-cream)/ Fruit, vegetables and nut products (fruit drinks, fruit products, fruit yoghurt, sugar-free jams and marmalades, low-calorie preserves)/ Other (oral cosmetics, pharmaceuticals, chewing gums, liquid concentrates, frozen and refrigerated desserts)
SYNONYMS:	As potassium salt: CFSAN = Acesulfame potassium/ CAS 5589-62-3/ Potassium acesulfame/ Acesulfame K/ Sunnette/ Acetosulfam/ Potassium 6-methyl-1,2,3-oxathiazin-4(3*H*)-1,2,2-dioxide/ HOE-095K General form: 6-Methyl-1,2,3-oxathiazin-4(3*H*)-1,2,2-dioxide/ 6-Methyl-3,4-dihydro-1,2,3-oxathiazin-4-1,2,2-dioxide/ Acetosulfam
FORMULA:	$C_4H_4NSO_4K$ (potassium salt); $C_4H_5NO_4S$ general form
MOLECULAR MASS:	163.15 (201.24 as potassium salt)
ALTERNATIVE FORMS:	Potassium salt
PROPERTIES AND APPEARANCE:	White crystalline solid. Potassium salt–colourless, odourless powder. Sweet taste (200 times sweeter than sucrose); very slight bitter/astringent aftertaste, noticeable at high concentrations. Rapid onset time; thin mouthfeel body
MELTING RANGE IN °C:	250 (potassium salt); begins to decompose at 225°C on slow heating
DENSITY AT 20°C (AND OTHER TEMPERATURES) IN g/l:	solid – 1.81 g/cm^3 bulk – 1.1 to 1.3 kg/dm^3
PURITY %:	99.0 after drying

HEAVY METAL CONTENT MAXIMUM IN ppm: 10

ARSENIC CONTENT MAXIMUM IN ppm: 3

SOLUBILITY % AT VARIOUS TEMPERATURE/pH COMBINATIONS:

in water: **@ 20°C** 31% weight/unit volume, or 360 g/l **@ 50°C** 830 g/l **@ 100°C** 100% weight/unit volume or 1300 g/l

in ethanol solution: **@ 20%** 221 g/l in 20 : 80 vol/vol ethanol/water solution at 23°C **@ 100%** 0.1% at 20°C or 1 g/l

FUNCTION IN FOODS: Non-nutritive sweetening agent (200 times sweetness of sucrose); flavour modifier

ALTERNATIVES: Aspartame (similar to acesulphame : thaumatin mixture)

TECHNOLOGY OF USE IN FOODS:

Very soluble in water, DMF, DMSO. Non-hygroscopic. Soluble in alcohol, glycerin-water. Stable for several months at pH 3 or higher; stable to pasteurisation and sterilisation at pH >3; stable to baking at temperatures >200°C.

Shelf-stable for more than 5 years in solid form; no hydrolysis of sterilised solution stored for one month at 40°C.

400–700 mg per litre produces a medium sweetness in drinks. Chemically and sensorily compatible with all sugars; best used in acidic foods; sweetness intensity not diminished in hot drinks.

No hydrolytic decomposition of stock solutions with pH >3 for several months.

When used in solution, adjust pH to 5.5 to 6.0 range using appropriate buffer system.

Can supplement sugar alcohols in sugar-free ice-cream without affecting melting and whipping properties. Use at 500 mg/kg. Can be used to mask bitter and other unpleasant taste characteristics of other products. Can be used alone in food products. Some benefits from combinations with bulk sweeteners; may be some cost advantage to combining with thaumatin with a taste result equivalent to aspartame.

Good solubility and stability in aqueous media. Sweetness in acid foods slightly higher than in neutral; heat does not cause reduction in sweetness relative to room temperature as it does in other sweeteners; can be added to liquid concentrates; can be processed in spray towers and in instant beverage powders due to its heat stability.

In combination with other sweeteners, best taste profile in soft drinks obtained when each sweetener contributes 50% of sweetness; pleasant taste results from mixtures with other high-intensity sweeteners, particularly aspartame and cyclamate.

Heat stability and good solubility aids in use in production of table-top formulations, solutions (should be adjusted to pH 5.5 to 6.0 with appropriate buffer systems) or spray-dried granular or powder preparations.

Disintegrant (e.g. low-viscosity carboxymethyl cellulose or polyvinyl pyrrolidine) required when compressing into tablets. In production of effervescent tablets, use sodium hydrogen carbonate (carbon dioxide donor), tartaric acid (acid medium) and small amounts of cold water-soluble gelatin. Result has a good shelf-life when stored in a dry place. Table-top powders can be produced by combining with inert substances or with citrates, tartrates, lactose and/or polyols. Calorie-free dustings can be produced by combining with pure cellulose.

Low-calorie preserves produced by combining with pectins and other gelling agents which provide bulk. More susceptible to microbiological spoilage than preserves containing sugar. Should be pasteurised or add 0.05 to 0.1% potassium sorbate preservative (if permitted). Add as an aqueous solution of 500 to 2500 mg/kg final product weight to aid even dispersion. Sugar-free jams and marmalades produced in combination with sorbitol. More susceptible to microbiological spoilage than preserves containing sugar. Should be pasteurised or add 0.05–0.1% potassium sorbate preservative (if permitted). Add as an aqueous solution of 500–2500 mg/kg final product weight to aid even dispersion. At a level of 1000–3000 mg/kg final product weight can be used to replace sugar in confectionery due to good heat stability. Combine with polydextrose, disaccharide alcohols, sorbitol or isomalt to provide bulk. At a level of 500–2000 mg/kg final product weight can be used to replace sugar in bakery products due to good heat stability. Combine with polydextrose, disaccharide alcohols, sorbitol or isomalt to provide bulk.

At a level of 500–600 mg/kg final product weight can replace sugar in desserts. At a level of 500–3000 mg/kg final product weight can replace sugar in chewing gum; 500 mg/kg final product weight may be added to sugar-free ice-cream to supplement polyols to achieve a well-balanced taste. Does not affect melting and whipping properties of the mix. Can be used to sweeten pharmaceuticals and oral hygiene products as it masks bitter and other unpleasant tastes.

At room temperature, 0.1% soluble in ethanol (0.1 g/100 ml), and more than 30% soluble in DMSO.

Solubility in water:

- 14 g/100 ml at 0°C
- 27 g/100 ml at 20°C (31%)
- 130 g/100 ml at 100°C (100%)

No browning reaction. Solubility in organic solvents poor; solubility increases in solvent : water mixtures.

Sweetness potency relative to sucrose decreases with increasing concentration; sweetness potency relative to sucrose varies with the medium in which the sweetener is being tested and the method used for quantifying sweetness; values range from 110 at 10% sucrose equivalence to 200 at 3% sucrose equivalence; taste profile considered to be superior to that of saccharin; sugar alcohols, maltol and ethyl maltol can be used to mask aftertaste.

Using acesulphame only: 600–800 mg/l appropriate for cola soft drinks; 550–750 mg/l appropriate for citrus-flavoured soft drinks. Blending aspartame gives a more acceptable soft-drink product.

Using 50 : 50 combination of acesulphame : aspartame: 160–170 mg/l appropriate for cola soft drinks; 140–150 mg/l appropriate for citrus soft drinks.

Appears to be non-reactive with other soft-drink ingredients. Adds potassium ions to beverage mixes, so care must be taken when selecting clouding agents and stabilisers.

HPLC may be used for quantitative analysis with detection in the UV range; quantitative analysis may be performed using thin-layer chromatography; methods using isotachophoretic techniques can be used to detect acesulphame-K, saccharin and cyclamate simultaneously; UV absorption in water: maximum 227 nm; fluorine: not more than 30 mg/kg.

Commercially successful combination of sucrose/acesulphame/aspartame – at 270 : 1.5 : 1.0 or 40% : 30% : 30% sweetness gives good sweetness at 33% of calories from sucrose alone. Used in blackcurrant, where fruitiness enhancement is important. Good stability in fruit-based soft drinks.

Solubiluty in water:
- 0°C = 150 g/l
- 10°C = 210 g/l
- 20°C = 270 g/l
- 30°C = 360 g/l
- 40°C = 460 g/l
- 50°C = 580 g/l
- 70°C = 830 g/l
- 100°C = 1300 g/l

Solubility at 20°C:
- Methanol = 10 g/l
- anhydrous ethanol = 1 g/l
- anhydrous glycerol = 30 g/l
- 80 : 20 glycerol : water (v/v) = 82 g/l
- 50 : 50 glycerol : water (v/v) = 162 g/l
- acetone = 0.8 g/l
- glacial acetic acid = 130 g/l

Solubility at 23°C:
- 80 : 20 ethanol : water (v/v) = 46 g/l
- 60 : 40 ethanol : water (v/v) = 100 g/l
- 40 : 60 ethanol : water (v/v) = 155 g/l
- 20 : 80 ethanol : water (v/v) = 221 g/l

SYNERGISTS: Sorbitol (1 : 150–200); sucrose (1 : 100–150); isomalt (1 : 250–300) – said to round up sorbitol's sweet taste; maltitol (1 : 150); up to 30% increase in sweetness intensity with cyclamate and aspartame; aspartame (1 : 1); sodium cyclamate (1 : 5); also with fructose, thaumatin; barely noticeable improvement in combination with saccharin

FOOD SAFETY ISSUES: Emits toxic fumes of SOx when heated to decomposition. Non-mutagenic. No adverse reactions have been found; not metabolised in the body, has no calorific value and suitable for diabetics; excreted rapidly and completely. Non-cariogenic with acute oral toxicity being extremely low. Metabolised by few microorganisms. Decomposes to acetoacetamide under certain conditions, but both acesulphame and acetoacetamide were found to be non-toxic

LEGISLATION:

USA:

FDA approved for ADI of 15 mg/kg body weight. Expected to be sufficient to allow almost complete replacement of all sugar in the diet of the average person.

Approved by US FDA in 1988 for use in dry beverage mixes, instant coffee, tea, gelatins, puddings, nondairy creamers, and chewing gum

CFR 21

PART 172–Food Additives Permitted for Direct Addition to Food for Human Consumption Subpart I – Multipurpose Additives

172.800 – Acesulfame potassium.

[53 FR 28382, July 28, 1988, as amended at 57 FR 57961, Dec. 8, 1992; 59 FR 61540, 61543, 61545, Dec. 1, 1994]

Acesulfame potassium (CAS Reg. No. 55589-62-3), also known as acesulfame K, may be safely used as a sweetening agent in food in accordance with the following prescribed conditions:

(a) Acesulfame potassium is the potassium salt of 6-methyl-1,2,3-oxathiazine-4(3*H*)-one-2,2-dioxide.

(b) The additive meets the following specifications:

(1) Purity is not less than 99% on a dry basis. The purity shall be determined by a method titled "Acesulfame Potassium Assay" (see reference). Copies are available from the Division of Food and Color Additives, Center for Food Safety and Applied Nutrition (HFF-330), Food and Drug Administration, 200 C St. SW., Washington, DC 20204, or available for inspection at the Office of the FEDERAL REGISTER, 800 North Capitol Street, NW., suite 700, Washington, DC 20408.

(2) Fluoride content is not more than 30 parts per million, as determined by method III of the Fluoride Limit Test of the *Food Chemicals Codex*, 3d edn (1981), p. 511 (see reference). Copies are available from the National Academy Press, 2101 Constitution Ave. NW., Washington, DC 20418, or available for inspection at the Office of the Federal Register, 800 North Capitol Street, NW., suite 700, Washington, DC 20408.

(c) The additive may be used in the following foods when standards of identity established under section 401 of the Federal Food, Drug, and Cosmetic Act do not preclude such use:

(1) Sugar substitute, including granulated, powdered, liquid, and tablet form.

(2) [Reserved]

(3) Chewing gum.

(4) Dry bases for beverages, instant coffee, and instant tea.

(5) Dry bases for gelatins, puddings, and pudding desserts.
(6) Dry bases for dairy product analogues.
(7) Confections, hard candy, and soft candy.
(8) Baked goods and baking mixes, including frostings, icings, toppings, and fillings for baked goods.
(9) Yoghurt and yoghurt-type products.
(10) Frozen and refrigerated desserts.
(11) Sweet sauces, toppings, and syrups.

(d) If the food containing the additive is represented to be for special dietary uses, it shall be labelled in compliance with part 105 of this chapter.

(e) The additive shall be used in accordance with current good manufacturing practice in an amount not to exceed that reasonably required to accomplish the intended effect.

UK and EUROPE:

FACC: 1983 (in UK)
JECFA assigned ADI of 0 to 9 mg/kg body weight in 1983
JECFA found was not mutagenic, carcinogenic, or any other toxicological properties
Approved for general use in Germany, Russia, Belgium and Denmark
Approved for use in toothpaste in Bulgaria and USSR
Regulatory Status: ADI level as of 1990 for use in soft drinks

– Belgium	600 mg/l
– Denmark	250 mg/l
– Finland	Not permitted
– France	360 mg/l
– East Germany	Not permitted
– West Germany	Permitted
– Greece	Permitted
– Ireland	Permitted
– Netherlands	600 mg/l (pending approval)
– Norway	Not permitted
– Spain	Not permitted
– Switzerland	Permitted
– Turkey	Permitted
– UK	Permitted

– USSR	Not permitted
– Yugoslavia	Not permitted

CANADA:

Canada CFR 67.30 – Table IX – Food additives that may be used as Sweeteners – Item A.01 (6 October 1994)

Permitted in or on	**Maximum level of use**
(1) Table-top sweeteners	Good Manufacturing Practice
(2) Carbonated beverages	0.025% in beverages as consumed
(3) Beverages; beverage concentrates, beverage mixes; dairy beverages; (except for any of these products for which standards are set out in these Regulations)	0.05% in beverages as consumed
(4) Desserts; dessert mixes; toppings; fillings; filling mixes; (except for any of these products for which standards are set out in these Regulations)	0.1% in products as consumed
(5) Chewing gum; breath-freshener products	0.35%
(6) Fruit spreads (except for any of these products for which standards are set out in these Regulations)	0.1%
(7) Salad dressings (except for any of these products for which standards are set out in these Regulations)	0.05%
(8) Confectionery	0.25%
(9) Bakery mixes; bakery products; (except for any of these products for which standards are set out in these Regulations)	0.1% in products as consumed

AUSTRALIA/PACIFIC RIM:

Approved for use in Australia
Regulatory Status: ADI level as of 1990 for use in soft drinks
– Australia: 3000 mg/kg
– Japan: Not permitted
– New Zealand: Pending approval

OTHER COUNTRIES:

Accepted as safe by WHO/FAO with an ADI of 0 to 9 mg/kg body weight
Approved for use in South Africa, Cyprus, and Egypt

No approval required for use in many countries because of its demonstrated safety.
Regulatory Status: ADI level as of 1990 for use in soft drinks

- Argentina: not permitted
- Brazil: 600 mg/l
- Israel: not permitted
- Kenya: not permitted
- Mexico: not permitted
- Saudi Arabia: not permitted
- South Africa: 1000 mg/l

REFERENCES:

Smith, J. (1991) *Food Additive User's Handbook.* Blackie Publishing, Glasgow.

Canadian Food and Drugs Act and Regulations (1994).

Mitchell, A. J. (1990) *Formulation and Production of Carbonated Soft Drinks.* Blackie Publishing, Glasgow.

McCue, N. (1996) *Showcase: Natural & High Intensity Sweeteners; Prepared Foods.* March, p. 84.

Chilton's Food Engineering Master '95. (1994) ABC Publishing Group. p. 10; 288–289.

New and Novel Foods; 484–485.

Minifie, B. W. (1989) *Chocolate, Cocoa, and Confectionery: Science and Technology.* 3rd edition. Van Nostrand Reinhold.

Matz, S. A. (1992) *Bakery Technology and Engineering.* 3rd edition. Van Nostrand Reinhold.

Food Chemistry. 2nd edition. (1985) Marcel Dekker, Inc.

Hicks, D. (1989) *Production and Packaging of Non-Carbonated Fruit Juices and Fruit Beverages.* Blackie Publishing, Glasgow.

Smoley, C. K. (1993) *Everything Added to Food in the United States.* US Food and Drug Administration, CRC Press, Inc.

Windholz, M., and Budavari, S. (1988) *Merck Index*, 10th edition. Merck and Co.

Ash, M., and Ash, I. (1995) *Food Additives, Electronic Handbook.* Gower.

Solutions CFR+ Database (1996).

Budavari, S. (1996) *The Merck Index.* 12th edition. Merck and Co.

Wong, D. W. S. (1989) *Mechanism and Theory in Food Chemistry.* Van Nostrand Reinhold.

ANY OTHER RELEVANT INFORMATION:

A potassium salt derived from acetoacetic acid. Discovered by accident in 1967 by employees of Hoechst AG. Currently marketed by Hoechst as Sunnett®

NAME:	**N-acetylglucosamine**
CATEGORY:	Sweetener
SYNONYMS:	CAS 7512-17-6/ EINECS 231-368-2
FORMULA:	$C_8H_{15}NO_6$
MOLECULAR MASS:	221.21
MELTING RANGE IN °C:	215
LEGISLATION:	**AUSTRALIA/PACIFIC RIM:** Japan: approved
REFERENCES:	Smoley, C. K. (1993) *Everything Added to Food in the United States.* U.S. Food and Drug Administration, CRC Press, Inc. Ash, M., and Ash, I. (1995) *Food Additives, Electronic Handbook.* Gower.

NAME:	**Arabinose**
CATEGORY:	Sweetener
SYNONYMS:	CFSAN L-Arabinose/ CAS 5328-37-0/ Pectin sugar
FORMULA:	$C_5H_{10}O_5$
MOLECULAR MASS:	150.13
MELTING RANGE IN °C:	157–160
SOLUBILITY % AT VARIOUS TEMPERATURE/pH COMBINATIONS:	
in water:	**@ 20°C** 1 g/ml
FUNCTION IN FOODS:	Used as a culture medium for some bacteria. Application in foods under investigation
LEGISLATION:	**AUSTRALIA/PACIFIC RIM:** Japan: approved
REFERENCES:	Smoley, C. K. (1993) *Everything Added to Food in the United States.* U.S. Food and Drug Administration, CRC Press, Inc. Ash, M., and Ash, I. (1995) *Food Additives, Electronic Handbook.* Gower. Budavari, S. (1996) *The Merck Index.* 12th edition. Merck and Co.
ANY OTHER RELEVANT INFORMATION:	Natural sweetener; can be found in plants in the form of a complex polysaccharide; can be found in mycobacteria

NAME:	Aspartame
CATEGORY:	Sweetener/ Nutritive additive/ Flavour enhancer
FOOD USE:	Baked goods (desserts, dessert toppings, dessert mixes, topping mixes, fillings, filling mixes)/ Bakery products (encapsulated aspartame)/ Bakery mixes (encapsulated aspartame)/ Dairy products (dry mixes for dairy products, frozen desserts)/ Beverages (beverage concentrates, tea beverages, beverage mixes)/ Cereals and cereal products/ Soft drinks/ Sugars, sugar preserves and confectionery (table use, confectionery, sugar-free syrups, jams and jellies, purées and sauces, confectionery glazes)/ Other (pharmaceutical tablets, chewable vitamins, chewing gum, emulsions, breath mints, breath-freshener products)/ Vinegar, pickles and sauces (salad dressings, condiments)/ Fruit, vegetable and nut products (peanut and other nut spreads)
SYNONYMS:	CFSAN Aspartame/ CAS 22839-47-0/ EINECS 245-261-3/ Aspartylphenylalanine methyl ester/ *N*-L-α-aspartyl-L-phenylalanine-1-methyl ester/ 3-Amino-*N*-(α-carboxyphenethyl)succinamic acid *N*-methyl ester/ SC 18862/ Tri-Sweet/ APM/ Usal/ Cauderal/ Pouss-Suc/ HSC Aspartame/ Canderel/ Equal/ NutraSweet/ Sanecta
FORMULA:	HOOC CH2 CH(NH2)CONH CH(CH2 C6H5)COO CH3
MOLECULAR MASS:	294.31
ALTERNATIVE FORMS:	α-(L)-Asp-(L)-Phe-OMe/ α-(L)-Asp-(L)-Met-OMe/ α-(L)-Asp-(L)-Phe-OEt[b]/ D,L-Ama-(L)-Phe-OMe[c]/ α-(L)-Asp-(L)-Tyr-OMe
PROPERTIES AND APPEARANCE:	White crystalline powder. Odourless. Clean sweet, agreeable taste (160 to 200 times sucrose); prolonged sweet aftertaste; no bitter or metallic aftertaste; fair mouthfeel
MELTING RANGE IN °C:	246–248
PURITY %:	98–102% assay

SOLUBILITY % AT VARIOUS TEMPERATURE/pH COMBINATIONS:		
in water:	**@ 20°C**	38% (pH 4.5–6.0 in 0.8% aqueous solution) 1.0 g/l
in vegetable oil:	Insoluble	
in ethanol solution:	**@ 100%**	0.4% at 25°C
FUNCTION IN FOODS:	Intense sweetener (160 to 250 times sweetness of sucrose); flavour enhancer, particularly in citrus drinks	

TECHNOLOGY OF USE IN FOODS:

Unstable in aqueous solution. 50% degraded after 36 days. 50–60% degraded at pH 3.5 after 36 days. 100% hydrolysed in pH 7.4 in 9 days. 5-year shelf-life in tightly closed container with sealed inner bag under cool, dry conditions.

Do not use in canned foods or products which will be baked or roasted. Can be used in acidified liquids and high-acid carbonated beverages. Not soluble in fats and oils.

Does not act the same as sucrose, so should not be used as a simple substitute; unstable under certain conditions; mixtures result in improved processing and shelf stability, and producing a balanced taste; blends well with other food flavours; interacts with other flavours differently than does sucrose so should not be used as a simple sucrose substitute; has flavour-enhancing properties, especially with citrus fruit drinks; overall acceptability of certain carbonated soft drinks remains high over a range of concentrations.

Can be used in HTST (high-temperature/short time) processes, allowing it to be used in dairy products and baking using this method. Not soluble in fats and oils. At pH 3.5 after 36 days, 50–60% degraded. At pH 7.4 after 9 days, fully hydrolysed. No browning reaction.

Solubility is adequate for most food applications; solubility increases in acid conditions and with increasing temperatures; sparingly soluble in solvents.

Has a similar taste profile to sucrose (one of the main reasons for its success); relative sweetness 180 at 10% sucrose equivalence, level most often used in soft drink preparations.

Flavour enhancement with fruit flavours, notably natural flavours. As sole sweetener: 500–600 mg/l appropriate for cola beverages; 400–600 mg/l appropriate for lemonade beverages.

Degrades in solution: hydrolysis of ester bond produces aspartyl-L-phenylalanine and methanol; at pH 5 and above, degrades to diketopiperazine (DKP) and methanol; then DKP hydrolyses to aspartyl-L-phenylalanine which can hydrolyse to aspartic acid and phenylalanine; pH, temperature, moisture and time dictate rate of decomposition; optimum pH range is 3.0 to 5.0 with maximum stability at pH 4.3.

Minimally affected by UHT aseptic processes; typical losses in the range of 0.5 to 5%.

More stable in ready-to-drink products than in post-mix or fountain syrups due to lower pH in concentrates. As aspartame concentration decreases, relative sweetness of the product increases; up to 40% can be lost before the product becomes unacceptable.

Stable for several years when properly stored in dry form, so ideal for powdered soft drinks; research into making aspartame more stable has resulted in several patents, most involving co-drying with acidulants and/or bulking agents or encapsulating the product. None is applicable to liquid situations.

When used in soft drinks, qualitative and quantitative spectroscopic analyses can be made through amino acid detection based on its reaction with ninhydrin; presence in soft drinks can be detected using HPLC methods, which may also allow for simultaneous detection of other product constituents; presence in soft drinks can be detected using a non-chromatographic method based on non-aqueous perchloric acid titration.

Granular form reduces dusting and increases flowability; encapsulated form recommended for baked goods.

Liquid provides quick dissolution and handling ease; cannot be used in canned or fried foods due to low thermal stability; can be thermally stabilised somewhat through encapsulation.

Slow to hydrolyse; susceptible to hydrolysis, other chemical interactions, and microbial degradation in aqueous system. At alkaline pH, solution reacts readily with vanillin, resulting in a loss of vanilla flavour.

Commercially successful formulations are:

- Aspartame
- Aspartame : saccharin – at ratio of 2 : 1 or 50% : 50% sweetness; gives good sweetness and saccharin stabilises sweetness to provide longer shelf-life
- Sucrose : acesulfame : aspartame – at ratio of 270 : 1.5 : 1.0 or 40% : 30% : 30% sweetness gives good sweetness and 33% of calories of using sucrose alone; used in blackcurrant, where fruitiness enhancement is important.

Solubility increases with decrease in pH.

Storage stability affected by instability in high-acid/low-pH beverages; cannot be used in in-pack pasteurised products; compatible with most flavours, and enhances fruit flavours.

SYNERGISTS:

Saccharin, acesulphame, cyclamates, sucrose, glucose, isomalt (isomalt steuroside), stevioside.

With saccharin, produces sweeter tastes than either alone; cost reduction by mixing with acesulfame-K, sodium saccharin, sodium cyclamate, glucose or sucrose: Chewing gum: sodium saccharinate-aspartame (1 : 2); cola drinks: sodium saccharin-aspartame (1 : 1); orange juice: aspartame-glycosyl stevioside (1–10 : 1); dry-mix: 1 g aspartame to 50 g sorbose; table-top sweetener: sodium saccharinate-aspartame (4 : 2 mg); table-top sweetener: sodium saccharinate-aspartame-sodium cyclamate (4 : 10 : 30 mg); sucrose substitute for diabetics: xylitol (44.184 g)-sorbitol (179 g)-aspartame (0.35 g).

Aspartame-cyclamate (1–2 : 6–1) and acesulfame K-aspartame (1 : 1) are more stable, improve flavour of sweeteners.

Cost can be reduced using mixtures with acesulfame K, sodium saccharin, sodium cyclamate, glucose or sucrose as they are synergistic.

Isomalt is synergistic with aspartame as it masks the metallic aftertaste.

ANTAGONISTS:

Conditions of pH, temperature and moisture cause decomposition of aspartame, resulting in loss of flavour. Loss of sweetness may result on prolonged exposure to high temperatures; loses flavour in neutral solutions such as dairy products. Sweetness loss from hydrolysis. In alkaline pH solutions, reacts readily with glucose to reduce aspartame's sweetness

FOOD SAFETY ISSUES:

Allergic dermatitis by ingestion (human systemic). Possible link to neural problems. Headaches, dizziness to those sensitive to chemicals.

Emits toxic fumes of NOx when heated to decomposition.

Not to be used by individuals with phenylketonuria (PKU) as it contains an amino acid the intake of which must be limited in these individuals. Label must state: "Phenylketonurics: contains phenylalanine".

One of the most thoroughly tested food additives; toxicity of components (amino acids, aspartic acid, phenylalanine) and metabolite (methanol) is dose-related. Expected levels of consumption are not expected to pose a risk.

Metabolised by the body into methanol and two amino acids, aspartic acid and phenylalanine, which are then further metabolised. All three are available from other foods, so no non-natural chemicals are being introduced into the body.

Anticariogenic.

Calorific value of 4 calories/g.

Substantial evidence lacking to link aspartame with claims of adverse health effects.

Virtually non-calorific due to intense sweetness.

Average consumption by Canadians 1.7 to 3.7 mg/kg body weight; upper estimate of Canadian intake 5.8 to 16.8 mg/kg body weight; current Canadian consumption levels, even among children, considered well within acceptable limits.

FDA Advisory Committee on Hypersensitivity to Food Constituents concluded some individuals have an unusual sensitivity to aspartame, although this does not represent a significant health risk.

Risk to pregnant women of aspartame consumption leading to children with PKU considered non-existent.

LEGISLATION:

USA:

Approved by FDA in 1974 for use as a sweetener, flavour enhancer and an ingredient in some dry food products.

Approved by FDA in 1983 for use in carbonated beverages and carbonated beverage syrups. Later approved for chewable multivitamin tablets.

Approved by FDA in 1987 for use in frozen juice drinks, frozen novelties, tea beverages and breath mints.
USFDA gave ADI of 50 mg/kg body weight.
Approved in US for use in soft drinks, desserts, dessert toppings and household use.
Regulatory Status: ADI level as of 1990 for use in soft drinks – Permitted
CFR 21
PART 172 – Food Additives Permitted for Direct Addition to Food for Human Consumption
Subpart I – Multipurpose Additives
172.804 – Aspartame.
The food additive aspartame may be safely used in food in accordance with GMP as a sweetening agent, or for an authorised technological purpose in foods for which standards of identity established under section 401 of the Act do not preclude such use under the following conditions:
(a) Aspartame is the chemical 1-methyl *N*-L-a-aspartyl-L-phenylalanine ($C14H18N2O5$).
(b) The additive meets the specifications of the *Food Chemicals Codex*, 3rd edition (1981) pp. 28–29 and First Supplement p. 5, which is incorporated by reference in accordance with 5 U.S.C. 552(a). Copies are available from the National Academy Press, 2101 Constitution Avenue NW., Washington, DC 20418, or for inspection at the Office of the Federal Register, 800 North Capitol Street, NW., suite 700, Washington, DC 20408.
(c) The additive may be used as a sweetener in the following foods:
(1) Dry, free-flowing sugar substitutes for table use (not to include use in cooking) in package units not exceeding the sweetening equivalent of 1 pound of sugar.
(2) Sugar substitute tablets for sweetening hot beverages, including coffee and tea. L-leucine may be used as a lubricant in the manufacture of such tablets at a level not to exceed 3.5 percent of the weight of the tablet.
(3) Breakfast cereals.
(4) Chewing gum.
(5) Dry bases for:
(i) Beverages.
(ii) Instant coffee and tea beverages.
(iii) Gelatins, puddings, and fillings.
(iv) Dairy product analogue toppings.
(6) Ready-to-serve non-alcoholic flavoured beverages, tea beverages, fruit juice-based beverages, and their concentrates or syrups.
(7) Chewable multivitamin food supplements.
(8) [Reserved]
(i) Fruit juice-based drinks (where food standards do not preclude such use).
(ii) Fruit flavoured drinks and ades.
(iii) Imitation fruit-flavoured drinks and ades.

(9) Frozen stick-type confections and novelties.
(10) Breath mints, hard and soft candy.
(11) and (12) are reserved categories
(13) Refrigerated ready-to-serve gelatins, puddings, and fillings.
(14) Fruit (including grape) wine beverages with ethanol contents below 7 percent volume per volume.
(15) Yoghurt-type products where aspartame is added after pasteurisation and culturing.
(16) Refrigerated flavoured milk beverages.
(17) Frozen desserts.
(18) Frostings, toppings, fillings, glazes, and icings for pre-cooled baked goods.
(19) Frozen, ready-to-thaw-and-eat cheesecakes, fruit, and fruit toppings.
(20) Frozen dairy and non-dairy frostings, toppings, and fillings.
(21) Fruit spreads, fruit toppings, and fruit syrups.
(22) Malt beverages of less than 7 percent ethanol by volume and containing fruit juice.
(23) Baked goods and baking mixes in an amount not to exceed 0.5 percent by weight of ready-to-bake products or of finished formulations prior to baking. Generally recognised as safe (GRAS) ingredients or food additives approved for use in baked goods shall be used in combination with aspartame to ensure its functionality as a sweetener in the final baked product. The level of aspartame used in these products is determined by an analytical method entitled "Analytical Method for the Determination of Aspartame and Diketopiperazine in Baked Goods and Baking Mixes," October 8, 1992, which was developed by the NutraSweet Co., and is incorporated by reference in accordance with 5 U.S.C. 552(a) and 1 CFR part 51. Copies are available from the Office of Premarket Approval, Center for Food Safety and Applied Nutrition, 200 C St. SW., Washington, DC 20204, or are available for inspection at the Office of the Federal Register, 800 North Capitol St. NW., suite 700, Washington, DC.

(d) The additive may be used as a flavour enhancer in chewing gum, hard candy, and malt beverages containing less than 3 percent alcohol by volume.

(e) To assure safe use of the additive, in addition to the other information required by the Act:

(1) The principal display panel of any intermediate mix of the additive for manufacturing purposes shall bear a statement of the concentration of the additive contained therein;

(2) The label of any food containing the additive shall bear, either on the principal display panel or on the information panel, the following statement:
PHENYLKETONURICS: CONTAINS PHENYLALANINE
The statement shall appear in the labelling prominently and conspicuously as compared to other words, statements, designs or devices and in bold type and on clear contrasting background in order to render it likely to be read and understood by the ordinary individual under customary conditions of purchase and use.

(3) When the additive is used in a sugar substitute for table use, its label shall bear instructions not to use in cooking or baking.

(4) Packages of the dry, free-flowing additive shall prominently display the sweetening equivalence in teaspoons of sugar.

(f) If the food containing the additive purports to be or is represented for special dietary uses, it shall be labelled in compliance with part 105 of this chapter.

[42 FR 14491, Mar. 15, 1977, as amended at 48 FR 31382, July 8, 1983; 49 FR 22468, May 30, 1984; 51 FR 43000–43002, Nov. 28, 1986; 53 FR 20837–20842, June 7, 1988; 53 FR 40879, Oct. 19, 1988; 53 FR 51273, Dec. 21, 1988; 54 FR 23647, June 2, 1989; 54 FR 31333, July 28, 1989; 57 FR 3702, 3703, 3704, Jan. 30, 1992; 58 FR 19771, Apr. 16, 1993; 58 FR 21097, 21098, 21099, Apr. 19, 1993; 58 FR 48598, Sept. 17, 1993]

UK and EUROPE:

JECFA gave ADI of 40 mg/kg body weight
UK gave Group A classification in Sweeteners in Food Regulations in 1983
Regulatory Status: ADI level as of 1990 for use in soft drinks

– JECFA	0 to 40 mg/kg body weight
– Belgium	750 mg/l (pending approval to increase from 500 mg/l)
– Denmark	500 mg/l
– Finland	500 mg/l
– France	600 mg/l
– East Germany	Not permitted
– West Germany	300 mg/l
– Greece	600 mg/l
– Ireland	Permitted
– Netherlands	750 mg/l (pending approval to increase from 700 mg/l)
– Norway	500 mg/l
– Spain	Permitted
– Switzerland	Permitted
– Turkey	600 mg/l
– UK	Permitted
– USSR	Permitted
– Yugoslavia	Not permitted

CANADA:

Approved for use in Canada in 1981

Table IX – Food additives that may be used as Sweeteners – Item A.1 (25 May 1993): Aspartame

Permitted in or on	**Maximum level of use**
(1) Table-top sweeteners	Good Manufacturing Practice
(2) Breakfast cereals	0.5%
(3) Beverages; beverage concentrates, beverage mixes; (except for any of these products for which standards are set out in these Regulations)	0.1% in beverages as consumed
(4) Desserts; dessert mixes; toppings; topping mixes; fillings; filling mixes; (except for any of these products for which standards are set out in these Regulations)	0.3% in products as consumed
(5) Chewing gum; breath freshener product	1.0%
(6) Fruit spreads; purées and sauces; table syrups; (except for any of these products for which standards are set out in these Regulations)	0.2%
(7) Salad dressings; peanut and other nut spreads; (except for any of these products for which standards are set out in these Regulations)	0.05%
(8) Condiments (except for any of these products for which standards are set out in these Regulations)	0.2%
(9) Confectionery glazes for snack foods; sweetened seasonings or coating mixes for snack foods	0.1%
(10) Confections and their coatings (except for any of these products for which standards are set out in these Regulations)	0.3%

Table IX – Food additives that may be used as Sweeteners – Item A.2 (25 May 1993): Aspartame, encapsulated to prevent degradation during baking

Permitted in or on	**Maximum level of use**
Bakery products and baking mixes (except for any of these products for which these standards are set out in these Regulations)	0.4% in product as consumed

Canada first country to allow use in soft drinks in 1981

acceptable for use in table-top sweeteners, ready-to-eat cereals, beverages, beverage concentrates and mixes, desserts, toppings, fillings and their mixes, chewing gum and breath fresheners.

Canadian ADI 40 mg/kg body weight which is greater than the 8.3 mg/kg that would be consumed on average if it replaced all sucrose in the diet or the 25 mg/kg if it replaced all carbohydrate in the diet

Item A.A3, Table III, pages 67–22 and 67–22a of Canada FDR-Update as of 1996

- sweetener and flavour enhancer in table-top sweeteners – Good Manufacturing Practice
- use as sweetener and favour enhancer in unstandardised food extended in 1995 to provide for use in beverages, beverage concentrates, beverage mixtures, confectionery glazes for grain, nut and corn-based snack foods, sweetened seasonings, or coating mixes for snack foods at a level of 0.1%
- in desserts, dessert mixes, toppings, topping mixes, fillings, filling mixes, confections and their coatings, candies, frostings and icings at a level of 0.3%

AUSTRALIA/PACIFIC RIM:

Regulatory Status: ADI level as of 1990 for use in soft drinks

- Australia: 1000 mg/l
- Japan: permitted
- New Zealand: permitted

OTHER COUNTRIES:

WHO – ADI 40 mg/kg

Now permitted for use in more than 50 countries

Permitted for use in soft drinks in 39 countries

Regulatory Status: ADI level as of 1990 for use in soft drinks

- Argentina: permitted
- Brazil: 750 mg/l
- Israel: 700 mg/l
- Kenya: permitted
- Mexico: permitted
- Saudi Arabia: not permitted
- South Africa: 1000 mg/l

REFERENCES:

Smith, J. (1991) *Food Additive User's Handbook.* Blackie Publishing, Glasgow.

Canadian Food and Drugs Act and Regulations (1994).

Mitchell, A. J. (1990) *Formulation and Production of Carbonated Soft Drinks.* Blackie Publishing, Glasgow.

McCue, N. (1996) *Showcase: Natural & High Intensity Sweeteners; Prepared Foods.* March, p. 84.

New and Novel Foods; p. 484

Nutritional Recommendations; *Health and Welfare Canada* (1990), pp. 190–193.

Minifie, B. W. (1989) *Chocolate, Cocoa, and Confectionery: Science and Technology.* 3rd edition. Van Nostrand Reinhold.

Matz, S. A. (1992) *Bakery Technology and Engineering.* 3rd edition. Van Nostrand Reinhold.

Food Additives Update. Food in Canada, April 1996.

Food Chemistry. 2nd edition. (1985) Marcel Dekker, Inc.

Hicks, D. (1989) *Production and Packaging of Non-Carbonated Fruit Juices and Fruit Beverages.* Blackie Publishing, Glasgow.

Ash, M., and Ash, I. (1995) *Food Additives, Electronic Handbook.* Gower.

Health & Human Services Publication No. (FDA) 85-2205.

Sweeteners: Nutritive and Non-Nutritive. Status summary by the IFT Expert Panel on Food Safety and Nutrition, August (1991).

Smoley, C. K. (1993) *Everything Added to Food in the United States.* US Food and Drug Administration, CRC Press, Inc.

Krutosikova, U. (1992) *Natural and Synthetic Sweet Substances.* Gilis Horwood.

Wong, D. W. S. (1989) *Mechanism and Theory in Food Chemistry.* Van Nostrand Reinhold.

Budavari, S. (1996) *The Merck Index.* 12th edition. Merck and Co.

ANY OTHER RELEVANT INFORMATION:

Manufacturers: Ajinomate, Browne & Dureau Intl., Calaga Food Ingredients fruitsource, Holland Sweetener, NutraSweet AG, Quimdis, Sanafi, Scan chem, Sweeteners Plus, FH Worlee Gmblt

Marketed by G.D. Searle as Nutrasweet®, Equal® and Canderel® (as table sweetener), a dipeptide methyl ester composed of two amino acids (phenylalanine and aspartic acid).

Marketed by Holland Sweetener Company of thc Netherlands as Sanecta®

Discovered in 1965 by J. Schlatter of G.D. Searle Laboratories.

First food ingredient marketed using a "branded ingredient" strategy under the brand name "NutraSweet" by G.D. Searle. All products sweetened with this ingredient branded with the NutraSweet logo.

NAME:	Cyclamate
CATEGORY:	Sweetener/ Flavour enhancer
FOOD USE:	Beverages (dry beverage mixes)/ Sugars, sugar preserves and confectionery (table-top sweeteners, jams and jellies, low-calorie frozen desserts)/ Fruit, vegetables and nut products/ Jams and jellies/ Soft drinks/ Vinegar, pickles and sauces (salad dressings)/ Other (chewing gum)
SYNONYMS:	Cyclamate: CFSAN Cyclamate/ CAS 977016-96-8 Cyclamic acid: CFSAN Cyclamic acid/ CAS 100-88-9/ Cyclohexanesulfamic acid/ Cyclohexylsulfamic acid/ *N*-cyclohexyl-sulphamic acid/ Hexamic acid Sodium cyclamate: Sodium cyclohexylsulfamate/ Cyclamate sodium/ Assugrin/ Sucryl sodium/ Sucrosa. Cyclamate is the generic name for cyclohexylsulphamate
FORMULA:	Sodium cyclamate: $C6\,H12\,N\,Na\,O3\,S$; cyclamic acid: $C6\,H13\,NO3\,S$
MOLECULAR MASS:	Cyclamic acid 179.24; sodium cyclamate 201.2
ALTERNATIVE FORMS:	Cyclamic acid/ Calcium cyclamate/ Sodium cyclamate
PROPERTIES AND APPEARANCE:	White crystalline powder. Good tasting and low in cost. In soft drinks, sweetness relative to sucrose is 30 to 40; sodium cyclamate – ‘chemical’ sweetness with no aftertaste; sweet taste has slow onset time; detectable sweet/sour aftertaste most noticeable at high concentrations; clean sweet taste of high intensity, close resemblance to sugar in sweetness; no bitter aftertaste; 30 to 60 times sweeter than sucrose; distinct off-taste noticeable at high concentrations. Characteristics of sweeteners (cyclamate alone): Sweetness intensity (at 10% sucrose): 33 Sweetness quality: slight chemical sweetness Time profile: slower and persistent Associated taste: off-taste at high concentrations Mouthfeel body: good Enhancement of fruitiness: good

Characteristics of sweeteners (1 : 10 saccharin : cyclamate)

Sweetness intensity (at 10% sucrose):	100
Sweetness quality:	Sugar-like
Time profile:	as sucrose
Associated taste:	none
Mouthfeel body:	good
Enhancement of fruitiness:	good

Cyclamic acid: Sweet-sour crystals

Sodium cyclamate:

Pleasantly sweet crystals

30 times as sweet as refined cane sugar

sweetness easily perceptible at dilution of 1 : 10,000 in water

MELTING RANGE IN °C: Cyclamic acid: 169–179

SOLUBILITY % AT VARIOUS TEMPERATURE/pH COMBINATIONS:

in water: **@ 20°C** Sodium salt 200 g/l Calcium salt 250 g/l

FUNCTION IN FOODS: Sweetener. Cyclamic acid has some flavour-enhancing capabilities at low levels. Cyclamate : aspartame also found to improve stability and taste profiles of diet soft drinks, dry beverage mixtures and chewing gum.

TECHNOLOGY OF USE IN FOODS:

Sodium and calcium salt forms most often used. 1/10th sweetness of equal weight of saccharin. 10 : 1 cyclamate: saccharin ratio masked the aftertaste of saccharin which boosted the low sweetness of cyclamates. Cyclamate : aspartame also found to improve stability and taste profiles of diet soft drinks, dry beverage mixtures and chewing gum. Sodium and calcium salts are soluble is water at room temperature.

Stable at pH 2 to 7 at normal process temperatures; in aqueous solution at pH 2.1 it hydrolyses to:

- 350 mg cyclohexylamine per kg cyclamate at 30°C
- 500 mg per kg at 44°C
- Both after 40 days.

Stable in tablet form for several years; in aqueous solution it hydrolyses slowly to sulphuric acid and cyclohexylamine.

Lower cost versus sucrose.

No browning reaction.

Decomposition is accelerated in the presence of amino acids and water-soluble vitamins at elevated temperatures.

Alone in soft drinks did not provide sweetness taste quality of the saccharin : cyclamate blend.

Sodium salt most commonly used.

Stable under conditions of soft drink manufacture such as pH 2 to 7, pasteurisation and UHT heat treatment.

Can be detected in soft drinks by spectrophotometric methods and titration followed by liquid chromatography.

Popular in soft drinks as a cyclamate : saccharin blend of 10 : 1; has been used in combination with saccharin to overcome bitterness and aftertaste; recommended for use in beverages, fruit juices, processed fruits, desserts, jellies, jams, toppings, salad dressings, and confections.

Fairly thermostable.

Commercially successful combination as cyclamate/saccharin at 10 : 1 or 50% : 50% sweetness gives clean sugar-like sweetness at low cost with good storage stability; often used mixed with other sweeteners.

Readily soluble in water; good stability in fruit beverages.

Compatible with a broad range of beverage ingredients.

Cyclamic acid: fairly strong acid, sparingly soluble in water, slowly hydrolysed by hot water.

Sodium cyclamate: freely soluble in water, practically insoluble in alcohol, ether, benzene, chloroform.

Sodium cyclamate: pH of 5.5 to 7.5 in 10% solution.

SYNERGISTS:

Synergistic with aspartame, saccharin, sucrose and acesulphame.

Sweetness quality improved by combining with other intense sweeteners.

Citric acid and other citrus products have a synergistic effect on sweetness.

FOOD SAFETY ISSUES:

Non-cariogenic.

Non-caloric as metabolism does not release any energy.

May be concerns of toxicity. At high enough doses, cyclamate found to be metabolised into cyclohexylamine which has been implicated in the occurrence of bladder tumours after two years. This based on studies of rats fed a 10 : 1 cyclamate : saccharin mixture. There are other concerns about its effects on genetic material. Due to low sweetness, quantities required would probably exceed ADI.

Most people metabolise only 1% of cyclamate intake, but 47% of the population can metabolise (in intestine) 20–60% to cyclohexylamine, a known carcinogen.

Recent studies do not support the claim of cyclamates being carcinogenic; some have linked the metabolite cyclohexylamine to high blood pressure, testicular atrophy and cancer promotion in rats.

In 1985, National Academic of Science (NAS) and National Research Council (NRC) decided that cyclamates act as carcinogen promoters or co-carcinogens in the presence of substances like saccharin.

LD_{50} of sodium cyclamate is 15.25–17.0 g/kg.

LEGISLATION:

USA:

Approved by USFDA in 1949. GRAS status by USFDA in 1957. Banned in US in 1969 for general-purpose foods based on studies linking bladder tumours in rats with repeated use of cyclamate : saccharin blend in rats. Banned for all foods in August 1970 by US FDA

CFR 21 PART 189: 189.135 – Cyclamate and its derivatives.

[42 FR 14659, Mar. 15, 1977, as amended at 49 FR 10114, Mar. 19, 1984; 54 FR 24899, June 12, 1989]

(a) Calcium, sodium, magnesium and potassium salts of cyclohexane sulfamic acid: (C6H12NO3S)2Ca, (C6H12NO3S)Na, (C6H12NO3S)2Mg, and (C6H12NO3S)K. Cyclamates are synthetic chemicals having a sweet taste 30 to 40 times that of sucrose, are not found in natural products at levels detectable by the official methodology, and have been used as artificial sweeteners.

(b) Food containing any added or detectable level of cyclamate is deemed to be adulterated in violation of the act based upon an order published in the FEDERAL REGISTER of October 21, 1969 (34 FR 17063).

(c) The analytical methods used for detecting cyclamate in food are in sections 20.162–20.172 of the *Official Methods of Analysis of the Association of Official Analytical Chemists*, 13th Ed. (1980), which is incorporated by reference. Copies may be obtained from the Association of Official Analytical Chemists, 2200 Wilson Blvd., Suite 400, Arlington VA 22201–3301, or may be examined at the Office of the Federal Register, 800 North Capitol Street, NW., suite 700, Washington, DC 20408.

ADI level as of 1990 for use in soft drinks: Not permitted

UK and EUROPE:

Still permitted in some applications in countries such as Spain, Germany and Switzerland.

JECFA approved cyclamates with an ADI of 11 mg/kg body weight in 1982. Accepted in UK in 1964. EEC Directive on Additives includes cyclamates on the positive list.

Regulatory Status: ADI level as of 1990 for use in soft drinks;

– JECFA – 0 to 11 mg/kg body weight

– Belgium	400 mg/l (pending approval for change from Not Permitted)
– Denmark	250 mg/l
– Finland	100 to 400 mg/l depending upon type of soft drink
– France	Not permitted
– East Germany	450 to 600 mg/l depending upon type of soft drink

– West Germany	800 mg/l
– Greece	Permitted
– Ireland	Permitted
– Netherlands	400 mg/l (pending approval to change from Not Permitted)
– New Zealand	1500 mg/l
– Norway	Permitted
– Spain	4000 mg/l
– Switzerland	800 mg/l
– Turkey	Not permitted
– UK	Not permitted
– USSR	Not permitted
– Yugoslavia	Permitted

CANADA:

Part E – Cyclamate and Saccharin Sweeteners

E.01.001 – (1) In this Part, "cyclamate sweetener" means

(a) cyclohexyl sulfamic acid or a salt thereof, or

(b) any substance containing cyclohexyl sulfamic acid or a salt thereof that is sold as a sweetener;

(2) Part B of these Regulations does not apply to any cyclamate sweetener or saccharin sweetener

E.01.002 – **(Sale)** No person shall

(a) sell a cyclamate sweetener or a saccharin sweetener that is not labelled as required by this Part

E.01.003 – **(Advertising)** No person shall, in advertising to the general public a cyclamate sweetener or saccharin sweetener, make any representation other than with respect to the name, price and quantity of the sweetener

E.01.004 – **(Labelling)**

(1) Every cyclamate sweetener that is not also a saccharin sweetener shall be labelled to state that such sweetener should be used only on the advice of a physician

E.01.005 – Commencing June 1, 1979, every cyclamate sweetener or saccharin sweetener shall be labelled to show

(a) a list of all the ingredients and, in the case of (i) cyclohexyl sulfamic acid, (ii) a salt of cyclohexyl sulfamic acid, the quantity thereof contained in the sweetener; and

(b) the energy value of the sweetener expressed in calories (i) per teaspoonful, drop, tablet or other measure used in the directions for use, and (ii) per 100 grams or 100 millilitres of the sweetener

ADI level as of 1990 for use in soft drinks: Not permitted

AUSTRALIA/PACIFIC RIM:

Regulatory Status: ADI level as of 1990 for use in soft drinks

– Australia	600 mg/l
– Japan	Not permitted
– New Zealand	1500 mg/l

OTHER COUNTRIES:

Lost GRAS status in 1969. Banned in other countries (e.g. UK, Canada, Japan) soon after US ban. Permitted for use in soft drinks in 25 countries.

Regulatory Status: ADI level as of 1990 for use in soft drinks

– Argentina	2000 mg/l
– Brazil	1600 mg/l
– Israel	193 mg/l
– Kenya	Not permitted
– Mexico	Not permitted
– Saudi Arabia	Not permitted
– South Africa	2500 mg/l

REFERENCES:

Smith, J. (1991) *Food Additive User's Handbook.* Blackie Publishing, Glasgow.

Mitchell, A. J. (1990) *Formulation and Production of Carbonated Soft Drinks.* Blackie Publishing, Glasgow.

New and Novel Foods, pp. 485–486.

Canadian Food and Drugs Act and Regulations (1994).

Minifie, B. W. (1989) *Chocolate, Cocoa, and Confectionery: Science and Technology.* 3rd edition. Van Nostrand Reinhold.

Food Chemistry. 2nd edition. (1985) Marcel Dekker, Inc.

Hicks, D. (1989) *Production and Packaging of Non-Carbonated Fruit Juices and Fruit Beverages.* Blackie Publishing, Glasgow.

Smoley, C. K. (1993) *Everything Added to Food in the United States.* US Food and Drug Administration, CRC Press, Inc.

Budavari, S. (1996) *The Merck Index.* 12th edition. Merck and Co.
Wong, D. W. S. (1989) *Mechanism and Theory in Food Chemistry.* Van Nostrand Reinhold.

ANY OTHER RELEVANT INFORMATION: Synthetic sweetener; discovered in 1937 by Michael Sveda of Abbott Laboratories in Chicago. Use as sweetener started in mid-1950s; popular in 1960s, instrumental in making diet products popular. Cyclamate : saccharin first commercial multiple sweetener. Use of saccharin : cyclamate blends ended with cyclamate ban in 1970

NAME:	**Glycyrrhizin**
CATEGORY:	Sweetener/ Nutritive additive/ Foaming agent
FOOD USE:	Baked goods/ Dairy products (frozen dairy desserts)/ Fruit, vegetables and nut products (hydrolysed vegetable protein (HVP), bean paste)/ Beverages (non-alcoholic beverages)/ Soft drinks (root beer)/ Sugars, sugar preserves and confectionery (confectionery manufacture, soft candy, confection, frosting, hard candy)/ Alcoholic drinks (liqueurs)/ Vinegar, pickles and sauces (soy sauce)/ Other (tobacco, chocolate, vanilla, medicines, gelatin, pudding, chewing gum)
SYNONYMS:	Glycyrrhizin: CFSAN Glycyrrhizin/ Ammoniated (*Glycyrrhiza* spp.)/ CAS 5395-04-0 Glycyrrhizinic acid: Glycyrrhetinic acid glucoside/ (3β,20β)-20-Carboxy-11-oxo-30-norolean-12-en-3-yl 2-O-β-D-glucopyranuronosyl-α-D-glucopyranosiduronic acid Monoammonium glycyrrhizin: CAS = 1407-03-0
FORMULA:	Ammonium glycyrrhizinate pentahydrate: $C_{42}H_{65}NO_{16}\cdot5H_2O$ Monoammonium glycyrrhizin: $C_{42}H_{61}O_{16}NH_4\cdot5H_2O$ Glycyrrhizic acid: $C_{46}H_{62}O_{16}$ Dipotassium glycyrrhizin: $C_{42}H_{60}K_2O_{16}$
MOLECULAR MASS:	822.94
ALTERNATIVE FORMS:	Calcium glycyrrhizinate/ Disodium glycyrrhizinate/ Monoammonium glycyrrhizinate/ Potassium glycyrrhizinate/ Magnesium glycyrrhizinate/ Ammonium glycyrrhizinate pentahydrate
PROPERTIES AND APPEARANCE:	Ammonium glycyrrhizine (AG): spray-dried brown powder; monoammonium glycyrrhizin (MAG): white crystalline powder. Slow to taste, but taste long-lasting; leaves strong, lingering licorice-like aftertaste. Ammoniated salt approximately 50 to 100 times sweeter than sucrose
MELTING RANGE IN °C:	Ammonium glycyrrhizinate pentahydrate: decomposes at 212–217
FUNCTION IN FOODS:	Foaming agent; nutritive sweetener; aromatisation

TECHNOLOGY OF USE IN FOODS:

Does not dissolve well in cold water, but is soluble in hot water or ethanol; practically insoluble in ether; ammoniated form soluble in hot or cold water and propylene glycol.

50 to 100 times sweeter than sucrose. Ammoniated salt generally used at the following levels:

– Baked goods	61 ppm
– Frozen dairy	91 ppm
– Non-alcoholic beverages	36 to 51 ppm
– Soft candy	1511 ppm
– Confection, frosting	625 ppm
– Gelatin, pudding	79 ppm
– Alcoholic beverages	59 ppm
– Hard candy	676 ppm
– Chewing gum	2278 ppm

Ammonium glycyrrhizin inactivated in acid media (pH below 4.5) due to precipitation of acid form. Relatively heat-stable, but flavour tends to deteriorate above 105°C; can precipitate at pH below 4.5.

Strong licorice flavour, so use in bakery products limited (a few confectionery products); applied mostly in tobacco and pharmaceutical products.

Has foam-enhancing properties which can be useful in beverage formulation.

SYNERGISTS: Glycyrrhizine potentiated to 100 times original sweetness in presence of sucrose

FOOD SAFETY ISSUES: Non-calorific

LEGISLATION:

USA:

GRAS flavouring agent
CFR 21: Part 184
184.1408 – Licorice and licorice derivatives.

(a) (1) Licorice (glycyrrhiza) root is the dried and ground rhizome and root portions of *Glycyrrhiza glabra* or other species of *Glycyrrhiza*. Licorice extract is that portion of the licorice root that is, after maceration, extracted by boiling water. The extract can be further purified by filtration and by treatment with acids and ethyl alcohol. Licorice extract is sold as a liquid, paste ("block"), or spray-dried powder.

(2) Ammoniated glycyrrhizin is prepared from the water extract of licorice root by acid precipitation followed by neutralisation with dilute ammonia. Monoammonium glycyrrhizinate ($C_{42}H_{61}O_{16}NH_4 5H_2O$, CAS Reg. No. 1407-03-0) is prepared from ammoniated glycyrrhizin by solvent extraction and separation techniques.

(b) The ingredients shall meet the following specifications when analysed:

(1) *Assay.* The glycyrrhizin content of each flavouring ingredient shall be determined by the method in the *Official Methods of Analysis of the Association of Official Analytical Chemists*, 13th Ed., §§ 19.136–19.140, which is incorporated by reference, or by methods 19.CO1 through 19.CO4 in the *Journal of the Association of Official Analytical Chemists*, 65:471–472 (1982), which are also incorporated by reference. Copies of all of these methods are available from the Association of Official Analytical Chemists, 2200 Wilson Blvd., Suite 400, Arlington, VA 22201-3301, or available for inspection at the Office of the Federal Register, 800 North Capitol Street, NW., suite 700, Washington, DC 20408.

(2) *Ash.* Not more than 9.5 percent for licorice, 2.5 percent for ammoniated glycyrrhizin, and 0.5 percent for monoammonium glycyrrhizinate on an anhydrous basis as determined by the method in the *Food Chemicals Codex*, 3rd edn. (1981), p. 466 (see reference). Copies are available from the National Academy Press, 2101 Constitution Ave. NW., Washington, DC 20418, or available for inspection at the Office of the Federal Register, 800 North Capitol Street, NW., suite 700, Washington, DC 20408.

(3) *Acid unsoluble ash.* Not more than 2.5 percent for licorice on an anhydrous basis as determined by the method in the *Food Chemicals Codex*, 3rd edn. (1981), p. 466 (see reference).

(4) *Heavy metals (as Pb).* Not more than 40 parts per million as determined by method II in the *Food Chemicals Codex*, 3rd edn. (1981), p. 512 (see reference).

(5) *Arsenic (As).* Not more than 3 parts per million as determined by the method in the *Food Chemicals Codex*. 3rd edn. (1981), p. 464 (see reference).

(c) In accordance with §181.1(b)(2), these ingredients are used in food only within the following specific limitations:

Category of Food	**Maximum level in food (percent glycyrrhizin content of food)**	**Functional Use**
Baked foods, §170.3(n)(1) of this chapter	0.05	Flavour enhancer, §170.3(o)(11) of this chapter; flavouring agent, §170.3(o)(12) of this chapter.
Alcoholic beverages, §170.3(n)(2) of this chapter	0.1	Flavour enhancer, §170.3(o)(11) of this chapter; flavouring agent, §170.3(o)(12) of this chapter; surface active agent, §170.3(o)(29) of this chapter.
Non-alcoholic beverages, §170.3(n)(3) of this chapter	0.15	Do.
Chewing gum, §170.3(n)(6) of this chapter	1.1	Flavour enhancer, §170.3(o)(11) of this chapter; flavouring agent, §170.3(n)(12) of this chapter.
Hard candy, §170.3(n)(25) of this chapter	16.0	Do.

Herbs and seasonings, §170.3(n)(26) of this chapter	0.15	Do.
Plant protein products, §170.3(n)(33) of this chapter	0.15	Do.
Soft candy, §170.3(n)(38) of this chapter	3.1	Do.
Vitamin or mineral dietary supplements	0.5	Do.
All other foods except sugar substitutes, §170.3(n)(42) of this chapter. The ingredient is not permitted to be used as a non-nutritive sweetener in sugar substitutes	0.1	Do.

(d) Prior sanctions for this ingredient different from the uses established in this section do not exist or have been waived.
[50 FR 21044, May 22, 1985, as amended at 54 FR 24899, June 12, 1989]

UK and EUROPE:

FEMA:
- No. 2630 (root)
- No. 2628 (extract)
- No. 2629 (extract powder)
- No. 2528 (ammoniated salt)

REFERENCES:

Burdock, G. A. (1995) *Fenaroli's Handbook and Flavour Ingredients*, Volume I, 3rd edition, CRC Press, pp. 173–175.

New and Novel Foods; 486.

Matz, S. A. (1992) *Bakery Technology and Engineering.* 3rd edition. Van Nostrand Reinhold.

Food Chemistry. 2nd edition. (1985) Marcel Dekker, Inc.

Sweeteners: Nutritive and Non-nutritive, Scientific Status Summary by the IFT Expert Panel on Food Safety & Nutrition, August (1986).

Smoley, C. K. (1993) *Everything Added to Food in the United States.* US Food and Drug Administration, CRC Press, Inc.

Krulosikova, Uher (1992) *Natural and Synthetic Sweet Substances*. Ellis Hardwood.

Ash, M., and Ash, I. (1995) *Food Additives Electronic Handbook.* Gower.

Wong, D. W. S. (1989) *Mechanism and Theory in Food Chemistry.* Van Nostrand Reinhold.

Budavari, S. (1996) *The Merck Index.* 12th edition. Merck and Co.

ANY OTHER RELEVANT INFORMATION:	Glucoside; extracted from licorice root; root extract contains both calcium and sodium salts; hydrolysis yields glycyrrhetinic acid and glucuronic acid. Licorice root contains 6–20% glycyrrhizin; commercial block juice contains 14–20% glycyrrhizin. Triterpenoid glycoside with two glyceronic units. Detectable at 1/50 the threshold taste level of sucrose.

NAME:	Isomalt
CATEGORY:	Sweetener/ Nutritive additive/ Flavour modifier/ Bulking agent
FOOD USE:	Baked goods/ Soft drinks/ Sugars, sugar preserves and confectionery (confectionery, coatings for hard-boiled and chewable candies, soft caramels, soft candies)/ Other (chewing gums)
SYNONYMS:	Hydrogenated isomaltulose/ Hydrogenated palatinose
MOLECULAR MASS:	368
PROPERTIES AND APPEARANCE:	White, odourless crystals. Pleasant sweet taste (sweetness intensity 0.45 compared to sucrose); no aftertaste. Available in crystalline form.
MELTING RANGE IN °C:	145–150
SOLUBILITY % AT VARIOUS TEMPERATURE/pH COMBINATIONS:	
in water:	**@ 20°C** 25
FUNCTION IN FOODS:	Bulking agent: adds texture and mouthfeel properties. Sweetening agent: functionally similar to sucrose, sweetness intensity of 0.45 compared to sucrose. Enhances shelf-life of hygroscopic products; increases chemical stability; increases affinity for water without altering sweetening power; reduces tendency to crystallise. Use in baking: reduces fermentability; increases resistance to non-enzymic browning reactions; decreases tendency to crystallise; increases hygroscopicity
ALTERNATIVES:	Sucrose; other polyols

TECHNOLOGY OF USE IN FOODS:

Solubility in water is a function of temperature. 25% solubility at 25°C and 55% solubility at 60°C. Solubility decreases linearly with the addition of alcohol. Highly stable against chemical and microbial breakdown; has no Maillard reaction so does not require browning inhibitors.

Changes colour slightly when held at 170°C for 60 minutes, but no further colour changes, unlike sucrose solutions.

Viscosity in aqueous solution comparable to sucrose solutions.

Less sweet than sugar, so require intense sweeteners as supplements to bring to equivalence with sucrose.

Does not produce a cooling sensation when dissolving; mixtures with saccharin and aspartame mask the metallic aftertaste; can be mixed with polydextrose in calorie-reduced foods.

Crystallises easily, so useful in simplifying coating of hard-boiled and chewable candies. With the addition of a crystallisation inhibitor such as HGS, can be used as a melt for manufacturing soft caramels, chewing gums and soft candies due to the high percentage of solids dissolved in the aqueous phase.

Low hygroscopicity; resistant to chemical and microbial breakdown

SYNERGISTS: Synergistic with other polyols (sorbitol, xylitol, HGS) and with intense sweeteners (saccharin, aspartame)

FOOD SAFETY ISSUES: Low cariogenicity. Suitable for inclusion in diabetic products because not dependent on insulin for metabolism and so results in no significant change in blood glucose. Low energy due to malabsorption in intestine (2 kcal/g). Has been proven to have no adverse health effects. Laxative effect at high doses, so warning labels are required in some countries (laxative effect 20–30 g/day)

LEGISLATION:

UK and EUROPE:

One of 12 sweeteners listed as permissible for use in the UK as of 1983
JECFA allocated and ADI of "Not Specified" in 1985

CANADA:

Canada FDA: 67.31 – Table IX – Food additives that may be used as Sweeteners
– Item I.1 (14 December 1994)
– Permitted on Unstandardised Foods at levels dependant upon Good Manufacturing Practice

REFERENCES:

Smith, J. (1991) *Food Additive User's Handbook.* Blackie Publishing.
Canadian Food and Drugs Act and Regulations (1994).
Matz, S. A. (1992) *Bakery Technology and Engineering.* 3rd edition. Van Nostrand Reinhold.
Smoley, C. K. (1993) *Everything Added to Food in the United States.* US Food and Drug Administration, CRC Press, Inc.

ANY OTHER RELEVANT INFORMATION: Polyol produced through hydrogenation of isomaltulose (also known as palantinose).
Marketed by Suddeutsche Zucher AG of Germany as Palatinit®.
Marketed by Tate and Lyle of UK as Lylose®.

NAME:	Lactitol
CATEGORY:	Sweetener/ Nutritive additive/ Bulking agent
FOOD USE:	Baked goods (bakery products, biscuit-making)/ Dairy products (ice-cream)/ Fruit, vegetables and nut products (jams and marmalades)/ Beverages (instant beverages)/ Soft drinks/ Sugars, sugar preserves and confectionery (confectionery, surface dustings for confectionery, table-top sweeteners, chocolate, hard and soft candies)/ Other (chewing gums, fruit gums, pastilles)
SYNONYMS:	CAS 81025-04-9 (monohydrate)/ 4-O-β-D-Galactopyranosyl-D-glucitol (monohydrate)/ β-Galactoside sorbitol/ Lactitol MC (monohydrate)/ Lactit/ Lactit M/ Lactite/ Lactobiosit/ Lactosit/ Lactositol
FORMULA:	$C_{12}H_{24}O_{11}$
MOLECULAR MASS:	Monohydrate: 362; dihydrate: 380; mol.wt. 344.32
ALTERNATIVE FORMS:	Lactitol monohydrate/ Lactitol dihydrate
PROPERTIES AND APPEARANCE:	Monohydrate: white, sweet, odourless, crystalline solid, non-hygroscopic. Dihydrate: white, sweet, odourless, crystalline powder. Pleasant sweet taste (sweetness intensity 0.35 compared to sucrose); no aftertaste
MELTING RANGE IN °C:	Monohydrate salt: 94–97, also reported as 120; dihydrate salt: 75
SOLUBILITY % AT VARIOUS TEMPERATURE/pH COMBINATIONS:	
in water:	**@ 20°C** Monohydrate salt: 150 g/100 ml 206 g/100 g solvent at 25°C dihydrate salt: 140 g/100 ml at 25°C **@ 50°C** Monohydrate: 512 g/100 g solvent
in ethanol solution:	**@ 100%** Monohydrate: 0.75 g/100 g solution at 25°C 0.88 g/100 g solution at 50°C

FUNCTION IN FOODS:	Bulking agent: adds texture and mouthfeel properties. Sweetener: sweetness intensity 0.35 compared to sucrose; functionally similar to sucrose. Texturising agent. Lowers freezing point of solutions in a manner similar to that of sucrose; increases chemical stability; increases affinity for water without altering sweetening power; reduces tendency to crystallise. When used in baking: reduces fermentability; increases resistance to non-enzymic browning reactions; decreases tendency to crystallise; increases hygroscopicity
ALTERNATIVES:	Sucrose; other polyols

TECHNOLOGY OF USE IN FOODS:

Good solubility in water which increases with temperature (at 25°C, 150 g of monohydrate salt or 140 g dihydrate salt will dissolve in 100 ml water)

Monohydrate: at 25°C	206 g/100 g solvent (water)
	0.75 g/100 g solvent (ethanol)
	0.4 g/100 g solvent (ether)
at 50°C	512 g/100 g solvent (water)
	0.88 g/100 g solvent (ethanol)
at 75°C	917 g/100 g solvent (water)

Partially converted to lactitan, sorbitol and other lower polyols at 179 to 240°C.

Viscosity in solution equal to that of sucrose (weight for weight).

Decomposition a function of temperature and acidity. In solution, stable at pH 3.0 to 7.5 and at temperatures up to 60°C for up 1 month. A 10% solution demonstrated 15% decomposition at pH 3.0 after 2 months; 10% solution displayed no decomposition at 105°C at pH 12.0 after 2 months.

Less sweet than sugar, so require intense sweeteners as supplements to bring to equivalence with sucrose. Generally aspartame or acesulfame-K is used. A 10% lactitol solution containing 0.03% of either is equivalent to a 10% sucrose solution.

Can replace sucrose as a texturising or bulking agent where the sweet taste is beneficial. Result considered to have equal palatability and no aftertaste.

Low hygroscopicity and low calorific value, so suitable for use as a bulking agent for intense sweeteners in table-top use and biscuit making; crispness maintained as a result.

Suitable for inclusion in low-calorie and sugarless products such as chewing gum, fruit gums and pastilles, chocolate, instant beverages and jams, but a crystallisation inhibitor such as HGS (Hydrogenated Glucose Syrup: e.g. Lycasin®) may be required.

In solution, good at pH 3.0 to 7.5 and at temperatures below 60°C for 1 month

FOOD SAFETY ISSUES:

Low cariogenicity. Contains 50% fewer calories than sugars; energy: 2 kcal/g.

Suitable for inclusion in diabetic products because not dependent on insulin for metabolism and results in no significant change in blood glucose.

Low energy due to malabsorption in intestine, and can also act as dietary fibre as it is fermented by the microflora of the large intestine and contributes to faecal mass.

Laxative effect at high doses, so warning labels are required in some countries (laxative effect 70–80 g/day)

LEGISLATION:

UK and EUROPE:

One of 12 sweeteners listed as permissible for use in the UK as of 1988

JECFA allocated and ADI of "Not Specified' in 1983

CANADA:

Canada FDR 67.31 – Table IX – Food additives that may be used as Sweeteners – Item L.1 (14 December 1994)

– Permitted in Unstandardised Foods to a maximum level of use dictated by Good Manufacturing Practice

REFERENCES:

Smith, J. (1991) *Food Additive User's Handbook.* Blackie Publishing.

Canadian Food and Drugs Act and Regulations (1994).

McCue, N. (1996) *Showcase: Natural & High Intensity Sweeteners; Prepared Foods.* March, p. 88.

Matz, S. A. (1992) *Bakery Technology and Engineering.* 3rd edition. Van Nostrand Reinhold.

Smoley, C. K. (1993) *Everything Added to Food in the United States.* US Food and Drug Administration, CRC Press, Inc.

Budavari, S. (1996) *The Merck Index.* 12th edition. Merck and Co.

ANY OTHER RELEVANT INFORMATION:

Polyol produced through hydrogenation of lactose. Lactitol dihydrate developed by CC Biochem of the Netherlands. Sold by Philpot Dairy Products in UK as Lacty®; Lacty is marketed by Purac Biochem

NAME:	**Maltitol**
CATEGORY:	Sweetener/ Nutritive additive/ Humectant
FOOD USE:	Baked goods (glazes for baked goods, muesli bars)/ Dairy products (frozen desserts, ice-cream)/ Fruit, vegetables and nut products (fruit fillings, jams, jellies, canned fruit, fruit toppings)/ Beverages/ Soft drinks/ Sugars, sugar preserves and confectionery (chocolate and compound coatings, confectionery (especially gloss coatings), hard-boiled candies, soft caramels)/ Vinegar, pickles and sauces (sauces)/ Other (chewing gum, chewing gum coatings, pharmaceuticals, dietary products, diabetic products, chocolate, gelatin gums and jellies, sugarless tablets)
SYNONYMS:	CAS 585-88-6/ EINECS 209-567-0/ Amalty/ Finmalt-L/ 4-*O*-α-glucopyranosyl-D-sorbitol/ 4-*O*-α-D-glucopyranosyl-D-glucitol
FORMULA:	C12H24O11
MOLECULAR MASS:	344.47
PROPERTIES AND APPEARANCE:	White crystalline powder, also available in liquid form. Crystalline form 0.8 to 0.9 relative to sucrose, liquid form 0.6 relative to sucrose. Pleasant, sweet taste with no aftertaste; low mouth-cooling effect
MELTING RANGE IN °C:	135–140
PURITY %:	maximum 92.5 assay (anhydrous)
WATER CONTENT MAXIMUM IN %:	1.5
HEAVY METAL CONTENT MAXIMUM IN ppm:	10
ARSENIC CONTENT MAXIMUM IN ppm:	2.5
SOLUBILITY % AT VARIOUS TEMPERATURE/pH COMBINATIONS:	
in water:	**@ 20°C** 60% weight/unit volume

FUNCTION IN FOODS:	Bulking agent, so adds texture and mouthfeel to properties. Functionally similar to sucrose; have an affinity for water without altering sweetening power. Highly hygroscopic, so reduced tendency to crystallise. Aids moisture retention in baked goods. For use in baking: reduced fermentability; increased resistance to non-enzymic browning reactions; decreased tendency to crystallise; increased hygroscopicity; Enhances and retains shelf-life of baked goods. Crystalline form stabilises colour and improves shelf-life of fillings. Bulk sweetening agent and glaze-former. Syrup form: excipient; humectant: anti-crystallisation agent.
TECHNOLOGY OF USE IN FOODS:	Functionally similar to sucrose. More chemically stable; heat stable, no loss of colour during boiling. Low fermentability by common moulds and bacteria. Less sweet than sugar, so require intense sweeteners as supplements to bring to equivalence with sucrose. Aids moisture retention in baked goods. Energy: 2 kcal/g No browning reaction; medium solubility in water at room temperature; chemically and thermally stable. Available in both crystalline and liquid forms. In liquid form, has applications in sugar-free confections, chewing gums, pharmaceuticals, sauces and variegates, baked goods and frozen desserts. Crystalline maltitol has applications in chocolate, confectionery, chewing gum, baked goods and fruit spreads. Amalty MR Grades crystalline maltitol: – % sulphates: 0.010 maximum – % chlorides: 0.005 maximum – % reducing sugars: 0.30 maximum – % residue on ignition: 0.1 maximum – Lead: 1 ppm maximum – Classification: polyhydric alcohol (polyol) – Mesh size: MR 20, 100% through #20 US sieve – Mesh size: MR 50, 100% through #20 US sieve

- Mesh size: MR 100, 80% maximum through #20 US sieve
- ERH at 20°C: 89%

Slightly soluble in alcohol; heat of solution: –5.5 to –16.3 cal/g; solubility at 37°C: 201 g/100 ml water.

Cooling effect (150 g powder/50 ml water at 37°C): –12

Sweetening power: 0.9

FOOD SAFETY ISSUES:

Low energy due to malabsorption in intestine, so suitable for inclusion in diabetic products because not dependent on insulin for metabolism. Does not significantly increase serum glucose or serum insulin levels after ingestion

Non-cariogenic as it is not metabolised by oral bacteria.

Laxative threshold of 100 g/day.

Material Safety Data Sheet for AMALTY Crystalline Maltitol

- pH: Neutral
- % volatile by volume: negligible
- Extinguishing media: water fog, alcohol foam, carbon dioxide, dry chemical, halogenated agents.
- Special fire-fighting protective equipment: self-contained breathing apparatus with full facepiece and protective clothing
- Unusual fire and explosive hazards: this product may form explosive dust clouds in air
- Incompatibility: oxidising agents
- Hazardous decomposition products: carbon dioxide, carbon monoxide
- Ingestion: LD_{50} in rats above 25 g/kg (2500 mg/kg). Classified as 'relatively harmless' by ingestion
- First-aid procedures:

Skin: wash off with plenty of soap and water. If redness, itching or a burning sensation develops, obtain medical attention

Eyes: immediately flush with plenty of water for at least 15 minutes. If redness, itching, or a burning sensation develops, have eyes examined and treated by medical personnel

Ingestion: do not induce vomiting. Give one or two glasses of water to drink and refer to medical personnel or take direction from either a physician or a poisons control centre. Never give anything to an unconscious person

Inhalation: remove victim to fresh air. If a cough or other respiratory symptoms develop, consult medical personnel

- In case material is released or spilled: sweep up and recover or mix material with a moist absorbent and shovel into a chemical waste container. Wash residue from spill area with water and flush to a sewer serviced by a permitted waste water treatment facility
- Container disposal: empty container retains product residue. Do not distribute, make available, furnish or reuse empty container except for storage and shipment of original product. Empty container, remove all product residue, puncture or otherwise destroy empty container before disposal
- TLV or suggested control value: minimise exposure in accordance with good hygiene practice.
- Ventilation: use ventilation adequate to maintain safe levels
- Respiratory protection: use MSHA-NIOSH approved respirator for organic vapours, dusts and mists.
- Protective clothing: impervious gloves and apron
- Eye protection: safety glasses with side shields
- Other protective equipment: eyewash station in work area.

LEGISLATION:

USA:

Not permitted for use in foods

UK and EUROPE:

Used in some European and Asian countries in dark and milk chocolate, hardboiled candies, soft caramels, arabic, gelatin gums and jellies, chewing gums and bubble gums, chocolate dragees, sugarless tablets, muesli bars, jams, and ice-cream

CANADA:

Crystalline Maltitol

Table IX – Food additives that may be used as Sweeteners: Item M.1 (25 May 1993)

Permitted in or on	**Maximum level of use**
Unstandardised foods	Good Manufacturing Practice

Maltitol syrup

Table IX – Food additives that may be used as Sweeteners: Item M.2 (25 May 1993)

Permitted in or on	**Maximum level of use**
Unstandardised foods	Good Manufacturing Practice

AUSTRALIA/PACIFIC RIM:

Japan: permitted for use in certain foods; has been used since before 1981

REFERENCES:	Smith, J. (1991) *Food Additive User's Handbook.* Blackie Publishing, Glasgow. Canadian Food and Drug Act and Regulations (1994). *Chilton's Food Engineering Master '95.* (1994) ABC Publishing Group. p. 204; 270. SPI Polyols technical specifications (1996). SPI Polyols material safety data sheet, 1 April (1996). Matz, S. A. (1992) *Bakery Technology and Engineering.* 3rd edition. Van Nostrand Reinhold. *Roquette technical bulletin*. (1996) As provided by Kingley & Keith (Canada) Inc. Smoley, C. K. (1993) *Everything Added to Food in the United States.* US Food and Drug Administration, CRC Press Inc. Wong, D. W. S. (1989) *Mechanism and Theory in Food Chemistry.* Van Nostrand Reinhold.
ANY OTHER RELEVANT INFORMATION:	Polyol produced through hydrogenation of maltose; Marketed as Malbit® Lycasin – Roquette America Inc., Maltitol Syrup SunMalt – Mitsubishi International Corporation, Fine Chemicals Department: crystalline maltose; maltose monohydrate carbohydrate sweetener Maltisorb – Roquette America Inc.: crystalline maltitol

NAME:	Mannitol
CATEGORY:	Sweetener/ Nutritive additive/ Bulking agent
FOOD USE:	Baked goods/ Sugars, sugar preserves and confectionery (sugar-free confections)/ Other (chewable tablets, chewing gum, vitamins, chocolates, effervescent powders, intravenous osmotic diuretics, oral intestinal transit enhancers)
SYNONYMS:	CFSAN D-Mannitol/ CAS 69-65-8/ EINECS 200-711-8/ E421/ Mannose sugar/ Mannite/ Manna sugar/ Cordycepic acid/ 1,2,3,4,5,6-Hexanehexol/ UniSweet MAN/ Manicol/ Mannidex/ Diosmol/ Osmitrol/ Osmosal/ Resectisol
FORMULA:	C6H14O6
MOLECULAR MASS:	182.17
PROPERTIES AND APPEARANCE:	Pleasant, sweet taste with no aftertaste (0.6 relative to sucrose). White crystalline, odourless, slightly sweet powder. Low mouth-cooling effect
BOILING POINT IN °C AT VARIOUS PRESSURES (INCLUDING 760 mm Hg):	290–295
MELTING RANGE IN °C:	165–169
ARSENIC CONTENT MAXIMUM IN ppm:	1
SOLUBILITY % AT VARIOUS TEMPERATURE/pH COMBINATIONS:	
in water:	**@ 20°C** 14.5 g/100 g solution **@ 50°C** 31 g/100 g solution
in ethanol solution:	**@ 100%** Very slightly soluble
FUNCTION IN FOODS:	Bulking agent, so adds texture and mouthfeel to properties. More chemically stable. Have an affinity for water without altering sweetening power. Reduced tendency to crystallise. Anti-adhesion agent. Excipient for use in baking: reduced fermentability; increased resistance to non-enzymic browning reactions; decreased tendency to crystallise; increased hygroscopicity

TECHNOLOGY OF USE IN FOODS:

Functionally similar to sucrose. Naturally occurring; low solubility in water. Useful as an anti-adhesion agent (inhibits crystallisation of other polyalcohols) in the manufacture of chewing gum.

Less sweet than sugar, so require intense sweeteners as supplements to bring to equivalence with sucrose. Generally used in chewing gums as a sugar substitute, supplemented by intense sweeteners, typically saccharin (added in a small quantity to keep saccharin's taste imperceptible). Claimed to require no intense sweetener to have near-sucrose equivalence in sweetness and flavour-release properties. Mainly used in crystalline form in sugar-free chewing gum. Also used in chewable pharmaceutical products because it is inert to most drug components.

Due to low solubility, not generally used in soft drinks or ice-cream.

Low hygroscopicity; no browning reaction; chemically stable.

Applications: sugar-free confections, chocolates, chewing gums, tablets, vitamins; use in chocolate, cocoa and confectioneries limited by low solubility; special use in sugarless chewing gum due to low solubility; special use in effervescent powders due to low hygroscopicity.

Solubility in water:

0°C = 9.1 g/100 g solution
10°C = 12.3 g/100 g solution
20°C = 14.5 g/100 g solution
30°C = 20 g/100 g solution
40°C = 25 g/100 g solution
50°C = 31 g/100 g solution
60°C = 37.5 g/100 g solution

Pharmaceutical applications: Excipient: ideal for dry formulations due to pleasant taste, high physico-chemical stability and lack of reducing power and low hygroscopicity; ideal freeze-drying carrier due to quick-drying characteristics. Results in enhanced drug stability; freeze-dried powders using this dissolve quickly; granulated form has excellent flow and compression properties.

Pharmaceutical applications: Active properties: due to special biological properties, can be used in applications such as intravenous osmotic diuretics and oral intestinal transit enhancers; mannitol hexanitrate has vasodilating properties similar to nitrated esters.

Food applications: recommended as sweetener in confectionery or in sugarless confectionery; may be used as an anti-crystallising agent in chewing gums or bubble gums; useful as dusting product due to low hygroscopicity; used as an aroma carrier; used for flow-improving; used as a mould-releasing additive.

Loss on drying: 0.3% max; reducing sugars: nil.

Specific optical rotation in borax solution: +23° to +24°; specific optical rotation in molybdic solution: +137° to +145°

Acidity (NaOH 0.02 N for 5 g): 0.3 ml max.

Chlorides: 50 ppm max; sulphates: 100 ppm max.

Solubility: 1 g/5.5 ml water; readily soluble in alkaline solutions; very slightly soluble in pyridine; very slightly soluble in alcohol (1 g dissolves in 83 ml alcohol at room temperature); practically insoluble in ether; soluble in glycerol (1 g dissolves in 18 ml glycerol at room temperature).

Compliance: USP XXII

B.P 88 French Pharmacopoeia. EEC Directive E421

Particle size: crystalline, standard grade – 10% max residue on 250 microns (60 mesh); crystalline, F grade – 2% max residue on 150 microns (100 mesh); crystalline, SF grade – 0.5% max residue on 100 microns (150 mesh); granular, Pearlitol® MG – 20% max residue on 500 microns (32 mesh); granular, Pearlitol® MG – 85% min residue on 100 mirons (150 mesh); granular, Pearlitol® GG2 – 0.5% max residue on 840 microns (20 mesh); granular, Pearlitol® GG2 – 90% min residue on 150 microns (100 mesh)

FOOD SAFETY ISSUES: Suitable for inclusion in diabetic products because not dependent on insulin for metabolism. Low energy due to malabsorption in intestine (<4 kcal/g). Useful for caries prevention. Laxative effect at high doses, so warning labels are required in some countries (20 g/day); has a low laxative threshold so is not recommended for diabetic products, and glucose is a breakdown product in the gut.

LEGISLATION:

USA:

CFR 21

PART 100 – General

Subpart G®Specific Administrative Rulings and Decisions 100.130 Combinations of nutritive and non-nutritive sweeteners in "diet beverages".

(4) To avoid confusion by diabetics, the label of a beverage containing sorbitol, mannitol, or other hexitol, must bear the statement "Contains carbohydrates, not for use by diabetics without advice of a physician". To further avoid confusion of these beverages with those sweetened solely with non-nutritive artificial sweeteners which have been marketed in containers bearing prominent statements such as "sugar free", "sugarless", or "no sugar", the labels of beverages containing hexitols must not bear these or similar statements.

PART 180®Food Additives Permitted in Food on An Interim Basis Or in Contact with Food Pending Additional Study

Subpart B®Specific Requirements for Certain Food Additives 180.25 Mannitol.

[42 FR 14636, Mar. 15, 1977, as amended at 49 FR 5610, Feb. 14, 1984]

(a) Mannitol is the chemical 1,2,3,4,5,6-hexanehexol (C6H14O6) a hexahydric alcohol, differing from sorbitol principally by having a different optical rotation. Mannitol is produced by the electrolytic reduction, or the transition metal catalytic hydrogenation, of sugar solutions containing glucose or fructose.

(b) The ingredient meets the specifications of the *Food Chemicals Codex*, 3rd edn. (1981), pp. 188–190, which is incorporated by reference. Copies may be obtained from the National Academy Press, 2101 Constitution Ave. NW., Washington, DC 20418, or may be examined at the Office of the Federal Register, 800 North Capitol Street, NW., suite 700, Washington, DC 20408.

(c) The ingredient is used as an anticaking agent and free-flow agent as defined in §170.3(o)(1) of this chapter, formulation aid as defined in §170.3(o)(14) of this chapter, firming agent as defined in §170.3(o)(10) of this chapter, flavouring agent and adjuvant as defined in §170.3(o)(12) of this chapter, lubricant and release agent as defined in §170.3(o)(18) of this chapter, nutritive sweetener as defined in §170.3(o)(21) of this chapter, processing aid as defined in §170.3(o)(24) of this chapter, stabiliser and thickener as defined in §170.3(o)(28) of this chapter, surface-finishing agent as defined in §170.3(o)(30) of this chapter, and texturiser as defined in §170.3(o)(32) of this chapter.

(d) The ingredient is used in food at levels not to exceed 98 percent in pressed mints and 5 percent in all other hard candy and cough drops as defined in §170.3(n)(25) of this chapter, 31 percent in chewing gum as defined in §170.3(n)(6) of this chapter, 40 percent in soft candy as defined in §170.3(n)(38) of this chapter, 8 percent in confections and frostings as defined in §170.3(n)(9) of this chapter, 15 percent in non-standardised jams and jellies, commercial, as defined in §170.3(n)(28) of this chapter, and at levels less than 2.5 percent in all other foods.

(e) The label and labelling of food whose reasonably foreseeable consumption may result in a daily ingestion of 20 grams of mannitol shall bear the statement "Excess consumption may have a laxative effect".

(f) In accordance with §180.1, adequate and appropriate feeding studies have been undertaken for this substance. Continued uses of this ingredient are contingent upon timely and adequate progress reports of such tests, and no indication of increased risk to public health during the test period.

(g) Prior sanctions for this ingredient different from the uses established in this regulation do not exist or have been waived.

UK and EUROPE:

One of 12 sweeteners listed as permissible for use in the UK as of 1983

Subject to certain restrictions, permitted as sweetener or food additive in UK, Belgium, Denmark, Greece, Spain, France, Germany, Switzerland and Sweden

CANADA:

Canada FDA: 67.31

Table IX – Food additives that may be used as Sweeteners – Item M.3 (25 May 1993)

Permitted in or on	**Maximum level of use**
Unstandardised foods	Good Manufacturing Practice

Page 67–25a of Canada FDR – 1996 update: sweetener and texture modifier in carbohydrate or calorie reduced foods meeting the requirements of B.24.004 and B.24.006 – Good Manufacturing Practice. Release agent and sweetener in confectionery – Good Manufacturing Practice.

AUSTRALIA/PACIFIC RIM:

Japan and Australia: subject to certain restrictions, permitted as sweetener or food additive

OTHER COUNTRIES:

South Africa: subject to certain restrictions, permitted as sweetener or food additive

REFERENCES:

Smith, J. (1991) *Food Additive User's Handbook.* Blackie Publishing, Glasgow.

Canadian Food and Drugs Act and Regulations (1994).

Chilton's Food Engineering Master '95. (1994) ABC Publishing Group. p. 270.

Minifie, B. W. (1989) *Chocolate, Cocoa, and Confectionery: Science and Technology.* 3rd edition. Van Nostrand Reinhold.

Matz, S. A. (1992) *Bakery Technology and Engineering.* 3rd edition. Van Nostrand Reinhold.

Food Additives Update. Food in Canada, April (1996).

Roquette technical brochure. (1996) Provided by Kingsley & Keith (Canada) Inc.

Smoley, C. K. (1993) *Everything Added to Food in the United States.* US Food and Drug Administration, CRC Press, Inc.

Wong, D. W. S. (1989) *Mechanism and Theory in Food Chemistry.* Van Nostrand Reinhold.

Budavari, S. (1996) *The Merck Index.* 12th edition. Merck and Co.

ANY OTHER RELEVANT INFORMATION:

Polyhydric alcohol produced through hydrogenation of mannose.

Developed in the 1970s; sold as Malbit® (90% maltitol and 5% maltotritol). Roquettes Frères applied for French patent in 1987 for compressible maltitol powder (85% maltitol); relatively expensive due to the process required to separate it from sorbitol and other contaminants. A hexahydric sugar alcohol; isomeric with sorbitol; occurs widely in nature – celery, larch and manna ash (*Fraxinus ornus*) are examples. Juice of manna ash called "manna"; found in high concentrations (15–20%) in certain varieties of mushrooms, marine algae (particularly of genus *Laminaria*).

NAME:	**Monellin**
CATEGORY:	Sweetener/ Nutritive additive
FORMULA:	Polypeptide chain of approximately 91 amino acids
MOLECULAR MASS:	11,500 daltons
PROPERTIES AND APPEARANCE:	Sweet taste (3000 times that of sucrose by weight). Slow onset time; lingering sweet aftertaste (for up to 1 hour)
FUNCTION IN FOODS:	Sweetening agent: up to 3000 times sweeter than sucrose
TECHNOLOGY OF USE IN FOODS:	Stable at pH 2 to 10. Sweetness irreversibly lost above 60°C. Unstable to heat. Complete loss of sweetness below pH 2 when held in solution at room temperature
REFERENCES:	Smoley, C. K. (1993) *Everything Added to Food in the United States.* US Food and Drug Administration, CRC Press, Inc. *New and Novel Foods*; p. 486. *Food Chemistry.* 2nd edition. (1985) Marcel Dekker, Inc. Budavari, S. (1996) *The Merck Index.* 12th edition. Merck & Co.
ANY OTHER RELEVANT INFORMATION:	Derived from the fruit of the *Dioscoreophyllum cumminisii* (also known as serendipity berries). Characteristics similar to thaumatin. Expensive

NAME:	**Saccharin**
CATEGORY:	Sweetener
FOOD USE:	Baked goods/ Fruit, vergetables and nut products (fruit drinks, preserves)/ Beverages (cola beverages, coffee)/ Soft drinks/ Sugars, sugar preserves and confectionery (table-top sweeteners, candies, preserves, chocolate products)/ Vinegar, pickles and sauces (salad dressings)/ Other (oral hygiene products, chewing gum, gelatin desserts, cocoa)
SYNONYMS:	Insoluble form: CFSAN Saccharin/ CAS 81-07-2/ EINECS 201-321-0/ EINECS 220-120-9/ Saccharin insoluble/ *O*-Benzoic acid sulfimide/ 3-Oxo-2,3-dihydro-1,2-benzisothiazole-1,1-dioxide/ 2,3-Dihydroxy-1,2-benzothiazolin-3-one-1,1-dioxide/ 1,2-Benzisothiazol-3(2*H*)-one-1,1-dioxide/ 2,3-Dihydro-3-oxobenzisosulfonazide/ 2,3-Dihydro-3-oxobenzisosulfonazole/ 1,2-Dihydro-2-ketobenzisosulfonazole/ 1,2-Benzisothiazolin-3-one-1,1-dioxide/ *O*-sulfobenzoic acid imide/ Benzoic sulfimide/ Sycal SDI/ Unisweet SAC/ Benzosulfimide/ Benzoic sulfimide/ Gluside/ Glucid/ Garantose/ Saccharinol/ Saccharinose/ Saccharol/ Saxin/ Sykose/ Hermesetas/ *O*-Sulfobenzimide. Ammonium salt: 1,2-Benzisothiazolin-3-one, 1,1-dioxide, ammonium salt/ CFSAN Saccharin, ammonium salt/ Sucline/ Daramin/ Saccharin ammonium/ CAS 6381-61-9/ EINECS 228-971-8. Calcium salt: CAS 6381-91-5/ 1,2-Benzisothiazolin-3(2*H*)-one, 1,1-dioxide, calcium salt/ CAS 6485-34-3/ EINECS 229-349-9/ Syncal CAS/ CFSAN Saccharin, calcium salt Sodium salt (dihydrate): Sodium saccharide/ Sodium benzosulfimide/ 1,1-Dioxide-1,2-benzisothiazol-3(2*H*)-one, sodium salt/ Sodium 2,3-dihydro-1,2-benzisothiazolin-3-one-1,1-dioxide/ Saccharin sodium/ Crystallose/ Dagutan/ Soluble saccharin/ CFSAN Saccharin, sodium salt/ CAS 128-44-9/ EINECS 204-886-1/ Saccharin soluble/ Sodium saccharine/ Sodium *O*-benzosulfimide/ Kristallose/ Sucaryl/ Sucromat Sodium salt (anhydrous): Sucredulcor
FORMULA:	Insoluble: $C_7H_5NO_3S$; sodium salt: $C_7H_4NNaO_3S$-$2H_2O$; ammonium salt: $C_7H_8N_2O_3S$
MOLECULAR MASS:	Sodium salt 205.16; insoluble 183.19
ALTERNATIVE FORMS:	Ammonium saccharin/ Calcium saccharin/ Sodium saccharin/ Insoluble saccharin

PROPERTIES AND APPEARANCE: Sodium saccharin: 200–700 times sweetness of sucrose.

White crystalline powder; bitter, astringent or metallic off-taste, particularly objectionable in delicately fruit-flavoured products and at high concentrations.

Characteristics of sweeteners (saccharin alone)

Sweetness intensity (at 10% sucrose):	350
Sweetness quality:	Slightly chemical sweetness
Time profile:	Slower and persistent
Associated taste:	Bitter/metallic
Mouthfeel body:	Thin
Enhancement of fruitiness:	Nil

Characteristics of sweeteners (1 : 10 saccharin : cyclamate)

Sweetness intensity (at 10% sucrose):	100
Sweetness quality:	Sugar-like
Time profile:	As sucrose
Associated taste:	None
Mouthfeel body:	Good
Enhancement of fruitiness:	Good

MELTING RANGE IN °C: Insoluble: 229–230

SOLUBILITY % AT VARIOUS TEMPERATURE/pH COMBINATIONS:

in water: **@ 20°C** saccharin: 3 g/l
sodium saccharin: 700 g/l (83%)
calcium saccharin: 400 g/l (67%)

in ethanol solution: **@ 100%** sodium saccharin at 20°C: 20 g/l
calcium saccharin at 20°C: 200 g/l

FUNCTION IN FOODS: Non-nutritive sweetener

TECHNOLOGY OF USE IN FOODS:
Highly stable and relatively inexpensive. Mixed with other products to mask taste, provide bulk and take advantage of synergy. May also be combined with cream of tartar, glucono-δ-lactone, sodium gluconate, glycols, gentian root, maltol, pectin, lemon flavour, ribonucleotides, adipic acid, aldohexuronic acid, and citric acid.

3:1 cyclamate:saccharin combination provided sugar-like sweetness in beverages until cyclamates were banned in the US in 1969.

Cyclamate combination replaced with a calcium chloride combined with cornstarch hydrolysate, lactose, sucrose, tartrates and fructose with gluconate salts.

High stability, even under extreme processing conditions. Only approved sweetener able to withstand heating, baking and high-acid media.

Has been used in soft drinks, candies, preserves, salad dressings, low-calorie gelatin desserts. Also combined with bulk sweeteners in baking for sugar-reduced products; used alone as table-top sweetener in tablet and liquid form, or in chewing gum.

In combination with other sweeteners, used as table-top sweetener; combined with sorbitol or aspartame, used in chewing gum; also popular for use in oral hygiene products.

Relative sweetness in soft drinks in range of 300 to 700 units; sodium saccharin sweetneess in soft drinks is 360 to 500 units.

Some people are more sensitive than others to the bitter/metallic aftertaste of saccharin; aftertaste can be masked using fructose, gluconates, tartrates, ribonucleotides, sugars, sugar alcohols (polyols) and other intense sweeteners.

1/20th price of sugar in terms of sweetness equivalency.

Stable in pH range 2 to 7; heat stable – unchanged after 1 hour at 150°C in pH 3.3 to 8.0; no browning reaction. In dry form, is stable for several years when stored appropriately. Stable under normal soft-drink processing conditions.

Lower cost versus sucrose.

When used alone in soft drinks did not provide sweetness taste quality of the saccharin:cyclamate blend.

Does not interact with other ingredients encountered during soft-drink manufacture. Concentrated soft-drink solutions can be stored. Detected in beverages using HPLC or spectrophotometric techniques.

Excellent heat and pH stability. Often blended with aspartame to reduce bitter aftertaste; has bitter, astringent or metallic off-taste particularly objectionable in delicately fruit-flavoured products; off-notes may be partially concealed in foods containing sucrose or corn sweeteners; off-notes less obvious in cola beverages, hot cocoa and other chocolate products, coffee, etc.

Stable under normal storage and preparation conditions; stable in acidic environments and under extended heat treatment.

Commercially successful in fruit drinks as:

- Sucrose/saccharin – 25–35% of sweetness from sacccharin with total cost 5% below that of using sucrose alone, dominated UK fruit drink market in 1960s and 1970s. Some consumers preferred this product to that with sucrose alone. Also marketed in a 500:1 ratio for "light" drinks in the US with 50% calorie reduction and good sweetness and fruitiness.
- Aspartame:saccharin – in ratio of 2:1 or 50%:50% sweetness gives good sweetness and saccharin stabilises total sweetness to extend shelf-life.
- Fructose/saccharin – strong sweetness intensity synergism and good fruitiness enhancement.
- Cyclamate/saccharin – at 10:1 ratio or 50%:50% sweetness gives clean, sugar-like sweetness at low cost with good storage stability.

Good stability in fruit juice beverages; readily soluble in fruit beverages as either sodium or calcium salt; compatible with other fruit beverage ingredients.

SYNERGISTS:

With isomalt, it masks the metallic aftertaste of saccharin or aspartame.

Synergistic with cyclamates, aspartame, sucrose, isomalt, stevioside, NHDC (Neohesperidin); synergism has been reported with fructose, sorbitol, xylitol and sucralose

FOOD SAFETY ISSUES:

May have anticariogenic properties. Is not metabolised by the body, so has no calorific value; is excreted unchanged.

Some research has shown an increase in bladder tumours in rats fed a saccharin : cyclamate blend; this was later confirmed to be caused by the saccharin component of the blend. Effect was determined to be species-specific as studies of those who consume large doses, such as diabetics, revealed no link between saccharin and bladder tumours in humans.

LEGISLATION:

USA:

FDA withdrew GRAS in 1972 and attempted to ban its use in 1977.
All products packaged in the US must bear a statement warning that saccharin has been shown to cause cancer in laboratory animals.
Only all-purpose non-nutritive sweetener approved for use in the US.
ADI level as of 1990 for use in soft-drinks: 12 mg/fluid ounce
CFR 21
PART 100 – General
Subpart G®Specific Administrative Rulings and Decisions
100.130 Combinations of nutritive and non-nutritive sweeteners in "diet beverages".
As a result of the removal of cyclamic acid and its salts from the list of substances generally recognised as safe (part 182 of this chapter) by an order published in the FEDERAL REGISTER of October 21, 1969 (34 FR 17063), the Commissioner of Food and Drugs has received inquiries as to the proper composition and labelling, from the standpoint of application of the Federal Food, Drug, and Cosmetic Act, of so-called "diet beverages" that will be made from mixtures of nutritive sweeteners and saccharin or its salts. The Commissioner concludes that:
(2) The label must bear a statement of the caloric content per fluid ounce, the carbohydrate content per fluid ounce, a statement of the percentage of saccharin or saccharin salt used, and the statement "Contains •• mg saccharin (or saccharin salt, as the case may be) per ounce, a non-nutritive artificial sweetener."
CFR 21
PART 100 – General
Subpart A – General Provisions

101.11 Saccharin and its salts; retail establishment notice.

[43 FR 8795, Mar. 3, 1978]

Each retail establishment (except restaurants) that sells food that contains saccharin shall display the following notice in the locations set forth in paragraph (b) of this section:

Each notice shall be displayed prominently, in a manner highly visible to consumers (e.g., not shielded by other store signs or merchandise displays) and set up to reduce the likelihood that a notice will be torn, defaced, or removed.

(a) The notice shall be printed in a combination of red and black ink on white card stock and be at least 11 by 14 inches. The background of the bold heading, "Saccharin Notice," and the boxed warning statement shall be bright red and the lettering, white. The remaining background shall be white with black ink. All lettering shall be in gothic typeface.

(b) Except as provided in paragraph (c) of this section, each retail establishment that sells food that contains saccharin shall display a notice in each of the following three locations:

(1) Near the entrance to the retail establishment and arranged so that consumers are likely to see the notice upon entering.

(2) Centrally located in the area of the retail establishment in which soft drinks containing saccharin are displayed. If there is more than one such place, then in the area where the greatest quantity of diet soft drinks are displayed.

(3) In the area in the establishment in which the largest quantity of saccharin-containing foods (including saccharin sold in package form as a sugar substitute) are displayed, other than the area where diet soft drinks are displayed.

(c) The following are exceptions to the requirements set forth in paragraph (b) of this section:

(1) A retail establishment with 3,200 square feet or less of floor space shall display at least one notice. The notice shall be located near the entrance to the retail establishment and arranged so that consumers are likely to see the notice upon entering.

(2) A retail establishment with more than 3,200 but less than 10,000 square feet of floor space shall display at least two notices. The first notice shall be located near the entrance to the retail establishment and arranged so that consumers are likely to see the notice upon entering. The second notice shall be centrally located in the area of the retail establishment in which soft drinks containing saccharin are displayed. If there is more than one such place, then in the area where the greatest quantity of diet soft drinks are displayed. If diet soft drinks are not sold, then in the area of the establishment in which the largest quantity of saccharin-containing foods (including saccharin sold in package form as a sugar substitute) are displayed.

(3) A large retail establishment, e.g., department store, whose primary business consists of selling nonfood items (i.e., the proportion of food sold is extremely small compared to other items) shall display at least one notice. The notice shall be located in the area of the establishment in which foods containing saccharin are displayed. If there is more than one such area, then a notice shall be displayed in each area.

(d) Each manufacturer of saccharin-containing food who customarily delivers his products directly to retail establishments shall make available at least three notices to each retail establishment in which his products are sold. Each manufacturer shall also arrange to supply additional notices to a retail establishment that asks for them.

(e) Manufacturers who do not customarily deliver their saccharin-containing food products directly to retail establishments may fulfill their obligation to provide notices either in the manner set forth in paragraph (d) of this section or by participating in, and performing the actions required by, a trade association coordinated program that meets the following requirements:

(1) The coordinating association shall have filed notice of the program with the Food and Drug Administration, including the association's name, mailing address, telephone number, and contact person.

(2) Each manufacturer participating in the program shall file notice of its participation with the coordinating association, including its name, mailing address, telephone number, and contact person.

(3) The association shall ensure that retail establishment notices, in the form specified in this section, are readily available to participating manufacturers.

(4) The association shall take affirmative steps to coordinate with retail establishments, their trade associations, and the trade press to disseminate information about the applicable requirements of the Saccharin Study and Labeling Act and these regulations, the existence of the association coordinated program, and the availability of notices through the program.

(5) Each manufacturer shall, in consultation with the association, communicate with its contacts in the distributional chain to inform them of the applicable requirements of the Saccharin Study and Labeling Act and these regulations, and the continued availability of notices.

(6) Each manufacturer shall ensure that notices are promptly provided on request to any retail establishment carrying its products.

(7) The association shall consult with participating manufacturers concerning the implementation and progress of the program and shall disseminate information to facilitate the conduct of the program based on such consultations or consultation with the Food and Drug Administration.

(8) The association shall, on request, permit the Food and Drug Administration to have access to the participation notices filed by manufacturers, samples showing the form of retail establishment notices made available, and typical communication materials used by the association in the course of the program.

UK and EUROPE:

Approved in UK by FACC in 1982, JECFA and EEC in 1984

MAFF in UK assigned a Group B status in Sweeteners in Food Regulations in 1983.

The SCF gave saccharin an ADI of 0 to 2.5 mg/kg body weight in December 1987 with proviso that this should come under review when further evidence became available.

JECFA ADI of 0 to 2.5 mg/kg

Regulatory Status: ADI level as of 1990 for use in soft drinks

– JECFA	0 to 2.5 mg/kg body weight
– Belgium	125 mg/l
– Denmark	75 mg/l
– Finland	30 to 70 mg/l depending upon type of soft drink
– France	100 mg/l depending upon type of soft drink

– East Germany	20 to 60 mg/l depending upon type of soft drink
– West Germany	200 mg/l
– Greece	Permitted
– Ireland	Permitted
– Netherlands	125 mg/l (pending approval to increase from 120 mg/l)
– Norway	50 to 120 mg/l depending upon type of soft drink
– Spain	200 mg/l
– Switzerland	Permitted
– Turkey	Permitted
– UK	80 mg/l
– USSR	Permitted
– Yugoslavia	180 mg/day

CANADA:

ADI level as of 1990 for use in soft drinks – Not permitted

Part E – Cyclamate and Saccharin Sweeteners

E.01.001 – (1) In this Part, "saccharin sweetener" means

(a) saccharin or a salt thereof, or

(b) any substance containing saccharin or a salt thereof that is sold as a sweetener

(2) Part B of these Regulations does not apply to any cyclamate sweetener or saccharin sweetener

E.01.002 – (**Sale**) No person shall

(a) sell a cyclamate sweetener or a saccharin sweetener that is not labelled as required by this Part; or

(b) commencing Jun 15, 1978, sell any saccharin sweetener to the general public except on the premises of a pharmacy

E.01.003 – (**Advertising**) No person shall, in advertising to the general public a cyclamate sweetener or saccharin sweetener, make any representation other than with respect to the name, price and quantity of the sweetener

E.01.004 – (**Labelling**)

(1) Every cyclamate sweetener that is not also a saccharin sweetener shall be labelled to state that such sweetener should be used only on the advice of a physician

(2) Commencing June 1, 1979, every saccharin sweetener shall be labelled to state that

(a) continued use of saccharin may be injurious to health; and

(b) it should not be used by pregnant women except on the advice of a physician.

E.01.005 – Commencing June 1, 1979, every cyclamate sweetener or saccharin sweetener shall be labelled to show
(a) a list of all the ingredients and, in the case of
(iii) a saccharin
(iv) a saccharin salt
the quantity thereof contained in the sweetener; and
(b) the energy value of the sweetener expressed in calories
(i) per teaspoonful, drop, tablet or other measure used in the directions for use, and
(ii) per 100 grams or 100 millilitres
of the sweetener

AUSTRALIA/PACIFIC RIM:

Regulatory Status: ADI level as of 1990 for use in soft drinks

– Australia	50 mg/l
– Japan	300 mg/l
– New Zealand	100 mg/l

OTHER COUNTRIES:

Has been assigned an ADI of 2.5 mg/kg body weight.
Used in more than 80 countries.
Permitted for use in approximately 75 countries.
Regulatory Status: ADI level as of 1990 for use in soft drinks

– Argentina	150 mg/l
– Brazil	500 mg/l
– Israel	44 mg/l
– Kenya	Permitted
– Mexico	400 mg/l
– Saudi Arabia	Not permitted
– South Africa	500 mg/l

determined safe by WHO

REFERENCES:

Smith, J. (1991) *Food Additive User's Handbook*, Blackie Publishing.
Mitchell, A. J. (1990) *Formulation and Production of Carbonated Soft Drinks*, Blackie Publishing.
New and Novel Foods, p. 484.

Canadian Food and Drugs Act and Regulations, 1994.

Minifie, B. W. (1989) *Chocolate, Cocoa, and Confectionery: Science and Technology*, 3rd edition, Van Nostrand Reinhold.

Matz, S. A. (1992) *Bakery Technology and Engineering*; 3rd edition, Van Nostrand Reinhold.

Food Chemistry, 2nd edition, Marcel Dekker Inc., 1985.

Hicks, D. (1989) *Production and Packaging of Non-Carbonated Fruit Juices and Fruit Beverages*, Blackie Publishing.

Smoley, C. K. (1993) *Everything Added to Food in the United States*, U.S. Food and Drug Administration, CRC Press Inc.

Wong, D. W. S. (1989) *Mechanism and Theory in Food Chemistry*, Van Nostrand Reinhold.

Budavari, S. (1996) *The Merck Index*; 12th edition; Merck and Co.

ANY OTHER RELEVANT INFORMATION:

Discovered accidentally by Fahlberg and Remsen in 1879 at Johns Hopkins University. First manufactured in 1884 as an antiseptic and preservative; first commercial patent for manufacture in 1885. Sold as a sweetener in 1900. Was banned for use in food and drinks in Germany in 1898 and in the US in 1912. Was reapproved during World War I due to a sugar shortage.

Synthesised commercially from toluene; one of the most widely used sugar substitutes in the world; was the first high-intensity sweetener to be marketed.

Use of saccharin : cyclamate blends ended with cyclamate ban in 1970.

NAME:	Sorbitol
CATEGORY:	Sweetener/ Nutritive additive/ Humectant/ Bulking agent/ Anti-caking agent
FOOD USE:	Baked goods (cookies, cakes, icings, fillings, biscuits)/ Sugars, sugar preserves and confectionery (confections, fondants, fudge, marshmallows, table-top sweetener, "boiled" sweets)/ Fruit, vegetables and nut products (preserves, jellies, fillings)/ Dairy products (ice-cream, frozen desserts, sorbets)/ Fish and seafood products (surimi, kamaboko, fish sausage, crab meat analogue)/ Vinegar, pickles and sauces (sauces, mayonnaise, fats)/ Alcoholic drinks (beer)/ Beverages (coffee, tea, chocolate drinks, flavoured drinks)/ Other (chewing gum, chocolate, caramels, toothpaste, liquid pharmaceuticals, solid pharmaceuticals, injectable pharmaceuticals, vitamins, mouthwash)/ Meat, poultry and egg products (meat-based products)
SYNONYMS:	CFSAN D-Sorbitol/ CAS 50-70-4/ EINECS 200-061-6/ E420i/ D-Glucitol/ D-Sorbite/ 1,2,3,4,5,6-Hexanehexol/ L-Gulitol/ Sorbit/ Cystosol/ Sorbilax/ Sorbol/ Sorbicolan/ Sorbo/ Sorbostyl/ Niuitin/ Cholaxine/ Karion/ Sionit/ Sionon/ Sorbilande/ Diakarmon/ Hydex 100 Coarse Powder/ Hydex 100 Coarse Powder 35/ Hydex 100 Granular 205/ Hydex 100 Coarse Powder 60/ Hydex tablet grade/ Hystar 7570/ Liponic Sorbitol Powder/ Liponic Sorbitol Solution 70% USP/ Unisweet 70/ Resulax/ Sorbitur
FORMULA:	CH2OH-(CHOH)4-CH2OH
MOLECULAR MASS:	182.17
PROPERTIES AND APPEARANCE:	High mouth-cooling effect. Imparts a cooling sensation due to heat absorption when dissolved. Pleasant, sweet taste with no aftertaste. Solution: clear, colourless syrup; sweet, bland taste; odourless. Crystalline form: white hygroscopic powder; slightly sweet odour. Sweet taste – sweetness 60% of sucrose
BOILING POINT IN °C AT VARIOUS PRESSURES (INCLUDING 760 mm Hg):	High mannitol solution and sorbitol solution: 105

MELTING RANGE IN °C:	Crystalline: 96–97. Anhydrous: 110–112
DENSITY AT 20°C (AND OTHER TEMPERATURES) IN g/l:	High mannitol solution: at 25°C, 10.8 lb/gal or 1,292.9 kg/l
PURITY %:	Grade USP/FCC (High mannitol solution): Minimum 64 D-Sorbitol (Assay) Grade NF/FCC (Non-crystallising): 45 to 55 D-Sorbitol (Assay) Grade USP/FCC (Solution): 69.5 in solution, 98.5 dry Grade NF/FCC (Crystalline): 91.0 to 100.5 Assay (anhydrous)
WATER CONTENT MAXIMUM IN %:	Solution: 28.5–31.5
HEAVY METAL CONTENT MAXIMUM IN ppm:	Solution: 5. Crystalline: 10
ARSENIC CONTENT MAXIMUM IN ppm:	Solution: 2.5. Crystalline: 3
ASH MAXIMUM IN %:	Solution: 0.1
SOLUBILITY % AT VARIOUS TEMPERATURE/pH COMBINATIONS:	
in water:	**@ 20°C** 68.7 g/100 g solution (maxium 70%) **@ 50°C** 83.3 g/100 g solution
FUNCTION IN FOODS:	Bulking agent, so adds texture and mouthfeel to properties. More chemically stable; has an affinity for water without altering sweetening power. High-mannitol solution: cryoprotectant. Anti-crystallisation agent: reduced tendency to crystallise. Tenderising agent. Imparts a cooling sensation due to heat absorption when dissolved; viscosity of sorbitol syrup useful in retarding crystallisation; sorbitol syrup could provide some humectant properties to confection as well as softening the texture as a result of controlling crystallisation. For use in baking: – acts as humectant and anticaking agent in baked goods; – reduced fermentability;

- increased resistance to non-enzymic browning reactions;
- decreased tendency to crystallise;
- increased hygroscopicity;
- moisture binding.

Used in fondants, fudges, marshmallows and caramels to retard sucrose crystallisation; this results in freshness and flavour being retained.

TECHNOLOGY OF USE IN FOODS:

Functionally similar to sucrose. Available in crystal form.

High-viscosity: at 20°C in aqueous solutions:

5% = 1.230 cP
10% = 1.429 cP
25% = 2.689 cP
50% = 11.09 cP
60% = 35.73 cP
70% = 185 cP
83% = more than 10,000 cP

Moisture binding:

- Intermediate water activity (0.6) = 30 g/100 g solids
- High water activity (0.95) = 485 g/100 g solids

Readily soluble in water; virtually insoluble in all organic solvents except alcohol.

Less sweet than sugar, so requires intense sweeteners as supplements to bring to equivalence with sucrose.

Generally used in chewing gums as a sugar substitute, supplemented by intense sweeteners, typically saccharin (added in a small quantity to keep saccharin's taste imperceptible).

Used in fondants, fudges, marshmallows and caramels to retard sucrose crystallisation. This results in freshness and flavour being retained.

Acts as humectant and anticaking agent in baked goods.

Used in table-top sweeteners, preserves, jellies and confectionery for diabetics when supplemented with an intense sweetener.

May replace sucrose in chocolate and ice-cream for diabetics, but products are noticeably different as a result.

HGS (Hydrogenated Glucose Syrup) may be combined with sorbitol as a crystallisation inhibitor.

Hygroscopicity: high (solution); low (powder).

No browning reaction.

Solubility in water: high (75 g/100 ml at room temperature).

Stability: stable to heat, chemically unreactive.
Price ratio relative to sugar: 1.4 (liquid); 2.3 (crystalline).
Provides moistness, sweetness and stability to baked goods, confections and ice-creams.
Crystalline sorbitol:
- cryoprotectant
- excipient
- applications: sugar-free chewing gum, tablets, vitamins, surimi, baked goods, frozen desserts.

Sorbitol solution, high-mannitol:
- composed chiefly of sorbitol
- small amount of mannitol added to provide high clarity in "sugar-free" boiled sweets
- viscosity at 25°C: 110 cP
- % mannitol: 3.0 to 4.0
- Specific gravity at 25/25°C: 1.285 minimum
- Refractive index at 20°C: 1.455 to 1.465
- % Residue on ignition: 0.1 maximum
- % Total sugars: 0.70 maximum
- % Reducing sugars: 0.21 maximum
- Sulphate: 80 ppm maximum
- Chloride: 35 ppm maximum
- Status: Meets all USP 23 and FCC requirements for Sorbitol Solution

Sorbitol solution, non-crystallising
- composed chiefly of sorbitol
- other polyhydric alcohols added to resist crystallisation
- recommended for dentrifices and sugar-free food applications
- useful as a humectant in a variety of food, pharmaceutical, cosmetic and industrial applications
- provides increased viscosity
- viscosity at 25°C: 190 cP
- % Total sugars: 9.5 to 14.5
- Specific gravity at 25/25°C: 1.290 to 1.320
- Status: complies with all compendial requirements of the US National Formulary (NF 18)

Sorbitol solution, USP/FCC:
- Consists primarily of D-sorbitol with a small amount of mannitol and other polyhydric alcohols
- Should be stored above 21°C to prevent crystallisation
- Can be used as a humectant in confections, cosmetics, tobacco and adhesives
- Contributes body and flavour to vehicles for pharmaceuticals and cosmetics, and to some beverages and foods
- Non-cariogenic, so used in "sugar-free" foods, dentrifices, and pharmaceuticals
- Used as a plasticiser, stabiliser and a raw material in the manufacture of sorbitan esters/polysorbates and polyurethanes
- Viscosity at 25°C: 110 cP
- Specific gravity at 25/25°C: 1.300
- % Residue on ignition: 0.01
- % Total sugars: 0.20
- % Reducing sugars: 0.02
- Status: Meets all compendial USP 23 and FCC requirements

Sorbitol Crystalline, NF/FCC:
- Slightly soluble in alcohol, methanol and acetic acid
- pH (5 g/100 ml water): 6.0 to 7.0
- Appearance: white, free-flowing powder and granules, essentially free of foreign matter
- Identification: Meets NF/FCC tests
- Loss on drying: maximum 1.0%
- Reducing sugars: maximum 0.30%
- Total sugars: maximum 1.0%
- Sulphate: maximum 0.010%
- Chloride: maximum 0.0050%
- Status: GRAS and complies with all the compendial requirements of the US National Formulary and Food Chemical Codex

Sorbitol Crystalline:
- Grade: NF/FCC
- Granulation/concentration: granular coarse powder, tablet type, 60 mesh
- Applications: used as tableting agent for making chewable and non-chewable tablets; replaces sugar in "sugar-free" gums; adds a pleasant cooling taste to candy products.

Sorbitol Solution:
- Grade: USP/FCC

– Granulation/concentration: 70% concentration
– Applications: used as humectant to preserve moistness; replaces sugar in “sugar-free” formulations.

Sorbitol solution, high mannitol:
– Grade: USP/FCC
– Granulation/concentration: 70% concentration
– Applications: used as humectant to preserve moistness; replaces sugar in “sugar-free” formulations; preferred in hard candy manufacture.

Sorbitol solution, non-crystallising:
– Grade: NF/FCC
– Granulation/concentration: 70% concentration

Applications: same as Sorbitol Solution, USP/FCC but non-crystallising; preferred in dentrifices

Valuable in production of diabetic chocolate and confectionery because it adds bulk as well as sweetness.

Purity and moisture content of crystalline form critical in use in chocolate and confectionery.

Liquid form used in many products, including confectionery and chewing gum, as a softener and prevents drying out.

Not often used in confectionery recipes as humectant as invert sugar produces the same results for less cost.

Viscosity of sorbitol syrup useful in retarding crystallisation.

Sorbitol syrup could provide some humectant properties to confection as well as softening the texture as a result of controlling crystallisation.

Sorbitol has a marked cooling effect when taken in solution.

Solubility in water:
- 0°C = 59.5 g/100 g solution
- 10°C = 64.3 g/100 g solution
- 20°C = 68.7 g/100 g solution
- 30°C = 73 g/100 g solution
- 40°C = 78.3 g/100 g solution
- 50°C = 83.3 g/100 g solution

Moisture-stabilising and textural properties used in confectionery, baked goods and chocolate to prevent drying and hardening, and maintain initial freshness during storage.

Chemically stable and unreactive.

Does not participate in Maillard (browning) reactions. Useful in production of cookies where no appearance of browning is desired.

Combines well with: sugars, gelling agents, proteins and vegetable fats.

Functions well in: chewing gums, candies, frozen desserts, cookies, cakes, icings and fillings, and oral care products such as toothpaste and mouthwash.

Useful in producing products listed as “reduced calorie” (25% reduction in calories) and “light” (1/3 reduction in calories) in US.

Heat of solution: –28 cal/g; crystalline form has heat of solution of –26.5 cal/g.

Solubility at 37°C: 334 g/100 ml water.

Cooling effect (150 g powder/50 ml water at 37°C): –22.

Thermally stable: resistant to heat. No yellowing or browning at temperatures up to 180°C and up to 200 to 220°C under certain conditions.

Chemically stable: resistant to acids and dilute alkalis.

Due to the absence of aldehyde groups, does not undergo Maillard or Cannizzaro reactions; esterification of sorbitol by fatty acids produces compounds with surface-active properties useful in foods, cosmetics, textiles and metallurgy; etherification of sorbitol by ethylene or propylene oxides produces polyethers useful in the synthesis of rigid polyurethane foams; reaction of sorbitol with aldehydes and ketones produces acetals such as dibenzylidene sorbitol, used primarily as a nucleating agent in polyolefins production.

Acts as a good chelating agent in alkaline media in the presence of certain metal ions such as aluminium, copper or iron. Particularly useful in chelating of alkaline-earth metal ions.

Low heat of solution and high solubility provides a pleasant cool taste.

Unfermentable by many microorganisms, particularly those actively involved in tooth decay.

Commercially available 70% aqueous solution has a pH between 6 and 7.

Addition of ethanol to an aqueous solution decreases solubility.

Slightly soluble in: methanol, isopropanol, butanol, cyclohexanol, phenol, acetone, acetic acid, dimethylformamide, pyridine, acetamide solutions.

Solubility decreases with increase in level of dissolved HGS.

Higher density than other polyols at the same concentration.

Higher viscosity than other polyols at the same concentration and temperature.

Refractive index increases linearly with dry solids concentration.

Boiling temperature increases with increase in concentration.

Decrease in freezing point of aqueous solutions with an increase in concentration.

Preserves protein fibres from denaturation during the freezing process, so useful as a cryoprotectant.

When placed in humid environments will absorb some of the moisture.

A fall in humidity levels will cause sorbitol to release some moisture to re-establish an equilibrium.

Can be used as a stabiliser by compensating for significant humidity changes.

Non-volatile, so is not affected by dry environments. Stabiliser content remains constant and permanent.

Can be used as a conditioning agent in creams or pastes.

Gives good product consistency and a high level of elasticity.

Heat of fusion: 43 to 45 cal/g.

High storage stability.

May be stored at 20°C without lumping at up to 68% relative humidity.

Has a dendritic surface and is readily compressible.

Humectant and stabiliser – good humectant under normal conditions:

- soft confectionery – at a level of 3 to 15% depending upon local legislative requirements. 10% use in fondant results in a 7% reduction in water activity
- sugar-based chewing gum and bubble gum – used as humectant and softening agent for gum base
- biscuits, cakes and pastries – at 3 to 15% prevents products from drying out. Aids in product appearance. Used in biscuits and cakes containing hazelnuts, coconuts and almonds to delay rancidity. Can replace all sugar in cake or biscuit recipes for diabetic products
- meat-based products, sauces, mayonnaise, fats – stabilise moisture content and reduce rancidity

Cryoprotective agent:

- Fish products – 3 to 8% level prevents low-temperature denaturation of fish proteins, as lower sweetness, and does not cause browning. Traditionally used in the manufacture of surimi and as a base for fish products such as kamaboko, fish sausage, and crab meat analogue.

Non-fermentable extract – beer and other beverages – non-fermentable by brewing yeasts. Improves flavour and body of alcohol-free, low-alcohol, and traditional beers.

Aroma carrier – in powder or liquid form products, preserves flavour and taste, and acts as a stable sugar-free carrier in products such as coffee, tea, chocolate and flavoured drinks.

Lowers freezing point – in ices, ice-creams and sorbets lowers the freezing point to make them softer and easier to scoop, and inhibits the recrystallisation of other sugars.

Some properties may be useful in animal nutrition.

Liquid pharmaceuticals (syrups, drinkable solutions and drops, elixirs, suspensions, mouth washes) – properties of sweetness, humectancy and bodying characteristics suitable for use as an excipient for the formulation of non-cariogenic liquid products. Prevents crystallisation around the cap of a bottle.

Solid pharmaceuticals (tablets, powders, capsules) – used in tablets produced either following wet granulation or direct compression.

Useful as a plasticiser in the production of capsules.

If used as an excipient in the production of tablets, results in white, smooth and shiny appearance, a hardness suitable for suckable lozenges and tablets, and a cooling effect.

Injectable pharmaceuticals – may be injected as a 5 to 10% solution with products used in parenteral nutrition associated with amino acids and vitamins.

Cosmetics – used in concentrations of 2 to 15% in lotions, body milks, moisturising creams, liquid soaps, shampoos, and shaving foams. Acts as a moisturiser, stabiliser, emollient and plasticiser, and is safe for the skin.

Toothpaste – stabilises the moisture content and gives good consistency and plasticity (used in a concentration of 20 to 60%).

Non-cariogenic, has a pleasant taste, and enhances the cooling effect of flavours used for this product.

Has a refractive index similar to silica used in toothpaste, so can be used to make a transparent product.

Chemical intermediary in the synthesis of L-ascorbic acid (vitamin C).

Chemical production of surfactants – sorbitol can undergo anhydration and esterification with fatty acids to sorbitan esters which themselves undergo ethoxylation to produce a range of non-ionic surfactants able to function as oil-in-water or water-in-oil emulsifying agents.

Used in chemical production to generate derivatives for products such as polyolefins (propylene/ethylene) and certain pharmaceuticals.

Used as a humectant and plasticiser in: tobacco; gouaches; cellulose films.
Complexing agent for metal ions such as iron, copper, and aluminium.
Enzyme stabiliser at concentrations of 45–50%.
Extends the shelf-life of suspensions of biological insecticide by inhibiting spore germination.
Used as media for the growth of *Claviceps*-type microorganisms.
Production of films and articles for food packaging.
Sweetness does not match sucrose in intensity.
Does not match mouthfeel body or fruitiness enhancement of sucrose.
Does not undergo browning reactions, so can improve shelf-life of fruit drinks.

SYNERGISTS: Isomalt

FOOD SAFETY ISSUES: Energy: 4 kcal/g.

High suitability for diabetics. Can cause osmotic diarrhoea if taken in quantities much above 25 g. Laxative effect at high doses (50 to 75 g/day), so warning labels are required in some countries.

High Mannitol Solution: 2.6 cal/g (dry basis)

Sorbitol Solution, Noncrystallising: 3.0 cal/g (dry basis).

1/3 calories of sucrose.

Suitable for inclusion in diabetic products because not dependent on insulin for metabolism; low energy due to malabsorption in intestine.

70% of orally ingested sorbitol converted to CO_2 without appearing as glucose in the blood. Metabolism causes only an insignificant rise in blood glucose. Metabolism may produce some glucose but as a retarding effect. Suitable for diabetics.

Non-cariogenic; useful for caries prevention.

98 percent taken in food digested, the remainder is excreted; D-sorbitol enters the glycogenolytic pathways without requiring insulin, so may replace D-glucose.

Maximum daily intake recommended as 60–80 g; individual doses should not exceed 10–20 g. No toxic symptoms.

LEGISLATION:

USA:

GRAS by US FASEB (Federation of American Societies for Experimental Biology)
FDA requires on label "Excess consumption may have a laxative effect"
GRAS by USFDA US
CFR 21
PART 100 – General
Subpart G®Specific Administrative Rulings and Decisions
100.130 Combinations of nutritive and nonnutritive sweeteners in "diet beverages".

(4) To avoid confusion by diabetics, the label of a beverage containing sorbitol, mannitol, or other hexitol, must bear the statement "Contains carbohydrates, not for use by diabetics without advice of a physician". To further avoid confusion of these beverages with those sweetened solely with nonnutritive artificial sweeteners which have been marketed in containers bearing prominent statements such as "sugar free", "sugarless", or "no sugar", the labels of beverages containing hexitols must not bear these or similar statements.

CFR 21
Part 182
182.90 Substances migrating to food from paper and paperboard products.

Substances migrating to food from paper and paperboard products used in food packaging that are generally recognized as safe for their intended use, within the meaning of section 409 of the Act, are as follows: Alum (double sulfate of aluminum and ammonium potassium, or sodium). Aluminum hydroxide. Aluminum oleate. Aluminum palmitate. Casein. Cellulose acetate. Cornstarch. Diatomaceous earth filler. Ethyl cellulose. Ethyl vanillin. Glycerin. Oleic acid. Potassium sorbate. Silicon dioxides. Sodium aluminate. Sodium chloride. Sodium hexametaphosphate. Sodium hydrosulfite. Sodium phosphoaluminate. Sodium silicate. Sodium sorbate. Sodium tripolyphosphate. Sorbitol. Soy protein, isolated. Starch, acid modified. Starch, pregelatinized. Starch, unmodified. Talc. Vanillin. Zinc hydrosulfite. Zinc sulfate.

[42 FR 14640, Mar. 15, 1977]
CFR 21
PART 184®DIRECT FOOD SUBSTANCES AFFIRMED AS GENERALLY RECOGNIZED AS SAFE
Subpart B®Listing of Specific Substances Affirmed as GRAS
184.1835 Sorbitol.

(a) Sorbitol is the chemical 1,2,3,4,5,6-hexanehexol (C6H14O6), a hexahydric alcohol, differing from mannitol principally by having a different optical rotation. Sorbitol is produced by the electrolytic reduction, or the transition metal catalytic hydrogenation of sugar solutions containing glucose or fructose.

(b) The ingredient meets the specifications of the *Food Chemicals Codex*, 3d edn. (1981), p. 308, which is incorporated by reference. Copies may be obtained from the National Academy Press, 2101 Constitution Ave. NW., Washington, DC 20418, or may be examined at the Office of the Federal Register, 800 North Capitol Street, NW., suite 700, Washington, DC 20408.

(c) The ingredient is used as an anticaking agent and free-flow agent as defined in §170.3(o)(1) of this chapter, curing and pickling agent as defined in §170.3(o)(5) of this chapter, drying agent as defined in §170.3(o)(7) of this chapter, emulsifier and emulsifier salt as defined in §170.3(o)(8) of this chapter, firming agent as defined in §170.3(o)(10) of this chapter, flavoring agent and adjuvant as defined in §170.3(o)(12) of this chapter, formulation aid as defined in §170.3(o)(14) of this chapter, humectant as defined in §170.3(o)(16) of this chapter, lubricant and release agent as defined in §170.3(o)(18) of this chapter, nutritive sweetener as defined in §170.3(o)(21) of this chapter, sequestrant as defined in §170.3(o)(26) of this chapter, stabilizer and thickener as defined in §170.3(o)(28) of this chapter, surface-finishing agent as defined in §170.3(o)(30) of this chapter, and texturizer as defined in §170.3(o)(32) of this chapter.

(d) The ingredient is used in food at levels not to exceed good manufacturing practices. Current good manufacturing practice in the use of sorbitol results in a maximum level of 99 percent in hard candy and cough drops as defined in §170.3(n)(25) of this chapter, 75 percent in chewing gum as defined in §170.3(n)(6) of this chapter, 98 percent in soft candy as defined in §170.3(n)(38) of this chapter, 30 percent in non-standardized jams and jellies, commercial, as defined in §170.3(n)(28) of this chapter, 30 percent in baked goods and baking mixes as defined in §170.3(n)(1) of this chapter, 17 percent in frozen dairy desserts and mixes as defined in §170.3(n)(20) of this chapter, and 12 percent in all other foods.

(e) The label and labeling of food whose reasonably foreseeable consumption may result in a daily ingestion of 50 grams of sorbitol shall bear the statement: "Excess consumption may have a laxative effect."

(f) Prior sanctions for this ingredient different from the uses established in this regulation do not exist or have been waived.

[42 FR 14653, Mar. 15, 1977, as amended at 49 FR 5613, Feb. 14, 1984]

UK and EUROPE:

One of 12 sweeteners listed as permissible for use in the UK as of 1983 – affirmed with GRAS status

UK Food Labelling Regulations 1984 states require label "best eat less than 25 g of sorbitol a day"

European Union Nutritional Labelling Directive states all polyols have a calorific value of 2.4 calories per gram

JECFA provides an ADI of "not specified" so no limits on use

EU set no limits for use in 1985

approved for use by European Union

Subject to certain restrictions, permitted as sweetener or food additive in Belgium, Denmark, Greece, Spain, France, Germany, Switzerland, Sweden, the Netherlands, Italy, Norway and Finland.

CANADA:

Canada FDR 67-29
– available as crystalline solid in fibre containers of 5 to 300 pounds
– available in solutions of 70% to 85% in one- and five-gallon cans, lined steel drums and tank cars
– soluble in water
– slightly soluble in alcohol
– insoluble in ether, fats or oils
– excessive quantities may have laxative effect.
– Mannitol is a chemical of similar structure and characteristics that is used for similar purposes
– Unstandardized foods: sweetener, to modify texture – good manufacturing practice
– confectionery: Sweetener, release agent – good manufacturing practice
– humectant for marshmallows, shredded coconut – good manufacturing practice
– to modify texture in a blend of prepared fish and prepared meat referred to in paragraph B.21.006 (n) at 3.5%

Table IX – Food additives that may be used as Sweeteners: Item S.1 (25 May 1993)

Permitted in or on	**Maximum level of use**
(1) A blend of prepared fish and prepared meat referred to in paragraph B.21.006(n) which states "B.21.006 – Prepared fish or prepared meat shall be whole or comminuted food prepared from fresh or preserved fish or meat respectively, may be canned or cooked, and may (n) in the case of a blend of prepared fish and prepared meat that has the appearance and taste of flesh of a marine or freshwater animal, contain filler, fish binder, whole egg, egg-white, egg-yolk, food colour, gelling or stabilizing agents, texture-modifying agents, natural or artificial flavouring preparations, pH-adjusting agents, sweetener and in a proportion not exceeding two per cent of the blend, a legume;"	3.5%
(2) Unstandardized foods	Good Manufacturing Practice

AUSTRALIA/PACIFIC RIM:

Japan and Australia: subject to certain restrictions, permitted as sweetener or food additive

OTHER COUNTRIES:

Codex: page 155
– Permitted for use in raisins at a maximum level of 5 g/kg.
– ADI not specified.

Artificial Sweeteners:
– Permitted for use in raisins

- Maximum level: 5 mg/kg
- ADI: Not specified

Subject to certain restrictions, permitted as sweetener or food additive in South Africa and Brazil.
The following ADM Sorbitol products are certified Kosher

- Sorbitol Crystalline
- Sorbitol Solution
- Sorbitol Solution, high mannitol
- Sorbitol Solution, non-crystallizing

REFERENCES:

Smith, J. (1991) *Food Additive User's Handbook.* Blackie Publishing.
Canadian Food and Drugs Act and Regulations, 1994.
Codex Alimentarius, Volume XIV, 1983.
McCue, N. (1996) *Showcase: Natural & High Intensity Sweeteners; Prepared Foods*, March, p. 84.
Chilton's Food Engineering Master '95. (1994) ABC Publishing Group. p. 10; 288–289.
ADM technical specifications, 1996.
ADM Product information, 1996.
Minifie, B. W. (1989) *Chocolate, Cocoa, and Confectionery: Science and Technology*. 3rd edition. Van Nostrand Reinhold.
Matz, S. A. (1992) *Bakery Technology and Engineering*. 3rd edition. Van Nostrand Reinhold.
Calorie Control Council informational brochure, 1995.
Food Additives Update, Food in Canada, April, 1996.
Roquette technical bulletin, as provided by Kingsley & Keith (Canada) Inc., 1996.
Roquette product information, as provided by Kingsley & Keith (Canada) Inc., 1996.
Hicks, D. (1989) *Production and Packaging of Non-Carbonated Fruit Juices and Fruit Beverages*. Blackie Publishing; Glasgow.
Smoley, C. K. (1993) *Everything Added to Food in the United States.* US Food and Drug Administration, CRC Press, Inc.
Wong, D. W. S. (1989) *Mechanism and Theory in Food Chemistry.* Van Nostrand Reinhold.
Budavari, S. (1996) *The Merck Index*. 12th edition. Merck and Co.

ANY OTHER RELEVANT INFORMATION:

Polyol produced through hydrogenation of glucose; first hydrogenated sugar produced in 1930s.
Naturally occurring; produced when crystalline form of lactitol heated to 179 to 240°C.
Used since 1920s by diabetics because metabolism causes only an insignificant rise in blood glucose.

Hexahydric sugar alcohol; isomeric with mannitol; naturally occurring in small quantities in many fruits.

Prepared commercially by catalytic hydrogenation of glucose.

Polyhydric alcohol: richest source is rowan or mountain ash berry; no natural supply is commercially important. Polymorphic and exists in three crystalline states. Only gamma form is stable; other two forms will transform to gamma form under moisture or heat.

Provided in US by Archer Daniels midland, Lonza Inc., Roquette America Inc., SPI Polyols, Inc. Marketed as NeoSorb – discovered in 1872 by French chemist B. J. Boussingault as a non-fermentable sugar present in rowan juice which he named sorbite, a linear hexitol.

Also present in high concentrations in: black grapes, apples, nectarines, peaches, apricots, plums, cherries, pears; formally named D-sorbitol.

Crystalline sorbitol: supplier Roquette America Inc.

Liquid Sorbitol: suppliers Roquette America Inc. and USP – ADM Corn Processing.

ADM sorbitol as marketed by ADM Food Additives Division; available as crystalline 70% solution, high-mannitol solution, and non-crystallising solution.

Monotropic polymorphism – able to crystallise in different forms (alpha, beta, gamma). Alpha and beta forms are metastable and may be converted to the stable gamma form under certain temperature and humidity conditions, causing sorbitol powder to cake. Impossible physically to extract sorbitol from natural sources due to its high water solubility.

NAME:	**Stevioside**
CATEGORY:	Sweetener/ Flavour enhancer/ Flavour modifier
FOOD USE:	Fruit, vegetable and nut products (fruit juices, fruit beverages)/ Beverages (fruit beverages, carbonated beverages)/ Sugars, sugar preserves and confectionery (cube sugar, table-top sweeteners)/ Others (sugarless chewing gum)
SYNONYMS:	Steviosin: (4α)-13-[(2-*O*-β-D-glucopyranosyl-α-D-glucopyranosyl) oxy] kaur-16-en-18-oic acid-β-D-glucopyranosyl ester/ Stevia Steviol: (4α)-13-Hydroxykaur-16-en-18-oic acid/ Hydroxydehydrostevic acid
FORMULA:	$C_{38}H_{60}O_{18}$
MOLECULAR MASS:	804.88
PROPERTIES AND APPEARANCE:	White hygroscopic powder (in pure form). Commercially available form ranges from cream to tan colour, depending upon purity. Licorice-like taste; slow taste onset; thin mouthfeel body. Sweet taste (300 times that of sucrose); lingering bitter-licorice aftertaste
MELTING RANGE IN °C:	Steviosin: 196–198. Steviol: 215
PURITY %:	90
SOLUBILITY % AT VARIOUS TEMPERATURE/pH COMBINATIONS:	
in water:	**@ 20°C** Pure form has a solubility in water of 1.2 g/l Commercial forms are more soluble, in the range of 300 to 800 g/l
in ethanol solution:	**@ 100%** Insoluble
FUNCTION IN FOODS:	Sweetening agent (300 times sweetness of sucrose); flavour modifier (used to suppress pungent flavours); flavour enhancer in fruit drinks

TECHNOLOGY OF USE IN FOODS:

Steviol is the aglucon of stevioside. On its own, produces an unacceptable liquorice-like taste in cola beverages. Combined with fructose to produce 50% calorie-reduced soft drinks; combined with polyols in sugarless chewing gums; combined with sucrose in calorie-reduced sugar cubes.

Gums using stevioside are considered highly acceptable by consumers.

Used in Japan in sugarless chewing gums, soft drinks, table-top sweeteners, juices and other products. Also used as a flavour modifier and to suppress pungent flavours in pickles, dried seafoods, fish, meat, bean pastes and soy sauce.

Stable at pH 3 to 9; withstands 100°C for 1 hour at pH 3 to 9. At room temperature, stable in citric acidified beverages for 3 months; stable in phosphoric acidified beverages for 5 months; also stable in the presence of salt. No browning reaction.

Stevioside has low solubility in alcohol.

In soft drinks, sweetness ranges from 140 to 280 times that of sucrose.

Not recommended as the sole sweetener due to its bitter/licorice taste.

Taste can be improved by increasing the fraction of rebaudioside-A or by combining with fructose, lactose, hitidine, chlorodeoxysugars, cyclodextrin, aspartame or cyclamate.

Stevia extracts are generally shelf-stable. In carbonated beverages, there is no reported degradation over 5 months at 22°C or lower; some breakdown of rebaudioside-A (36%) and stevioside (25%) at 37°C over 4 months.

Stevia extracts do not interact with other food components in soft drinks.

Stevia detection in soft drinks best using HPLC, but can be performed using GLC or colorimetric methods.

Currently used in Japan by over 50 companies in soft drinks, instant juices, and fruit-flavoured drinks; compatible in fruit beverages. Sweetness intensity at 10% sucrose is 150. Soluble in dioxane

SYNERGISTS:	Glycyrrhizin (mixture is commercially available in Japan)/ Aspartame/ Cyclamate/ Acesulfame-K/ Sucrose/ Glucose/ Fructose
ANTAGONISTS:	Degradation in carbonated beverages greater with phosphoric acid than citric acid systems. Rebaudioside-A breaks down under UV light.
FOOD SAFETY ISSUES:	No significant mutagenic or genotoxic activity. Question of whether stevioside and rebaudioside are degraded in the bowel to steviol, which has biological risks. Unclear whether stevioside is excreted unmetabolised or is reduced to steviol (this has been observed in rats, but not in humans). There have been reports linking stevioside with anti-hormonal effects and infertility, but studies have not confirmed this. Non-cariogenic

LEGISLATION:

USA:

Not approved for use in US as of 1990
Regulatory Status: ADI level as of 1990 for use in soft drinks – Not permitted

UK and EUROPE:

SCF indicated in December 1987 that stevia was not toxicologically acceptable.
Regulatory Status: ADI level as of 1990 for use in soft drinks

– JECFA	None specified	– Ireland	Not permitted
– Belgium	Not permitted	– Netherlands	Not permitted
– Denmark	Not permitted	– Norway	Not permitted
– Finland	Not permitted	– Spain	Not permitted
– France	Not permitted	– Switzerland	Not permitted
– East Germany	Not permitted	– Turkey	Not permitted
– West Germany	Not permitted	– UK	Not permitted
– Greece	Not permitted	– Yugoslavia	Not permitted

CANADA:

Regulatory Status: ADI level as of 1990 for use in soft drinks – Not permitted

AUSTRALIA/PACIFIC RIM:

Natural product, so no clearance required in Japan
Permitted for use in Japan since 1970
Used in Japan in sugarless chewing gums, soft drinks, table-top sweeteners, juices and other products
In Japan, used as a flavour modifier and to suppress pungent flavours in pickles, dried seafoods, fish, meat, and bean pastes, and soy sauce.
Permitted for use in China and South Korea
Currently used in Japan by over 50 companies in soft drinks, instant juices, and fruit-flavoured drinks.
– Regulatory Status: ADI level as of 1990 for use in soft drinks

– Australia	Not permitted
– Japan	Permitted
– New Zealand	Not permitted

OTHER COUNTRIES:

Permitted for use in Brazil and Paraguay
Brazil's Health Ministry approved its use in diet drinks in November 1988
Regulatory Status: ADI level as of 1990 for use in soft drinks

– Argentina	Not permitted
– Brazil	750 mg/l
– Israel	Not permitted
– Kenya	Not permitted
– Mexico	Not permitted
– Saudi Arabia	Not permitted
– South Africa	Not permitted

REFERENCES:

Smith, J. (1991) *Food Additive User's Handbook.* Blackie Publishing, Glasgow.

Mitchell, A. J. (1990) *Formulation and Production of Carbonated Soft Drinks.* Blackie Publishing.

New and Novel Foods, p. 486.

Matz, S. A. (1992) *Bakery Technology and Engineering.* 3rd edition. Van Nostrand Reinhold.

Food Chemistry. 2nd edition. (1985) Marcel Dekker Inc.

Hicks, D. (1989) *Production and Packaging of Non-Carbonated Fruit Juices and Fruit Beverages.* Blackie Publishing, Glasgow.

Smoley, C. K. (1993) *Everything Added to Food in the United States.* US Food and Drug Administration, CRC Press Inc.

Wong, D. W. S. (1989) *Mechanism and Theory in Food Chemistry.* Van Nostrand Reinhold.

Budavari, S. (1996) *The Merck Index.* 12th edition. Merck and Co.

ANY OTHER RELEVANT INFORMATION:

Discovered in 1905. Structurally related to glycyrrhizic acid. One of the only sweeteners extracted and refined from plant sources without chemical or enzymic modification; the others are sucrose and thaumatin.

Sweet diterpene glycosides extracted from the leaves of the *Stevia rebaudiana* Bertoni, which is a variety chrysanthemum found in Paraguay and Brazil. The plant has been successfully cultivated in Japan, Korea, South Africa, South America, Taiwan and China.

Commercially available in Japan as crude extract, 50% pure and 90% pure or higher grades.

Taste profile improves with purity.

Rebaudioside-A is a constituent of stevioside which has less aftertaste and is currently under study. Stevioside and rebaudioside-A are the two extracts from the *Stevia* plant which are of commercial importance. Three other extracts of lesser importance named rebaudioside-C to rebaudioside-E.

In Japan, stevia marketed through a consortium of 11 major companies called the Stevia Konwakior (Stevia Association).

NAME:	**Sucralose**	
CATEGORY:	Sweetener/ Flavour enhancer	
FOOD USE:	Baked goods (baking mixes, bakery products, desserts, dessert mixes, toppings, topping mixes, fillings, filling mixes)/ Dairy products (dry milk products, dairy beverages, dairy desserts, frozen desserts, puddings, pudding mixes)/ Cereals and cereal products/ Fruit, vegetable and nut products (fruit spreads, processed fruit and vegetable products)/ Beverages (dairy beverages, beverage concentrates, beverage mixes, still beverages, carbonated beverages)/ Sugars, sugar preserves and confectionery (table-top sweeteners, confections, confection coatings, confectionery glazes, table syrups)/ Alcoholic drinks/ Vinegar, pickles and sauces (salad dressings, condiments)/ Other (chewing gum, breath freshener, sweetened seasonings, coatings mixes for snack foods)	
SYNONYMS:	CAS 56038-13-2/ 4,1′,6′-Trichlorogalactosucrose/ 1,6-Dichloro-1,6-dideoxy-β-D-fructofuranosyl-4-chloro-4-deoxy-α-D-galactopyranoside/ 4,1′,6′-Trichloro-4,1′,6′-trideoxy-galacto-sucrose/ Trichlorogalactosucrose/ TGS	
FORMULA:	C12 H19 C13 O8	
MOLECULAR MASS:	397.64	
PROPERTIES AND APPEARANCE:	White crystalline powder. Thin mouthfeel body; sweet (400 to 800 times that of sucrose); delayed onset of flavour; lingering sweet aftertaste; intensely sweet taste; taste similar to sucrose	
MELTING RANGE IN °C:	Anhydrous form: 130; pentahydrate: 36.5	
SOLUBILITY % AT VARIOUS TEMPERATURE/pH COMBINATIONS:		
in water:	**@ 20°C**	28% weight/unit volume
in ethanol solution:	**@ 100%**	Soluble
FUNCTION IN FOODS:	Sweetener (400 to 800 times that of sucrose)/ Flavour enhancer	

TECHNOLOGY OF USE IN FOODS:

Stable in solution at low pH; as stable in solution at high temperatures as sucrose.

Can be stored for several years in liquids. Resistant to enzymic hydrolysis; no browning reaction with proteins, gums, tannins and other carbohydrates.

Relative sweetness increases with a decrease in pH; in cola beverages, sweetness enhanced using aspartame.

Stable under most conditions of food processing; stable over pH range found in carbonated soft drinks.

Hydrolyses to 4-chloro-D-galactosucrose and 1, 6-dichloro-D-fructose in extreme pH and temperature situations.

Does not interact with most food ingredients except some iron salts.

Stable in dry form when stored correctly; slow decomposition of dry form at elevated temperatures resulting in colour change from white to brown; resistant to enzymic hydrolysis.

Stable at pH 3 to 7.5; good stability and maintains integrity during baking; good stability in acidic drinks; compatible with fruit beverage ingredients.

Sweetness intensity at 10% sucrose is 450

SYNERGISTS: Aspartame; cyclamate; saccharin; stevioside. High levels of synergism in tripartite blends with cyclamate and aspartame or acesulphame and saccharin

ANTAGONISTS: Negatively synergistic with aspartame in bipartite blends

FOOD SAFETY ISSUES: Non-toxic; non-carcinogenic; non-teratogenic; non-mutagenic; non-cariogenic. LD_{50} in rats >16 g/kg. Non-caloric: not metabolised by mammals, and poorly absorbed by the body

LEGISLATION:

USA:

Under FDA review as of 1992 for use in 15 food categories, including baked goods

UK and EUROPE:

JECFA ADI of 0 to 3.5 mg/kg

CANADA:

Canada FDR 67.31A – Table IX – Food additives that may be used as Sweeteners – Item S. 2 (25 May 1993)

Permitted in or on	**Maximum level of use**
(1) Table-top sweeteners	Good Manufacturing Practice
(2) Breakfast cereals	0.1%
(3) Beverages; beverage concentrates; beverage mixes; dairy beverages; (except for any of these products for which standards are set out in these Regulations)	0.025% in beverages as consumed

(4) Desserts; dessert mixes; toppings; topping mixes; dairy desserts; frozen desserts; fillings; filling mixes; (except for any of these products for which standards are set out in these Regulations)	0.025% in products as consumed
(5) Chewing gum; breath freshener products	0.15%
(6) Fruit spreads (except for any of these products for which standards are set out in these Regulations)	0.045%
(7) Salad dressings; condiments; (except for any of these products for which standards are set out in these Regulations)	0.04%
(8) Confections and their coatings; confectionery glazes for snack foods; sweetened seasonings or coatings mixes for snack foods; (except for any of these products for which standards are set out in these Regulations)	0.07%
(9) Baking mixes; bakery products; (except for any of these products for which standards are set out in these Regulations)	0.065% in products as consumed
(10) Processed fruit and vegetable products (except for any of these products for which standards are set out in these Regulations)	0.015%
(11) Alcoholic beverages (except for any of these products for which standards are set out in these Regulations)	0.07%
(12) Puddings; pudding mixes	0.04% in products are consumed
(13) Table syrups (except for any products for which standards are set out in these Regulations)	0.15%

Amendment to Canada FDR – Item S.15a, Table VIII as of 1996 – to provide for use of sucralose as a sweetener and flavour enhancer

REFERENCES:

Smith, J. (1991) *Food Additive User's Handbook.* Blackie Publishing, Glasgow.

Canadian Food and Drugs Act and Regulations (1994).

Mitchell, A. J. (1990) *Formulation and Production of Carbonated Soft Drinks.* Blackie Publishing, Glasgow.

New and Novel Foods, p. 486.

Matz, S. A. (1992) *Bakery Technology and Engineering.* 3rd edition. Van Nostrand Reinhold.

Food Additives Update; Food in Canada; April 1996.

Hicks, D. (1989) *Production and Packaging of Non-Carbonated Fruit Juices and Fruit Beverages.* Blackie Publishing, Glasgow.

Smoley, C. K. (1993) *Everything Added to Food in the United States.* US Food and Drug Administration, CRC Press Inc.

Budavari, S. (1996) *The Merck Index.* 12th edition. Merck and Co.

ANY OTHER RELEVANT INFORMATION:	Trichloro derivative of sucrose; discovered in 1976. Produced using selective chlorination of sucrose, process patented by Tate & Lyle of London, England. Jointly developed by Tate & Lyle Specialty Sweeteners (UK) and McNeil Specialty Products Company (US). Marketed by Tate & Lyle in UK and Europe. Johnson & Johnson market the product in the US.

NAME:	**Thaumatin**
CATEGORY:	Sweetener/ Flavour enhancer/ Flavour modifier
FOOD USE:	Baked goods (desserts)/ Dairy products (milk powders, ice-cream, iced milk)/ Beverages/ Soft drinks/ Others (pet foods, animal feeds, chewing gum, salt substitutes, coffee-flavoured products, toothpaste, mouthwash, MSG replacement, medication)
SYNONYMS:	CFSAN Thaumatin/ CAS 53850-34-3/ Talin/ Katemfe
MOLECULAR MASS:	21,000–22,000 daltons
ALTERNATIVE FORMS:	Thaumatin I/ Thaumatin II
PROPERTIES AND APPEARANCE:	Cream-coloured powder. Lingering licorice aftertaste; odourless. Sweet taste (750 to 1600 times that of sucrose by weight, 1300 to 2500 times that of sucrose in 5% to 10% sucrose range, 30,000 to 100,000 times sweeter than sucrose on a molar basis); delayed sweetness onset
MELTING RANGE IN °C:	172–174
SOLUBILITY % AT VARIOUS TEMPERATURE/pH COMBINATIONS:	
in water:	**@ 20°C** 60% weight/unit volume
in propylene glycol:	**@ 20°C** 5% weight/unit volume
FUNCTION IN FOODS:	Sweetener: 750 to 1600 times that of sucrose by weight; 1300 to 2500 times that of sucrose in 5% to 10% sucrose range; primarily in beverages and desserts. Flavour enhancer (primarily used at below sweet-taste threshold); flavour potentiator: flavour masking agent (in medicines); aroma enhancer; taste modifier
ALTERNATIVES:	Aspartame (equivalent to thaumatin: acesulphame)

TECHNOLOGY OF USE IN FOODS:

Protein loses sweetness on heating and in pH <2.5.

Primarily used as a flavour enhancer at levels below sweet-taste threshold due to problems with stability, taste profile and compatibility.

When combined with acesulfame-K, provides a less costly alternative to aspartame with equivalent taste in some products.

Due to licorice taste and delayed sweetness, applications are limited.

More commonly used as a partial sweetener in combination with other rapid-tasting sweeteners.

Flavour potentiation and slow onset of sweetness beneficial in products such as toothpaste, mouthwash and chewing gum.

Boosts low sweetness of bulk sweeteners added to sugarless gums.

Powerful flavour enhancer, magnifies spearmint, cinnamon, wintergreen and peppermint up to 10 times.

Enhances and improves flavour of coffee and milk products so used in coffee-flavoured products, ice-cream, iced-milk drinks-on-sticks, and spray-dried milk powders.

Enhances savoury flavours; can be used to replace MSG (mono-sodium glutamate) when combined with nucleotides, spices and/or other flavours.

Solubility at room temperature: good solubility in ethyl alcohol, isopropyl alcohol, glycerol, propylene glycol and higher polyols such as sorbitol; insoluble in ether, acetone, toluene and triacetin.

Stable in pH 1 to 9 at ambient temperatures; heat-stable in pH 2.7 to 6.0, optimum at pH 2.8 to 3.5. At pH <5.5, withstands 100°C for several hours.

Able to be stored indefinitely in dry form; able to be stored for several years in chemically preserved solutions at ambient temperatures.

Licorice aftertaste can be reduced slightly through combination with arabinogalactan, glucuronic acid, several types of sugars, or one of polyols.

Does interact with some food constituents by forming salts with suitable negatively charged compounds when they are present in excess.

Interaction with several types of gums and stabilisers such as CMC, xanthan, pectin, locust bean gum, and alginates.

In soft drinks, precipitation may occur with some synthetic colours which may be prevented by adding low levels of gum arabic.

Talin® used in UK in chewing gum as a flavour-potentiator at sub-threshold levels. Limited use world-wide in soft drinks.

Very soluble in cold water. Solubility: readily soluble in cold water to produce solutions in excess of 60% weight: volume; can produce solutions of 3–5% in 60% ethanol; can produce solutions of up to 5% in propylene glycol; soluble in other aprotic solvents.

Remains stable indefinitely in freeze-dried or spray-dried form if stored under ambient conditions; remains stable at 120°C in canning operations and under pasteurisation and UHT conditions; stable under acid conditions to below pH 2.0.

Has masking effect on bitter, metallic ions such as sodium, iron, potassium. Can be used to mask aftertaste of saccharin, added minerals, or pharmaceutical products.

Masks bitter components of natural citrus flavours in products containing citrus fruits or juices.

Addition of 10 ppm thaumatin allows a 30% reduction in the level of aspartame required for a product of the same sweetness.

Used with non-nutritive sweeteners to mask synthetic taste, provide body, all allow a reduction in the amount of sweetener required.

The addition of less than 1 ppm of thaumatin in carbonated beverages dramatically reduces sugar or sweetener inclusion.

Due to licorice-like taste when used at high levels, applications limited to where sweetness requirement is less than the equivalent of 10% sucrose.

Unsuitable for use in fruit-flavoured beverages

SYNERGISTS: Saccharin (masks its aftertaste at low levels); acesulfame-K; stevioside.

Aluminium ions known to increase perceived sweetness by a factor of 2.

Interaction with taste receptors causes it to heighten response to sweeteners and certain flavour compounds.

Flavour-enhancing properties used in "aggressive" flavours such as peppermint, ginger, cinnamon and coffee, as well as ability to reduce their fiery, peppery or bitter elements.

In savoury products, enhances flavour and synergises with flavour-enhancers such as MSG and 5′-nucleotides, creating the "Umami" (delicious) flavour.

Taste increased by trivalent salts.

ANTAGONISTS:

Taste reduced by mono- and divalent salts. Denatured by metaphosphoric and phytic acids at pH 2.9. Loss of sweetness with xanthan, CMC, pectin and alginate. Incompatible with carrageenans; incompatible with certain beverage ingredients such as synthetic colours. Undergoes denaturation when exposed to extreme pH and high temperatures, resulting in a loss of sweetness (this denaturation may be reversible)

FOOD SAFETY ISSUES:

Contributes the same number of calories as sucrose (4 kcal/g), but use is measured in ppm so it is essentially non-calorific in food products.

Is digested and metabolised completely by humans and animals.

No adverse effects found during toxicological studies; non-cariogenic.

In 1987, JECFA recorded no mutagenic, teratogenic or allergenic effects; concluded that the only dietary effect was an insignificant increase in normal protein intake. Generally accepted as a safe, natural substance.

LEGISLATION:

USA:

US granted GRAS for use in chewing gum in October 1984.
Approved for general use in US.
Has been reviewed and listed as GRAS (Substance No. 3732) by FEMA.
May be used in US in products labelled as "natural".

UK and EUROPE:

UK permitted use in foods, drinks and dietary products except baby foods by Sweeteners in Food Regulations with Group A status in 1983.
JECFA declared ADI "not specified" in 1985

Regulatory Status: ADI level as of 1990 for use in soft drinks

– Belgium	Not permitted
– Denmark	Permitted
– Finland	Not permitted
– France	Not permitted
– East Germany	Not permitted
– West Germany	Not permitted
– Greece	Not permitted
– Ireland	Not permitted
– Netherlands	Permitted
– Norway	Permitted
– Spain	Not permitted
– Switzerland	Not permitted
– Turkey	Not permitted
– UK	Permitted
– USSR	Not permitted
– Yugoslavia	Not permitted

CANADA:

Canada FDR s. 67.31B – Table IX – Food additives that may be used as Sweeteners – Item T.1 (25 May 1993)

Permitted in or on	**Maximum level of use**
(1) Chewing gum; breath fresheners products	500 ppm
(2) Salt substitutes	400 ppm
(3) (naming the flavour) Flavour referred to in section B.10.005;	100 ppm

Unstandardized flavouring preparations

B.10.005 (naming the flavour) Flavour

(a) shall be a preparation, other than a flavouring preparation described in section B.10.003 as sapid or odorous principals, or both, derived from the aromatic plant after which the preparation is named;

(b) may contain a sweetening agent, food colour, Class II preservative, thaumatin, Class IV preservative or emulsifying agent; and

(c) may have added to it the following liquids only
  (i) water;
  (ii) any of, or any combination of, the following: benzyl alcohol; 1,3-butylene glycol, ethyl acetate, ethyl alcohol, glycerol, glyceryl diacetate, glyceryl triacetate, glyceryl tributyrate, isopropyl alcohol, monoglycerides and diglycerides; 1,2-propylene glycol or triethylcitrate;
  (iii) edible vegetable oil; and
  (iv) brominated vegetable oil, sucrose acetate isobutyrate or mixtures thereof, when such flavour is used in citrus-flavoured or spruce-flavoured beverages

B.10.003 – (naming the flavour) Extract or (naming the flavour) Essence shall be a solution in ethyl alcohol, glycerol, propylene glycol or any combination of these, of sapid or odorous principles, or both, derived from the plant after which the flavouring extract or essence is named, and may contain water, a sweetening agent, food colour and a Class II preservative or Class IV preservative.

Chagnes to page 67–29 of Canada FDR as of 1996

- sweetener and flavour enhancer for chewing gum and breath freshener products – 500 ppm
- bitterness masking agent in salt substitutes – 400 ppm
- flavour enhancer in (naming the flavour) flavour (division 10) and unstandardized flavouring preparations – 100 ppm

AUSTRALIA/PACIFIC RIM:

Permitted as a natural food in Japan since June 1979
Appproved for use as a sweetener and flavour enhancer in Australia
Regulatory Status: ADI level as of 1990 for use in soft drinks

– Australia	Permitted
– Japan	Permitted
– New Zealand	Permitted

OTHER COUNTRIES:

The only natural high-intensity sweetener so products containing thaumatin do not require to be labelled "artificially sweetened".
Approved for use as a sweetener and flavour enhancer in Mexico.
Regulatory Status: ADI level as of 1990 for use in soft-drinks. May be permitted as a sweetener or flavour enhancer

– Argentina	Not permitted
– Brazil	Not permitted
– Israel	Not permitted
– Kenya	Not permitted
– Saudi Arabia	Not permitted
– South Africa	Permitted

REFERENCES:	Smith, J. (1991) *Food Additive User's Handbook.* Blackie Publishing, Glasgow. Canadian Food and Drug Act and Regulations (1994). Mitchell, A. J. (1990) *Formulation and Production of Carbonated Soft Drinks.* Blackie Publishing, Glasgow. New and Novel Foods, p. 486. Minifie, B. W. (1989) *Chocolate, Cocoa, and Confectionery: Science and Technology.* 3rd edition. Van Nostrand Reinhold. Matz, S. A. (1992) *Bakery Technology and Engineering.* 3rd edition. Van Nostrand Reinhold. Food Additives Update; Food in Canada; April 1996. Thaumatin – The Sweetest Substance Known to Man Has a Wide Range of Food Applications; *Food Technology*; January 1996; p. 74–75. *Food Chemistry*, 2nd edition (1985) Marcel Dekker Inc. Hicks, D. (1989) *Production and Packaging of Non-Carbonated Fruit Juices and Fruit Beverages.* Blackie Publishing. Smoley, C. K. (1993) *Everything Added to Food in the United States.* US Food and Drug Administration, CRC Press Inc. Budavari, S. (1996) *The Merck Index.* 12th edition. Merck and Co. Wong, D. W. S. (1989) *Mechanism and Theory in Food Chemistry.* Van Nostrand Reinhold.
ANY OTHER RELEVANT INFORMATION:	Thaumatin I (TI) and Thaumatin II (TII) isolated by Van der Wel at Unilever in 1972 as two major sweet-tasting proteins. Mixture of sweet-tasting proteins extracted from the fruit of the *Thaumatococcus danielli* Benth or Katemfe plant, found in West Africa. One of the only sweeteners extracted and refined from plant sources without chemical or enzymic modification (the others are sucrose and stevioside). Fruit of plant source has been used for centuries in West Africa. Five soluble proteins designated a, b, c, I and II, the latter two being the sweetest. Made up of 207 "normal" amino acid residues linked by eight disulphide bridges. Interacts with the majority of taste receptors; this function is shared only by MSG and nucleotides. Contains the following amino acids: glycine, threonine, alanine, half-cystine, serine, aspartic acid, proline, arginine, phenylalanine, lysine, asparagine, valine, leucine, iso-leucine, tyrosine, glutamic acid, glutamine, tryptophan, methionine

100,000 times as sweet as sucrose on a molar basis.

Marketed by Tate & Lyle in UK and Hayes Ingredients in US, selling it as a thaumatin-aluminium product under the name Talin.

Listed in the *Guinness Book of World Records* as the sweetest substance known.

Primarily sold in Japan.

NAME:	**Xylitol**
CATEGORY:	Sweetener/ Nutritive additive/ Humectant/ Bulking agent
FOOD USE:	Sugars, sugar preserves and confectionery (confections, icings, chocolate products, fondants, mints, caramels)/ Baked goods (dry mixes, cream fillings)/ Other (dietary supplement, diabetic foods, dietetic foods, pharmaceuticals, anti-caries products, oral and intravenous nutrients, chewing gum, children's chewable vitamins)
SYNONYMS:	CFSAN Xylitol/ CAS 87-99-0/ EINECS 201-788-0/ Xylite/ Xylit/ Eutrit/ Kannit/ Klinit/ Kylit/ Newtol/ Torch/ Xyliton/ Xylitol C/ Xylo-pentane-1,2,3,4,5-pentol/ Xylisorb
FORMULA:	$C_5H_{12}O_5$
MOLECULAR MASS:	152.15
PROPERTIES AND APPEARANCE:	White crystal or crystalline powder; practically odourless. Sweet taste with cooling sensation; same sweetness as sucrose; no aftertaste
MELTING RANGE IN °C:	Stable form: 93–94.5; metastable form: 61–61.5
PURITY %:	98.5% to 110% assay
SOLUBILITY % AT VARIOUS TEMPERATURE/pH COMBINATIONS:	
in water:	**@ 20°C** 62.7 g/100 g solution **@ 50°C** 80 g/100 g solution
in ethanol solution:	**@ 100%** Sparingly soluble in ethanol (1.2 g/100 g solution at 25°C)
FUNCTION IN FOODS:	Add texture and mouthfeel (bulking agent). Reduced tendency to crystallise. Sweetener (sweetness 1.0 relative to sucrose); humectant; non-fermentable. In baking, results in: reduced fermentability; increased resistance to non-enzymic browning reactions; decreased tendency to crystallise; increased hygroscopicity.
ALTERNATIVES:	Sorbitol (less sweet)

TECHNOLOGY OF USE IN FOODS:

An affinity for water without altering sweetening power. Functionally similar to sucrose; sweetening power twice sorbitol and glucose syrup. pH 5–7 in aqueous solution. Chemically stable.

Storage: 1-year stability in original sealed package; store below 25°C and <65% relative humidity.

High hygroscopicity: 6.0 g/100 g methanol; 1.2 g/100 g ethanol; 62.2 g/100 g water.

Sweetness varies slightly with temperature, pH and concentration.

Used alone, with other polyols, or with polydextrose in sugarless confectionery products. Polydextrose is a bulking agent with low laxative properties, low calorific value (1 kcal/g) and is tolerated by diabetics, but is not very sweet.

Primarily used as a sweetener in sugarless chewing gum.

In 50:50 combination with sorbitol in milled, blended or compressed form may have potential for use in mints and children's chewable vitamin tablets. There are problems of poor flow properties. Direct compression of xylitol with no more than 4% sorbitol preferable.

Good choice for replacement of sucrose in chocolate.

Conching can take place at temperatures up to 55°C.

Can be combined with sorbitol to provide syrup phase of fondants, and with mint and chocolate flavours; Can also be used in pectin and gelatin jellies, but extra gelling agent is required as xylitol reduces gel strength.

No browning (Maillard) reaction; resistant to heat; does not react with amino acids.

Solubility in water:

10°C = 58 g/100 g solution
20°C = 62.7 g/100 g solution
30°C = 68.5 g/100 g solution
40°C = 74.5 g/100 g solution
50°C = 80 g/100 g solution
60°C = 86 g/100 g solution

FOOD SAFETY ISSUES:

LD_{50} of 22 g/kg (25.7 g/kg orally; 3.77 g/kg intravenous). Moderately toxic by intravenous route; mildly toxic by ingestion.

Low cariogenicity; several studies have demonstrated its caries-inhibiting ability; is not metabolised by cariogenic bacteria.

Is slowly but completely absorbed in the intestine.

Laxative effect at high doses (50–70 g/day), so warning labels are required in some countries.

Has been proven to have no adverse health effects, but abnormalities arise when exceptionally large quantities are taken.

	Suitable for diabetic products because metabolism results in no significant change in blood glucose. ADI not specified. On heating to decomposition, emits acrid smoke and irritating fumes.
LEGISLATION:	**USA:** FDA clearance in 1978 for sweetening of Special Dietary Foods CFR 21 PART 172 – Food Additives Permitted for Direct Addition to Food for Human Consumption Subpart D – Special Dietary and Nutritional Additives 172.395 – Xylitol. Xylitol may be safely used in foods for special dietary uses, provided the amount used is not greater than that required to produce its intended effect. **UK and EUROPE:** Permissible for use in UK, EEC, Scandinavia. One of 12 sweeteners listed as permissible for use in the UK as of 1983. **CANADA:** Canada FDR s. 67.31B – Table IX – Food additives that may be used as Sweeteners – Item X.1 (25 May 1993). Permitted in or on Unstandardized foods at maximum Level of Use specified by Good Manufacturing Practice 1996 Addition: Page 67–29 of Canadian FDR – Table IX Permitted for use as a sweetener in chewing gum at a maximum level specified by Good Manufacturing Practice **OTHER COUNTRIES:** WHO/FAO clearance (1978) for sweetening of Special Dietary Foods Not GRAS approved Permitted for use in more than 40 countries as of 1991 including EEC, North America and Scandinavia

REFERENCES:

Smith, J. (1991) *Food Additive User's Handbook.* Blackie Publishing, Glasgow.

Canadian Food and Drug Act and Regulations (1994).

McCue, N. (1996) *Showcase: Natural & High Intensity Sweeteners; Prepared Foods.* March p. 84.

Minifie, B. W. (1989) *Chocolate, Cocoa, and Confectionery: Science and Technology.* 3rd edition. Van Nostrand Reinhold.

Matz, S. A. (1992) *Bakery Technology and Engineering.* 3rd edition. Van Nostrand Reinhold.

Food Additives Update; Food in Canada; April 1996.

Roquette technical bulletin, as provided by Kingley & Keith (Canada) Inc.; 1996.

Smoley, C. K. (1993) *Everything Added to Food in the United States.* US Food and Drug Administration, CRC Press Inc.

Ash, M., and Ash, I. (1995) *Food Additives: Electronic Handbook.* Gower.

Windholz, M., and Budavari, S. (1988) *Merck Index*, 10th edition. Merck and Co.

Solutions CFR+ Database (1996).

Budavari, S. (1996) *The Merck Index.* 12th edition. Merck and Co.

Wong, D. W. S. (1989) *Mechanism and Theory in Food Chemistry.* Van Nostrand Reinhold.

ANY OTHER RELEVANT INFORMATION:

Pentahydric alcohol or sugar alcohol; hydrogenated form of xylose; naturally occurring in raspberries, strawberries, plums, some vegetables, mushrooms, greengages and cauliflowers.

Manufacturers: American Roland, American Xyrofin, Automergic Chemicals, F. R. Benson, Cerestar International, Food Additives & Ingredients, Forum Chemicals, Fruitsource, Melida, Penta Manufacturing, Raquette U.K., Scanchem Xyratin.

Developed in the 1970s; relatively expensive, but appealing due to other beneficial qualities. Produced commercially by enzymatic or microbial conversion of xylose-rich precursors such as birchwood chips.

Xylisorb® meets requirements of USPXXII, NF XVII, Japanese Ph XI, and DAC 90.

Strongly negative heat of solution, combined with high solubility at body temperature results in refreshingly cool effect to the taste. Heat of solution: –34.8 cal/g versus water –4.3 cal/g. Cooling effect (150 g powder/50 ml water at 37°C): –20 versus –9 for sugar

Index